新生物学丛书

生物燃料：可再生能源、农业生产和技术进步对全球的影响（影印版）

Biofuels: Global Impact on Renewable Energy, Production Agriculture, and Technological Advancements

〔美〕D. Tomes 〔澳〕P. Lakshmanan 〔美〕D. Songstad 编著

科学出版社

北京

图字：01-2013-1375 号

内 容 简 介

生物燃料领域是生物技术相关学科中被讨论得最多的研究领域。本书艺术地总结了生物燃料的经济现状、农业生产力和可持续发展，以及全球视角。本书很好的把生物燃料的知识分隔成了不同的章节和特定知识点，使之更易阅读和理解。此外，每章的参考文献都是进一步研究的宝贵资源。

本书的涉及范围广泛，一方面可以在实验室和教学中向学生介绍最新的技术进展，另一方面也可以让关注于生物燃料在国际范围内的经济分析的企业家们感到兴趣。

Reprint from English language edition: *Biofuels*
by Dwight Tomes, Prakash Lakshmanan and David Songstad
Copyright © 2011 Springer New York
Springer New York is a part of Springer Science+Business Media All Rights Reserved

This reprint has been authorized by Springer Science & Business Media for distribution in China Mainland only and not for export therefrom.

图书在版编目(CIP)数据

生物燃料：可再生能源、农业生产和技术进步对全球的影响=Biofuels: global impact on renewable energy, production agriculture, and technological advancements：英文/(美)托姆斯(Tomes, D.)等编著. —影印本. —北京：科学出版社，2013.3
 (新生物学丛书)
 ISBN 978-7-03-036997-0

Ⅰ.①生… Ⅱ.①托… Ⅲ.①生物燃料-英文 Ⅳ.TK6

中国版本图书馆 CIP 数据核字(2013)第 045384 号

责任编辑：岳漫宇 / 责任印制：钱玉芬 / 封面设计：北京美光制版有限公司

科学出版社出版
北京东黄城根北街16号
邮政编码：100717
http://www.sciencep.com

源海印刷有限责任公司 印刷
科学出版社发行　各地新华书店经销
*
2013 年 3 月第 一 版　　开本：787×1092　1/16
2013 年 3 月第一次印刷　　印张：23 1/4
　　　　　　　　　　　字数：540 000
定价：120.00 元
（如有印装质量问题，我社负责调换）

《新生物学丛书》专家委员会成员名单

主　任：蒲慕明

副主任：吴家睿

专家委员会成员（按姓氏汉语拼音排序）

昌增益	陈洛南	陈晔光	邓兴旺	高　福
韩忠朝	贺福初	黄大昉	蒋华良	金　力
李家洋	林其谁	马克平	孟安明	裴　钢
饶　毅	饶子和	施一公	舒红兵	王　琛
王梅祥	王小宁	吴仲义	徐安龙	许智宏
薛红卫	詹启敏	张先恩	赵国屏	赵立平
钟　扬	周忠和	朱　祯		

《新生物学丛书》丛书序

当前，一场新的生物学革命正在展开。为此，美国国家科学院研究理事会于2009年发布了一份战略研究报告，提出一个"新生物学"（New Biology）时代即将来临。这个"新生物学"，一方面是生物学内部各种分支学科的重组与融合，另一方面是化学、物理、信息科学、材料科学等众多非生命学科与生物学的紧密交叉与整合。

在这样一个全球生命科学发展变革的时代，我国的生命科学研究也正在高速发展，并进入了一个充满机遇和挑战的黄金期。在这个时期，将会产生许多具有影响力、推动力的科研成果。因此，有必要通过系统性集成和出版相关主题的国内外优秀图书，为后人留下一笔宝贵的"新生物学"时代精神财富。

科学出版社联合国内一批有志于推进生命科学发展的专家与学者，联合打造了一个21世纪中国生命科学的传播平台——《新生物学丛书》。希望通过这套丛书的出版，记录生命科学的进步，传递对生物技术发展的梦想。

《新生物学丛书》下设三个子系列：科学风向标，着重收集科学发展战略和态势分析报告，为科学管理者和科研人员展示科学的最新动向；科学百家园，重点收录国内外专家与学者的科研专著，为专业工作者提供新思想和新方法；科学新视窗，主要发表高级科普著作，为不同领域的研究人员和科学爱好者普及生命科学的前沿知识。

如果说科学出版社是一个"支点"，这套丛书就像一根"杠杆"，那么读者就能够借助这根"杠杆"成为撬动"地球"的人。编委会相信，不同类型的读者都能够从这套丛书中得到新的知识信息，获得思考与启迪。

《新生物学丛书》专家委员会

主　任：蒲慕明

副主任：吴家睿

2012年3月

In Vitro Cellular and Developmental Biology - Plant

"A Message from the Editor-in-Chief, Dwight T. Tomes"

'Biofuels: Global Impact on renewable energy, production agriculture, and technological advancements'

This comprehensive volume developed under the guidance of guest editors Prakash Lakshmanan and David Songstad features broad coverage of the topic of biofuels and its significance to the economy and to agriculture. These chapters were first published by In Vitro Cellular and Developmental Biology In Vitro Plant in 2009 and consists of 15 chapters from experts who are recognized both for their scientific accomplishments and global perspective in their assigned topics.

The subject of biofuels is multi faceted and must be considered in the context of economic, agricultural productivity and sustainability as well as global perspectives. The first section is devoted to a combination of a historical perspective (David Songstad) and the current research of the DOE Bioenergy Science Center (Russ Miller).

In the popular press, biofuels are portrayed as a competitor to food production and the reason for food price inflation. The primary factors for food price increases such as oil prices, emerging economies, weather and currency fluctuations are outlined in detail (Greg Phillips). Furthermore, the economics of biofuels (Andy Aden) lead to more realistic conclusions of the positive contribution of biofuels to agriculture and energy sustainability. The feed stocks for biofuels are varied depending on location and availability. These feedstocks can arise from multiple monocot species (Jose Gonzalez-Hermandez), dedicated energy crops (Russel Jessup), grasses (Katrin Jocob), or woody short-rotation crops (Hinchee). The international perspective from Brazil with their mature biofuels practices with sugarcane (Paulo Arruda) and from the emerging economies of India (Mambully Gopinathan) and China (Chun-Zhao Liu) adds to the understanding of biofuels from a global perspective.

Essentially, biofuel production is an application of fermentation technology and a major determinant of the feasibility of different biofuel products. This section contains the most up to date treatment of plant modification of the lignin pathway

(Z.Y. Wang), the use of plant expressed enzymes for cellulosic ethanol (Manuel Sainz), integration of biorefineres and high-value co-products (W. Gibbons and S. Hughes), and biodiesel production (Bryan Moser).

This book is an excellent resource of diverse biofuels knowledge organized by major areas and specific topics in an easily accessible and readable format. The broad scope of this book has application in the lab and classroom for the latest technological advances and also for the entrepreneur for the economic analysis and international scope of biofuels. Furthermore, the references within each chapter are a valuable resource for further study.

Contents

1 **Historical Perspective of Biofuels:**
   Learning from the Past to Rediscover the Future..................... 1
   David Songstad, Prakash Lakshmanan, John Chen,
   William Gibbons, Stephen Hughes, and R. Nelson

2 **The DOE BioEnergy Science Center-A U.S.**
   Department of Energy Bioenergys Research Center.................. 9
   Russ Miller and Martin Keller

3 **Drivers Leading to Higher Food Prices:**
   Biofuels are not the Main Factor... 19
   Paul Armah, Aaron Archer, and Gregory C. Phillips

4 **The Economics of Current and Future Biofuels**..................... 37
   Ling Tao and Andy Aden

5 **A Multiple Species Approach to Biomass Production**
   from Native Herbaceous Perennial Feedstocks........................ 71
   J.L. Gonzalez-Hernandez, G. Sarath, J.M. Stein,
   V. Owens, K. Gedye, and A. Boe

6 **Development and Status of Dedicated Energy**
   Crops in the United States... 97
   Russell W. Jessup

7 **Genetic Improvement of C4 Grasses as Cellulosic**
   Biofuel Feedstocks... 113
   Katrin Jakob, Fasong Zhou, and Andrew H. Paterson

8 **Short-Rotation Woody Crops for Bioenergy**
   and Biofuels Applications.. 139
   Maud Hinchee, William Rottmann, Lauren Mullinax,
   Chunsheng Zhang, Shujun Chang, Michael Cunningham,
   Leslie Pearson, and Narender Nehra

9 **The Brazilian Experience of Sugarcane Ethanol Industry** 157
   Sizuo Matsuoka, Jesus Ferro, and Paulo Arruda

10 **Biofuels: Opportunities and Challenges in India** 173
   Mambully Chandrasekharan Gopinathan and Rajasekaran Sudhakaran

11 **Biofuels in China: Opportunities and Challenges** 211
   Feng Wang, Xue-Rong Xing, and Chun-Zhao Liu

12 **Genetic Modification of Lignin Biosynthesis
   for Improved Biofuel Production** 223
   Hiroshi Hisano, Rangaraj Nandakumar, and Zeng-Yu Wang

13 **Commercial Cellulosic Ethanol:
   The Role of Plant-Expressed Enzymes** 237
   Manuel B. Sainz

14 **Integrated Biorefineries with Engineered Microbes
   and High-value Co-products for Profitable
   Biofuels Production** 265
   W. R. Gibbons and S. R. Hughes

15 **Biodiesel Production, Properties, and Feedstocks** 285
   Bryan R. Moser

Erratum E1

Index 349

Chapter 1
Historical Perspective of Biofuels: Learning from the Past to Rediscover the Future

David Songstad, Prakash Lakshmanan, John Chen, William Gibbons, Stephen Hughes, and R. Nelson

Abstract This issue of *In Vitro* Plant is dedicated to various aspects of biofuel research and development. The editors have sought the experts in this field and solicited manuscripts for this special issue publication from various academic institutions, government (USDA, DOE), industry (Mendel, Alellyx, Canavilas, Syngenta, Monsanto), and various countries (USA, China, Brazil, India, and Australia). This has resulted in state-of-the-art articles describing ethanol and also biodiesel research. These publications highlight the status of biofuel research across the globe and also focus on private, public, and government interests. This is especially noteworthy in that President Barack Obama has stated that renewable energy is a pivotal aspect of his policy for the USA. The objective of this introduction is to provide the reader with the pertinent background information relative to the biofuel efforts within the private sector, academia, and government laboratories. In particular, the history of biofuel research and commercialization is provided as well as a summary of the various crop systems available for biofuel production.

Keywords Biofuel • Bioethanol • Biodiesel • History

History of Bioethanol

Those not familiar with the legacy of bioethanol might believe that this is a late twentieth century development. Those that experienced the Arab oil embargos of the 1970s remember this period as the first vocal call for a domestic source of renewable energy to counter the rapid escalation in oil prices that has continued through the early twenty-first century. However, the reality is that ethanol was developed as an alternative fuel before the discovery of petroleum by Edwin Drake

---

D. Songstad (✉)
Monsanto, 800 N. Lindbergh Blvd., St. Louis, MO, USA
e-mail: david.d.songstad@monsanto.com

in 1859 (Kovarik 1998). Prior to this year, the energy crisis revolved around finding a replacement for the diminishing supply of whale oil, which was commonly used as a lamp oil. Other lamp oils derived from vegetables and animals were also used, but whale oil was preferred. By the late 1830s, ethanol blended with turpentine (refined from pine trees) was used to replace the more expensive whale oil.

During this time period, Samuel Morey invented the internal combustion engine (US Patent 4378 Issued April 1, 1826) which was fueled by a combination of ethanol and turpentine. Morey's patent was entitled "Gas or Vapor Engine" and was signed by President John Quincy Adams and Secretary of State Henry Clay (Hodgson 1961). Morey was able to use this engine to power a boat at speeds of 7 to 8 mph. Unfortunately, Morey was unable to find an investor to further develop the internal combustion engine. However, in 1860, the German inventor Nicholas Otto rediscovered the internal combustion engine and also used an elixir containing ethanol as fuel. He was denied a patent but was able to secure initial funding from Eugen Langen, who interestingly owned a sugar refining company and most likely had associations with ethanol markets in Europe (Kovarik 1998).

The next era where biofuels became prominent was early in the twentieth century and was linked with the discovery of the automobile. Henry Ford envisioned automobiles that relied on ethanol as their fuel source (Kovarik 1998). Keen interest in ethanol as a fuel for automobiles was described in a 1906 New York Times article entitled "Auto Club Aroused Over Alcohol Bill" (Anonymous 1906). This article described concerns over the influence of ethanol on the gasoline industry in the quote "Gasoline is growing scarcer, and therefore dearer, all the time." This might be the first indication of competition between the petroleum and ethanol industries. Furthermore, Alexander Graham Bell was quoted in a 1917 National Geographic interview (Anonymous 1917) stating "Alcohol can be manufactured from corn stalks, and in fact from almost any vegetable matter capable of fermentation. Our growing crops and even weeds can be used. The waste products of our farms are available for this purpose and even the garbage of our cities. We need never fear the exhaustion of our present fuel supplies so long as we can produce an annual crop of alcohol to any extent desired." Dimitri and Effland (2007) commented that the $2.08 tax per gallon of ethanol, which was initially imposed in 1861 as a means to fund the Civil War, continued well past the war's conclusion and was finally removed in 1906 (this would equivalent to about $35/gallon ethanol tax in 2007 http://www.icminc.com/timeline). This obviously made ethanol more expensive than gasoline and favored use of gasoline for internal combustion engines. However, when this tax was removed, it was difficult for the ethanol infrastructure to compete with gasoline, the now accepted fuel for automobiles. However, it became apparent that "previously established industrial partnerships" emerged in the 1920s blocking use of ethanol as the solution to engine knock, in favor of tetraethyl lead (Dimitri and Effland 2007).

During the 1920s and 1930s, a new movement called Chemurgy emerged and was promoted by Henry Ford (Finlay 2004). Chemurgy can be thought of an early version of the agricultural "new uses" initiative, and it put pressure on the USDA to focus crop utilization on bio-based materials. Ethanol was prominent on the Chemurgy

1 Historical Perspective of Biofuels: Learning from the Past to Rediscover the Future

Fig. 1.1 Photo from April, 11, 1933 of a bioethanol fueling station in Lincoln, Nebraska. (Copyright permission to use photo in this publication by the Nebraska History Museum)

agenda, as evident for the need to produce synthetic rubber at the onset of World War II. In 1942 Congress appointed a "rubber czar" to over see the production of synthetic rubber only from agricultural resources and in 1943, nearly 77% of the synthetic rubber produced in the USA was derived from ethanol (Finlay 2004).

The World War years of 1917–1919 and 1941–1945 also witnessed an increase in ethanol demand in the USA due to the rationing of raw materials and natural resources, including gasoline. Ethanol was produced as an alternative to gasoline for domestic use. During the World War I years, ethanol production increased to 60 million gallons and this further increased during the World War II years to 600 million gallons (http://www.icminc.com/timeline). Proof of interest in bioethanol during this timeframe is evident in the photo provided in Fig. 1.1 (Credit—Nebraska History Museum) which shows a gasoline filling station of Earl Coryell in Lincoln, Nebraska, promoting 10% ethanol. This photo, taken in April, 1933, is an example of several gasoline filling stations throughout the midwestern region of the USA that were testing ethanol as a gasoline additive.

The 1974 Arab oil embargo also resurrected interest in ethanol production throughout the world. As before, much of the early interest in the USA was in the Midwestern states. In 1978, the College of Agriculture and Biological Sciences at South Dakota State University provided start-up funds and a campus building for Dr. Paul Middaugh (Fig. 1.2) of the Microbiology Department to begin work on

Fig. 1.2 Dr. Paul Middaugh, Professor of Microbiology at South Dakota State University, 1964–1980. Developed first dry mill ethanol production plant from which much of the ethanol industry has been established; photo taken in 1974 (SDSU Photo Archives)

farm-scale ethanol production plant. The East River Rural Electric Cooperative provided $30,000 in funding as well. In 1979, Dr. Middaugh had established the first operating dry mill ethanol plant in the USA and operated it on a demonstration basis for about 18 mo. In 1980, Dr. Middaugh received a competitive research grant of $90,000 from USDA to determine yields, costs, and energy balances in this facility, while a $20,000 DOE grant was obtained to develop educational/training materials for ethanol plant operation. A team of researchers from the Departments of Plant Science, Microbiology, Agricultural Engineering, Mechanical Engineering, Animal Science, Dairy Science, and Economics will spend the next 4 yr establishing the technical and economic baseline for farm-scale ethanol production and developing many of the innovations still in practice today (Middaugh 1979). Dr. Middaugh's legacy continues through his students, two of which are coeditors of this Special Issue publication, Dr. William Gibbons and Dr. David Songstad.

Two of the first students on Dr. Middaugh's team were Paul Whalen and Dennis Vander Griend. Paul Whalen, a microbiology graduate student, established the processing parameters for the dry mill plant. Dennis, an undergraduate mechanical engineer, and his brother, Dave, were recruited by Dr. Middaugh to design, construct, and operate the distillation columns. The Vander Griend brothers continued their focus on ethanol, and Dave founded ICM in 1995 as a company that designs and constructs ethanol plants. Dave continues as the President and CEO of ICM in Colwich, KS (suburb of Wichita), while Dennis is the lead process engineer.

Other leading companies in the ethanol industry also have their roots in the Midwest. Jeff Broin, owner of POET, traces the beginnings of his company to the family farm in Minnesota where in 1986, they built an ethanol plant capable of producing 100,000 gal/yr of ethanol. In this same year, he purchased an existing ethanol plant in Scotland, SD and continues to use this plant to test engineering and design improvements that have been applied to the construction of 26 ethanol plants in seven states. The combined ethanol production of POET facilities now exceeds one billion gallons annually (Gabrukiewicz 2009).

VeraSun Energy was founded in 2001 by Don Endres of Brookings, SD. Through construction of new plants and acquiring existing plants through a merger with US Bioenergy, VeraSun became the largest ethanol producer in the USA, operating 16 ethanol plants across eight states with an average potential ethanol production capacity of 1.64 billion gallons per year (http://www.verasun.com/About/) VeraSun became a publically traded company on the New York Stock Exchange on June 14, 2006. However, VeraSun declared Chapter 11 bankruptcy in October, 2008 and announced in February, 2009, that some of the company's ethanol plant assets will be sold and all assets were auctioned on March 16, 2009. However, it is important to realize that ethanol production from these plants continues, only under new corporate management.

Recently, bioethanol has been criticized in the feed, food, fuel debate. In this special publication, Armah et al. described the economics associated with bioethanol and biodiesel and identified rising input costs (e.g., crude oil, labor, and transportation costs) as the main drivers associated with the rising price of food. This report, based on factual data, is contrary to the sensational stories often reported in the popular media. Furthermore, a recent report by Darlington (2009) reported that increasing US corn ethanol production to 15 billion gallons by 2015 will not result in significant conversion of non-agricultural lands into corn production in the USA or abroad. Rather, this will be possible due to the increased rate of growth of yield of corn as the result of advances in breeding and novel traits delivered through biotechnology.

History of Biodiesel

The term "biodiesel" was first coined in 1988 (Wang 1988), but the history of using vegetable oil in place of diesel as a fuel dates back to 1900. The roots of what eventually became known as "biodiesel" extend back to the discovery of the diesel engine by Rudolf Diesel. The first demonstration of the diesel engine was at the 1900 World's Fair in Paris. Knothe (2001), in a book chapter entitled "Historical perspectives on vegetable oil-based diesel fuels," describes that the diesel engine built by the French Otto Company was tested at this event using peanut oil. It is not totally clear if Rudolf Diesel had the idea to use peanut oil because apparently, Diesel gives credit to this to the French government. Knothe (2001) also relates that the French government was interested in vegetable oil fuels for diesel engines because of its availability in their colonies in Africa, thereby eliminating the need to import liquid fuels or coal.

Knowledge that vegetable oils could be used to fuel the diesel engine gave a sense of energy self-sufficiency to those countries producing oil crops, especially for those countries in Africa in the 1940s (Knothe 2001). This was especially the case during the World War II yr, where even in Brazil, export of cottonseed oil was prohibited so it could be used as a substitute for diesel (Anonymous 1943). In China, tung oil and other vegetable oils were used to produce a version of gasoline and kerosene (Cheng 1945; Chang and Wan 1947). Furthermore, prompted by fuel shortages during World War II, India conducted research on conversion of a variety of vegetable oils to diesel (Chowhury et al. 1942). This interest in biodiesel was also evident in the USA where research was performed to evaluate cottonseed oil as a diesel fuel (Huguenard 1951).

Related to this are the efforts of automobile entrepreneur Henry Ford and the development of the "soybean car" in 1941. Mr. Ford was a true visionary and was motivated by combining the strength of the automobile industry with agriculture. According to the Benson Ford Research Center, there was a single experimental soybean car built, and this was suspended due to World War II (http://www.thehenryford.org/research/soybeancar.aspx#). This car weighed 2,000 lbs, which was 1,000 lbs less than the all steel cars in production in 1941. This lightweight car would certainly be more fuel efficient than its heavier counterpart. The genius of Mr. Ford is clearly evident in a car made in part with soybean and propelled by ethanol derived from corn. However, production of the soybean car did not resume after the end of World War II, and this does illustrate the lesson on how difficult it can be to sustain innovation (Young 2003). Photos of the soybean car are available at the website listed above.

Since the 1950s, interest in converting vegetable oils into biodiesel has been driven more by geographical and economic factors than by fuel shortages. For instance, the USA is a top producer of soybean oil, whereas Europe produces large amounts of canola oil, and this essentially determines which oil is used for biodiesel within these geographies. Also, for those remote geographic locations to which fossil fuel refining and distribution are problematic, vegetable oil-based biodiesel is a sustainable and practical means to meet the fuel energy demands. Furthermore, sources for biodiesel have been expanded to include spent vegetable oil from the food service industry as well as animal fats from slaughterhouses (Knothe 2001). However, additional research is required to identify new oil crops to meet the increasing demand for biodiesel. A variety of tools including plant breeding, molecular breeding, and biotechnology are needed to increase oil production from conventional crops such as soybean and to develop new oil crops for specific regions.

Conclusion

This introduction to the history of biofuels was intended to give the reader a deeper appreciation for the achievements of those previously that set the stage for the role of renewable biofuels in the twenty-first century. It is the intent that this overview will help establish a legacy of those that saw the vision of bioethanol and biodiesel yr ago, and it is on their shoulders that we now stand.

This introduction is dedicated to memory of Dr. Paul Middaugh (1920–2003).

References

Anonymous (1906) Auto club aroused over alcohol bill. New York Times, April 26, 1906.
Anonymous. National geographic 31: 131; 1917.
Anonymous Brazil uses vegetable oil for diesel fuel. *Chem. Metall. Eng* 50: 225; 1943.
Chang C. -C.; Wan S. -W. China's motor fuels from tung oil. *Ind. Eng. Chem* 39: 1543–1548; 1947. doi:10.1021/ie50456a011.
Cheng F. -W. China produces fuels from vegetable oils. *Chem. Metall. Eng* 52: 99; 1945.
Chowhury D. H.; Mukerji S. N.; Aggarwal J. S.; Verman L. C. Indian vegetable fuel oils for diesel engines. *Gas Oil Power* 37: 80–85; 1942.
Darlington, T. (2009) Land Use Effects of U.S. Corn-Based Ethanol. http://www.ethanolrfa.org/objects/documents/2192/land_use_effects_of_us_corn-based_ethanol.pdf
Dimitri C.; Effland A. Fueling the automobile: An economic exploration of early adoption of gasoline over ethanol. *J. Agriculture & Food Industrial Org* 5: 1–17; 2007.
Finlay M. R. Old efforts at new uses: A brief history of chemurgy and the American search for biobased materials. *J. Industrial Ecol* 7: 33–46; 2004. doi:10.1162/108819803323059389.
Gabrukiewicz, T. (2009) Ethanol innovator driven to replace oil. Sioux Falls Argus Leader, January 4, 2009 (http://www.argusleader.com/article/20090104/NEWS/901040311/1001).
Hodgson A. D. Samuel morey: Inventor extraordinary. Historical Fact Publications, Oxford; 1961.
Huguenard, C. M. Dual fuel for diesel engines using cottonseed oil, M.S. Thesis, Ohio State University; 1951.
Knothe G. Historical perspective on vegetable oil-based diesel fuels. *AOCS Inform* 12: 1103–1107; 2001.
Kovarik, B. (1998) Henry Ford, Charles F. Kettering and the Fuel of the Future. Automotive History Review 32:7–27. http://www.radford.edu/~wkovarik/papers/fuel.html
Middaugh, P. R. (1979) Gasohol USA; Vol 4: 22–27. Interview: Dr. Paul Middaugh, alcohol professor. OSTI ID: 6350203 http://www.osti.gov/energycitations/product.biblio.jsp?osti_id=6350203
Wang R. Development of biodiesel fuel. *Taiyangneng Xuebao* 9: 434–436; 1988.
Young A. L. Biotechnology for food, energy, and industrial products. New Opportunities for bio-based products. *Environ. Sci & Pollut. Res* 10: 273–276; 2003.

Chapter 2
The DOE BioEnergy Science Center-A U.S. Department of Energy Bioenergys Research Center

Russ Miller and Martin Keller

Abstract The BioEnergy Science Center, a nationally and internationally peer-reviewed center of leading scientific institutions and scientists, is organized and in operation as a U.S. Department of Energy Bioenergy Research Center. This Oak Ridge National Laboratory-led Center has members from top-tier universities, leading national labs, and private companies organized as a single project team, with each member chosen for its significant contributions in the Center's research focus areas. The recalcitrance of cellulosic biomass is viewed as (1) the most significant obstacle to the establishment of a cellulosic biofuels industry, (2) essential to producing cost-competitive fuels, and (3) widely applicable, since nearly all biofuels and biofeedstocks would benefit from such advances. The mission of the BioEnergy Science Center is to make revolutionary advances in understanding and overcoming the recalcitrance of biomass to conversion into sugars, making it feasible to displace petroleum with ethanol and other fuels.

Keywords Cellulosic biomass • Biofuels • Recalcitrance • Ethanol • Consolidated bioprocessing • Poplar • Switchgrass

The submitted manuscript has been authored by a contractor of the U.S. Government under Contract No. DE-AC05-00OR22725. Accordingly, the U.S. Government retains a non-exclusive, royalty-free license to publish or reproduce the published form of this contribution, or allow others to do so, for U.S. Government purposes.

R. Miller (✉)
Partnerships and Technology Transfer, BioEnergy Science Center, Oak Ridge National Laboratory, Oak Ridge, TN 37831-6169, USA
e-mail: millerrr@ornl.gov

Background and History

In December 2005, two U.S. Department of Energy (DOE) programs working together convened a Biomass to Biofuels Workshop (the Office of Biological and Environmental Research in the Office of Science, and the Office of Biomass Programs in the Office of Energy Efficiency and Renewable Energy) for the purpose of defining barriers and challenges to the rapid expansion of the production of ethanol from cellulose. A core barrier was identified: the resistance, or "recalcitrance," of cellulosic biomass to the conversion into sugars which can then be fermented into ethanol. The two programs articulated a joint research agenda involving the application of modern tools of biological research. The result of this workshop was the publication of a roadmap for speeding the research related to cellulosic ethanol production (DOE/SC-0095 2006).

In the past few years, biological research has seen a major transformation with the emergence of an approach known as "systems biology." Arising from human genome research, this approach involves the application of advanced techniques, tools and high-throughput technologies, computational modeling, and also the adoption of team science. Systems biology is at the core of the DOE Genomics:GTL (GTL) program (http://genomicsgtl.energy.gov/benefits/index/shtml). GTL (formerly known as "Genomes to Life") is a program of the Office of Biological and Environmental Research of the U.S. Department of Energy Office of Science.

The GTL research program is focused on developing technologies that can be used to understand plants and microbes and to enable biological solutions to challenges in energy, environment, and climate. DOE's "mission challenge" related to energy is described as understanding the principles of the structural and functional design of plants and microbes, and developing the capability of modeling, prediction, and engineering of optimized enzymes and microorganisms for the production of such biofuels as ethanol and hydrogen (http://genomicsgtl.energy.gov/benefits/index.shtml).

To focus the research resources of GTL on the biological challenges of biofuel production, DOE announced in June 2007 the establishment of three new Bioenergy Research Centers (BRCs). Each center is projected to receive $135 million over a 5-yr period to pursue the basic research related to high-risk, high-return biological solutions for bioenergy applications. The new scientific knowledge generated by these three BRCs is expected to lay a foundation of methods and tools for the emerging biofuel industry. The purpose of this paper is to provide insight into the BRC which is being lead by Oak Ridge National Laboratory (ORNL).

A competitive, peer-review process resulted in the selection of Oak Ridge National Laboratory in Tennessee to lead the DOE BioEnergy Science Center (BESC); the University of Wisconsin-Madison to lead the DOE Great Lakes Bioenergy Research Center (GLBRC); and DOE's Lawrence Berkeley National Laboratory to lead the DOE Joint BioEnergy Institute (JBEI). Each of these three centers represents a multidisciplinary partnership with research expertise in the physical and biological sciences, including genomics, microbial and plant biology, analytical chemistry, and computational biology.

The BESC team consists of ten institutional partners and individual investigators selected for their extensive experience in biomass research at seven other institutions. The new Joint Institute for Biological Sciences systems biology research facility at ORNL serves as the central hub for coordinating research activities among the various BESC partners. Through developing the scientific basis for understanding biomass recalcitrance, BESC's researchers hope to lay the foundation for the re-engineering of plants, microbes and enzymes at the system, cellular and molecular levels to improve the productivity of bioenergy crops, and to reduce the cost of conversion.

BESC Partners

DOE's Oak Ridge National Laboratory (ORNL), Oak Ridge, Tennessee. DOE's largest science and energy laboratory, ORNL is a world leader in poplar genome research, with strong programs in comparative and functional genomics, structural biology, and computational biology and bioinformatics. The ORNL Spallation Neutron Source and supercomputers at the ORNL National Leadership Computing Facility are used in BESC for investigating the activity of enzyme complexes.

Georgia Institute of Technology, Atlanta. Georgia Tech's Institute for Paper Science and Technology provides BESC with expertise in biochemical engineering and instrumentation for high-resolution analysis of plant cell walls.

National Renewable Energy Laboratory (NREL), Golden, Colorado. NREL has over 30 yr of experience in biomass analysis and biofuel research, and houses premiere facilities for analyzing biomass surfaces. NREL also has a long history of establishing biofuel pilot plants and working with industry for commercial development of technologies.

University of Georgia, Athens, Georgia (UGA). UGA's Complex Carbohydrate Research Center maintains state-of-the-art capabilities in mass spectrometry, NMR spectroscopy, chemical and enzymatic synthesis, computer modeling, cell and molecular biology, and immunocytochemistry for studying the structures of complex carbohydrates and the genes and pathways controlling plant cell-wall biosynthesis.

University of Tennessee, Knoxville, Tennessee (UT). UT conducts successful programs in bioenergy-crop genetic and field research (particularly switchgrass) and biotechnological applications of environmental microbiology.

Dartmouth College, Hanover, New Hampshire. Dartmouth's Thayer School of Engineering is a leader in the fundamental engineering of microbial cellulose utilization and consolidated bioprocessing approaches.

ArborGen, Summerville, South Carolina. ArborGen is a world leader in forest genetics research, tree development, and commercialization.

Verenium Corporation, Cambridge, Massachusetts. Verenium is a biofuels-focused biotechnology company and developer of specialty enzymes found in diverse natural environments and optimized for targeted applications.

Mascoma Corporation, Lebanon, New Hampshire. Mascoma is a leader in developing microbes and processes for economical conversion of cellulosic feedstocks into ethanol.

The Samuel Roberts Noble Foundation, Ardmore, Oklahoma. This non-profit research foundation is devoted to improving agricultural production and to advancing the development of switchgrass and other grasses through genomic research. Their activities are conducted by the Agricultural Division and the Foundation's Plant Biology and Forage Improvement programs.

Individual researchers from the University of California, Riverside; DOE's Brookhaven National Laboratory (Upton, New York); Cornell University (Ithaca, New York); Virginia Polytechnic Institute and State University (Blacksburg, Virginia); North Carolina State University (Raleigh, North Carolina); University of Minnesota (St. Paul, Minnesota); and Washington State University (Pullman, Washington) specialize in biomass pretreatment, characterization of plant-associated microbes, cellulose and enzyme modeling, consolidated bioprocessing, and lignin biochemistry.

BESC focuses on the fundamental understanding of the resistance of cellulosic biomass to enzymatic breakdown into sugars. This is thought to be the single-greatest obstacle to cost-effective production of biofuels (Lynd et al. 2008) (Fig. 1). BESC's approach to making biomass easier to degrade involves (1) designing plant cell walls for rapid deconstruction and (2) engineering a multi-talented microbe tailor-made for converting plants into biofuel in a single step, an approach known as consolidated bioprocessing. In this way, the problem is approached from two directions; both from the perspective of being able to design optimized biofuel feedstock crops and also from the "deconstruction" part of the problem. In BESC, these two research areas are not being pursued independently of each other, but are coupled in a single integrated plan.

The BESC Research Strategy

The unifying theme of BESC is to understand and overcome the recalcitrance of lignocellulosic biomass for conversion to fermentable sugars. This is a complex problem since the cell walls of plants are largely made up of cellulose and hemicellulose (polysaccharides that make up most of the mass of plant cell walls) and lignin that cross-links them together. This structure contributes mechanical strength and support to elements of the plant body. In addition, the lignin has properties which interfere with enzymatic conversion of the polysaccharides (DOE/SC-0095 2006). The researchers now working on these challenges in BESC were among the first to identify this biomass recalcitrance as a major obstacle to biomass conversion processes (Lynd et al. 1999) and are responsible for making fundamental advances in a wide range of related science (Lynd et al. 2005; Himmel et al. 2007).

BESC will pursue the following goals in three research focus areas (Fig. 2.2):

Focus Area 1. *Biomass formation and modification*—Goals: develop a thorough understanding of the genetics and biochemistry of plant cell-wall biosynthesis so the process can be modified to reduce biomass recalcitrance.

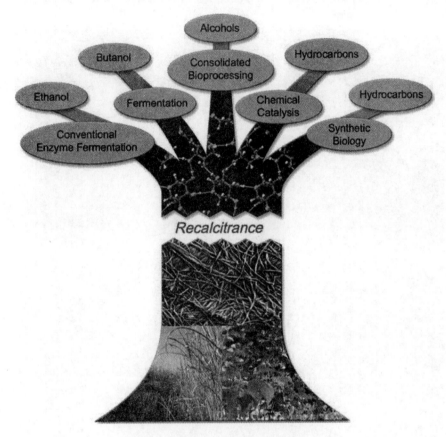

Fig. 1.1 Access to the sugars in lignocellulosic biomass is the current critical barrier. Solving this problem will cut processing costs significantly for cellulosic biofuels

Focus Area 2. *Biomass deconstruction and conversion*—Goals: develop an understanding of enzymatic and microbial biomass deconstruction; characterize and mine biodiversity; and use this knowledge to develop superior biocatalysts for consolidated bioprocessing (CBP).

Focus Area 3. *Characterization and modeling*—Goals: develop a high-throughput pretreatment and characterization pipeline for the systematic study of the structure, composition, and deconstruction of biomass. In a cross-cutting approach, integrate the scientific knowledge gained using chemical, spectroscopic, and imaging methods and computational modeling and simulation.

Focus Area 1. Biomass Formation and Modification. BESC's biomass formation and modification research involves working directly with two potential bioenergy crops—switchgrass and poplar—to develop varieties that are easier to break down into fermentable sugars yielding ultimately improved biofuel yields. Currently, little is known about how cellulose and hemicelluloses are synthesized; distributed within cell

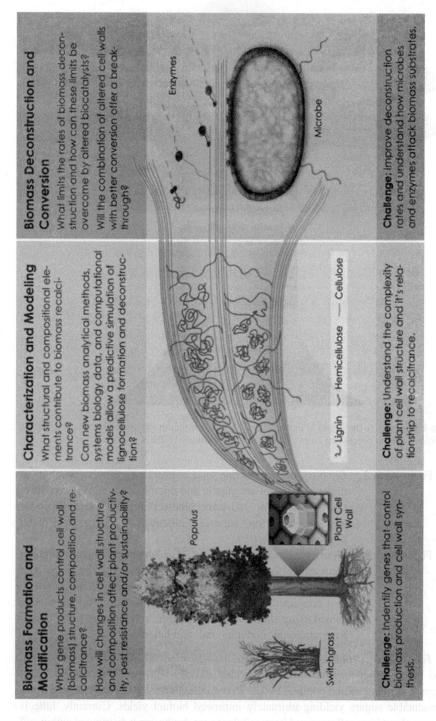

Fig. 2.2 The BESC scientific approach integrates three science focus areas

walls; and attached to each other, to lignin, or to cell-wall proteins (DOE/SC-0095 2006). Molecular, genetic, genomic, biochemical, chemical, and bioinformatics tools are being used by BESC researchers in this Focus Area to advance the understanding of cell-wall biosynthesis, cell-wall structure, and biological consequences of plant cell-wall modification. In addition, BESC is developing computational models of cell-wall biosynthesis that will help in poplar and switchgrass.

One major focus in plant cell-wall biosynthesis is the identification of genes that affect biomass production and to use this information to generate plants with reduced recalcitrance. Specifically:

- To identify and characterize gene products involved in cell-wall biosynthesis and structure, and to establish which genes alter biomass recalcitrance;
- To determine how changes in cell-wall composition and structure affect plant recalcitrance and to begin to develop a systems biology understanding of the genetic basis for cell wall structure and recalcitrance.

To accomplish the above, BESC is conducting fundamental research on the enzymes involved in plant cell-wall polysaccharide (cellulose, hemicellulose, and pectin) and lignin synthesis, a plant cell-wall biosynthesis gene database will be established and curated, and biosynthetic and regulatory genes encoding proteins involved in specific polysaccharide and lignin synthesis pathways will be investigated. Genes will be under- and over-expressed in *Populus* (*Populus trichocarpa*) and switchgrass (*Panicum virgatum*), via both stable and transient expression systems, and biomass from the resulting stable genetic variants will be characterized through various screening methods to evaluate any impact on cell-wall structure.

Focus Area 2. Biomass Deconstruction and Conversion. Two key hypotheses drive BESC's biomass deconstruction and conversion research: (1) microorganisms can be engineered to enable consolidated bioprocessing (CBP), a one-step, microbe-mediated strategy for directly converting plant biomass into ethanol; and (2) enzymes and microbial biocatalysts can be engineered to take advantage of recalcitrance-reducing plant modifications to achieve better biomass deconstruction. CBP will enable fundamental process cost savings through the establishment of a much simplified industrial process (Lynd et al. 1999). One model organism being studied in BESC for CBP development is *Clostridium thermocellum*, a bacterium that rapidly degrades pure cellulose and then ferments the resulting sugars into ethanol. This microbe's strategy for combined biomass deconstruction and conversion employs cellulosomes—multifunctional enzyme complexes that specialize in degrading cellulose (Brown et al. 2007). Understanding cellulosomes is important for rapid improvement in the deconstruction of more complex plant cell walls. BESC is studying the structures and activities of these multi-enzyme complexes to design new variants with better cell-wall-deconstruction capabilities. In addition to working with *C. thermocellum*, BESC researchers are investigating samples from hot springs at Yellowstone National Park to identify heat-tolerant enzymes and microbes with superior biomass-degrading functions that can be used to discover additional strategies for new CBP microorganisms.

In this Focus Area, we are exploring several fundamental frontiers: (1) mining the natural diversity of biomass-degrading enzymes and microbes; (2) studying how different biomass features affect the activities of enzymes and microbes; (3) examining the relationship between enzyme structure and function; (4) investigating how enzymes and microbes interact with pretreated cell walls; and (5) testing strategies for using microbial cultures for biomass deconstruction and conversion. An over-arching goal is to integrate information obtained from these various investigations into comprehensive conceptual and computational models of cellulose deconstruction in both natural and engineered environments.

Focus Area 3. Characterization and Modeling. Advancing BESC goals to develop improved plant materials and CBP methods that facilitate cost-effective conversion of biomass to fermentable sugars will require detailed knowledge of (1) the chemical and physical properties of biomass that influence recalcitrance, (2) how these properties can be altered by engineering plant biosynthetic pathways, and (3) how biomass properties change during pretreatment and how such changes affect biomass–biocatalyst interactions during deconstruction by enzymes and microorganisms.

BESC researchers are applying modern analytical technologies to examine chemical and structural changes that occur to the plant cell walls that have been genetically modified. Switchgrass and poplar samples generated in BESC will be catalogued, bar coded, and analyzed in detail for chemical composition. This biomass undergoes pretreatment and enzyme-digestibility studies followed as needed by detailed chemical, physical, and imaging analysis. The resulting data is being used to build relationship models between biomass structure and recalcitrance. Characterizing and modeling cell-wall synthesis pathways, biomass structure and composition, and microbe–enzyme interactions with biomass surfaces is expected to provide knowledge which will stimulate the coordinated development of (1) CBP microbes and (2) new generations of switchgrass and poplar optimized for deconstruction. The combination of characterization, modeling, and data sharing will help define the genomic and physical basis of plant cell-wall recalcitrance and deconstruction.

A strength of BESC is the crosscutting integration of diverse experimental, theoretical, and computational approaches. For example, high-throughput physical characterization of biomass will provide the basis for subsequent data mining that can reveal previously unknown correlations between recalcitrance and biomass structure.

Translation of BESC Science into Commercial Applications

BESC is supported by a governing board and a scientific advisory board populated with internationally known academic and bioenergy industry leaders. This structure is similar to what might be found in a biotech startup company, and is intended to provide the perspective that a corporate board might have, as well as the insight and objectivity that an external science advisory board brings. In addition, BESC has formed a "commercialization council" of technology-transfer and intellectual property management professionals from partner institutions to evaluate the commercial potential of new inventions arising from BESC research and to promote and facilitate the transfer of new discoveries to industry. New BESC inventions are posted on the program's website (http://www.bioenergycenter.org/licensing). The BESC inventions disclosed at this point in the program (in the middle of year #2) come from eight of our 17 partners, and arise from research in all three focus areas. The subject matter of these early inventions includes plant and microbial genetic transformation techniques, special methods of microscopy, and innovations in biomass sample handling.

BESC has among its goals the effective, coordinated commercialization of these technologies through licensing to companies pursuing biofuels development. The translation of BESC research results into applications testing and potential commercial deployment is an important step toward reaching DOE's bioenergy objectives. Further, toward our goal of forming external relationships, dissemination of results of the research, and generating information about commercial opportunities, BESC offers companies the opportunity to become BESC Industry Affiliates.

BESC's home base is less than 40 mi from a pre-commercial switchgrass-to-ethanol demonstration plant which represents a significant addition to our ability to move BESC scientific discoveries into commercial use. The $40-million facility funded by the Tennessee Biofuels Initiative will be an innovative pilot-scale biorefinery and state-of-the-art research and development facility for cellulosic ethanol or ethanol from non-food sources (http://www.utbioenergy.org/TNBiofuelsInitiative). The pilot-scale biorefinery is expected to be a catalyst for a new biofuel industry for Tennessee. The facility will produce cellulosic ethanol as a transportation fuel from two different non-food biomass feedstocks, corn stover (cobs and fiber) and switchgrass.

Education and Outreach

By leveraging successful educational and training programs already in place at partner academic institutions, BESC will offer students, postdoctoral staff, and scientists interdisciplinary research opportunities that cut across a broad range of biofuel-related fields. To extend the reach of BESC science to diverse locations and communities, collaborative workshops for training students and scientists and an open seminar series reporting scientific progress will be held at partner institutions. BESC also will provide opportunities each year for non-BESC scientists to participate in research at one or more partner sites.

In addition to our efforts to prepare graduate students and postdocs, our center has taken a novel approach in our education efforts. In collaboration with the Creative Discovery Museum in Chattanooga, we have developed lesson plans aimed at grades 4–6 to educate and inform students about the basics of energy production and utilization. These lessons include basic concepts such as the carbon cycle, lignocellulosic biomass as substrate for the production of biofuels, and technical and economic obstacles to a biobased fuel economy. The hands-on activities and guided questions are also designed to meet educational objectives for these grades. This program has been piloted in a hundred classrooms in North Georgia and Tennessee and will be made available to schools nationwide. We have also begun to pilot interactive "science night" programs offered to students and the general public through local schools, museums, and community centers.

Acknowledgement The BioEnergy Science Center (BESC) is a U.S. Department of Energy Bioenergy Research Center supported by the Office of Biological and Environmental Research in the DOE Office of Science.

References

BioEnergy Science Center. Technologies available for licensing. http://www.bioenergycenter.org/licensing. Cited 30 Jan 2009.

Brown S. D.; Raman B.; McKeown C. K.; Kale S. P.; He Z.; Mielenz J. R. Construction and evaluation of a *Clostridium thermocellum* ATCC 27405 whole-genome oligonucleotide microarray. *Appl. Biochem. Biotech.* 137–140(1–12): 663–674; 2007.

DOE/SC-0095. Breaking the biological barriers to cellulosic ethanol: a joint research agenda. U.S. Department of Energy. http://genomicsgtl.energy.gov/biofuels/b2bworkshop.shtml; 2006.

Himmel M.; Ding S. Y.; Johnson D. K.; Adney W. S.; Nimlos M. R.; Brady J. W.; Foust T. D. Biomass recalcitrance: engineering plants and enzymes for biofuels production. *Science* 315: 804–807; 2007.

Lynd L. R.; Laser M. S.; Bransby D.; Dale B. E.; Davison B.; Hamilton R.; Himmel M.; Keller M.; McMillan J. D.; Sheehan J.; Wyman C. E. How biotechnology can transform biofuels. *Nature Biotech.* 26: 169–172; 2008.

Lynd L. R.; van Zyl W. H.; McBride J. E.; Laser M. et al. Consolidated bioprocessing of cellulosic biomass: an update. *Curr. Opin. Biotech.* 16(5): 577–583; 2005.

Lynd L. R.; Wyman C. E.; Gemgross T. U. Biocommodity engineering. *Biotech. Prog.* 15(5): 777–793; 1999.

Tennessee Biofuels Initiative. Recent News Oct 14, 2008, Jul 23, 2008. Office of Bioenergy Programs at the University of Tennessee, Knoxville, TN. http://www.utbioenergy.org/TNBiofuelsInitiative. Cited 30 Jan 2009.

U.S. Department of Energy Office of Science. Genomics:GTL. http://genomicsgtl.energy.gov/benefits/index.shtml. Cited 30 Jan 2009.

Chapter 3
Drivers Leading to Higher Food Prices: Biofuels are not the Main Factor

Paul Armah, Aaron Archer, and Gregory C. Phillips

Abstract Since 2007, the overall rise in food prices in the USA was twice that of the overall inflation rate, led by inflation for poultry and dairy products in particular. Prominent studies have indicated that the main drivers associated with the rise in food prices in the past 3 yr are the increased energy costs (and the trickle-down impact on farm input costs) and the devaluation of the US dollar. However, currently, the impact of crude oil as one of the primary drivers in food prices has waned significantly, as crude oil prices have fallen dramatically since late 2008. The data reviewed here debunk the popular myth that food producers, particularly farmers growing corn or soybean for biofuel feedstocks, are making huge profits with high commodity prices. Producers continue to face extraordinary risks in their farming operations, and profit margins have narrowed considerably because of the high input prices driven by high energy costs. One of the primary solutions to the food price inflation is to increase the supply of commodity crops in a responsible way that is both respectful and sustainable regarding economic, social, and environmental aspects. At the center of this solution is implementing and developing new technologies to increase crop yields and nutritional values without increasing the amount of fossil-based inputs used in agriculture. Conventional breeding, molecular breeding, genomics, and biotechnology are pivotal technologies for increasing crop yields to meet these supply needs, combined with the impacts of other new technologies and best management practices in agricultural production.

Keywords Biofuels • Food price inflation • Biotechnology • Soybean • Corn • Crude oil prices • Consumer price index

---

G.C. Phillips (✉)
College of Agriculture & Technology, Arkansas State University, State University, AR 72467, USA
e-mail: gphillips@astate.edu

Introduction

The price of food has increased over the past 3 yr. This outcome was forecast in a CAST Commentary (QTA 2006-3) by Cassman, Eidman, and Simpson in 2006. In 2007, the overall rise in food prices was twice that of the overall inflation rate, led by inflation for poultry (meat and eggs) and dairy products in particular (Henderson 2008a). Because of the current high agricultural commodity prices, many people believe that farm producers are earning more net income and are in part to blame for rising food prices. Nothing could be farther from the truth, because the input costs for food production have increased just as dramatically as the commodity prices (FAPRI 2008), resulting in consistently narrow profit margins for the producers (Henderson 2008b).

A number of recent reports suggest factors associated with food price inflation throughout the world (Elobeid et al. 2007; Kruse et al. 2007; Abbott et al. 2008; Henderson 2008a, b; Trostle 2008). These prominent studies have indicated that the main drivers associated with the rise in food prices are the increased energy costs (and the trickle-down impact on farm input costs, especially for fertilizers and chemical control products), correlated with the cost of crude oil. These reports indicate that between 2007 and 2008, soaring energy prices led to a surge in US farm input costs, especially energy-based inputs. For example, Henderson (2008b) reported that by July 2007, US farm input costs had surpassed expectations, rising 20% above 2006 levels. The largest price increases were for fertilizer (derived from natural gas), which more than doubled 2006 levels. Fuel prices also nearly doubled. Seed prices rose 30% and chemical prices went up 12%. Other drivers associated with the rise in food prices were reported to include devaluation of the US dollar, global weather conditions, high energy demand from emerging economies (especially India and China), and the use of commodity crops (especially corn and soybean) in the production of biofuels. Each of these factors has a direct bearing on food price inflation, as illustrated in the following paragraphs.

Energy costs. Surging energy costs are linked with the price escalation of foods (Abbott et al. 2008; Henderson 2008a, b; Trostle 2008). Trostle (2008) specifically indicated that prices of all commodities (food and nonfood) have increased along with the price of oil as shown in Fig. 3. 1. In the last two decades, grain production for biofuels has increased by a factor of ten (Delgado and Santos 2008). Since that time, biofuels have started to play a relevant role in agricultural commodity markets. Because the amount of agricultural commodities (corn and soybean) needed to produce biofuels is large, this translates into pressing demand for inputs. In particular, US ethanol will account for one third of the country's corn crop in the 2009–2010 season (USDA 2008). As crude oil prices have risen over the years, Fig. 3.1 shows a similar path for commodity prices. Indeed, prices of all grains at the end of August 2008 were more than double the prices that prevailed at the beginning of 2007 (McCalla 2009).

The important issue emerging from these studies (Abbott et al. 2008; Henderson 2008a, b; Trostle 2008) is that biofuels today account for a very small fraction of the global liquid fuels supply (1.5%), but their share is rapidly increasing. Furthermore,

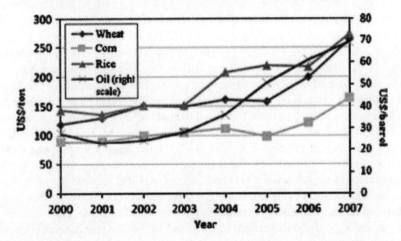

Fig. 3.1 Rising commodity prices, 2000–2007. Source: adapted from von Braun (2008)

since the amount of agricultural commodities (corn and soybean) needed as inputs to produce biofuels is large, this translates into high demand for these commodities (Delgado and Santos 2008). Consequently, the upward pressure on food prices exerted by biofuels not only operates directly via higher food crop demand but also indirectly via food prices and competition for other resources including energy and land. It is clear that biofuels affect food prices, as they constitute an additional source of demand, but this remains a minor contribution to the equation.

In addition, there is a direct link between cost of energy and price of fertilizers, agricultural chemicals, propane, and diesel used in production agriculture (Abbott et al. 2008). To illustrate this point, the Arkansas State University (ASU) Farm experienced a 251% increase in fuel costs to operate the farm machinery comparing the 2007–2008 actual expenditures to those of 2005–2006 (W. Pendegraft and M. Johnson, Arkansas State University, personal communication). During the same time period, fertilizer costs increased by 329% at the ASU Farm. Livestock feed prices increased by 266% in that period. It is noteworthy in this example that the trickle-down impact of increased cost of fertilizers is dramatically higher than the increased cost of energy itself, while the increased cost of commodities used in livestock feed tracks closely to the cost of energy. Early review of the 2008–2009 ASU Farm operation expenditures indicate that these input costs are not increasing at similar rates this year, due to the recent fall in crude oil prices. However, there does seem to be a lag effect, that is, fertilizer and chemical input costs are not declining as rapidly as crude oil prices. We predict that crude oil prices will rise once again as the global economic recovery progresses. Agriculture is very susceptible to fluctuating prices, both for inputs as well as for products.

Devaluation of the US dollar. The devaluation of the US dollar has contributed to the rise in the price of food. Abbott et al. (2008) indicate that the impact of the US dollar on the price of food has been underestimated by several previous studies and that the simultaneous movement of these variables (US dollar and price of oil) supports this trend over the past several years. Because food commodities are traded globally, the decreased value of the US dollar has resulted in worldwide food price inflation. As a result of the devaluation of the dollar relative to foreign currencies since 2002, the cost to other countries of importing agricultural commodities denominated in US dollars, such as grains, food, and oil, has declined. Consequently, countries whose currencies have appreciated relative to the US dollar imported the same amount of agricultural commodities at a lower cost, or imported more at the same cost. This increased global demand as a result of the depreciation of the US dollar has put upward pressure on commodity prices (Trostle 2008). Furthermore, US concerns about recession in 2008 led to successive cuts in nominal and real interest rates that reduced the price of the dollar and encouraged buying and holding of real commodities. These together helped to drive up all real international commodity prices (Frankel 2008). For example, in 2008, the domestic rice prices ranged 6%, 9%, and 11% of the international price rise, respectively, in the Philippines, India, and Vietnam, each of which is a net food exporter. In contrast, countries which import some foods and export other foods, such as Bangladesh, Indonesia, and China, experienced price rises of 43%, 53%, and 64%, respectively (McCalla 2009).

Climate change. Global weather conditions have impacted the price of foods. In 2007, adverse weather events affected yields worldwide, documented by Trostle (2008) as follows: "Floods during harvest in Northern Europe; drought in Southern Europe, Russia, Ukraine, Australia and Northwest Africa; late freeze followed by drought in Argentina; hot dry growing conditions in Canada; and late hard freeze that affected the red winter wheat crop in the United States." The study by Abbott et al. (2008) affirms that adverse weather conditions made a bad situation for food crops even worse during the last few years. Between 2005 and 2007, drought in large wheat-producing countries such as Australia and Ukraine, floods in Asia and dry weather in the USA, the European Union and Canada negatively affected world agricultural production. Several years of weather impacts including Europe in 2006, North America in 2006 to 2007, and a continuing severe drought in Australia 2006–2008 drew global stocks down to critical lows between 2005 and 2008 (McCalla 2009). In 2008, the late spring and early summer floods in Iowa, Missouri, Illinois, Arkansas, and elsewhere had a direct impact on the prices of food commodity crops (Carey 2008). In Australia and Canada, for instance, adverse weather led to a combined fall of over 20% in production per hectare of wheat and other grains between 2005 and 2007. Then in 2006 and again in 2007, adverse weather in a number of major producing countries reduced global production of major grains. All of these weather-related factors together resulted in declining global agricultural stocks that ultimately contributed to higher prices for agricultural commodities. Importers faced declining market supplies and many countries experienced politically sensitive increases in domestic food prices which led some to contract aggressively for future imports, even at world record prices (Trostle 2008; McCalla 2009). The general public often forgets that the risk of bad weather at the wrong times of the growing season makes farming the riskiest business in the world. Approximately half of producers experience a loss on income with one or more crops every 4 or 5 yr (T. Griffin, University of Arkansas, Division of Agriculture, Cooperative Extension Service, personal communication).

Emerging economies. A relevant driving factor that has dramatically caused a change in agricultural commodity prices is the increasing food demand arising from unprecedented economic growth in highly populated emerging countries such as China and India over the last decade. Increased imports of food commodities such as meat products and soybean to Asia have clearly increased over the past decade, and forecasts indicate continued demand and diet diversification in these regions (Henderson 2008a, b; Trostle 2008). China, for instance, has had annual economic growth rates above 10% in most of the last 15 yr. This rapid growth in highly populated emerging economies has translated into both a rise in demand for major agricultural commodities and a change toward a meat-based diet. From 1980 to 2002, total meat consumption in developing countries nearly tripled (Mullins 2008). As a result of the move toward a more meat-based diet, demand for feed has multiplied in these rapid economic-growth countries. Annual per capita meat consumption in China, for instance, rose from 34 to 49 kg between 1997 and 2007, implying that around 20% of the worldwide increase in grain and oilseed demand between 1997 and 2007 is attributable to the switch to meat in China (Gould and Villarreal 2006; Delgado and Santos 2008; Streifel 2009). Soybean consumption in China has been growing at about 15% in recent years, while soy oil and palm oil have been surging at more than 20% and 25%, respectively (Streifel 2009). In fact, more than 80% of the increase in Chinese grain and oilseed demand stems from diet changes (Delgado and Santos 2008; Streifel 2009). Like China, India continues to consume a sizeable share of many agriculture commodities because of its large population and economic growth. India is the largest consumer of tea and sugar, and the second largest consumer of wheat, rice, cotton, and palm oil. Also, soybean oil consumption in India has experienced more rapid growth (Streifel 2009). Similar to China, India's growth in the consumption of agricultural commodities is also likely to increase due to changing eating patterns and rising income (Gould and Villarreal 2006; Delgado and Santos 2008; Streifel 2009). Therefore, the continuous growth in the middle classes of many emerging countries including India and China will continue to shift toward a meat-based diet, thus sustaining the upward pressure on feed commodity prices.

These high levels of imports of agricultural commodities into Asia and other emerging economies will continue for the foreseeable future, because the depreciated US dollar makes US agricultural commodities (soybean and meat) cheaper on the international export market. Coordinated with this growing economy is the concomitant rise in imports of nonfood products into Asia—most notably, oil and steel. Also noteworthy is that the net importation of rice, wheat, and corn into India and China has not changed significantly over the past several years (Abbott et al. 2008), primarily due to government mandates to become self-sufficient for these staples.

Biofuels. Current demand for biofuels in the USA has been driven largely by high crude oil prices, dependence on unstable and hostile countries for crude oil, environmental concerns, the weak US dollar, and the initiatives by the federal government and several states to expand the use of biofuels. The Renewable Fuels Standard, part of the national 2005 Energy Policy Act, requires blending of renewable fuels with gasoline to reach 7.5 billion gallons by 2012 (U.S. EPA 2006). The Energy Independence and Security Act of 2007 (The White House 2007) compels fuel producers to use no less than 36 billion gallons of biofuels by 2022. These stimuli have fueled renewable energy production in the USA in recent years and have created an alternative market for agricultural commodities used as feedstocks, such as corn used for ethanol production and soybean used for biodiesel production. Consequently, new construction and expansions of existing biofuels production facilities in the USA have accelerated dramatically. Currently, new capacity (new builds plus the expansion of existing facilities) will increase ethanol output by nearly 50% between 2006 and 2008. With respect to biodiesel, new facilities will boost capacity by nearly 200% in the same period (Havran 2006; Koplow 2006; Smith 2007). According to the Renewable Fuels Association, there are 119 ethanol biorefineries in the USA, which have the capacity to produce more than 6.1 billion gallons a year. In addition, 86 ethanol refineries are under construction and these would add more than 6.4 billion gallons annually. Today, 26 states have ethanol plants, up from 17 states at the start of 2000 (Renewable Fuels Association 2007; Smith 2007).

In general, increased demand for biofuels has caused the prices of agricultural commodities used as feedstocks and other competing crops to increase because of the direct impact of biofuels production on agricultural commodities and markets, with both direct and indirect implications for food prices and allocations of rural agricultural land (IEA 2004; Baker and Zahniser 2006; Donnelly 2006; Hill et al. 2006; von Lampe 2006; IMF 2007; Sauser 2007; Secchi and Babcock 2007; Smith 2007; Tokgoz et al. 2007; United Nations 2007). While there is consensus on the impact of biofuel production on the prices of agricultural commodities, it is unclear whether the increased commodity prices have resulted in positive incomes for rural farmers. For example, results from some studies on public support for biofuel production and their impact on rural development argue that the production of biofuels is itself fuel intensive, obviating many of the benefits of growing agricultural commodities as feedstocks for biofuels, and suggest that there are less expensive options for both greenhouse gas mitigation and rural development (Farrell et al. 2006; Hill et al. 2006; U.S. EPA 2006). However, a recent report from the Congressional Budget Office (2009) suggests that this is a worse-case scenario that only applies to conversion of forested land to agricultural use in order to produce bioenergy crops. We have yet to see the definitive studies that may settle this issue.

The use of corn and soybean as biofuel crops has generated increased demand for these commodities. In the case of corn, much attention from the public media has been placed on the use of corn for ethanol production and that this alternate demand for corn has been the primary reason for increased food prices. In fact, the 2007 corn harvest was a record 13.1 billion bushels and only 2.3 billion bushels were used in the production of ethanol—17.6% of the corn harvest (National Corn Growers Association Statistics). A portion (estimated to be 23.75%) of the increase in the price of corn is due to increased demand for biofuels. However, data gleaned by Abbott et al. (2008) indicate that the rise in the price of oil from $40 to $120 per barrel was mirrored by the rise of corn from $2 to $6 per bushel; out of the

$4 increase in the price of corn, $3 is attributed to the price of oil and $1 to ethanol. Furthermore, foods that use corn as an ingredient make up less than one third of the retail food spending, and a 50% rise in the price of commodity corn would result in less than a 1% rise above the normal rate for food price inflation (Leibtag 2008). A recent Congressional Budget Office (2009) report concluded that the increased use of biofuels accounted for only 10–15% of total food price increases between April 2007 and April 2008, whereas higher energy costs accounted for 36% of the overall food price increase.

Ethanol production has been growing steadily in major corn producing states such as Iowa, Indiana, Illinois, and Minnesota over the last two decades. Acreage planted with corn in the USA increased from 79.5 million in 1997 to 93.6 million in 2007 (National Corn Growers Association 2008). Many of the studies cited above have focused on the corn-based ethanol industry; however, comparatively little attention has been focused on the soybean-based biodiesel industry. The row crop intensive agricultural area in eastern Arkansas (#1 US producer of rice, #2 US producer of cotton, #10 US producer of soybean which is commonly used as a second crop with either rice or wheat) is very competitive in the production of soybean, the main biodiesel feedstock. As biodiesel consumption in the USA has risen dramatically since 1999, its production in Arkansas has been aided by public support and advantageous production conditions. The state of Arkansas also provides incentives in the form of tax credit for biodiesel production to complement established federal incentives. All these factors have accelerated biodiesel production in Arkansas in recent years. While no soybean-based biodiesel was produced in Arkansas until 2006, total production has since increased from 11 million gallons in 2006 to 15 million gallons in 2008 (Archer 2008). Since 2006, three new biodiesel plants have been constructed at Stuttgart, Dewitt, and Batesville, AR, causing the estimated annual production capacity to jump by over 1,500% between 2007 and 2008. When the proposed fourth biodiesel plant is completed in 2009, annual production will be increased to more than 63 million gallons.

The recent biodiesel plant constructions in Arkansas have become an increasingly important and rapidly growing outlet market for Arkansas produced soybean when soybean prices are favorable. As the pace of this industry growth accelerates, it is useful to examine issues related to its impact on the agricultural economies where the key raw material (soybean) is produced. The following sections will assess the Arkansas soybean industry, as an example, regarding the factors driving soybean commodity prices, production costs, producer incomes, and the relationship of these factors to food prices generally.

Impact of crude oil prices on soybean prices. As the crude oil price has risen in recent years, biodiesel production in Arkansas has increased and the price of soybean used as the main biodiesel feedstock has soared as well. To determine the impact of the high crude oil prices and the resultant biodiesel production on soybean prices, a "causality" model was developed (Archer 2008). This model, based on the cross-correlation functions, was used to test the price relationships between soybean, crude oil, and Consumer Price Index (CPI), as well as the direction of causality relationship between the prices.

Table 3.1 shows that the cross-correlation coefficients, T statistics, and P values are significant at the 1% level at lags of +3, +2, +1, 0, −1, −2, −3 yr. We originally suspected direct leads and lags at low values, say 0–3 yr, and feedback, if it existed, to be running from CPI and crude oil prices to soybean prices at maximum lags of about 1 to 3 yr. Thus, analysis of the "lead/lag" causality model and cross-correlations were made with lags of one to three periods in each direction. This statistical significance suggests that there is some evidence of two-way or feedback causality; specifically, the soybean price series are influenced by crude oil prices and CPI. Moreover, the prominence of the highly significant positive coefficients at lag zero for all pairs of price series suggests an instantaneous relationship between soybean prices, CPI, and crude oil prices. Soybean prices are highly correlated with inflation rates (82%) and crude oil prices (77%; see the respective coefficients at 0 yr lag/lead in Table 3.1). Furthermore, all the coefficients at positive and negative lags have nonzero values. These results indicate that soybean producers employ past and future information on crude oil prices and inflation rates in forming their production decisions and that the production of soybean is directly influenced by past and future crude oil prices and inflation rates (CPI).

Further analysis of the impact of crude oil prices and CPI on soybean prices was performed using a "lead/lag" causality model (the regression used here allowed autoregressive structure in data errors) calculated at lead/lags of 1, 2, and 3 yr to capture the direction of relationships. Results of the causality analysis in Table 3.2 (coefficients and R^2) indicate that there are strong bidirectional causality relationships between crude oil prices, CPI, and soybean prices.

Table 3.2 also shows that a rise of $1 in crude oil prices raises soybean prices by a little more than $0.06 per bushel (see coefficient of 0.064 for crude oil at 0 yr lead/lag). These results imply that the rising global demand for crude oil and its corresponding high prices have directly stimulated demand for soybean feedstock for the expanding biodiesel industry, increasing soybean prices and thereby creating higher incentives for increased soybean production in Arkansas.

The value of the US dollar has also impacted soybean prices in recent years. Generally, a weak US dollar creates inflation in the economy. However, since crude

Table 3.1 Cross-correlation of soybean prices, CPI, and crude oil prices: 1977–2007

Lead/lag[z] years	Crude oil Coefficients	T test results T statistics	P value	CPI Coefficients	T test results T statistics	P value
3	0.46	5.05	<0.0001	0.55	3.08	0.0023
2	0.54	5.51	<0.0001	0.63	3.38	0.0008
1	0.61	5.87	<0.0001	0.71	3.57	0.0004
0	0.77	5.9	<0.0001	0.82	3.83	0.0002
−1	0.63	5.64	<0.0001	0.72	4.16	<0.0001
−2	0.57	5.6	<0.0001	0.66	4.62	<0.0001
−3	0.51	5.62	<0.0001	0.57	4.66	<0.0001

All price series were prefiltered using an ARIMA filter to remove substantial autocorrelation.
[z] Negative values imply that crude and CPI lead soybean prices; positive values imply they lag soybean prices.

Table 3.2 Lead/lag regression of soybean prices on past and future crude oil prices and CPI

Lead/lag years	Crude oil				CPI			
	Coefficient	R^2	F statistics	Standard error	Coefficient	R^2	F statistics	Standard error
3	0.041	53.92	16.98	0.027	0.022	48.66	18.33	0.042
2	0.053	67.01	21.48	0.027	0.026	63.31	24.46	0.044
1	0.059	78.28	25.48	0.028	0.031	71.46	27.24	0.041
0	0.064	89.42	26.33	0.028	0.036	82.33	29.46	0.032
−1	0.056	73.96	24.04	0.029	0.033	68.72	25.62	0.052
−2	0.048	73.78	23.98	0.029	0.027	58.69	24.73	0.053
−3	0.039	75.63	24.63	0.029	0.023	41.01	24.66	0.053

Prices are from 1977 to 2007. Negative values imply crude oil prices and CPI lead soybean prices; positive values imply they lag soybean prices.

oil and most agricultural commodities are priced in US dollars, any depreciation of the dollar raises soybean prices. Between 2002 and 2007, the US dollar depreciated by about 45%. Over the same period, soybean prices increased from $5.12 to $11.20 per bushel, an increase of 119%. At a recent peak price of over $12, soybean would not be exportable. Despite this, international soybean export continues to rise because the depreciated US dollar makes US soybean cheaper on the international export market. The devalued dollar has therefore raised soybean prices through export demand. Furthermore, Table 3.2 shows that changes in CPI have influenced soybean prices—a rise in inflation rate of $1 raises soybean prices by $0.04 (see CPI coefficient at 0 yr lead/lag). Thus, high crude oil prices, changes in CPI, and the devalued US dollar have all contributed to the recent spike in soybean prices.

Impact of high crude oil prices on soybean production costs. As high crude oil prices have spurred demand for biodiesel, the resultant increase in soybean production has stimulated demand for major farm inputs such as fertilizer, pesticides, diesel, and chemicals which themselves are fossil fuel intensive or use crude oil (fossil-based fuel) as key ingredients. Prices for fossil-based inputs for soybean production have risen considerably since 2002 due to the "demand pull" from soybean being used in the biodiesel industry as well as the "cost push" from fossil-based products used as key ingredients in producing agricultural inputs.

In evaluating the impact of high crude oil prices on the cost of producing soybean, per acre soybean production costs were cross-correlated with crude oil prices and CPI (Archer 2008). Figure 3.2 shows that crude oil prices, CPI, and soybean production costs move together. Between 1997 and 2007, crude oil prices increased from $19.04 per barrel to $67.93 per barrel, an increase of $48.89. During the same period, soybean production costs increased by $111.52 per acre.

Table 3.3 also shows that crude oil prices, CPI, and soybean production costs are highly correlated. The correlation coefficient between soybean production cost and crude oil prices is 0.92. Table 3.3, however, reveals that the correlation coefficient between soybean production costs and CPI is only 0.52. The implication is that changes in crude oil prices have greater relationships with soybean production costs than do changes in CPI; in other words, crude oil prices impact soybean

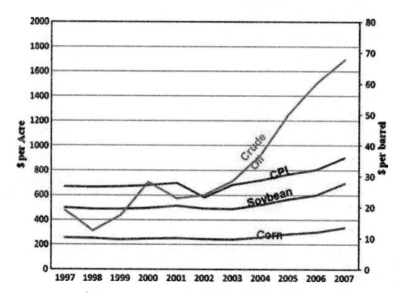

Fig. 3.2 Production costs of soybean and corn, crude oil prices, and CPI. CPI used is the ratio of prices received to prices paid (%). Ratio uses the most recent prices paid index. Source: Drawn from ERS data—http://www.ers.usda.gov/Data/CostsAndReturns/testpick.htm

Table 3.3 Per acre production costs, crude oil prices, and CPI

	Corn	Soybean	Crude oil	CPI
Years				
1997	258	241.47	19.04	169.5
1998	252	237.99	12.52	173.4
1999	241	250.80	17.51	177.0
2000	250	249.24	28.26	181.3
2001	251	263.31	22.95	186.1
2002	245	247.07	24.1	90.5
2003	241	250.77	28.53	193.2
2004	260	263.26	36.98	196.6
2005	287	280.44	50.24	200.9
2006	302	300.51	60.24	205.9
2007	341	352.99	67.93	210.7
Correlation coefficients between corn, soybean, crude oil prices, and CPI				
Crude oil	0.91	0.92		
CPI	0.51	0.52		

Source: extracted from ERS data—http://www.ers.usda.gov/Data/CostsAndReturns/testpick.htm.

production costs more than inflation rates. Therefore, both inflation rates and crude oil prices must be factored in when assessing the recent spike in soybean production costs.

A regression model was also used to examine the impact of crude oil prices and CPI on soybean production cost. Table 3.3 shows that in the past 10 yr (since 1997),

Table 3.4 Regression of per acre soybean production costs (average per acre production cost for Arkansas soybean producers) to crude oil prices and consumer CPI, 1997–2007

	Coefficient	R^2	Standard error
Crude oil prices	1.437^z	0.91	0.28
Consumer Price Index	0.394^z	0.87	0.17
β^3	0.757^z	0.46	0.11
Intercept	64.37^z	0.68	12.33

β^3 captures autoregressive dependence in errors.
^{z}Indicates significance at the 1% level.

soybean production costs in Arkansas have increased from $241.47 to $352.99 per acre, an increase of $111.52. The regression results in Table 3.4 show that crude oil prices contributed $70.26 (i.e., $48.89 crude oil price increase over the 10 yr × $1.437 cost increase per $1 crude oil price increase) and inflation $16.23 (i.e., 41.2% increase in index × $0.394 cost increase per 1% CPI increase), respectively, to the $111.52 increase in the soybean production cost over the 10-yr period. Therefore, as crude oil price continues to rise, soybean production costs will also rise in relation to inflation, which will ultimately lead to higher market prices if soybean farmers are to survive. The implication is that both inflation rates and high crude oil prices, to a large extent, have made soybean production more expensive by raising the cost of inputs such as fertilizers, pesticides, and transportation.

Impact of high crude oil prices on incomes of soybean producers. The recent rise in commodity prices and the resulting rise in commodity-based food prices have created the perception that farmers producing soybeans for the biodiesel industry are making huge profits. This perception, often repeated in the media, has been represented as "fact", and the public often accepts it uncritically. In particular, the alleged high profits being made by soybean farmers as a result of high crude oil and food prices is often based on inaccurate or incomplete information. These perceptions of unusually high profits also fail to take into account the rising input/production costs and inflation, to the dismay of soybean farmers.

Figure 3.3 shows that crude oil prices from 1997 through 2007 move in different directions with operating and net incomes of soybean producers, except for the period 2000 through 2003 when crude oil prices and incomes of soybean producers all increased. However, as crude oil prices rose since 2003, incomes of soybean producers have continued to fall. Indeed, Fig. 3.3 shows that soybean producers have actually lost money (negative net incomes) since 2003 when crude oil prices have been skyrocketing. This trend disproves the general belief that soybean farmers are making high profits by producing soybeans for the biodiesel industry, as a result of the recent rises in crude oil and commodity prices.

Further analysis of the impact of crude oil prices on Arkansas soybean producers' incomes was made by correlating per acre net and operating incomes with crude oil prices and CPI from 1997 through 2007. The results in Table 3.5 show that net and operating incomes of soybean producers have low correlation coefficients with both crude oil prices (0.13 and 0.26, respectively) and CPI (0.03 and −0.01, respectively). The implication is that the recent rise in crude oil prices and inflation rates have not

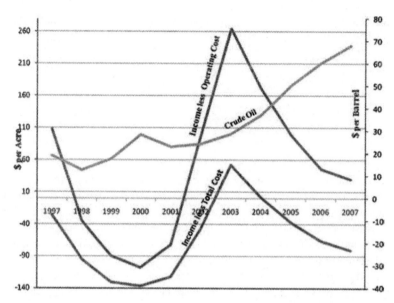

Fig. 3.3 Soybean production income and crude oil prices—Arkansas (1997–2007). Source: extracted from ERS Production Cost Data—http://www.ers.usda.gov/Data/CostsAndReturns/testpick.htm

Table 3.5 Per acre income (US $) for Arkansas produced soybean (soybean income per planted acre, excluding government payments)

	Income less total cost[z]	Income less operating cost[x]	Crude oil price	Consumer Price Index
1997	−25.09	133.50	19.04	169.5
1998	−94.49	60.07	12.52	173.4
1999	−130.05	40.59	17.51	177
2000	−136.44	28.76	28.26	181.3
2001	−122.36	50.11	22.95	186.1
2002	−46.84	144.22	24.1	90.5
2003	52.44	212.38	28.53	193.2
2004	1.84	169.49	36.98	196.6
2005	−37.06	136.35	50.24	200.9
2006	−65.82	112.08	60.24	205.9
2007	−80.88[y]	110.99[y]	67.93	210.7
Correlation coefficients between soybean income, crude oil prices, and CPI				
Crude	0.13	0.26		
CPI	0.03	−0.01		

Source: extracted from ERS data—http://www.ers.usda.gov/Data/CostsAndReturns/testpick.htm.
[z] Income less total cost (net income) is gross value of production less operating cost (e.g., fertilizer, chemicals, fuel, etc.) and overheads (e.g., hired labor, equipment, insurance, etc.).
[y] Estimates.
[x] Income less operating cost refers to gross value of production less fertilizer, chemicals, fuel, seeds, etc.

raised soybean producers' incomes but have actually reduced their income. The major factor contributing to the falling incomes is the rising farm level input costs which themselves are fossil fuel intensive or use crude oil as key ingredients.

Impact of crude oil prices on overall food prices. Many studies, including those from the United Nations (2007), indicate that the significant use of soybean for biodiesel production has resulted in higher costs of food. Indeed, a recent report by the International Monetary Fund states that: "Higher prices of corn and soybean oil will also likely push up the price of partial substitutes, such as wheat and rice, and other edible oils, and exert upward pressure on meat, dairy, and poultry prices by raising animal rearing costs, given the predominant use of corn and soy meal as feedstock, particularly in the United States (more than 95 percent)." (IMF 2007).

Generally, the world population has been growing and demanding more and different types of foods. Furthermore, increased economic growth in many emerging and developing countries has provided rapid increases in incomes resulting in higher consumer purchasing power. These factors have generated a rising demand for nontraditional staples and toward higher-value foods such as meat and milk. This shift in "food ways" has caused increased demand for grains used to feed livestock. As the production cost of grains has increased in recent years due to higher crude oil prices, higher input costs, overall inflation, and conversion of agricultural commodities to biofuel production, food prices have also soared.

The recent increases in food prices have made many in the media conclude that commodity farmers, including soybean and corn producers, are profiting while consumers are paying much higher prices for food necessities. However, as previously explained, high crude oil prices and its trickle-down effects have ultimately raised the cost of production at the farm level. These have caused the prices of food products, in general, and those that use soybean and corn as ingredients, in particular, to rise.

Solutions and policy implications. One of the primary solutions to the food inflation problem is to increase the supply of commodity crops in a responsible way that is sustainable regarding economic, societal, and environmental aspects. At the center of this solution is implementing and developing new technologies to increase crop yields without increasing the amount of land and fossil-based inputs used in agriculture.

A successful track record can be seen in that, historically, the introduction of new technologies has increased the rate of gain for crop yields in the past. For example, corn yields have increased in the USA since the introduction of hybrid corn in the mid-1930s. However, this yield increase has not been kept at a constant rate. In both the mid-1930s and again in the mid-1950s, the introduction of new hybridization technology increased the rate of gain for corn yields. The introduction of double cross hybrids in the 1930s raised the rate of gain from 0 to around 1 bu ac^{-1} yr^{-1}, and the introduction for single cross hybrids around 1955 further increased the rate of gain with an average gain of about 1.8 bu ac^{-1} yr^{-1} (CAST 2006; Troyer 2006; Fig. 3.4). Improvements in fertilization, agronomic practices and farm machinery certainly contributed to this improvement, but 50% to 60% of this gain in output is attributed to plant breeding, with the remainder coming from improvements in fertilization, agronomic practices, and farm machinery (Russell 1991; Duvick 1992, 1999).

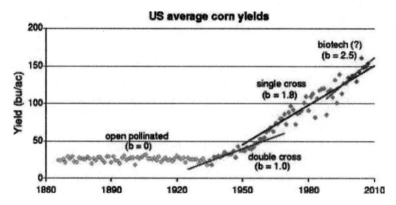

Fig. 3.4 Rate of yield gain for periods characterized by different corn breeding methods. Rate of yield gain (b) is in bu ac^{-1} yr^{-1}. Source: Troyer (2006) with updated data from USDA-NASS (2007)

More recently, marker-assisted breeding has further increased the rate of yield gain for corn (Johnson and Mumm 1996; Stuber et al. 1999; Johnson 2002; Troyer 2006). Marker-assisted breeding takes advantage of emerging molecular genetics tools, and its implementation has doubled the rate of gain for yield relative to conventional breeding in both commercial corn and soybean breeding programs (Eathington et al. 2007). Nearly all corn grown commercially today comes from hybrids produced in private sector breeding programs (Duvick and Cassman 1999; Hurley 2005; Aguiba 2007; ERS 2008). This means that forward-looking yield estimates should include as much information as possible from commercial breeding programs. Applying this technology to breeding programs throughout the world is expected to translate into increased rates of gain in commodity yield averages.

Biotechnology, including direct gene transfer and genomics methodologies, is another new technology that has increased corn yields (Stuber et al. 1999). Corn protected from corn borer was first introduced on a large scale in the USA in 1997. Over the next 10 yr, corn carrying biotechnology traits has increased to more than 35.2 million hectares (approximately 87 million acres) in 2007 (James 2007). Monsanto's introduction of "Roundup Ready 2 Yield$^{(TM)}$ soybean technology" is projected to increase soybean yield by 7% to 11% in 2009 (Reuter News 2008). Furthermore, the benefits from biotech traits have been friendly to the environment through the use of insect control traits that result in less pesticide usage and due to the dramatic implementation of conservation tillage (no-till) through the use of herbicide tolerance traits. In particular, no-till farming results in less erosion, conservation of nutrients and soil organic matter, and reduced input costs, primarily in diesel fuel savings (Carpenter et al. 2002).

Biotechnology traits currently on the market act to protect corn yields through either insect pest protection or herbicide tolerance. Recent reports from the field indicate the additive effect on yield of stacking biotech traits for above ground protection (herbicide tolerance and Lepidopteran insect control) and below ground protection (Coleopteran insect control; Edgerton 2009). In addition, second generation biotech traits currently under development (such as drought stress tolerance) have also shown yield increases in testing programs (James 2007). Biotechnology tools are being used to develop soybeans with improved nutritional value and greater resistance to disease, herbicides, and drought (Soper et al. 2003). Biotech traits in soybean offer the promise of crop protection benefits, higher yield, and grain value enhancement through oil and protein modification (Soper et al. 2003). Therefore, it is likely that stacking biotech yield traits will also show an additive benefit, especially when combined with superior genetics identified through molecular or marker-assisted breeding. Together, marker-assisted breeding and biotechnology are expected to provide a significant enhancement in the rate of corn yield gain at least though the early part of the 2020s (Edgerton 2009).

Another recent technological innovation that is resulting in dramatic change in agricultural production and management is the use of geographic information systems (GIS) technology, also referred to as spatial or geospatial technology, which enables a higher degree of precision farming. This technology recognizes that variability exists not only between fields and locations but also within fields. Important measurable variables such as soil type, nutrient level, moisture level, pest or pathogen pressure, as they vary across the field, are incorporated into the decision software for spatial analysis (Berry et al. 2008). GIS is now used to map the health of crop fields and prescribe variable application rates for factors such as seed planting density, fertilizer, and irrigation water (Rickman et al. 2003). This strategy may save producers on input costs by application of inputs only where needed, and the goal is to achieve more uniform quality and harvest operations. While it is still too early to quantify the financial impact of this new technology in agricultural production, it is becoming rapidly adopted for the benefits of precision and uniformity. Geospatial technology is touted as one of the three "mega-technologies" of the twenty-first century, along with biotechnology and nanotechnology (Berry et al. 2008). Two of the three mega-technologies (geospatial and biotechnology) have already been adopted by agricultural producers, and the third will likely impact agricultural production in the foreseeable future, so it is fair to conclude that US agricultural producers are amenable to rapid adoption of new technologies.

Conclusions. We predict that the next 15 to 20 yr will see the development and implementation of a wide variety of new technologies in agricultural production. Second- and third-generation biotech traits are under development. There is current demand for increased food traceability, food quality assurance, and food safety, and these stimuli will inevitably lead to the development and adaptation of new technologies to gain greater precision and accuracy in these areas. The impacts of current discussions related to climate change will also result in improved management practices to recycle nutrients and carbon in a more efficient and environmentally sustainable manner. Therefore, the continued adoption and implementation of agricultural technology advancements will result in a continued increase in the rate of yield gain with the same or fewer inputs and improve the supply of commodity grain to ease demand. The costs associated with such technological developments are small compared to the fundamental driver of energy cost as related to food prices.

We urge the public to become more engaged in critically examining the information they are provided through popular media sources and to strive to understand the complexity and global nature of many of our current agricultural issues. Farmers should be commended for taking on production risks that no other industry will tolerate, rather than being unfairly blamed for rising food prices! How can we help? We support the use of sound scientific analysis to critically examine the issues, and we support investment in research and development of the next-generation efficient technologies to tackle the ever increasing demands on food production.

References

Abbott, P. C.; Hurt, C.; Tyner, W. E. What's driving food prices? Farm Foundation Issue Report, July 2008, Farm Foundation, Oak Brook, IL, 2008.

Aguiba, M. Corn hybridization rate seen to rise to 35%. Manila Bulletin; Tuesday, July 31 2007; http://www.articlearchives.com/agriculture-forestry/agriculture-crop-production-grain/56716-1.html (Accessed on April 04, 2009)

Archer, A. Economic impact of biofuel feedstock on rural agricultural economy—the case of rural Arkansas. M.S. thesis, College of Agriculture & Technology, Arkansas State University, Jonesboro; 2008.

Baker, A.; Zahniser S. Ethanol reshapes the corn market. *Amber Waves* 42: 30–35; 2006.

Berry, J. K., Delgado, J. A.; Khosla, R. Precision farming advances agricultural sustainability. V1 Magazine, Sunday, February 10, 2008. http://www.vector1media.com/article/feature/precision-farming-advances-agricultural-sustainability/ (Accessed on April 07, 2009); 2008.

Carey, N. Midwest farmland flooding boosts food prices. Reuters, June 17, 2008 article idUSN1346134820080617; 2008

Carpenter, J.; Felsot, A.; Goode, T.; Hammig, M.; Onstad, D.; Sankula S.Comparative environmental impacts of biotechnology-derived and traditional soybean, corn and cotton crops. Council for Agricultural Science and Technology, Ames, Iowa, 2002.

Congressional Budget Office (CBO). The impact of ethanol use on food prices and greenhouse-gas emissions. Congress of the United States, Pub. No. 3155, April 2009. http://www.cbo.gov/ftpdocs/100xx/doc10057/04-08-Ethanol.pdf (Accessed on April 10, 2009); 2009.

Council for Agriculture Science and Technology (CAST). Convergence of agriculture and energy: Implications for research and policy. CAST Commentary QTA 2006-3/. CAST, Ames, IA; 2006.

Delgado, J.; Santos, I. The new food equation: Do EU policies add up?" In: Bruegel Policy Brief. Issue 2008/06, July 2008. http://www.bruegel.org/Public/fileDownload.php?target=/Files/media/PDF/Publications/Policy%20Briefs/pb-2008-06_FINAL.pdf (Accessed on April 01, 2009); 2008.

Dobermann, A.; Cassman, K. G. Plant nutrient management for enhanced productivity in intensive grain production systems of the United States and Asia. *Plant Soil* 247: 153–175; 2002.

Donnelly, J. Farmers could get crop change: A call for fuel may bring cut in grassy field. Boston Globe, October 2, 2006; 2006.

Duvick, D. N. Genetic contributions to advances in yield of U.S. maize. *Maydica.* 37: 69–79; 1992.

Duvick D. N. Heterosis: Feeding people and protecting natural resources. In: Coors J. G.; Pandey S. (eds) The genetics and exploitation of heterosis in crops. American Society of Agronomy, Inc., Crop Science Society of America, Inc., Soil Science Society of America, Inc., Madison, WI, pp 19–29; 1999.

Duvick, D. N.; Cassman, K. G. Post-Green Revolution trends in yield potential of temperate maize in the North-Central United States. *Crop Sci* 39: 1622–1630; 1999.

Eathington, S. R.; Crosbie, T. M.; Edwards, M. D.; Reiter, R. S.; Bull, J. K. Molecular markers in a commercial breeding program. *Crop Sci* 47: S–154–S-163; 2007.

Economic Research Service of USDA Adoption of genetically engineered crops in the U.S. USDA ERS, Washington, DC. 2008. http://www.ers.usda.gov/Data/BiotechCrops/ExtentofAdoptionTable1.htm (Accessed on April 04, 2009).

Edgerton, M. Increasing crop productivity to meet global needs for feed, food, and fuel. *Plant Physiol* 149: 7–13; 2009.

Elobeid, A.; Tokgoz, S.; Hayes, D. J.; Babcock, B. A.; Hart, C. E. The long-run impact of corn-based ethanol on the grain, oilseed, and livestock sectors with implications for biotech crops. AgBioForum 101: 11–18; 2007.

Farrell, A. E.; Plevin, R. J.; Turner, B. T.; Jones, A. D.; O'Hare, M.; Kammen, D. M. Ethanol can contribute to energy and environmental goals. *Science* 311: 506–508; 2006.

Food and Agricultural Policy Research Institute—FAPRI. Agricultural Outlook. www.fapri.missouri.edu/outreach/publications/2008/OutlookPub2008.pdf 2008.

Frankel, J. An explanation of soaring commodity prices. VoxEU.org (Accessed on March 25, 2008); 2008.

Gould, B. W.; Villarreal, H. J. An assessment of the current structure of food demand in urban China. *Agric. Econ.* 34: 1–16; 2006.

Havran, N. Maintenance as a profit center in a fuel ethanol plant. Ethanol Producer Magazine, March 2006, pp. 60–62; 2006.

Henderson, J. What is driving food price inflation? *The Main Street Economist* 31: 1–5; 2008a.

Henderson, J. Are energy prices threatening the farm boom? In AgDM Newsletter, November, 2008, Iowa State University. http://www.extension.iastate.edu/AGDm/articles/others/HenNov08.html 2008b.

Hill, J.; Nelson, E.; Tilman, D.; Polasky, S.; Tiffany, D. Environmental, economic, and energetic costs and benefits of biodiesel and ethanol biofuels. *Proc. Nat. Acad. Sci. (U.S.A.)* 10330: 11206–11210; 2006.

Hurley T. M. Bt resistance management: Experiences from the U.S.. In: Wesseler J. H. H. (ed) Environmental costs and benefits of transgenic crops in Europe. Wageningen UR Frontis Series, Vol. 7. Springer, Dordrecht, pp 81–93; 2005.

IEA. World energy outlook 2004. International Energy Agency, Paris; 2004.

IMF—International Monetary Fund. World economic outlook 2007. April 2007, p. 44. http://www.imf.org/external/pubs/ft/weo/2007/01/pdf/text.pdf 2007.

James, C. Global status of commercialized biotech/GM crops: 2007. ISAAA Brief No. 37. ISAAA, Ithaca, NY, pp 1–125; 2007.

Johnson, G. R. Process for predicting the phenotypic trait of yield in maize. U.S. Patent 6,455,758; 2002.

Johnson, G. R.; Mumm, R. H. Marker assisted maize breeding. Amer. Seed Trade Assoc. Proc. 51st Annual Corn and Sorghum Industry Research Conf. 51:75–84; 1996.

Koplow D. Biofuels at what cost?: Government support for ethanol and biodiesel in the United States. The Global Subsidies Initiative (GSI) of the International Institute for Sustainable Development (IISD), Geneva, Switzerland; 2006.

Kruse, J.; Westhoff, P.; Meyer, S.; Thompson, W. Economic impacts of not extending biofuel subsidies. *AgBioForum* 102: 94–103; 2007.

Leibtag, E. Corn prices near record high, but what about food costs? *Amber Waves* 61: 11–15; 2008.

McCalla, A. World food prices: Causes and consequences. *Can. J. Agric. Econ.* 571: 23–34; 2009.

Mullins, L. What is pushing up crop price. US News and World Report, January 24, 2008. http://www.usnews.com/Topics/tag/Author/l/luke_mullins/index.html (Accessed on November 15, 2008); 2008.

National Corn Growers Association. Corn Production Trends 1991–2007, http://www.ncga.com/corn-production-trends (Accessed on April 04, 2009); 2008.

Renewable Fuels Association. News Release, April 2, 2007, http://www.ethanolrfa.org/media/press/rfa/2007/view.php?id=1054 (Accessed on February 15, 2008); 2007.

Reuter News. Farmer excitement building around soybean economic and yield opportunities. Tuesday, April 22, 2008, 6:04pm EDT. http://www.reuters.com/article/pressRelease/idUS256592+22-Apr-2008+PRN20080422. (Accessed on October 07, 2008); 2008.

Rickman, D.; Luvall, J. C.; Shaw, J.; Mask, P.; Kissel, D.; Sullivan, D. Precision agriculture: Changing the face of farming. Geo*times*, November 2003. http://www.geotimes.org/nov03/feature_agric.html (Accessed on October 10, 2008); 2003.

Russell, W. A. Genetic improvement of maize yields. *Adv. Agron.* 46: 245–298; 1991.

Sauser, B. Ethanol demand threatens food prices. Technology Review. February 13, 2007. http://www.technologyreview.com/Energy/18173/ (Accessed on February 15, 2007); 2007.

Secchi, S., Babcock, B. A. Impact of high crop prices on environmental quality: A case of Iowa and the Conservation Reserve Program. Working Paper 07-WP 447. Ames, IA: Center for Agricultural and Rural Development, Iowa State University; 2007.

Smith, F. B. Corn-based ethanol: A case study in the law of unintended consequences. Competitive Enterprise Institute, June 2007; 2007.

Soper, J.; Judd, D.; Schmidt, D.; Sullivan, S. The future of biotechnology in soybeans. *AgBioForum* 61&2: 8–10; 2003.

Streifel, S. Impact of China and India on global commodity markets focus on metals and minerals and petroleum. Development Prospects Group World Bank. http://siteresources.worldbank.org/INTCHIINDGLOECO/Resources/ChinaIndiaCommodityImpact.pdf (Accessed on April 1, 2009); 2009.

Stuber, C. W.; Polacco, M.; Senior, M. L. Synergy of empirical breeding, marker-assisted selection, and genomics to increase crop yield potential. *Crop Sci.* 39: 1571–1583; 1999.

The White House. The Fact Sheet—Energy Independence and Security Act of 2007. http://www.whitehouse.gov/news/releases/2007/12/20071219-1.html (Accessed on March 18, 2008); 2007.

Tokgoz, S.; Elobeid, A.; Fabiosa, J. F.; Hayes, D. J.; Babcock, B. A., Yu, T. H., Dong, F., Hart C. E., Beghin, J. C. Emerging biofuels: Outlook of effects on U.S. grain, oilseed, and livestock markets. Food and Agricultural Policy Research Institute (FAPRI) Publications 07-sr101, Ames, IA: FAPRI, Iowa State University; 2007.

Trostle, R. Global agricultural supply demand: Factors contributing to the recent increase in food commodity prices. WRS-0801, USDA Economic Research Service; 2008.

Troyer, F. A. Adaptedness and heterosis in corn and mule hybrids. *Crop Sci.* 46: 528–543; 2006.

United Nations. Sustainable energy: A framework for decision makers. UN Energy, May 8, 2007; 2007.

U.S. Department of Agriculture. USDA Agricultural Projections to 2017. Long-term Projections Report OCE-2008-1; 2008.

U.S. EPA (Environmental Protection Agency). Renewable Fuel Standard Program: Draft Regulatory Impact Analysis. Document No. EPA420-D-06-008, Washington, DC: U.S. Environmental Protection Agency, September 2006; 2006.

von Braun, J. Poverty, climate change, rising food prices, and the small farmers. PowerPoint presentation given at a meeting of the International Fund for Agricultural Development, April 22, 2008, Rome. Washington, DC: International Food Policy Research Institute; 2008.

von Lampe, M. Agricultural market impact of future growth in the production of biofuels. Directorate for Food, Agriculture and Fisheries Committee for Agriculture—Working Party on Agricultural Policies and Markets; OECD; AGR/CA/APM(2005)24/FINAL. February 2006; 2006.

Chapter 4
The Economics of Current and Future Biofuels

Ling Tao and Andy Aden

Abstract This work presents detailed comparative analysis on the production economics of both current and future biofuels, including ethanol, biodiesel, and butanol. Our objectives include demonstrating the impact of key parameters on the overall process economics (e.g., plant capacity, raw material pricing, and yield) and comparing how next-generation technologies and fuels will differ from today's technologies. The commercialized processes and corresponding economics presented here include corn-based ethanol, sugarcane-based ethanol, and soy-based biodiesel. While actual full-scale economic data are available for these processes, they have also been modeled using detailed process simulation. For future biofuel technologies, detailed techno-economic data exist for cellulosic ethanol from both biochemical and thermochemical conversion. In addition, similar techno-economic models have been created for *n*-butanol production based on publicly available literature data. Key technical and economic challenges facing all of these biofuels are discussed.

Keywords Biofuel • Biodiesel • Biobutanol • Process economics • Techno-economic analysis • Transportation fuel • Ethanol

Introduction

Biofuels production worldwide is continuing to grow at a very rapid pace. In the USA, ethanol production has more than tripled over the past 5 yr (RFA 2008), from 2.8 billion gallons in 2003 to over 9 billion gallons in 2008. This

A. Aden (✉)
National Renewable Energy Laboratory, 1617 Cole Blvd., Golden, CO 80401, USA
e-mail: andy.aden@nrel.gov

is used primarily for fuel oxygenate in blend percentages (10% or E10); however, some ethanol is sold as dedicated (85% or E-85) fuel for flexible fuel vehicles. Brazil is the second largest producer worldwide with approximately 5 billion gallons in 2007 (FO Lichts, Kent, UK), but uses the fuel in higher percentage blends, either 20–25% or as high as 100% (E-100). Biodiesel is produced in the USA at lower volumes than ethanol, but is also growing rapidly. Production has increased over tenfold from approximately 25 million gallons in 2004 to an estimated 700 million gallons in 2008 (NBB 2009). Energy policy in the form of a renewable fuel standard (EISA 2007) has helped to maintain strong markets for both of these fuels. Under the Energy Independence and Security Act of 2007 (EISA 2007), 36 billion gallons of renewable fuel are mandated by 2022 of which 15 billion gallons are corn-based ethanol, 16 billion are "cellulosic biofuels", and at least 1 billion gallons are biodiesel. Tax credits for both ethanol and biodiesel are also in place to encourage blenders to use these alternative fuels.

However, recent market price fluctuations for fuels and feedstocks have been quite dramatic. Petroleum prices have ranged from $20 per barrel in 2002 to record highs over $140 per barrel in 2008. Average corn prices during this time have also ranged from $2 per bushel to $4.20 per bushel (USDA 2008) with spot prices rising over $8 per bushel. Costs of steel, natural gas, and fertilizer have also experienced dramatic increases during the past 5 yr. In light of these immense price shifts, it becomes increasingly important to understand the production economics of these and other potential biofuels.

For today's commercial biofuels technologies, actual production cost data are available through a variety of sources. The economics of each existing biofuel production will differ from one another depending on a variety of specific and local market and technological variations. For example, ethanol plants will pay differing rates for corn, natural gas, and enzymes based on company hedging practices, negotiated contract prices, and regional market conditions. In addition, detailed process and economic models (also called techno-economic models) also exist for these processes and can be used to demonstrate specific economic impacts of certain key parameters.

Production cost data for nonexisting or noncommercial biofuels processes (cellulosic ethanol, corn-based butanol) are less readily available and come largely from detailed techno-economic models and evaluations. Many companies are currently developing these technologies and oftentimes may present cost data through press releases or other simplified formats. However, this type of cost data must be taken at limited value because the assumptions, level of costing detail, and other bases behind the values are oftentimes absent or unvalidated. The assumed years-dollars are also important because inflationary increases can make year to year comparisons difficult. True and meaningful comparisons between fuels and technology variations can only be made when production economic data are presented clearly, transparently, and in sufficient level of detail to be reproducible. Therefore, we applied detailed techno-economics models and evaluations to several current and future biofuel processes.

Development of Concept Design Methodologies and Models for Process Economics

Developing process economics of this sort requires significant amounts of experimental data, modeling toolsets, and engineering expertise. The first step is the development of a conceptual-level design. Conceptual design refers to the engineering decision making that is required to conceive a process from information on the products, feeds, and chemical and physical steps proposed as a basis (Douglas 1989). The effective use of conceptual design methods in the early stages of process design can have a large impact on overall process design and development. There are frequently many process alternatives that can be chosen and many design variables to specify. Conceptual design approaches practically always include both assumptions and heuristics along with numerical modeling and optimization techniques. The conceptual designs for the future biofuels processes described here were largely developed by National Renewable Energy Laboratory (NREL) in collaboration with a number of research partners. The graphical depiction of such designs is referred to as "process flow diagrams" (PFDs).

The material and energy balance and flow rate information for such designs are then generated using process simulation software packages. For these particular applications, Aspen Plus (Aspen Plus 2006) was used. This software contains physical property and thermodynamic data on a large number of chemical compounds. NREL has further developed customized physical property data for biomass constituents such as cellulose, lignin, and xylan (Wooley and Putsche 1996). The material and energy balance data generated by these models are fed into spreadsheets built for capital and operating cost estimation. As documented (Wooley et al. 1999; Aden et al. 2002; Phillips et al. 2007), capital costs are developed for each piece of equipment using a number of sources, including vendor quotations (for more specialized equipments), costing software estimates (for simpler equipment such as pumps and tanks), and engineering company database information. Installation factors are derived from a number of sources as well.

Using published engineering methodology (Peters and Timmerhaus 1991), a discounted cash flow rate of return analysis is generated using capital and operating cost data. The minimum ethanol selling price (MESP; $/gallon) is the minimum price that the ethanol must sell for in order to generate a net present value of zero for 10% internal rate of return (IRR). This makes the MESP slightly higher than a true cost of production.

It should be emphasized that a certain percentage of uncertainty exists around conceptual cost estimates such as these. These values are best used in relative comparison against technological variations or process improvements. Use of absolute values without detailed understanding of the basis behind them can be misleading. Process economics requires a wide range of detailed studies, including conceptual level of process design to develop detailed process flow diagram (based on research data from NREL and other data sources), rigorous materials and energy balance calculations (via commercial simulation tools such as Aspen Plus), capital and project cost estimation (via in house model using spreadsheets), discounted cash flow economic model (via in house model using spreadsheets), and then a final calculation of minimum biofuel selling price.

Process Economics and Comparative Analysis

Commercialized biofuels process economics   Biodiesel is produced worldwide from a variety of feedstocks. Most of the ethanol produced in the USA is derived from corn grain. Brazil uses sugarcane and European Union (EU) countries use wheat and sugar beets as feedstocks for commercial ethanol production. Many parts of Asia use palm oil feedstocks for biodiesel production, while the EU uses oil derived from rapeseed (canola variety) and the USA uses soybean oil and other oil-bearing crops. In this paper, three commercial processes are discussed. Corn ethanol and sugarcane ethanol process economics are discussed in details as representative examples for commercialized ethanol production. For biodiesel production, a soybean-derived biodiesel process is selected as one example for biodiesel process economics.

Corn ethanol

Process description

There are two general types of processing: wet milling and dry grind. Dry-grind ethanol plants are much more prevalent; greater than 80% of existing ethanol plants in the USA are dry grind (RFA 2008). Dry-grind processes are less capital and energy intensive than their wet mill counterparts. However, they also produce fewer products. Dry-grind plants produce ethanol and animal feed, known as distillers dried grains (DDG or DDGS). Wet mills, on the other hand, are structured to produce a number of products, including starch, high fructose corn syrup, ethanol, corn gluten feed, and corn gluten meal. As a result, ethanol yields from wet mills are slightly lower (2.5 gal per bushel) than from dry-grind processes (2.8 gal per bushel).

General process diagrams are shown in Fig. 4.1 for these processes. In a dry-grind process, corn grain is milled and slurried with water and amylase enzymes. The mixture is cooked and mixed with additional enzyme to complete starch hydrolysis to glucose. The subsequent glucose sugars are fermented to ethanol, CO_2, and other minor by-products using various yeast strains. The ethanol is concentrated and purified through a series of distillation and molecular sieve dehydration steps. The by-product solids are dewatered and dried through a series of centrifugation, evaporation, and drying steps.

In wet milling, the corn kernels are first soaked in a mixture of water and SO_2 through a process known as "steeping" in order to allow for separation of the kernel

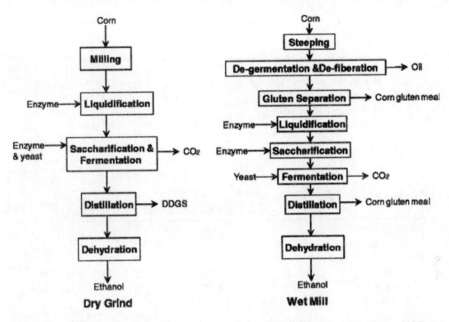

Fig. 4.1 Schematic process flow diagram from corn to ethanol (Bothast and Schlicher 2005)

components. Germ, fiber, gluten, and starch are separated from one another through a series of screens, cyclones, presses, and other equipment. Oil can be further extracted from the germ. Enzymes are added to the starch stream for hydrolysis to sugars, and the sugars can be fermented to ethanol, similarly to dry-grind processing.

Process economics

Several sources of data on corn processing economics exist. In 2002, the US Department of Agriculture (USDA 2002) published a "cost of production" survey (Shapouri et al. 2006) where 21 dry-grind corn ethanol plants were asked to estimate their production costs, including both variable and capital expenses. Total average costs ranged from $0.92/gal to $0.99/gal. Of that total, $0.80/gal was the cost for purchasing corn grain feedstock. These averaged cost data for both years 2002 and 1998 are shown in Table 4.1.

In 2007, FO Lichts conducted a worldwide survey of ethanol production costs (FO Lichts 2007). For the USA during 2006/2007, net production costs of approximately $400/m^3 (~$1.50/gal) were documented for corn purchased at $3.35 per bushel (Table 4.2). Net production costs for wet mills and dry-grind processes were within 3% of each other. However, the increased cost of capital in 2006 *vs.* 2002 was evident because the net feedstock costs were closer to 60% of the overall cost as opposed to 80% or greater in 2002.

Table 4.1 Ethano production cost in USDA 2002 ethanol production-of-cost survey from US ethanol manufacturers

US average	2002	1998
Feedstock costs ($/gal)	0.8030	0.8151
By-product credit ($/gal)	−0.2580	−0.2806
Net feedstock cost ($/gal)	0.5450	0.5345
Operating cost ($/gal)	0.4124	0.4171
Total cost ($/gal)	0.9574	0.9516

Table 4.2 Ethanol production cost in USA, Brazil, and EU (source: FO Lichts 2007)

	2004/2005	2005/2006	2006/2007
Maize ($/bushel)	2.05	2.01	3.35
Maize ($/tonne)	80.71	79.13	131.89
Net feedstock cost ($/gal)			
Dry mill	0.55	0.55	0.90
Wet mill	0.37	0.50	0.72
Net production cost ($/gal) [in USA]			
Dry mill	1.06	1.10	1.54
Wet mill	1.05	1.77	1.50
Net production cost ($/gal) [in Brazil]			
Sugarcane ethanol	0.66	0.92	1.14
Net production cost ($/gal) [in EU]			
Grain-based ethanol	1.54	1.65	2.19
Beet-based ethanol	1.40	2.69	1.83

Dry-grind and wet mill techno-economic models also exist as developed by USDA researchers (Kwiatkowski et al. 2006; Ramirez et al. 2009). These models, in general, are good representations of corn ethanol economics. NREL was given a copy of the 2007 version of the dry-grind model by USDA and had used it for several biofuels research projects. For this particular biofuels report, NREL used this model to develop comparative economics between corn ethanol and other biofuels technologies. Baseline assumed values for several key parameters are shown in Table 4.3. The model was run using these assumed parameter values, which resulted in the modeled costs for ethanol shown in Table 4.4. The assumed plant design capacity was 45 MM gallons per year ethanol. The operating costs are shown in both $/gallon ethanol and $MM/yr. The corn feedstock cost is the largest single cost contributor as expected. Coproduct DDGS contributes significantly to the net production cost as well, as does natural gas (utilities). Operating costs are based on a yield of 115 gal ethanol yield per dry ton corn. Depreciation of capital is included in the overall operating costs for all models in this paper. The resulting total project investment (TPI) is summarized with other biofuel processes in Table 4.15.

4 The Economics of Current and Future Biofuels

Table 4.3 Baseline economic assumptions for corn dry mill ethanol

Economic parameter/prices	Baseline value
Installation factor	3.0 (across all equipment)
Equipment life/depreciation time	20 yr (straight line)
Equipment scaling exponent	0.6
Online percentage	96% (350 d)
Corn price	$3.35/bushel
DDGS price	$95/ton
Caustic price	$0.055/lb ($110/ton)
Enzyme price	$1.03/lb
Gasoline (denaturant) price	$2.00/gal
Sulfuric acid price	$0.05/lb ($100/ton)
Lime price	$0.04/lb ($80/ton)
Makeup water price	$0.00002/lb ($0.167/1,000 gal)
Urea price	$0.10/lb ($200/ton)
Yeast price	$0.85/lb
Electricity	$0.05/kWh
Natural gas	$7.50/MMBtu

Table 4.4 Operating cost of dry mill corn ethanol plant with 45 MM gal/yr production rate

Corn ethanol 45 MM gal/yr	
Operating costs ($/gal ethanol)	
Shelled corn	$1.21
Denaturant	$0.04
Other raw materials	$0.05
Utilities	$0.30
Labor, supplies, and overheads	$0.11
Depreciation	$0.14
Coproduct credits	−$0.32
Production cost per gallon	$1.53
Operating costs ($/yr)	
Shelled corn	$57,777,726
Denaturant	$2,055,982
Other raw materials	$2,524,366
Utilities	$14,267,437
Labor, supplies, and overheads	$5,819,600
Depreciation	$6,570,000
Coproduct credits	−$15,037,000
Total production cost	$73,978,110

Production cost per gallon is $1.53

Sugarcane ethanol

Process description

Brazil's production accounts for 42% of world sugarcane production in 2005 (BNDES 2008). Brazilian cane biorefineries began as sugar mills, but had grown to include simultaneous production of ethanol as well. It has been noted that the low cost of Brazilian sugar is largely related to the development of agricultural and industrial technology associated with the expansion of bioethanol production (BNDES 2008). Because both sugar and ethanol coproduction are so closely integrated within these biorefineries, it is difficult, if not impossible, to completely decouple the economics of sugar production from ethanol production.

Sugarcane ethanol is a sugar-based production process that requires one less step than starch-based ethanol production. For corn, the starch must be hydrolyzed to glucose before it can be used by yeast. The sucrose extracted from sugarcane can be used without such a hydrolysis step. Sugarcane cannot be stored for more than a few days, and mills subsequently operate during the harvest period only (6–7 mo) and perform maintenance while down the rest of the year.

A general process diagram for sugarcane ethanol production is shown in Fig. 4.2. The production process consists of three general steps: sugar extraction from the cane, sugar fermentation to ethanol, and ethanol separation and purification. Sugar extraction is a common step shared by both sugar production and ethanol production trains. Once in the mill, sugarcane is generally washed and sent to preparation and

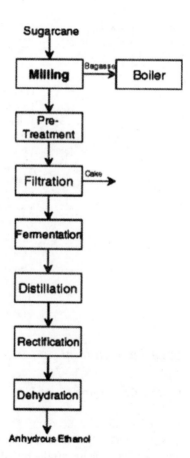

Fig. 4.2 Schematic process flow diagram from sugarcane to ethanol. Sugar production sections not shown

extraction. Extraction is made by roll mills that separate the sugarcane juice from the bagasse, which is sent to the mill's power plant for use as fuel. The bagasse is further pressed through a drying roller which reduces moisture for more efficient combustion in the boilers. The extracted juice is treated and sent through a rotary vacuum filter. The resulting filter cake is applied back on fields as fertilizer, and the sugar is recovered from the decanted slurry. The juice is sent to either sugar production or ethanol production. For ethanol production, the juice is fermented by yeast (*Saccharomyces cerevisae*). Fermentation times range from 8 to 12 h, generating beer with ethanol concentrations from 7% to 10%. Yeast is recovered, treated, and reused. Beer is sent to distillation where ethanol is initially recovered in a hydrated form. This nearly azeotropic hydrated ethanol can further be dehydrated using a number of existing technologies. These include adsorption with molecular sieves and/or extractive distillation with monoethylene glycol (Seabra 2007; BNDES 2008). However, the most common commercial technology in Brazil is extractive distillation using cyclohexane as the entrainer. The decanted cyclohexane-rich stream is recycled internally.

Process economics

FO Lichts data for 2006/2007 (Table 4.2) put sugarcane ethanol costs at $1.14/gal ethanol, which is lower than corn ethanol produced in the USA. While the USA does not currently produce any ethanol from sugarcane, a USDA 2006 study (Shapouri et al. 2006) estimated what those costs might be. Shapouri and his coworkers estimated sugarcane ethanol production costs in the USA as high as $2.40/gal based on 2003/2004 sugarcane market prices and estimated processing costs. Feedstock costs were estimated at $1.48/gal of ethanol, representing 62% of the total ethanol production cost. However, sugarcane feedstock costs in Brazil could be as low as $0.30/gal of ethanol (Shapouri et al. 2006), which makes sugarcane ethanol more economical in Brazil.

Delivered costs of sugarcane to the biorefinery are estimated at $21.08/ton by Rodriguez (2007) and Seabra (2007). For yields of 22.7 gal ethanol per ton cane, the feedstock contributes $0.93/gal ethanol. Other baseline assumptions for production costs are shown in Table 4.5. The full operating costs for a 45-MM gal/yr process are shown in Table 4.6 and are largely derived from two economics studies (Rodrigues 2007; Seabra 2007). These studies examined the Brazilian manufacturing process in great detail. Though these biorefineries do not operate year-round, the economics shown assume year-round production for better comparison against other biofuels processes.

Also note that electricity surplus in these processes is not large and is therefore not included in the cost data. This is changing, however, as Brazilian biorefineries continue development and improvement in the overall power management of these facilities. Biomass residue and/or bagasse combustion feeds the cogeneration section to optimize the overall power, steam, and electric demands of the mills.

The TPI for the 45 million gallon per year sugarcane ethanol plant is shown in Table 4.15 alongside the other biofuels processes under comparison. Even though

Table 4.5 Baseline economic assumptions for sugarcane ethanol process

Economic parameter/prices	Baseline value
Installation factor	3.0 (across all equipment)
Equipment life/depreciation time	20 yr (straight line)
Equipment scaling exponent	0.6
Online percentage	96% (350 d)
Sugarcane price	$28.5/ton
Other chemicals	$0.078/gal ethanol
Gasoline (denaturant) price	$0.043/gal ethanol

Table 4.6 Operating cost of sugarcane ethanol, with 45 MM gal/yr production

Sugarcane 45 MM gal/yr	
Operating costs ($/gal ethanol)	
Sugarcane	$0.93
Denaturant	$0.04
Other raw materials	$0.08
Water	$0.02
Fuels and lubricants	$0.02
Labor, maintenance	$0.15
Depreciation	$0.05
Production cost per gal	$1.29
Operating costs ($/yr)	
Sugarcane	$42,150,000
Denaturant	$1,953,801
Other raw materials	$3,539,760
Water	$756,800
Fuels and lubricants	$899,560
Labor, maintenance	$6,904,080
Depreciation	$2,185,000
Total production cost	$58,389,000

decoupling the economics of ethanol from sugar production is difficult, simplifying assumptions were made for the common areas of feedstock handling juice extraction. It is normal in Brazil for roughly half of the juice to go to sugar and half to ethanol so 50% of the capital in the common area was attributed to ethanol production. Because of this, combined with higher feedstock cost in Seabra and his coworker's study (Rodrigues 2007; Seabra 2007), we estimated sugarcane cost at $1.29 (Fig. 4.6) which is higher than FO Lichts (2007)'s prediction.

Soybean biodiesel

A diesel engine gets its name from its inventor, Rudolf Christian Karl Diesel. Diesel fuel has traditionally been derived from petroleum refining. Biodiesel, on the other hand, is defined by American Society for Testing and Materials standards as a fatty acid methyl ester produced through transesterification of triglycerides (found in vegetable oils, fats, and greases) with alcohols such as methanol. Biodiesel offers lower emissions than the petrodiesel, which makes it an attractive alternative transportation fuel. Common feedstocks for biodiesel production include soy, canola, camolina, corn, rapeseed, and palm. Peanut, mustard seed, sunflower, and cotton seed present additional potential feedstocks.

The cost of biodiesel varies depending largely on feedstock cost. For instance, biodiesel from soybean can be more than twice as expensive as petrodiesel, although the capital investment for a biodiesel plant is not that significant. The high value of soybean oil makes production of a cost competitive biodiesel fuel very challenging. New feedstocks are undergoing detailed research, such as beef tallow, waste cooking oil, pork lard, yellow grease, etc. (Demirbas 2005). A 10 million gallon per year plant was estimated to cost $2.15/gal biodiesel based on feedstock cost at $0.25/lb (Graboski and McCormick 1998) of soybean oil and the capital cost was also estimated at about $2/gal using 20% discounted cash flow and 100% equity financing assumptions. Four continuous process configurations with HYSYS simulation were studied in detail for biodiesel capital and production cost comparison by Zhang et al. (2003) using waste oils, with cost ranging from $644 to $884/ton (~$2.10–2.90/gal). At a value of $0.52/kg ($0.24/lb) for feedstock soybean oil, Hass et al. (2006) predicted biodiesel production cost at $2.00/gal using Aspen simulation and cash flow calculation. Although previous researchers have used different plant scale, the soybean biodiesel production cost over the past years lies in $2.00–2.50/gal range with similar feedstock pricings.

Process description

The most common way to make biodiesel is by transesterification, which refers to a catalyzed chemical conversion involving vegetable oil and an alcohol to yield fatty acid alkyl esters (i.e., biodiesel) and glycerol as by-product. Methanol is the most commonly used alcohol. Transesterification reactions can be alkali, acid, or enzyme catalyzed. Among these three technologies, alkali catalyzed is the most common. The production process contains the following major steps: raw material handling, transesterification, methanol recovery and recycle, biodiesel and glycerol separation, purification of both biodiesel and by-product glycerol, and wastewater treatment. The simplified PFD is shown in Fig. 4.3.

Process economics

The soybean oil pricing is the dominant factor for the biodiesel cost as shown in Table 4.8. Even when soybean oil is priced below $0.20/lb, the feedstock still dominated more than 70% of overall biodiesel cost. The coproduct glycerin of 80% purity is assumed to be roughly equivalent to $0.15/gal biodiesel produced based

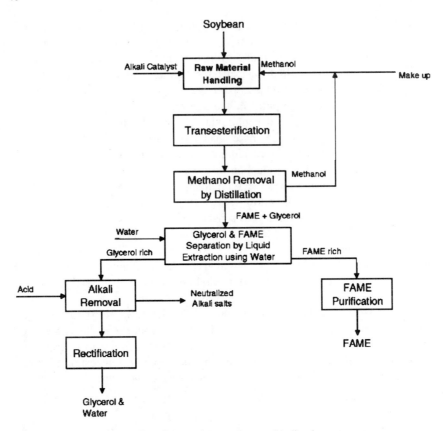

Fig. 4.3 Schematic process flow diagram from soybean to biodiesel

on Energy Information Administration (EIA) data (EIA 2004); however, the price of glycerin can often offset the high feedstock price to a certain degree.

Baseline economics assumptions are listed in Table 4.7, together with feedstock and coproduction glycerin pricings. The IRR after tax is assumed 10% and equity of total investment is 100%. The operating cost of soybean biodiesel is calculated in the model to be $2.55/gal for a 45-MM gal biodiesel plant, shown in Table 4.8. The TPI of a soybean biodiesel process is also listed in Table 4.15. From the cost analysis of soybean biodiesel, it is obvious that the feedstock cost of oil is the largest single component of biodiesel production costs. To make biodiesel cost competitive to petrodiesel, low cost feedstocks and their availability are the key. It is necessary to understand the process economics from the variety feedstocks, in addition to soybean biodiesel. Shown in Fig. 4.4, soybean prices from $0.10–0.88/lb contribute 75% to 95% of overall operational cost.

Going forward—economical challenges Commercial biofuels to date have been very successful in garnering larger and larger shares of the fuel markets. However, many economic challenges still remain for these fuels. Feedstock availability is one issue. While many million acres of US cropland are devoted to soybean and corn

4 The Economics of Current and Future Biofuels 49

production, much of these are used in competing processes such as animal feeding and/or livestock. It may prove difficult for corn ethanol, for example, to grow to more than 15 billion gallons per year without substantially impacting exports or other markets. Another issue facing conventional biofuels is the decoupling of feedstock costs and biofuel prices. Because these biofuels are largely blended into petroleum fuels at lower concentrations, their prices track those of gasoline or diesel.

Table 4.7 Baseline economic assumptions for soybean diesel process

Economic parameter/prices	Baseline value
Installation factor	3.0 (across all equipment)
Equipment life/depreciation time	20 yr (straight line)
Equipment scaling exponent	0.6
Online percentage	96% (350 d)
Soybean oil price ($/lb)	$0.30
80% glycerin price ($/gal biodiesel)	$0.15
Methanol price ($/lb)	$0.13
Sulfuric acid price	$0.05/lb ($100/ton)
Sodium methoxide catalyst price ($/lb)	$0.45
Makeup water price	$0.00002/lb ($0.167/1,000 gal)
Electricity	$0.05/kWh
Natural gas	$7.50/MMBtu
IRR	10%
Equity	100%

Table 4.8 Operating cost of 45 MM gal/yr soybean biodiesel plant

Soybean biodiesel 45 MM gal/yr	
Operating costs ($/gal biodiesel)	
Feedstock (raw soybeans)	$2.21
Raw materials	$0.26
Utilities	$0.06
Labor supplies	$0.01
General works	$0.14
Subtotal operating costs	$2.70
Coproduct credits	−$0.15
Gross operating costs	$2.55
Operating costs ($/yr)	
Feedstock (raw soybeans)	$99,506,040
Raw materials	$11,673,531
Utilities	$596,000
Labor supplies	$6,389,064
General works	$121,455,000
Subtotal operating costs	$26,826,018
Coproduct credits	$94,628,982
Gross operating costs	$121,455,000

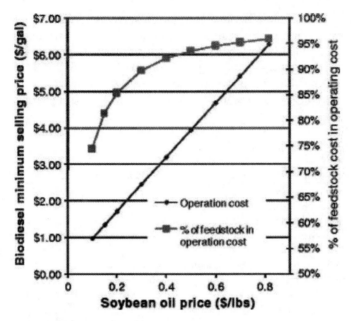

Fig. 4.4 Soybean oil price is the dominant factor for soybean biodiesel price

However, feedstock prices track with other commodity grain prices. Therefore, history has presented several instances of high feedstock costs coupled with low ethanol prices, which can cause producers to have negative margins and lose money.

A final issue facing commercial biofuels is one of perceived value and sustainability. Enormous debate over the potential benefits of biofuels has taken place around concepts of net energy, water use and quality, food *vs.* fuel issues, and most recently, land-use change impacts. Fervor on these topics are likely to only increase as time goes on, but new developments in cellulosic biofuels and advanced biofuels hold promise in this discussion.

Future biofuels process economics The future biofuels process economics shown here encompass both future fuels (butanol) as well as new feedstocks (lignocellulosic biomass). Various lignocellulosic materials, such as wood, agricultural residues, and energy crops, have the potential to be valuable raw materials for economical and compatible biofuel processes. Agricultural residues include wheat straw, sugar cane bagasse, and corn stover. Forest residues include sawdust, and dedicated energy crops include salix, switchgrass, and miscanthus. The lignocellulosic materials and subsequent conversion processes are sufficiently abundant and have potential to significantly reduce environmental impacts compared to today's technologies. Transportation biofuels such as hydrocarbons or cellulosic ethanol, if produced from low-input biomass grown on agriculturally marginal land or from waste biomass, could provide much greater supplies and environmental benefits than food-based biofuels (Hill et al. 2006). Therefore, various process design studies have been proposed over the last decade on ethanol production from lignocellulosic materials.

However, process economics data are still relatively limited. In this paper, two representative lignocellulosic ethanol processes were selected for techno-economic analysis: biochemical conversion and thermochemical conversion.

Advanced biofuel development is as important as the development of alternative feedstocks. Advanced fuels, meaning nonethanol fuels, are high energy and high performance biofuels that include higher molecular weight alcohols (such as butanol and isobutanol) as well as infrastructure compatible hydrocarbons produced from either thermochemical (Fischer–Tropsch fuels) or biochemical (fermentation-derived hydrocarbons) pathways. Biobutanol will likely be the first of these biofuels to reach global markets (Spake 2007). One specific partnership active in this area is the collaboration between BP and DuPont (BP News 2006). Active and rapid technology development by these groups has led to initial construction of a US $400 million world-scale biofuels plant in Hull, UK. Initially, the plan will have the capacity to produce 420 million liters (110 million US gallons) of bioethanol annually from locally grown wheat but could subsequently be converted to biobutanol production. Several other companies are active in this area, including EEI (D. Ramey), Cobalt, Gevo, and Tetravitae. In this paper, process economics for *n*-butanol production from corn grain are shown based on Aspen Plus modeling and detailed economic analysis.

Cellulosic ethanol via biochemical conversion route

Process description

The biochemical design and cost estimates are based on an updated version of the NREL design report (Aden et al. 2002) for conversion of corn stover to ethanol. This design uses dilute acid pretreatment followed by enzymatic hydrolysis and cofermentation with recombinant *Zymomonas mobilis*. The corn stover is first treated with dilute sulfuric acid catalyst at a high temperature (190°C) for a short time (average at 2 min), liberating the hemicelluloses sugars and other compounds. Before going into enzymatic hydrolysis, proper conditioning is required for acid neutralization and detoxification prior to the biological portions of the process. Detoxification is only applied to the liquor fraction of the pretreated biomass and not the solids. Solids from pretreatment will be internally washed before remixed with the detoxified liquor for saccharification and fermentation. A purchased cellulase enzyme is added to the hydrolyzate at an optimized temperature for enzyme activity. If saccharification and fermentation steps are conducted at different temperatures, a cooling step is required to ensure growth of fermenting organism *Z. mobilis* at anaerobic condition. Between 3 and 7 d are required to convert most of the cellulose and xylose to ethanol. The "beer" liquor with ~4–8 wt.% of ethanol is then sent to recovery and purification, which uses standard adsorption technology. The solids after fermentation are separated and combusted in a fluidized bed combustor to produce high pressure steam for electricity credits and process heat. This is very similar to systems at Brazilian sugar mills and North American pulp and paper mills.

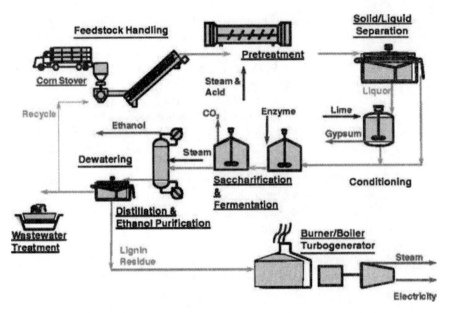

Fig. 4.5 Schematic process flow diagram from corn stover to ethanol (Aden et al. 2002)

The process design presented (Fig. 4.5) reflects the best available estimates for performance of an enzyme-based process with the current status of NREL research efforts. The process includes four general steps:

(a) Co nversion of feedstocks to sugar. Corn stover is the model feedstock in this study. The approach used to accomplish this step normally distinguishes different process technologies (or configurations) for the overall process design, since it is the most critical and challenge step or steps. This includes the areas of pretreatment and/or hydrolysis both chemical and enzymatic.
(b) Fermentation of sugars to ethanol. The fermenting organism *Z. mobilis* recombinant bacterium is used to co-ferment all sugars simultaneously to ethanol.
(c) Ethanol recovery area. High purity ethanol (99.5 wt.%) is produced in the recovery area, using distillation columns, a molecular sieve unit, and water evaporation units. Although it is an energy intensive operation, the temperature gradient from the first distillation column (beer column) can be used as a driving force for energy intensification in the area. Recycled water streams with reasonable levels of impurities are introduced into different areas such as pretreatment of feedstocks, in order to optimize the water and/or steam usage.
(d) Residue use. The solids from distillation (largely lignin), the concentrated syrup from the evaporator, and the biogas from anaerobic digestion are combusted in a fluidized bed combustor to produce high pressure steam for electricity production and provide process steam or heat. Excess electricity credits are included in the cost analysis.

Process economics

The biochemical (Aden et al. 2002) and thermochemical (Phillips et al. 2007) ethanol design reports from NREL were completed 5 yr apart. As a result, costs for much of the equipment and raw material inputs (chemicals, etc.) were developed using different year's dollars. Because of inflation and other factors, the time value of money for each year changes. Therefore, it was necessary to put all costs in the same year's dollars in order to make them all more directly comparable. This was done using a factored indexing approach for capital costs, chemical costs, and labor costs as described by Aden and coworkers and Phillips and coworkers using well-known cost indices. The cost indices for nonlabor costs have risen significantly since 2003 due to a variety of international demands for materials and increased energy costs. For this comparison, all costs are shown using year-$2007. Similar approach has been applied to the other four processes discussed in the paper. Operating costs were calculated (Table 4.9) based on the following assumptions of 45 MM gal ethanol production per year plant size with ethanol yield of 90 gal per dry ton feedstock (corn stover). The internal rate of return after tax is assumed 10% and equity of total investment is 100%. The assumed plant operates 350 d per year for a total of 8,400 h. Total project investment is listed in Table 4.15 and is depreciated in 20 yr. Detailed baseline economics assumptions can be found by referring to the Aden et al. 2002 study. The costs are further broken down into respective process areas as shown in Fig. 4.6. In the biochemical process, the lignin residue boiler and turbogenerator section represent a huge portion of the overall capital costs; however, much of this are offset by the heat and power generated for the process. Therefore, the net cost of this process area is diminished. The next largest net cost area of the biochemical process is pretreatment and conditioning, accounting for $0.25/gal (or 19%) of the overall cost. Much of this is due to the high capital cost of the pretreatment reactors needed for dilute sulfuric acid. The largest single operating cost, however, remains the feedstock cost itself.

Cellulosic ethanol via thermochemical conversion route (Phillips et al. 2007)

Process description

The thermochemical design is illustrated in Fig. 4.7 and uses a different feedstock than the biochemical design. Wood chips instead of corn stover are converted to ethanol and other higher alcohols though a series of solid-phase and gas-phase reactions. Wood chips are brought to the plant and then screened, milled, and dried. The wood is then gasified using a circulating fluidized bed indirect gasification system. Biomass char and a small slipstream of unreformed synthesis gas are

Table 4.9 Conversion cost of biochemical conversion of corn stover to ethanol

Cellulosic ethanol biochem 45 MM gal/yr	
Operating costs ($/gal ethanol)	
Feedstock	0.51
CSL	0.03
Cellulase	0.10
Other raw materials	0.11
Waste disposal	0.02
Electricity	−0.07
Fixed costs	0.16
Capital depreciation	0.18
Average income tax	0.13
Average return on investment	0.32
Production cost per gal	1.48
Operating costs ($/yr)	
Feedstock	23,000,000
CSL	1,400,000
Cellulase	4,400,000
Other raw material costs	2,500,000
Waste disposal	700,000
Electricity	−3,000,000
Fixed costs	$7,000,000
Capital depreciation	$8,000,000
Average income tax	$5,900,000
Average return on investment	$14,300,000
Total operating costs ($/yr)	$64,200,000

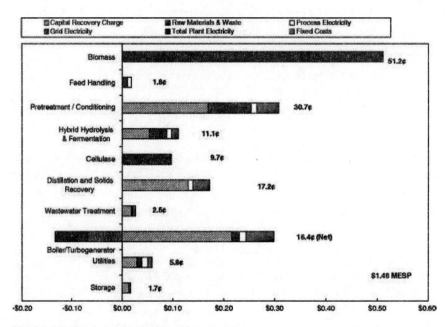

Fig. 4.6 Biochemical costs distributed by process area

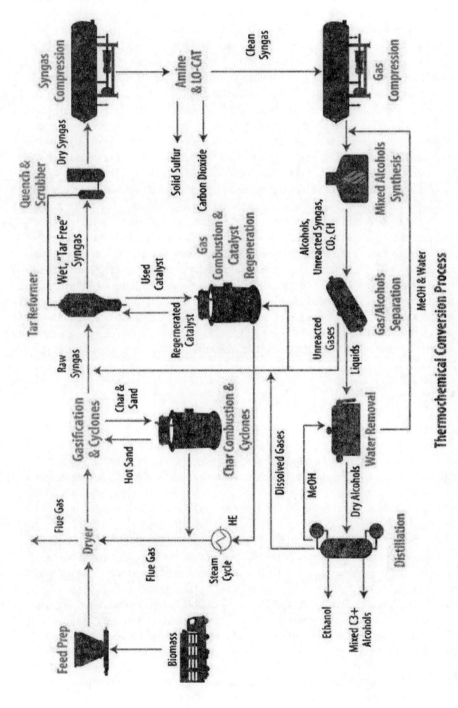

Fig. 4.7 Schematic process flow diagram from corn stover to ethanol (Phillips et al. 2007)

combusted and this heat is transferred to the gasifier through the circulation of hot sand (olivine) between the two process vessels. The crude synthesis gas (syngas) is primarily composed of CO, H_2, CO_2, CH_4, tars, and water. The tars (including benzene) are then reformed into useful syngas using a fluidize-able tar reforming catalyst. Deactivated reforming catalyst is separated from the effluent syngas and regenerated online similarly to fluidized catalytic cracking technology used in petroleum refining. The hot syngas is cooled through a series of heat exchange and water scrubbing steps. The scrubber removes impurities such as particulates and residual tars. This scrubber water undergoes primary treatment onsite to recover a portion of the quench water, while the rest is sent offsite for further wastewater treatment. The cooled syngas is compressed to 435 psi before it enters an amine unit to remove a majority of the acid gases (CO_2, H_2S) present. The CO_2 is vented to the atmosphere and the sulfur is captured in its elemental form using a Klaus-like unit called LO-CAT.

The cleaned and conditioned syngas is further compressed to the required synthesis pressure and sent through a fixed-bed molybdenum sulfide-based catalyst to synthesize a variety of mixed alcohols. After synthesis, the alcohols are cooled and condensed away from the unconverted syngas. The unconverted syngas is recycled to the tar reformer while the condensed alcohols undergo distillation and purification to recover pure ethanol. Methanol is recovered and recycled to the synthesis reactor in order to boost ethanol and higher alcohol yields. The C3+ alcohols are then sold as a coproduct based on an assumed fuel value ($1.15/gal). In this design, the steam cycle is integrated throughout the process while driving compressors and generating electricity as well.

Process economics

Operating costs were calculated for a 45 million gallon per year facility and are shown in Table 4.10 The internal rate of return after tax is assumed 10% and equity of total investment is 100%. The plant operates 350 d per year for a total of 8,400 h. The TPI of this size plant is shown in Table 4.15. Detailed baseline assumptions can be found in Phillips et al. 2007.

For both cases of biochemical and thermochemical conversion, the single largest cost component is the biomass feedstock, though the absolute contribution in $/gallon differs because of the different ethanol process yields. For the thermochemical process, the largest cost area is the tar reforming and syngas conditioning section, which accounts for $0.38/gal (or 28%) of the overall cost (Fig. 4.8). Reductions in costs for pretreatment and syngas cleanup and conditioning will lead to significant reductions in the overall cost of ethanol for each process. This also helps to depict why each of these respective areas is a key focus of process R&D.

A direct comparison of the costs (Table 4.11) is shown for biochemical and thermochemical ethanol processes. Process alcohol yields are also shown. Note that

4 The Economics of Current and Future Biofuels

Table 4.10 Conversion cost of thermochemical conversion of wood chips to ethanol

Cellulosic ethanol thermochem 45 MM gal/yr	
Operating costs ($/gal product)	
Feedstock	$0.57
Catalysts	$0.00
Olivine	$0.01
Other raw materials	$0.02
Waste disposal	$0.01
Fixed costs	$0.24
Coproduct credits	−$0.21
Capital depreciation	$0.19
Average income tax	$0.14
Average return on investment	$0.34
Total operating costs (cents/gal)	$1.32
Operating costs ($/yr)	
Feedstock	$25,800,000
Catalysts	$100,000
Olivine	$300,000
Other raw material costs	$200,000
Waste disposal	$200,000
Fixed costs	$10,900,000
Coproduct credits @ $1.15/gal	−$9,300,000
Capital depreciation	$8,400,000
Average income tax	$6,400,000
Average return on investment	$15,400,000
Total operating costs ($/yr)	$58,400,000

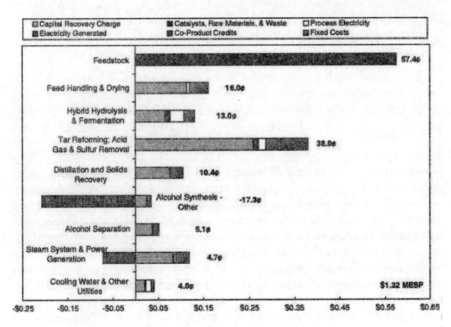

Fig. 4.8 Thermochemical costs distributed by process area

Table 4.11 Cost comparison between biochemical and thermochemical ethanol production from biomass

	Biochemical	Thermochemical
Year-dollars ($)	$2007	$2007
Feedstock	Corn stover	Wood chips
MESP ($/gallon)	$1.48	$1.32
Total installed cost ($MM)	$92	$121
Total project investment ($MM)	$183	$241
Delivered feedstock cost ($/dry ton)	$46	$46
Coproduct credit	Electricity ($0.05/kWh)	Higher alcohols (C3+) $1.15/gallon)
Coproduct credit/gallon EtOH	$0.09	$0.21
Ethanol production (MM gal/yr)	45	45
Total alcohol production (MM gal/yr)	45	52.8
Ethanol yield (gal/dry ton feed)	89.7	80.1
Total alcohol yield (gal/dry ton feed)	89.7	94

while the feedstocks are different for each process, identical delivery costs ($46/dry ton) are assumed. The MESP ($/gallon) is slightly lower for the thermochemical process despite having lower ethanol yields than the biochemical process. One primary reason for this is that the total alcohol yield from thermochemical production is slightly higher than for biochemical production, and the higher alcohols coproduct return, a larger coproduct credit than the power credit the biochemical process receives. Subtracting the coproduct credits from the respective MESP values results in ethanol costs that are almost identical, especially within the margin of uncertainty for these models.

Corn butanol

Fermentation of sugar-containing substrates to acetone, butanol, and ethanol (ABE) is well known. These solvents were produced by fermentation at commercial quantities in the early-to-mid-1900s using *Clostridium acetobutylicum*, a solvent-producing gram-positive bacteria. At times, acetone was the product of focus and butanol was a by-product. After the mid-1950s, petroleum became a more cost-preferred feedstock for producing butanol. The ABE plants simply could not compete and were forced to shut down. However, with the onset of rising petroleum prices and new biotechnology, biological butanol production may become the more cost-effective technology once again. In the long run, the butanol production via biomass may be more economical than petrochemical industry as a transportation fuel with much larger scale than in the past.

The old fermentation processes were batch processes with low productivity and solvent concentration. Molasses or corn mash was generally used as feedstocks. More recent research has been aimed at developing a continuous process with high product yield, concentration, and productivity. From this research, initial economics of ABE produced from corn starch hydrolysate were compared against propylene-based butanol. Under certain scenarios, the economics for butanol looked promising at the time. Another well-known strain, *Clostridium beijerinckii* (or *C. butylicum*) produces solvents in approximately the same ratio as *C. acetobutylicum*, but isopropanol is produced in place of acetone. These strains are spore formers and obligate anaerobes with relatively simple growth requirements. Blaschek et al. (2005) have significant research experience with "hyper-producing" versions of such strains. Ramey and Yang (2004) have also been successful in producing higher butanol yields by separating the fermentation into two separate steps: an acidogenesis phase and a solventogenesis phase, where different organisms are used during each phase.

Butanol's higher energy density and transportation advantages compared to ethanol provide incentive for investment from industry. In addition to BP and DuPont collaborations on butanol isomers, Cobalt Biofuels announced it raised $25 million in equity to accelerate the commercialization of *n*-butanol, an advanced biofuel (Cobalt News 2008). Another startup company, Gevo, is currently focused on the development of advanced biofuels and renewable chemicals that are based on isobutanol and its hydrocarbon derivatives (Gevo News 2009) using a microorganism recently licensed from Cargill. TetraVitae Bioscience's focus is the production of biobutanol using a proprietary fermentation process and enhanced microorganism platform and claims making significant improvements over conventional approaches (TetraVitae Bioscience 2009). UK-based Green Biologics and Mumbai-based Laxmi Organic Industries have signed an agreement to build a commercial-scale biobutanol plant in India and the demonstration plant is expected to produce 1,000 metric tons (2.2 million pounds) of butanol a year starting in 2010 (Cleantech 2008). In general, technological advances are being focused on engineering strains for higher butanol productivity and yield through minimization or elimination of the metabolic pathways that lead to acetone and ethanol production. Engineering advances are also being employed to help overcome the toxicity challenges that these systems have seen in the past.

Process description

While several designs and process variations have been explored in literature, very little modeling has been done using rigorous physical property models. Simple assumptions of how components are expected to fractionate and separate, for example, can often be proven wrong through such an approach. After analyzing several conceptual designs, NREL developed its own conceptual design for ABE production. The USDA corn dry mill ethanol model was used as a basis for modeling. The ethanol fermentation and downstream recovery was replaced with a butanol fermentation using *Clostridium*. A complex sequence of distillation steps was then added to separate and purify the acetone, butanol, and ethanol components from the water. Nonrandom two-liquid modeling package was used to more accurately predict the liquid/liquid interactions that would take place in a system such as this. As with corn ethanol, the by-product from this process is animal feed. While DDGS from yeast-based production is common, bacterial-based DDGS would have to undergo animal feeding trials. H_2 is also produced from clostridial strains of this sort. This is collected and purified using commercial pressure-swing adsorption (PSA) technology. In this fashion, five total products are produced: ethanol, hydrogen, DDGS, acetone, and butanol.

The process is depicted in Fig. 4.9, adopted initially from the corn dry mill process. Corn is milled first then sent to liquefaction. After liquefaction, microorganisms are added to ferment the glucose to acetone, butanol, and ethanol mixture. Total residence time in the fermentors is 72 h. The gas stream from the fermentors goes through a PSA unit to recover hydrogen, which is sold as coproduct. The whole

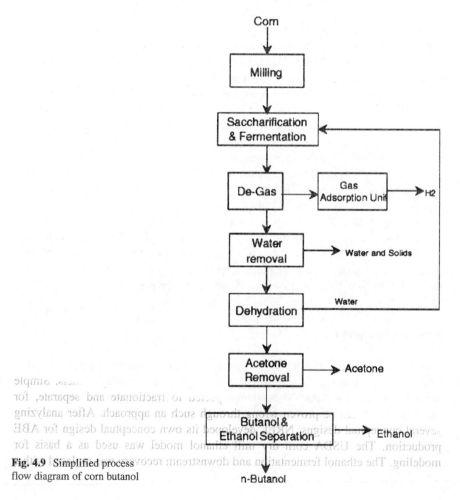

Fig. 4.9 Simplified process flow diagram of corn butanol

beer is degassed and sent to distillation systems. About 85% water is removed from the dehydration column then recycled back to liquefaction. Downstream distillation columns further separate acetone, ethanol, and n-butanol from residual water, combined with process options of extractive distillations, molecular sieve, or pervaporation membrane units.

Process economics

Qureshi and Blaschek published ABE butanol production costs at $1.56/gal based on $1.80 per bushel corn feedstock cost (Qureshi and Blaschek 2000). If the feedstock is raised to $3.35/bushel (FO Lichts 2007), their estimated butanol cost is about $2.10/gal, which is similar to what is estimated in the model reported here.

The economics (annualized cost of production, not cash flow analysis) for corn ethanol production are calculated in an Excel spreadsheet. The material and energy balance data obtained from the Aspen Plus simulation is used in the spreadsheet to size and cost the specified capital equipment. It is also used to calculate the fixed (labor and supplies) and variable (raw materials) operating costs of the plant. The economics have been updated to reflect year-$2007. The specific costing data sources are originally gathered by USDA and represent a combination of vendor specifications, costing program results, and chemical and utility list prices. Overall economic process assumptions are listed in Table 4.12.

The resultant process economics are shown in Table 4.13. For a 45-MM gal/yr butanol plant using $3.35/bushel corn, the production cost of $1.96/gal n-butanol results and a yield of 1.36 gal/bushel (57 gal/dry ton corn) are calculated. With a total installed cost of $138 MM, this equates to just over $6.10/annual gallon, shown in Table 4.15. These results are a fairly accurate depiction of the economics during early 2007. The comparison of process economics of corn ethanol and corn butanol is listed in Table 4.14. Using almost identical baseline economic assumptions, the yield of corn ethanol is almost twice as to corn butanol. However acetone, ethanol, and hydrogen coproducts in corn butanol process contribute more significantly to the net production cost than coproducts do to the corn ethanol process. These results in a production cost of $1.96/gal butanol compared with $1.53/gal ethanol produced from corn. Also note that the TPI is also doubled for the corn butanol process compared to corn ethanol, simply due to low feedstock yield, and a more complicated separation of liquid mixtures of acetone, ethanol, butanol, and water from the fermentation beer with low solvent concentrations.

Cellulosic butanol potentials Using corn grain to make butanol is a logical technology progression because much of the technology is already commercial. However, further increases in biofuel production (ethanol and butanol) to meet the goals of the renewable fuels standard can be produced from cellulosic biomass, such as corn stover, corn fiber, wheat straw, barley straw, and energy crops like switchgrass and miscanthus. While *C. beijerinckii* (Parekh et al. 1999) and *C. acetobutylicum* (Lin and Blaschek 1983) have been used primarily for butanol production, *C. beijerinckii* BA101 has also been used to treat corn fiber hydrolysate

Table 4.12 Baseline economic assumptions for corn butanol process economics

Economic parameter/prices	Baseline value
Installation factor	3.0 (across all equipment)
Equipment life/depreciation	20 yr (straight line)
Equipment scaling exponent	0.6
Online percentage	96% (350 d)
Corn price	$3.35/bushel
Acetone price	$0.45/lb
Ethanol price	$0.35/lb
Hydrogen price	$1.30/lb
Caustic price	$0.055/lb ($110/ton)
Enzyme price	$1.03/lb
Sulfuric acid price	$0.05/lb ($100/ton)
Lime price	$0.04/lb ($80/ton)
Makeup water price	$0.00002/lb ($0.167/1,000 gal)
Urea price	$0.10/lb ($200/ton)
Yeast price	$0.85/lb
Electricity	$0.05/kWh
Natural gas	$7.50/MMBtu

Table 4.13 Butanol cost from a 45-MM gal/yr plant from corn

Corn Butanol 45 MM gal/yr	
Operating costs ($/gal butanol)	
Shelled corn	$2.46
Denaturant	$0.00
Other raw materials	$0.11
Utilities	$0.97
Labor, supplies, and overheads	$0.44
Depreciation	$0.43
Coproduct credits	−$2.45
Production cost per gallon	$1.96
Operating costs ($/yr)	
Shelled corn	$110,829,994
Denaturant	$0
Other raw materials	$5,036,590
Utilities	$43,457,545
Labor, supplies, and overheads	$19,959,364
Depreciation	$19,262,217
Coproduct credits	−$110,443,187
Total production cost	$88,102,522

Table 4.14 Economic results comparison for corn dry mill ethanol and n-butanol

	Corn to ethanol	Corn to butanol
Annual ethanol production cost	$1.53/gal	$1.96/gal
Plant size (fuel production, MM gal/yr)	45.00	45.00
Yield (gal/bushel)	2.77	1.36
Capital costs—installed ($MM)		
Feed handling and milling	$4.05	$5.99
Liquefaction and saccharification	$3.56	$5.31
Fermentation	$10.97	$19.99
Distillation	$13.56	$22.60
Water recycle and solids handling	$25.85	$63.08
Storage	$1.53	$7.32
Utilities and other	$6.23	$9.21
Total	$65.75	$133.51
Operating costs ($/gallon)		
Shelled corn	$1.21	$2.46
Denaturant	$0.04	$0.00
Other raw materials	$0.05	$0.11
Utilities	$0.30	$0.97
Labor, supplies, and overhead	$0.11	$0.44
Depreciation	$0.14	$0.43
Coproduct credits	−$0.32	−$2.45
Total	$1.53	$1.96

(Qureshi et al. 2008) and soy molasses (Qureshi et al. 2001) for butanol production. *C. beijerinckii* P260 was used to wheat straw by Qureshi et al. 2007. *C. acetobutylicum* P262 has been used to corn stover (Parekh et al. 1988) and wheat straw (Marchal et al. 1984). *Clostridium saccharoperbutylacetonicum* has been used for bagasse and rice straw (Soni et al. 1982). Newly engineered *Escherichia coli* strains as the microbial cultures have been developed to produce isobutanol (Atsumi et al. 2007, 2008; Hanai et al. 2007), in addition to n-butanol production. It should be noted that butanol-producing cultures are able to use a wide variety of carbohydrates, such as cellobiose, sucrose, glucose, fructose, mannose, lactose, dextrin, starch, xylose, and arabinose (Qureshi and Thaddens 2008). There is no reported cost data for cellulosic butanol production available yet.

Comparison on process economics The TPIs for the six biofuel processes discussed in the paper are listed in Table 4.15 for comparison at the plant scale of 45 MM gal production/yr. For TPI per gallon biofuel production, corn butanol costs more than any of the other processes due to both low yield and high installed equipment cost under current technologies.

Once converted to a production cost with energy equivalence to gasoline (Table 4.16), the modeled butanol production cost is more similar to the food-based commercial processes' production cost. Although both cellulosic ethanol processes have not achieved commercial scale yet and are still in the development stages, the process economics with higher TPI still look very promising.

Table 4.15 Total project investment of all six processes

Total capital investment ×1,000						
45 MM gal/yr biofuel	Corn ethanol	Sugarcane ethanol	Soybean diesel	Corn butanol	Corn stover ethanol (biochem)	Wood chips ethanol (thermochem)
Total installed equipment cost	$65,745	$43,700	$11,681	$138,317	$91,901	$120,637
Warehouse and site development	$6,575	$4,589	$1,168	$13,832	$9,190	$12,064
Total installed cost	$72,320	$48,290	$12,849	$152,148	$101,091	$132,701
Indirect costs						
Field expenses	$14,464	$9,658	$2,570	$30,430	$20,218	$26,540
Home office and construction fee	$18,080	$12,072	$3,212	$38,037	$25,273	$33,175
Project contingency	$14,464	$9,658	$2,570	$30,430	$20,218	$26,540
Total capital investment	$119,328	$79,676	$21,201	$251,045	$166,800	$218,956
Other costs (startup, permits, etc.)	$11,933	$7,968	$2,120	$25,104	$16,680	$21,896
Total project investment	$131,261	$87,644	$23,321	$276,149	$183,480	$240,852
TPI per gal biofuel ($/gal)	$2.92	$1.95	$0.52	$6.14	$4.08	$5.35

Process intensification Numerous ways of improving the overall efficiencies and intensities of biofuels processes are being explored for both current and future biofuels. In addition, the colocation of cellulosic facilities with existing conventional facilities is being examined. As part of an on-going collaboration between USDA-ARS and NREL, several conceptual options for integrating these technologies were explored and summarized in a report (Wallace et al. 2005). Biomass in corn ethanol boiler systems in place of natural gas is being used in Minnesota. Not only does this help to begin integrating biomass into existing systems but it also avoids high natural gas prices and fluctuations. If gasifier systems, such as one being implemented by Chippewa Valley Ethanol Co., Benson, MN, are used, they provide the next step toward integrating cellulosic ethanol production into conventional systems.

Similarly, within cellulosic biofuels processes themselves, opportunity exists for integrating biochemical and thermochemical technologies to increase the overall efficiency of a biorefinery (Fig. 10). This has been demonstrated through the Role of Biomass in America's Energy Future project. While capital costs for these systems will likely be large, the increased energy efficiency of such systems can lead to

4 The Economics of Current and Future Biofuels

Table 4.16 The process economics comparison of all six processes

45 MM gal/yr biofuel	Corn ethanol	Sugarcane ethanol	Soybean diesel	Corn butanol	Corn stover ethanol (biochem)	Corn stover ethanol (thermochem)
Total production cost ($M/yr)	$74	$58	$121	$88	$64	$58
Total project investment ($M)	$131	$88	$23	$276	$183	$241
Production cost ($/gal)	$1.53	$1.29	$2.55	$1.96	$1.48	$1.32
Energy density (BTU/gal)	76,330	76,330	119,550	99,837	76,330	76,330
Production cost with energy equivalent to gasoline ($/gal)	$2.33	$1.95	$2.48 equivalent to gasoline; $2.74 equivalent to diesel	$2.28	$2.25	$2.00

Energy density data from Argonne GREET 1.8 model (http://www.transportation.anl.gov/modeling_simulation/GREET/index.html).

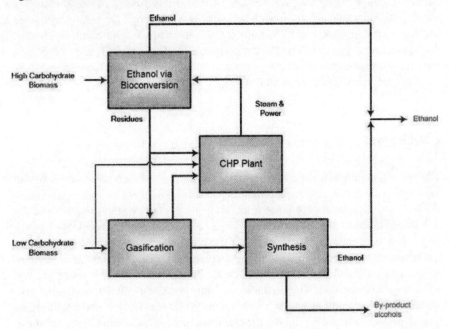

Fig. 4.10 Conceptual configuration for a combined biochemical/thermochemical biorefinery

attractive economics for certain modeled scenarios. Specifically, the thermochemical section can provide combined heat and power production by gasifying the lignin residue from a biochemical process while simultaneously producing additional liquid fuels. The overall yields of alcohols and the total process efficiencies have been shown to increase beyond current designs.

While the benchmark biochemical cellulosic ethanol design uses purchased cellulase enzymes, the option for including onsite enzyme production is also a viable technological consideration. Realistically, enzyme cost targets in the range of $0.30/gal at the commercial scale should be achievable in the near future by avoidance of transportation and formulation costs (Merino and Cherry 2007). In such scenario, on-site or near-site enzyme production is essential, where enzymes are produced using reduced-cost feedstocks, transported short distances, and not stored for extended periods of time. The least expensive alternative in this situation involves the direct use of whole fermentation broth (including cell mass) to circumvent expensive cell removal and enzyme formulation steps. To investigate this possibility, they compared the use of whole fermentation broth and cell-free broth as catalysts; results strongly indicated that there is no difference in terms of enzymatic performance for both preparations with equivalent dose. Most current commercial cellulase products are sold as cell-free stabilized concentrates. Although these formulations meet market needs with regard to application and cost, for many enzyme products, it is uncommon for the recovery cost and formulation costs to be a major portion of the overall cost. It then follows that no postfermentation processing has the lowest cost. Performance in saccharification was indistinguishable among fresh whole fermentation broth, fresh fully recovered and formulated product, and 28-d-old fermentation broth. These results suggest that typical recovery and formulation costs can be eliminated for use in biorefinery operations, especially in an integrated plant that both makes and uses the cellulase enzymes (Dean et al. 2006).

Conclusions

The process economics for commercial biofuels processes are presented here and are fairly well understood throughout the industry. For these processes, the feedstock cost comprises a very large fraction of the overall production cost, and the overall capital costs are not particularly large compared to other processes or industries. Techno-economic models exist for these processes that appropriately depict the proper process behavior, results, and economics. This knowledge in techno-economic analysis of biofuels can also be applied to future biofuels processes that are not yet commercial. This includes not only the cellulosic processes but also advanced biofuels processes, such as butanol. While feedstock costs are the single largest portion of the overall cost for cellulosic processes, capital costs are much higher for these processes because of the increased difficulty presented by the deconstruction and use of these materials. Significant opportunity for improving both existing and future biofuels economics will come from a variety of sources including better process intensity and integration of multiple technologies.

Acknowledgment NREL would like to thank the US Department of Energy (DOE) Office of the Biomass Program (OBP) for its continued leadership, support, and collaboration in the biofuels arena.

References

Aden, A.; Ruth, M.; Ibsen, K.; Jechura, J.; Neeves, K.; Sheehan, J.; Wallace, R. Lignocellulosic biomass to ethanol process design and economics utilizing co-current dilute acid prehydrolysis and enzymatic hydrolysis for corn stover. NREL report NREL/TP-510-32438, http://www.nrel.gov/docs/fy02osti/32438.pdf; 2002.

Aspen Plus™ Release 2006.5. Aspen Technology, Cambridge, MA2006.

Atsumi S.; Cann A. F.; Connor M. R.; Shen C. R.; Smith K. M.; Brynilden M. P.; Chou K. J. Y.; Hanai T.; Liao J. C. Metabolic engineering of *Escherichia coli* for 1-butanol production. *Metabolic Eng* 10: 305–311; 2008.

Atsumi S.; Hanai T.; Liao J. C. Non-fermentative pathway for synthesis of brached-chain higher alcohols as biofuels. *Nature* 451: 86–89; 2007. doi:10.1038/nature06450.

Blaschek H. P.; Ezeji T. C.; Qureshi N. Continuous butanol fermentation and feed starch retrogradation: butanol fermentation sustainability using *Clostridium beijerinckii* BA101. *J. Biotechnol.* 115: 179–187; 2005. doi:10.1016/j.jbiotec.2004.08.010.

Bothast R. J.; Schlicher M. A. Biotechnological processes for conversion of corn into ethanol. *Appl. Microbiol. Biotechnol.* 67: 19–25; 2005. doi:10.1007/s00253-004-1819-8.

BNDES. Sugarcane-based bioethanol: energy for sustainable development/coordination BNDES and CGEE. www.sugarcanebioethanol.org; 2008

BP Biofuels News. Advanced biofuels, working together, two global leaders are creating the next generation of biofuels. http://www.bp.com/sectiongenericarticle.do?categoryId=9021783&contentId=7041026; 2006.

Cleantech News. UK firm plans biobutanol plant in India. http://www.cleantech.com/news/3550/uk-firm-plans-biochemical-plant-india; 2008

Cobalt Biofuel News. Cobalt biofuels raises $25 million to commercialize next generation biofuel—biobutanol. http://www.cobaltbiofuels.com/news/news-item/series-c-round; 2008

Dean B.; Dodge T.; Valle F.; Chotani G. Development of biorefineries—technical and economic considerations. Biorefineries—industrial process and products. In: Kamm B.; Gruber P. R.; Kamm M. (eds) Status quo and future directions, vol 1. Wiley-VCH, Weinheim, pp 67–83; 2006.

Demirbas A. Biodiesel production from vegetable oils via catalytic and non-catalytic supercritical methanol transesterification methods. *Prog. Energy Combust. Sci.* 31: 466–487; 2005.

Douglas J. M. Conceptual design of chemical processes. McGraw-Hill, New York. 1989.

EIA. Biodiesel performance, costs and use. http://tonto.eia.doe.gov/FTPROOT/environment/biodiesel.pdf; 2004

EISA. EISA of 2007 calls for additional production of biofuels. http://www.renewableenergyworld.com/rea/partner/stoel-rives-6442/news/article/2008/01/eisa-of-2007-calls-for-additional-production-of-biofuels-51063; 2007

FO Lichts Ethanol production costs: a worldwide survey, a special study from FO Lichts and Agra CEAS Consulting. Agra Informa, Tunbridge Wells, Kent; 2007.

Gevo News. Gevo, Inc secures cellulosic technology to make advanced biofuel. http://www.gevo.com/news_Cargill-pr_022609.php; 2009.

Graboski M. S.; McCormick R. L. Combustion of fat and vegetable oil derived fuels in diesel engines. *Prog. Energy Combust. Sci.* 24: 125–164; 1998.

Hanai T.; Atsumi S.; Liao J. C. Engineered synthesis pathway for isopropanol production in *Escherichia coli. App. Environ. Microbiol.* 73: 7814–7818; 2007.

Hass M. J.; McAloon A. J.; Yee W. C.; Foglia T. A. A process model to estimate biodiesel production cost. *Biores. Technol.* 97: 671–678; 2006.

Hill J.; Nelson E.; Tilman D.; Polasky S.; Tiffany D. Environmental, economic, and energetic costs and benefits of biodiesel and ethanol biofuels. *PNAS* 10330: 11206–11210; 2006.

Kwiatkowski J. R.; McAloon A. J.; Taylor F. Modeling the process and costs of fuel ethanol production by the corn dry-grind process. *Ind. Crops Prod.* 233: 288–296; 2006.

Lin Y.; Blaschek N. P. Butanol production by butanol-tolerant strain of *Clostridium acetobutylicum* in extruded corn broth. *Appl. Envirol. Microbiol.* 45: 966–973; 1983.

Marchal R.; Rebeller M.; Vandecasteele J. P. Direct bioconversion of alkali-pretreated straw using simultaneous enzymatic hydrolysis and acetone butanol production. *Biotechnol. Lett.* 6: 523–528; 1984.

Merino S. T.; Cherry J. Progress and challenges in enzyme development for biomass utilization. *Adv Biochem Engin Biotechnol* 108: 95–120; 2007.

NBB. Estimated US biodiesel production by fiscal year. http://www.biodiesel.org/pdf_files/fuelfactsheets/Production_Graph_Slide.pdf; 2009

Parekh M.; Formanek J.; Blaschek H. P. Pilot-scale production of butanol by *Clostridium beijerinckii* BA101 using a low-cost fermentation medium based on corn steep water. *Appl. Microbiol. Biotechnol.* 51: 152–157; 1999.

Parekh S. R.; Parekh R. S.; Wayman M. Ethanol and butanol production by fermentation of enzymatically saccharified SO2-prehrdolysed lignocellulosics. *Enzyme Microb. Technol.* 10: 660–668; 1988.

Peters M. S.; Timmerhaus K. D. Plant design and economics for chemical engineers. 4th ed. McGraw-Hill, New York; 1991.

Phillips, S.; Aden, A.; Jechura, J.; Dayton, D. Thermochemical ethanol via indirect gasification and mixed alcohol synthesis of lignocellulosic biomass. National Renewable Energy Laboratory Golden CO. NREL report no TP-510-41168. http://www.nrel.gov/docs/fy07osti/41168.pdf; 2007.

Qureshi N.; Blaschek N. P. Economics of butanol fermentation using hyper-butanol producing *Clostridium beijerinckii BA 101*. *Food Bioprod. Process.* 78: 152–167; 2000.

Qureshi N.; Lolas A.; Blaschek H. P. Soy molasses as fermentation substrate for production of butanol using *Clostridium beijerinckii BA 101*. *J. Ind. Microbiol. Biotechnol.* 26: 290–295; 2001.

Qureshi N.; Saha B. C.; Cotta M. A. Butanol production from wheat straw hydrolyzate using *Clostridium beijerinckii*. *Bioprocess Biosys. Eng.* 30: 419–427; 2007.

Qureshi N.; Thaddeus C. E. Butanol, a superior biofuel production from agricultural residues (renewable biomass): recent progress in technology. *Biofuels Bioproducts Biorefining* 2: 319–330; 2008.

Qureshi N.; Thaddeus C. E.; Ebener J.; Dien B. S.; Cotta M. A.; Blaschek H. P. Butanol production by *Clostridium beijerinckii*. Part I: use of acid and enzyme hydrolyzed corn fiber. *Bioresour. Technol.* 99: 5915–5922; 2008.

Ramey, D.; Yang, S. H. Production of butyric acid and butanol from biomass final report. Work performed under contract no: DE-F-G02-00ER86106 USDOE Morgantown WV; 2004.

Ramirez. E.; Johnston D.; Mcaloon A. J.; Singh V. Enzymatic corn wet milling: engineering process and cost model. *Biotechnol. Biofuels* 2: 2; 2009.

RFA. Changing the climate, ethanol industry outlook. RFA, Washington, DC; 2008.

Rodrigues, A. P. Participação dos fornecedores de cana na cadeia do açúcar e álcool. Congresso Internacional de Tecnologias na Cadeia Produtiva, Concana Uberaba, MG, março de; 2007.

Seabra, J. E. A. Technical-economic evaluation of options for whole use of sugar cane biomass in Brazil. Campinas Faculdade de Engenharia Mecânica Universidade Estadual de Campinas 274p PhD thesis (In portuguese); 2007.

Shapouri, H.; Salassi, M.; Fairbanks, N. The economics feasibility of ethanol production from sugar in the United States. USDA report; 2006

Soni B. K.; Das K.; Ghose T. K. Bioconversin of agrowastes inot acetone butanol. *Biotechnol. Lett.* 4: 19–22; 1982.

Spake, A. DuPont develops world's first advanced biofuel, biobutanol will be a high-energy petroleum alternative. http://www.america.gov/st/washfile-english/2007/September/20070919163628ndyblehs0.6094019.html; 2007

TetraVitae Bioscience. http://www.tetravitae.com; 2009

USDA. USDA's 2002 ethanol cost-of-production survey. Agricultural Economic Report Number 841; 2002

USDA. USDA soybean projections, 2008–17. http://www.ers.usda.gov/briefing/soybeansoilcrops/2008baseline.htm#ussoybean; 2008

Wallace, R.; Ibsen, K.; McAloon, A.; Yee, W. Feasibility study for co-locating and integrating ethanol production plants from corn starch and lignocellulosic feedstocks. http://www.nrel.gov/docs/fy05osti/37092.pdf; 2005

Wooley, R.; Putsche, V. Development of an ASPEN PLUS physical property database for biofuels components. National Renewable Energy Laboratory, Golden CO. NREL report no. MP-425-20685; 1996.

Wooley, R.; Ruth M.; Glassner D.; Sheehan J. Process design and costing of bioethanol technology: a tool for determining the status and direction of research and development. *Biotechnol. Prog.* 155: 794–803; 1999.

Zhang Y.; Dube M. A.; McLean D. D.; Kates M. Biodiesel production from waste cooking oil: 2. economic assessment and sensitivity analysis. *Bioresour. Technol.* 90: 229–240; 2003.

Chapter 5
A Multiple Species Approach to Biomass Production from Native Herbaceous Perennial Feedstocks

J.L. Gonzalez-Hernandez, G. Sarath, J.M. Stein, V. Owens, K. Gedye, and A. Boe

Abstract Due to the rapid rate of worldwide consumption of nonrenewable fossil fuels, production of biofuels from cellulosic sources is receiving increased research emphasis. Here, we review the feasibility to produce lignocellulosic biomass on marginal lands that are not well-suited for conventional crop production. Large areas of these marginal lands are located in the central prairies of North America once dominated by tallgrass species. In this article, we review the existing literature, current work, and potential of two native species of the tallgrass prairie, prairie cordgrass (*Spartina pectinata*), and little bluestem (*Schizachyrium scoparium*) as candidates for commercial production of biofuel. Based on the existing literature, we discuss the need to accelerate research in the areas of agronomy, breeding, genetics, and potential pathogens. Cropping systems based on maintaining biodiversity across landscapes are essential for a sustainable production and to mitigate impact of pathogens and pests.

Keywords Biomass • Herbaceous perennial feedstocks • Marginal lands • Little bluestem • Prairie cordgrass

Introduction

This review evaluates the rationale for using a variety of native perennial herbaceous species for the emerging lignocellulosic fuel sector and considers potential biotic threats that could impact this venture. Recent and previous reviews on herbaceous feedstocks have had a distinctly switchgrass flavor, as this crop (*Panicum virgatum* L.) had emerged as a lead perennial herbaceous candidate from a large body of study that was funded by the US-Department of Energy (Bouton 2007; McLaughlin and Kszos 2005; Parrish and Fike 2005; Sarath et al. 2008; Vogel et al.

J.L. Gonzalez-Hernandez (✉)
Department of Plant Sciences, South Dakota State University, Brookings, SD 57007, USA
e-mail: jose.gonzalez@sdstate.edu

2002). As a result of this interest, several important strides have been made in the development of genomic and other resources for switchgrass (Bouton 2007; Martinez-Reyna and Vogel 2008; Missaoui et al. 2005; Mitchell et al. 2002; Tobias et al. 2005, 2006, 2008). In addition, recent studies have demonstrated that switchgrass can be effectively transformed to produce polyhydroxybutyrate (Somleva et al. 2008), potentially opening the way for future engineering of quality parameters in this species. Taken together, these studies of switchgrass have provided a robust framework to evaluate other potential herbaceous perennial feedstocks and point out the detailed agronomic, genomic, breeding, and management infrastructures that will have to be developed for each species.

Vogel and Jung (2001) have made a number of research recommendations for developing herbaceous bioenergy crops. These recommendations centered on the need for efficient and rapid analyses of biomass quality and the need for greater understanding of the genetics and linkage parameters affecting traits of interest in these species. However, underlying assumptions included that these perennial crops would be grown on marginal land, could be cost effectively harvested, stored, and converted into a liquid fuel such as ethanol. Although conversion of biomass into liquid fuels, especially ethanol, has received significant interest and funding (Cardona and Sanchez 2007; Chen and Dixon 2007; Dien et al. 2006; Schmer et al. 2008; Simpson et al. 2008), several alternate platforms exist for converting biomass into energy including the production of syngas, pyrolysis oils, and through burning in coal-fired power plants (Demirbas 2008; Gaunt and Lehmann 2008; Keshwani and Cheng 2009). In each of these cases, biomass quality parameters can be important. As an example, low lignin biomass is probably most suitable in an ethanol biorefinery (Chen and Dixon 2007; Sarath et al. 2008) whereas biomass with higher levels of aromatics would possess greater energy content and potentially be a better fit for direct-conversion strategies. Thus, to a certain extent, ideal biomass quality will be driven by the conversion platform it is ultimately intended for, although all herbaceous species raised for bioenergy purposes (at least in the continental USA) will share several common traits.

First, an herbaceous bioenergy crop should not compete directly with food crops for land or other resources. Second, it must have fairly broad environmental adaptation with sustainable yields while requiring limited inputs. Third, it must be a perennial with good regrowth potential and tolerance to both biotic and abiotic stresses. Finally, it must be amenable for improvement through traditional, molecular, or/and engineered breeding. Given the edaphic and environmental variations in the continental USA, no one crop will exhibit all of these attributes across all growing regions, indicating the need to evaluate and develop multiple species with good adaptation to specific climatic and edaphic zones. These considerations suggest that native plants that have evolved within a specific plant adaptation region would be good targets for improvement (Tobias et al. 2005). Finally, it may be possible to use a mix of native species to create low-input high diversity (LIHD) bioenergy landscapes (Tilman et al. 2006).

Selecting the appropriate native species does present some challenges. Even after a specific species is selected, it will require several years of yield trials across different environments to adequately document its utility as a bioenergy crop. These initial agronomic studies are critical to the future success in deploying herbaceous crops and requires the needed commitment in fiscal, land, and human resources. Among the key factors during initial evaluation of a species is its ability to thrive and produce on marginal land. For our purposes, marginal land is assumed to be farmable, but would not normally support the economic production of food crops. Most of these lands are expected to be classified within the Conservation Reserve Program (CRP) of the USDA.

The question of marginal lands is an important one and requires some discussion. Over the course of the last few years, many studies have raised the food *versus* fuel debate (Cassman and Liska 2007; Gomez et al. 2008) as well as a range of potential environmental aspects of cultivating dedicated bioenergy crops (Fargione et al. 2008; Schmer et al. 2008; Schnoor 2006; Searchinger et al. 2008). Marginal lands in most producers' fields are attractive for bioenergy purposes since they normally adjoin cropped land and present an added opportunity for producers to obtain value. Placing such land in perennial plants can be expected to maintain most of conservation benefits, such as lowered soil erosion, enhanced soil C-sequestration, minimizing run-off, and providing wildlife habitat and yielding biomass for biorefineries, while allowing producers to get revenue. If the right crop or blend of crops (Schmer et al. 2008; Tilman et al. 2006) are selected, it should be possible to realize both environmental and fuel requirements. There is also the potential to manage noncrop marginal lands for bioenergy uses. An example would be soils and/or environment that favor the growth of a particular native perennial that could be used as a source of lignocellulosic biomass. In fact, part of the current push for herbaceous perennials as a fuel source has its origins from the abundant native tallgrass prairies of the USA.

Prior to the arrival of European immigrants, large tracts of the US Midwest had been under perennial vegetation, dominated by herbaceous grasses comprising both the native tall and short grass prairies. These ecosystems were generally characterized by deep soils containing abundant amounts of soil carbon and enough biomass production to support vast herds of bison and other large herbivores. These rich soils have since provided much of the US cereal grains and other crops, most notably soybeans, maize, and wheat. However, continued annual cropping has led to numerous problems associated with modern agriculture, including soil loss from wind and water erosion, leading to reduced yields and/or loss of arable acres. In addition to the significant resources which have been allocated to deal with these environmental issues, it has become apparent that we could also return to the tallgrass native prairie model to generate biomass for biofuels.

As mentioned earlier, the first plant selected for bioenergy use has been switchgrass. Although switchgrass is not the dominant species in tallgrass prairies, it possesses many useful agronomic traits such as ease of seed harvest, cleaning, and planting. In contrast, the two other more dominant tallgrass prairie species, big bluestem (*Andropogon gerardii* Vitman) and Indian grass (*Sorghastrum nutans* (L.) Nash; both chaffy-seeded species), have not received the same level of research interest, although they have yield potentials possibly greater than switchgrass.

Similarly, a number of other warm- and cool-season grasses have good potential to yield biomass under specific environmental conditions. These include prairie cordgrass (*Spartina pectinata* Bosc ex Link), reed canary grass (*Phalaris arundinacea* L.), bermuda grass (*Cyanodon dactylon* L.), napiergrass (*Pennisetum purpureum* Schumach), and the giant reed (*Arundo donax* L.; Anderson et al. 2008; Dien et al. 2006). Research on several of these species is needed in addition to the development of switchgrass as a primary model and possibly the first-generation dedicated bioenergy species.

Numerous studies have touted introduced species with good potential as a bioenergy feedstock. The most dominant of these species has been *Miscanthus* (*Miscanthus x giganteus*). *Miscanthus* appears to possess many of the traits critical for success as a bioenergy feedstock (Christian et al. 2008; Heaton et al. 2004, 2008). It currently suffers from a major limitation; it is a sterile hybrid which requires transplantation of rhizomes. It is possible that this limitation might be overcome through research and agronomic adaptation of transplanting machinery; however, another significant concern is that *Miscanthus* is a nonnative perennial with broad environmental adaptation. Therefore, it carries some risk of becoming an invasive weed like other such perennial species (Barney and Ditomaso 2008). Although this is not a focus of this review, it should be borne in mind that most of the cultivated crops are exotic introductions and have not posed a major environmental threat when managed appropriately. Given the significant resources that have been committed to the development of *Miscanthus* (for example http://miscanthus.illinois.edu/), this plant will likely become a member of the herbaceous bioenergy suite of plants.

Based on the switchgrass model, several promising plants need development of genetic, physiological, agronomic, and conversion resources. The development of such resources will allow successful cross-pollination of ideas and lead to discovery of common and species-specific mechanisms that can be deployed for continued improvement of herbaceous feedstocks for bioenergy purposes.

The Tallgrass Prairie of North America: A Model Biomass Production System

A number of herbaceous perennial feedstocks have been evaluated for their potential to provide high quality lignocellulose to biorefineries of the future. Much of the work has been conducted on switchgrass (*P. virgatum* L.), a tall-growing, C_4 species, native to much of the eastern USA (Hitchcock 1935). In 1991, switchgrass was selected by the US Department of Energy (DOE) as a "model" herbaceous feedstock because it is (1) widely adapted, (2) perennial, (3) grows well on land not highly suited to row crop production, (4) established from seed, and (5) beneficial for soil conservation and wildlife habitat (McLaughlin and Kszos 2005; Parrish and Fike 2005; Sanderson et al. 2006). The benefits of switchgrass are also found in a number of other species of the tallgrass prairie, including prairie cordgrass (*S. pectinata* Link.; Boe and Lee 2007; Potter et al. 1995), big bluestem (*A. gerardii* Vitman; Mulkey et al. 2008), *M.* x *giganteus* (Heaton et al. 2008), and reed canarygrass (*P. arundinacea* L.; Sanderson and Adler 2008). Because multiple species are of potential use as feedstocks, knowledge of local environmental conditions, potential limitations due to biotic and abiotic stresses, and other pertinent factors should also play a role in the choice of species for each region.

Tillman et al. (2006) suggested the use of LIHD mixtures of native grasses. They found that LIHD mixtures produced more potential energy on agriculturally degraded and nitrogen-poor sandy soil with no inputs than any of the individual species grown alone. However, biomass production of the LIHD mixture (about 4 Mg ha^{-1}) averaged well below the DOE goal of 22 Mg ha^{-1} (U.S. DOE 2006). In on-farm trials in the Great Plains, USA, switchgrass produced 540% more renewable than nonrenewable energy consumed (Schmer et al. 2008). In addition, switchgrass managed for high yield produced 93% more biomass than estimates from the human-made prairie established by Tilman et al. (2006). Switchgrass was a key component in maintaining yields above 5 Mg ha^{-1} at three locations in the northern Great Plains, USA (Owens et al. 2009, unpublished data). In their trial, switchgrass, big bluestem, and indiangrass were grown in monocultures and all two- and three-way mixtures. These were highly diverse mixes, but one used three of the dominant warm-season grasses of the tallgrass prairie. While switchgrass helped maintain yields, the inclusion of big bluestem improved soil cover.

Russelle et al. (2007) refuted some of the claims made by Tilman et al. (2006), noting that (1) minerals may have been replaced in the system since most biomass was burned after removing small areas within the plots, (2) the establishment difficulties associated with many native prairie species were ignored, (3) the experiment was conducted at only one location and yet results were extrapolated to the entire planet, and (4) the use of corn stover, as well as the grain, would improve the net energy of corn. On the other hand, annual fire events on the Konza Prairie in eastern Kansas, USA increased aboveground net primary productivity (ANPP) of native grasses 8 of 12 yr while ANPP for forbs increased only 4 of 12 yr (Knapp et al. 1998). Precipitation was a key component in increased in ANPP, i.e., in years with adequate precipitation, ANPP was much more likely to increase with burning. It is apparent from these examples that a diversity of species and management approaches will be needed to meet the needs of biorefineries. The decision to grow a monoculture or a mixture of several species will largely be determined by the requirements of the feedstock "consumer" (e.g., cellulosic ethanol facilities), local environmental conditions, types of government subsidies, as well as how society begins to value such diversity.

Many proponents of lignocellulosic biomass note the capability to produce feedstocks on marginal lands. In the USA, land capability classes were developed to help technical managers describe soils and their suitability for cultivation (Hockensmith and Steele 1950). Essentially, soils are given a class rating from I to VIII with possible subclasses describing specific limitations within each class. Land capability classes I–IV are suitable for cultivation with varying limitations. Land capability classes V–VIII are unsuitable for cultivation due to characteristics such as flooding, slope, stones, and roughness. However, they may be suitable for grazing or forestry. Because of this, proponents of perennial biomass crops suggest that these species be grown on land rated unsuitable for cultivation (V–VIII), noting that this will help mitigate the food–fuel debate. On the other hand, some authors have noted that conversion of marginal lands to biomass production may have a negative impact on forage–livestock production as pastureland and hayland are pushed to even more marginal lands (Ceotto 2008; Sanderson and Adler 2008). Nonetheless, the focus of this review is on perennial species adapted to marginal land such as those in land capability classes II–IV for which moisture or erosion are the primary limitations to cultivation, land capability class V on which flooding may frequently occur, and land capability classes VI–VIII where cultivation is not possible (Doolittle et al. 2002).

Class V soils may be affected by salinity or exhibit other problems associated with inherent wetness. There is approximately 13.2 million-hectare class V land within the contiguous states of the USA (USDA-NRCS 2000). A number of species native to the tallgrass prairie, such as big bluestem, little bluestem (*Schizachyrium scoparium* [Michx.] Nash), and prairie cordgrass are well-adapted to some or most of the conditions of class V land. Studies using these and other species have been conducted to evaluate their performance on purported marginal land including native, warm-season grass mixtures composed of switchgrass, big bluestem, and indiangrass (*S. nutans* [L] Nash; Mulkey et al. 2008); switchgrass on land enrolled in the CRP (Mulkey et al. 2006), on reclaimed land (Al-Kaisi and Grote 2007), on multiple farms in the Great Plains (Schmer et al. 2008), and on dryland in eastern Nebraska, USA (Varvel et al. 2008); prairie cordgrass (Boe and Lee 2007); tall fescue (*Festuca arundinacea* Schreb.) (Wells et al. 2003); and a number of monoculture cool- and warm-season grasses and legumes on reclaimed mine land in Virginia, USA (Evanylo et al. 2005). While the cited work was done on marginal land, it is not clear whether or not it was done on class V land.

More research is needed on class V land to ascertain yield, persistence, and management practices to optimize performance of specific species adapted to this environment. By the final years of a 4-yr study, Boe and Lee (2007) reported that two prairie cordgrass populations averaged 9.3 Mg ha^{-1} compared to 2.0 Mg ha^{-1} for "Cave-in-Rock" switchgrass and 4.8 Mg ha^{-1} for "Summer" and "Sunburst" switchgrass. In a 10-yr multilocation trial in the upper Great Plains, USA (Tober et al. 2008) reported big bluestem yields from 0.1 to 8.8 Mg ha^{-1} depending on years and location. However, nitrogen fertilizer was not applied in either of the aforementioned multiyear studies. Mulkey et al. (2008) found that yields of mixtures containing big bluestem, indiangrass, and switchgrass responded some years to N rates up to 112 kg ha^{-1}, but addition of N also increased the percentage of weedy species in the mix.

Basic management information (e.g., establishment, fertilizer practices, harvest timing, harvest methods) is severely lacking for species such as prairie cordgrass and little bluestem, two species adapted to the type of marginal land described in this review. This is particularly true when these species are considered for biomass energy. General guidelines regarding warm-season grass establishment (Masters et al. 2004) and management for biomass (Sanderson et al. 2004) can be found in the literature. However, specific production practices and effects of pests such as weeds, insects, and diseases at diverse environments are severely lacking for all potential feedstock species and especially for the less-studied species like prairie cordgrass, little bluestem, and big bluestem.

Breeding North America Prairies Native Species: Little Bluestem and Prairie Cordgrass

Little bluestem. Little bluestem [*S. scoparium* (Michx.) Nash] was the most important dominant of uplands in the central area of the tallgrass prairie. On drier soils, it comprised 50% to 90% of the vegetation (Weaver and Fitzpatrick 1932). It was also the dominant species on shallow slopes of the northern region of the tallgrass prairie that extended from southern Manitoba through the eastern Dakotas and western Minnesota. Weaver (1960) studied 63 typical prairies throughout 100,000 km^2 in the central Missouri Valley region and determined little bluestem composed 58% of the grassland compared with 38% for big bluestem (*A. gerardii* Vitman). In the driest areas of the region where annual precipitation ranged from 61 to 66 cm, percentages were 69% for little bluestem and 28% for big bluestem. Currently, little bluestem is dominant in many areas of the mixed-grass prairie in the northern Great Plains (Johnson and Larsen *1999*), the Nebraska Sandhills, and central Kansas, Oklahoma, and Texas (Sims and Risser 2000).

The adaptation of little bluestem to dry and shallow soils across wide latitudinal and longitudinal ranges suggests it may have potential as a dedicated lignocellulosic biomass feedstock for biofuels production on land not suitable for annual crops or other perennial grasses, such as switchgrass (*P. virgatum* L.), which are best adapted to more mesic midslopes and lowlands (Weaver and Fitzpatrick 1932). Gilbert et al. (1979) considered little bluestem to be a more stable forage producer than switchgrass under fluctuating climatic conditions.

Breeding work on little bluestem began during the late 1930s. Evaluation of strains from North Dakota to Texas in common gardens in the central Great Plains revealed ecoclinal variation for maturity, height, leafiness, and disease resistance (Anderson and Aldous 1938; Cornelius 1947). More than ten cultivars, selected populations, and source-identified populations of little bluestem have been released, with a collective range of adaptation extending from Kansas to southern Canada (Boe et al. 2004). As is the case for switchgrass and big bluestem, a strong relationship exists between latitude of origin and biomass production and winter hardiness in little bluestem. Cornelius (1947) evaluated 16 ecotypes of little bluestem with origins from Towner, ND; Vernon, TX; and Manhattan, KS. Individual spaced-plant biomass averaged across 2 yr ranged from 130 g for the North Dakota ecotype to 1,563 g for the Texas ecotype. Spaced plants of all of the ecotypes survived the winters at Manhattan, but severe winter injury occurred on Texas and Oklahoma ecotypes in seeded plots. Phan and Smith (2000) evaluated 14 ecotypes of little bluestem from Manitoba at Winnipeg and found biomass was the highest for the southern ecotypes (138 g plant^{-1}) and the lowest for the northern ecotypes (50 g plant^{-1}). Camper, with origins in Nebraska and Kansas, has shown high winter survival in seeded trials and spaced-plant trials in the northern Great Plains (Jacobson et al. 1984; Boe and Bortnem 2009).

In a recent genetic study of biomass production of spaced plants of 22 genotypes from "Camper" little bluestem in eastern South Dakota (Boe and Bortnem 2009), differences were found among genotypes for biomass, number of phytomers per tiller, tillers plant^{-1}, mass tiller^{-1}, mass of the primary axis, and mass of the axillary branches. This indicated genetic variation within Camper for biomass and yield components of biomass. Variation among genotypes for number of phytomers per tiller was likely due to within-cultivar variation in maturity as a result of multiple sources from Nebraska and Kansas being used in the development of Camper. Fu et al. (2004) reported that >91% of the total amplified fragment length polymorphism (AFLP) variation of six natural populations of little bluestem from Manitoba and Saskatchewan was within the populations. Since natural stands of little bluestem on marginal land in dry areas are highly bunchy (Weaver 1960), research to determine the impact of variation in interplant spacing on axillary branching, biomass production, seed production, water and nutrient use efficiency, and persistence is needed (Boe and Bortnem 2009).

Prairie cordgrass. Prairie cordgrass (*S. pectinata* Link.) is a tall, rhizomatous, perennial, warm-season species found predominantly in marshes, wet meadows, potholes, and drainage ways throughout Canada to 60° N latitude and throughout the continental USA, with the exceptions of Louisiana to South Carolina in the Southeast, and California, Nevada, and Arizona in the West (Hitchcock 1950; Mobberley 1956; Stubbiendieck et al. 1982). However, Mobberley (1956) frequently found prairie cordgrass in open dry prairie and on high ground along railroad rights-of-way in the Midwestern USA. The genus *Spartina* has the most northerly distribution of any of the C4 perennial grasses (Potter et al. 1995).

Prairie cordgrass is recognized for tolerance to salinity and value for wetland revegetation, stream-bank stabilization, wildlife habitat, and forage. It is adapted to soils that are too wet and not sufficiently aerated for big bluestem (*A. gerardii* Vitman) and switchgrass (*P. virgatum* L.), grows more rapidly than other tallgrass prairie dominants, and is conspicuously taller than big bluestem and switchgrass where their distributions overlap (Weaver 1954). Weaver and Fitzpatrick (1932) noted that during the early twentieth century when the uplands and big bluestem lowlands of the tallgrass prairie were broken for growing maize (*Zea mays* L.), some of the soils were too wet for growing maize and were returned to growing prairie cordgrass and left mostly intact for hay production and control of soil erosion.

Although cultivars of smooth cordgrass (*Spartina alterniflora* Loisel.) and salt-meadow cordgrass [*S. patens* (Ait.) Muhl.] have been developed for coastal marsh revegetation (Alderson and Sharp 1994), the development of improved populations of prairie cordgrass for inland wetland revegetation has been very limited. "Red

River Natural Germplasm" is a selected class release of prairie cordgrass from the USDA-NRCS Bismarck Plant Materials Center.

Evaluations of prairie cordgrass for biomass production at 52°N latitude in Europe (Potter et al. 1995), in southwestern Quebec (Madakadze et al. 1998), and in eastern South Dakota (Boe and Lee 2007) have indicated its high potential for biomass production in short-season areas, relative to switchgrass and other warm-season grasses. Its high tolerance for soil salinity and early season waterlogged soils has also been demonstrated (Montemayor et al. 2008).

In the semiarid northern Great Plains, water is the major factor controlling the growth of perennial grasses (Willis et al. 1983). As such, the highest yields of biomass from perennial grasses in the northern Great Plains would be expected to come from C_4 species that establish a photosynthetically active canopy early during the growing season in an environment where soil moisture is rarely deficient. Species in the genus *Spartina* develop photosynthetically active canopies earlier than most other warm-season grasses (U.S.D.O.E. 2006) and are well-adapted to soils that are wet throughout the growing season (e.g., land capability class V). On the other hand, the other native tallgrass prairie C_4 grasses are not well-adapted to those types of soils (Weaver and Ernest 1954). Therefore, prairie cordgrass, as pointed out by Weaver and Ernest (1954), should be superior to switchgrass and big bluestem for biomass production in low wet areas where it is a dominant component of the potential natural vegetation (USDA-SCS 1981). The recent 90-Billion Gallon Biofuel Deployment Study completed by Sandia National Laboratories and General Motors concluded that expanding feedstock production from dedicated biomass crops should target lands requiring little or no irrigation (DOE/Sandia National Laboratories, ScienceDaily, 11 Feb. 2009; http://HITECtransportation.org/news).

The feasibility of a billion-ton supply of biomass annually proposed by US Department of Energy (Perlack et al. 2005) assumes a high-yield scenario of an average of 18 Mg of dry matter per hectare for perennial grass crops. In the semi-arid northern Great Plains, it is highly unlikely that level of biomass production can be reached on rain fed marginal uplands. However, those levels might be attainable from high-yielding grasses on poorly drained soils (i.e., land capability class V). In South Dakota alone, there are more than 210,000 ha of land capability class V, which although too wet for conventional crop production, are generally regarded as the highest grass-producing soils in the State (D. Malo 2009, personal communication).

Maximizing biomass production in prairie cordgrass will require development of populations with the capacity to produce a high frequency of reproductive tiller across a range of environments. Mature reproductive tillers of a synthetic cultivar developed by one cycle of selection among and within seven populations from eastern South Dakota (Boe and Lee 2007) were found to exceed 2.8 m in height and weigh more than 30 g at Brookings, SD. Some of these populations out-yielded elite switchgrass cultivars in long-term studies (Fig. 5.1; Boe and Lee 2007). Preliminary results (unpublished data, 2008) from a biomass trial at Brookings indicated this improved population can produce >18 Mg ha^{-1}. The national goal set by US DOE for perennial grass biomass crops is at least 22 Mg ha^{-1}. A stand of

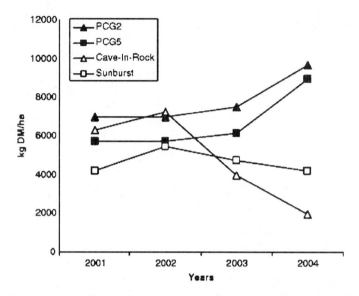

Fig. 5.1 Mean biomass production for two natural populations of prairie cordgrass (PCG2 and PCG5) and two cultivars of switchgrass (Cave-In-Rock and Sunburst) harvested during early October 2001 thru 2004 at Aurora, SD (adapted from Boe and Lee 2007)

well-developed reproductive tillers with a mean individual weight of 30 g would meet that desired goal with a density of only 75 tillers m^{-2}.

Weaver and Fitzpatrick (1932) reported reproductive culms of prairie cordgrass exceeded 3 m in height and 10 mm in diameter in wet areas in the tallgrass prairie. While few estimates exist for biomass production of prairie cordgrass in cultivated or natural stands, natural stands in the Konza prairie in northeastern Kansas produced >15 Mg ha^{-1} (Johnson and Knapp 1996) and natural stands recovering from long periods of drought produced >9 Mg ha^{-1} in eastern Nebraska (Weaver and Ernest 1954).

Weaver and Ernest (1954) noted that natural stands of prairie cordgrass were not damaged from three hay harvests during a growing season, and Boe and Lee (2007) found no detrimental effects from a single harvest at the end of the growing season for transplanted stands over a 4-yr period in eastern South Dakota. Therefore, we do not expect negative impacts on mature stands from an annual single harvest during autumn for biomass production. However, studies are needed to determine the effects of multiple harvests during a growing season and N and P fertilizer on (1) reproductive tiller frequency and development, (2) biomass production, and (3) persistence of mature stands of prairie cordgrass.

Biomass production of seven populations of prairie cordgrass over 4 yr on marginal gravelly upland in the northern Great Plains was less than one third of the production on prime land (Boe 2009, unpublished data). This reduction in biomass production was slightly greater than the reduction that occurred for Sunburst switchgrass (Fig. 5.1; Boe and Lee 2007). Although Mobberley (1956) considered open dry prairie and gravelly railroad embankments as primary habitats for prairie cordgrass in the Midwestern USA (as opposed to marshes, sloughs, and floodplains in the eastern USA and Canada) biomass production potential of prairie cordgrass, as well as that of switchgrass, was severely depressed on a gravelly soil in east central South Dakota. Comparisons between prairie cordgrass and switchgrass, big bluestem, little bluestem, *Miscanthus*, and other stress-tolerant grasses along environmental gradients are needed to determine amplitudes of adaptation and optimum landscape positions for species that have potential for use in an integrated multispecies approach to the establishment and maintenance of plant communities with simultaneous roles in biomass production, carbon sequestration, soil conservation, and as wildlife habitat.

Genetics of North America's Prairies Perennial Grasses: A Work in Need of Progress

Prairie cordgrass. Based on the limited amount of information available concerning the management and potential biomass productivity of prairie cordgrass and little bluestem, it is no surprise that our knowledge of genome organization and applied genetics for both species is also minimal. For example, a search of the National Center for Biotechnology Information (NCBI) database (Table 5.1) reveals that most of the genetic research involving either species consists of phylogenetic studies on the grasses where a limited number of DNA sequences were studied.

Prairie cordgrass has 40 chromosomes ($2n=40$, $x=10$) and is considered to be an autotetraploid (Marchant 1963, 1968), although there are no reports of detailed investigation on the mode of inheritance. A later investigation also reported octoploid accessions in the western states of the USA (Reeder 1977). This may indicate the existence of at least two different ecotypes with different ploidy levels, similar to switchgrass (Hultquist et al. 1996). The evolutionary history of the genus has been characterized by recent hybridizations and polyploidizations events (Baumel et al. 2002; Ainouche et al. 2004, b). For example, it is believed that two new hybrid species have originated within the last 150 yr (Ainouche et al. 2004). Since diploid species are unknown in the *Spartina* genus, prairie cordgrass might have the simplest genome in the genus.

Based on DNA content values ($1C=5.45$ pg) for the dodecaploid *Spartina angelica* ($2n=122$, $1C=5,341$ Mb; Bennett and Leitch 2005), prairie cordgrass would have an estimated DNA content of 2.8 pg. The average chromosome in the *Spartina* genus would have 87.6 Mb, and by extrapolation, the genome of *S. pectinata* is approximately 1,751 Mb in size. This is equivalent to approximately four times the rice genome. At present, there is no linkage map available for any species of the *Spartina* genus. If we assume that prairie cordgrass has similar recombination frequencies to switchgrass, based on similar DNA content, prairie cordgrass would have a similar recombinational genetic map length of ~4,500 cM (Missaoui et al. 2005).

Table 5.1 Number of entries in NCBI database (as on February 9, 2009) for *S. pectinata*, *S. scoparium*, and related species within both genera

Species	DNA	EST	Protein
Schizachyrium scoparium	8		3
Schizachyrium tenerum	4		
Schizachyrium malacostachyum	2		
Schizachyrium sanguineum	2		
Schizachyrium semitectum	2		
Schizachyrium brevifolium	2		
Schizachyrium neomexicanum	1		
Schizachyrium gaumeri	1		
Spartina alterniflora	112	1,255	5
Spartina densiflora x Spartina foliosa	60		
Spartina densiflora	20		4
Spartina pectinata	18		8
Spartina foliosa	18		3
Spartina anglica	17		7
Spartina gracilis	10		
Spartina patens	10		2
Spartina maritima	10		4
Spartina bakeri	6		1
Spartina cynosuroides	5		1
Spartina argentinensis	4		2
Spartina arundinacea	4		1
Spartina spartinae	3		
Spartina versicolor	2		
Spartina alterniflora x Spartina densiflora	2		
Spartina ciliata	1		
Spartina x townsendii	1		

Very little information is available on the genome organization in *Spartina* spp. The only molecular marker work reported on this genus is based on randomly amplified polymorphic DNA (RAPDs), AFLPs, and intersimple sequence repeats (ISSRs) as an approach to study genetic diversity in natural populations of *S. angelica* (Ainouche et al. 2004) and *S. alterniflora* (Ryan 2003; Travis and Hester 2005). A clonal germplasm collection with samples from the central prairies of North America is in development at South Dakota State University. At this time, this germplasm collection consists of clones from 128 sampling locations distributed thru South Dakota, North Dakota, Minnesota, Nebraska, Iowa, and Kansas. Future collections trips will target Missouri, Oklahoma, Texas, Arkansas, and Louisiana. An initial genetic diversity study using AFLPs on samples from the northern plains revealed extensive genetic diversity (Gonzalez-Hernandez 2009, unpublished data).

The development of molecular markers in prairie cordgrass has been nonexistent. Recent work has been done to develop a small number of simple sequence repeats (SSRs) (<50) on *S. alterniflora* and *Spartina foliosa* (Blum et al. 2004;

Sloop et al. 2005). When tested on *S. pectinata*, only a relatively low number of them amplified DNA. This observation, together with phylogenetic evidence-based DNA sequence analysis, may suggest the existence of allopolyploidy in the genus. SSR markers for prairie cordgrass are under development using a genomic library enrichment approach, consisting of enriched libraries for CA, GA, AAG, and CAG SSR repeats. An initial screening revealed an average enrichment of 65%. To date, over 1,300 SSR primer pairs have been designed with most of the loci being simple repeats. We are also in the process of mapping up to 500 of them in one or two mapping populations.

The development of a linkage map for autotetraploid species like prairie cordgrass and little bluestem presents unique challenges due to the complexities of tetrasomic inheritance, namely (a) multiplex segregation, (b) double reduction, and (c) mixed bivalent and quadrivalent pairings among homologous chromosomes. These complexities translate into higher number of alleles and greater number of genotype combinations. Double reduction is a phenomenon in which sister chromatids end in the same gamete because of homologous chromosomes forming a quadrivalent, followed by crossing over between the locus and the spindle attachment. Few strategies have been developed for linkage analysis in autopolyploids. In some species such as potato (*Solanun* spp.; Freyre et al. 1994) and alfalfa (*Medicago sativa*; Diwan et al. 2000; Kalo et al. 2000) mapping can be done in diploid derivatives. The limitation of this approach stems from the fact that polyploidization and subsequent evolution is a dynamic process (Song et al. 1995); therefore, approximating a polyploidy genome to its diploid relative may not be appropriate. Another strategy is based on the segregation analysis of single-dose loci. Segregation ratios (1:1, absence *versus* presence) and recombination fraction of these loci in coupling are equivalent to those from disomic species. This strategy has been used with restriction fragment length polymorphisms (RFLPs) in species such as sugarcane (*Saccharum spontaneum*; Da Silva et al. 1993), alfalfa, and switchgrass (Missaoui et al. 2005). More recent statistical approaches have been developed assuming solely random bivalent pairing of the four homologous chromosomes (Luo et al. 2001; Cao et al. 2005). The inclusion of codominant markers, such as SSRs or RFLPs, facilitate the linkage map construction because more information about marker dosage can be derived from these markers than from dominant markers such as AFLPs which require that dosage information be inferred from the segregation ratios (Luo et al. 2001; Wu et al. 2001; Wu and Ma 2005; Luo et al. 2006). These approaches avoid some analytical complexities, although simultaneously ignoring essential features in tetrasomic inheritance. Currently, a methodology is being developed to solve the challenge of linkage analysis in autotetraploids (Luo et al. 2004, 2006).

A preliminary characterization of the transcriptome of prairie cordgrass is revealing a great level of similarity with other grasses in general and sorghum in particular. Taking advantage of recent sequencing technologies developments, we conducted a transcriptome analysis using a normalized pooled full-length cDNA library from four tissues which yielded approximately 26,000 contigs representing a little over 10,000 unigenes (Gonzalez-Hernandez 2009, unpublished data).

We have discussed in previous ns the importance of using soils currently out of intensive production to exploit salinity tolerance of prairie cordgrass. This trait is shared by all *Spartina* species. In fact, most species in the genus are salt marsh species. This observation opens the question on how some *Spartina* species such as *S. pectinata* and *Spartina gracilis* (alkali cordgrass) colonized the interior prairies of North America. Recent studies (Baisakh et al. 2006; Baisakh et al. 2008) in *S. alterniflora* have provide evidence that, in addition to several previously unknown genes, genes involved in ion transport, osmolyte production, and housekeeping functions may play an important role in the primary responses to salt stress in this halophyte grass. The study of these genes in prairie cordgrass is the logical starting point to study the salt tolerance of this species and could have a significant impact in the development of highly productive varieties.

Little bluestem. Investigations into the genome of little bluestem is also limited. It is considered a tetraploid with the chromosome number and ploidies were determined as $2n=40$ (Bruner 1983). The utilization of molecular markers to determine genetic diversity among populations of little bluestem has occurred, although it was proposed that due to the outcrossing mode of reproduction of this species, resulting in highly heterogeneous populations, ascertaining an accurate representation of the genetic diversity of little bluestem would be difficult (Huff et al. 1998; Fu et al. 2004). It is postulated that the phenotype of little bluestem is independently and continuously variable across its range of habitation without discernable regional delineation (Wipff 1996).

Huff et al. (1998) examined RAPD variation among four populations which were derived from soils of high and low fertility with two geographic locations: New Jersey (a forest biome) and Oklahoma (a grassland biome). This study examined the genetic diversity in reference to using little bluestem for restoration and revegetation as concern was raised that commercially produced cultivars of little bluestem would have a limited germplasm base and have a restricted applicability for revegetation. Ten plants were sampled from each of the four populations and tiller material was collected from each plant for DNA extraction (Huff et al. 1998). Four RAPD markers were analyzed. Although this study is limited by the small number of RAPD markers used, analysis of molecular variance of polymorphic RAPDs indicated that there was significant genetic differences between the four populations ($p<0.05$), with the greatest genetic distance between the New Jersey high fertility population and the Oklahoma low fertility population (Huff et al. 1998). However, Huff et al. (1998) observed a relatively low level of population differentiation among the four little bluestem populations in comparison to other outcrossing species. It is assumed that a high level of genetic variability caused this low level of differentiation, maintained by frequent outcrossing and the ability of the species to have very long distance pollen and seed dispersal.

A second study into the genetic diversity of little bluestem was performed by Fu et al. (2004), utilizing AFLPs. In this study, the population examined was from the extreme north of the geographic range of bluestem, with all samples collected north of the 49° parallel in western Canada. From six locations, 30 plants were randomly selected for tissue collection (Fu et al. 2004). One of the goals of this research, similar to that of Huff et al. (1998), was to assess the genetic diversity of natural populations, with congruent concerns about improved plant materials not maintaining a level of genetic diversity requisite for adaptation to a new environment (Fu et al. 2004). Tissue for DNA extraction was collected from the plants and analyzed with five primer combinations, creating 158 scorable polymorphic bands (Fu et al. 2004). Results from the analysis indicated that a small yet significant ($p<0.01$) amount of genetic variation was discernable among the six populations (Fu et al. 2004). However, the genetic diversity was not correlated with geographical distances of the collection sites ($r=0.02$, $p=0.5244$; Fu et al. 2004).

Consistent between the studies performed by Huff et al. (1998) and Fu et al. (2004) was the considerable difference between within and among population genetic diversity. Huff et al. (1998) found 95% of total genetic variation resided within a population, while only 5% resided among the examined populations. Similarly, Fu et al. (2004) found 92.8% of the total genetic variation resided within a population, while 7.2% resided among the six populations examined. Both these results confirm the highly outcrossing nature of little bluestem.

Little bluestem has been successfully cultured *in vitro* (Songstad et al. 1986). Young inflorescence from three genotypes were used to develop callus culture on RM medium (Linsmaier and Skoog 1965) supplemented with 5 mg l^{-1} of 2,4-D, and the transfer of the callus tissue onto hormone-free media induced organogenesis (Songstad et al. 1986). Plant development was found to have been via somatic embryogenesis, and the explants were transplanted into pots in the greenhouse where they grew to maturity in 24 wk (Songstad et al. 1986). While no reports of transformation of little bluestem have been found, the results of Songstad et al. (1986) indicate that viable embryogenic callus, suitable for transformation, can be grown as in many other grasses.

Potential Pathogens and Pests of Perennial Feedstocks

As with all of the previously discussed topics, there is very little information currently available concerning the known or potential pathogens and pests of herbaceous perennial feedstocks, especially prairie cordgrass and little bluestem. As a result, much of what will be discussed will be conjecture based on historical surveys and research in related species. What is well-documented is that pathogens and pests can have significant lasting impacts on biomass productivity and stand survival in crops harvested for biomass, such as alfalfa (*M. sativa* L.; Nutter et al. 2002), and therefore, researchers need to understand the potential impact of these organisms on all potential feedstock species. There is little doubt that economical management strategies need to be developed in order to mitigate any losses from feedstock pathogens and pests. Failure to do so could jeopardize the sustainability of this emerging industry.

The reason that pathogens and pests are a potential concern for perennial feedstock crops can be summarized by one word: monoculture. When little bluestem, prairie cordgrass, switchgrass, and most other feedstock species were found in their native ecosystem(s), there was usually some level of heterogeneity in terms of both species diversity and genetic uniformity within each species (Fu et al. 2004; Kittelson and Handler 2006; Polley et al. 2007). That is, individuals of one species were not usually the only plants within a given area, and if they were, it was common that multiple genetically unique individuals were present. This diversity impacted the occurrence and importance of pathogens and pests because reproduction and spread of these organisms is often dependent on the density and uniformity of the host (Young 1995; Garrett and Mundt 1999; Mitchell et al. 2002).

For example, a single uredinium (pustule) of the rust fungus *Puccinia sparganioides* (Ash-Cordgrass rust) can produce thousands of urediniospores and these propagules can easily become airborne. However, the probability that any given spore will land on a susceptible host (namely prairie cordgrass) is inversely proportional to the density and diversity of the host species in a given area. In other words, if another susceptible host is too far away from the infected source plant, then the chances of the epidemic progressing to a substantial level are minimal. In contrast, monocultures are composed of single plant species and are often very genetically uniform. This greatly increases the chance that pathogens and pests will be able to spread within a given location and more importantly that such organisms can become highly adapted to a given host genotype and impact the crop even more dramatically. The loss of genetic variability through the process of selection and varietal improvement and the subsequent planting of crop monocultures are the primary reasons why pathogens and pests should be study extensively in perennial feedstock crops.

A variety of fungi and water molds, bacteria and mollicutes, protozoa, and nematodes can be pathogenic to plants and there are also several virus and viroid groups that utilize plants as a host. Similarly, many different insect and mite species can feed on plants and cause damage. The potential effects of diseases and pests on herbaceous perennial feedstock crops like prairie cordgrass and little bluestem can be organized into three general groupings based on what the final impact(s) will be (a) photosynthetic capacity, (b) plant–water relations, and (c) seed production and viability. It should be noted that these groupings are not mutually exclusive but are simply used in this review to partition impacts into concise categories.

The most obvious impact in terms of economic importance to feedstock crops is the disruption of photosynthetic capacity. Specifically, if the ability of a plant to photosynthesize is reduced, then the potential to produce aboveground biomass will be limited. Reductions can occur through physical means, such as when a pest feeds on a leaf. A prime example would be locust swarms where complete or near-complete defoliation of plants is known to occur, often over enormous regions (Stewart 1997; Todd et al. 2002; Ceccato et al. 2007). Reductions may also occur when a plant is parasitized by a pathogen and both prairie cordgrass and little bluestem have been documented to be susceptible to multiple foliar pathogens (Mankin 1969; Farr and Rossman 2009), some of which are highly specific (e.g., *P. sparganioides* on prairie cordgrass). In addition to pathogens, natural stands of prairie cordgrass have also been found to be heavily infested with insects, such as the lygaied *Ischnodemus falicus* (Johnson and Knapp 1996; Boe and Stein 2008, unpublished data). This piercing-sucking bug reduced biomass production of natural stands of prairie cordgrass by 40% in Kansas (Johnson and Knapp, 1996) and could become economically important as the feedstock industry develops.

Reductions may also be physiological such as when plants are infected by a virus or phytoplasma and it has long been known that leaves showing mosaic symptoms of viral infection (i.e., yellowing) usually have reduced photosynthetic rates (Diener 1963; Balachandran et al. 1997) and therefore lowered photosynthetic capacity. No viruses have been found in prairie cordgrass and only one virus has been identified in little bluestem to date, maize streak monogeminivirus (Brunt et al. 1996). However, the lack of known viruses in these species does not necessarily mean that none are present; instead, it is quite possible that additional viruses do exist but have yet to be found because no one has looked. For example, Spartina mottle rymovirus has been documented in related *Spartina* species in Europe (Brunt et al. 1996) and during the aforementioned germplasm collection trips, we found prairie cordgrass plants with leaf mottling symptoms at approximately 10% of the locations (Stein 2009, unpublished data). When these samples were submitted for commercial virus testing, they all came back negative for a number of common grass viruses (Stein 2009, unpublished data). *Spartina* mottle rymovirus was not part of the panel and therefore, further investigations in this topic are planned.

In addition to strictly physical and physiological disruptions, there are a number of plant pathogens and pests that initially cause physiological impacts (localized chlorosis) but ultimately cause tissue death (blighting). Many of these organisms produce toxins that initially impact cell physiology but can eventually kill the affected tissue (Orolaza et al. 1995; Young et al. 1992). This type of damage is functionally equivalent to direct tissue removal and can have similar impacts on biomass accumulation. Prairie cordgrass has at least two fungal pathogens of this type: *Septoria spartinae* and *Stagonospora spartinicola* and little bluestem one: *Pyrenophora tritici-repentis*. These organisms produce leaf spot/blight symptoms and have the potential to cause economic losses in production (Farr and Rossman 2009; Mankin 1969). As with the viruses, more species will likely be found as research efforts in the area are expanded (Leslie et al. 2004).

As with photosynthetic capacity, the disruption of plant–water relations by pathogens or pests can have profound impacts on biomass production in a feedstock crop. Symptoms can include stunting or wilting of the plant, nutrient deficiency in the foliage, and even lodging (breakage of the stem), all of which reduce harvestable biomass. The first and most obvious is the loss or disruption of root function through pest feeding or rot. Many fungi and insect genera are known to cause significant problems in other grasses but few have been documented in prairie cordgrass and little bluestem. Plant parasitic nematodes, on the other hand, have been known to be important in native prairie ecosystems for some time (Smolik 1973). For example, the application of a nematicide to experimental plots containing little bluestem resulted in a significant increase in ANPP for two seasons

(Ingham and Detling 1990). No nematode pests have been noted in prairie cordgrass, but others have been found to be important in other *Spartina* spp. (Plantard et al. 2007). Based on historical evidence and occurrences of nematodes in other cropping systems (Smiley et al. 2006), it is highly likely that they would become important under intensive production systems with little bluestem and possibly even prairie cordgrass.

Unlike nematodes, only a few primary root rotting fungal pathogens have been found in little bluestem (Farr and Rossman 2009; Leslie et al. 2004; Mankin 1969) and none have been confirmed to date in prairie cordgrass (Farr and Rossman 2009; Mankin 1969), although several fungal species that are important in other cropping systems (such a *Bipolaris*) have been isolated from root tissue of the latter (Stein 2009, unpublished data). As before, the lack of known pathogens and pests does not necessarily mean that none will become important in the future; instead, further research needs to be conducted in this topic area because these organisms limit productivity. For example, Fusarium root and crown rot was found to reduce wheat grain yields by 25% under field conditions in the Pacific Northwest (Smiley et al. 2006) and similar losses have been noted from rootworm (*Diabrotica* spp.) damage in maize (Riedell et al. 1992). Successful implementation of biotechnology tools has resulted in production of maize resistant to corn rootworm, which producers prefer stacked with biotech traits for control of European corn borer and herbicide tolerance—referred together as a triple stack (James 2008). Systemic viral infections are also known to disrupt root function and may impact biomass accumulation, but their important is unknown since few have been found to infect prairie cordgrass and little bluestem (Brunt et al. 1996), as noted above.

A similar mechanism of disruption involves water movement in the plant. Specifically, there are several pathogen and pest groups that colonize stems and impede water translocation (upward movement). For example, prairie cordgrass has been documented to have several stem boring moths that are specialists on the species (Reed 1996). During our recent summer disease surveys (in 2007 and 2008), evidence of stem borer activity was found at nearly all of the sampling sites and was quite severe in one large patch of prairie cordgrass. At this location, nearly 50% of the tillers were lodged due to excessive damage (Stein 2009, unpublished data); however, this was late in the growing season and so it is unknown if a significant reduction in biomass accumulation occurred.

The final impact of plant pathogens and pests on feedstock crops is seed production and viability. This is the least obvious because the ability of a feedstock crop to make viable seed is usually not important in terms of biomass productivity; however, for fertile species such as prairie cordgrass and little bluestem, the most economical way to establish crops will be through seeding. Therefore, a source of uniform viable seed will need to be maintained to support this industry. Seed production is highly dependent on overall plant health and therefore, the previous list of impacts to biomass production would also apply in this case. There are also pathogens and pests that can be directly limiting in this area. For example, several types of insects and also some birds are known to feed on the seeds of these species.

This is an obvious direct loss in productivity in terms of a seed crop. Smut/bunt fungi in other grass species are also well-documented to impact seed production and in some cases plant biomass (Gravert et al. 2000). Prairie cordgrass does not have any known smut/bunts pathogens; however, *Sorosporium ellissii* has been identified in little bluestem (Farr and Rossman 2009; Mankin 1969) and might become important in a seed production situation. The disease ergot, caused by the fungus *Claviceps* spp., has also been documented in both prairie cordgrass and little bluestem (Alderman et al. 2004) and might be sporadically important in seed production.

Plant pathogens and pests could severely impact the establishment, productivity, and even sustainability of herbaceous feedstock crops such as prairie cordgrass and little bluestem. It is vital that research be expanded in this area so that breeders, pathologists, agronomists, and growers are better prepared to deal with such issues as varieties are developed, released, and large-scale production begins. The first and probably most important is the continuation of disease and pest surveys over large geographical areas so that we can get a more thorough understanding of which organisms could become important in the future. This should include prairie cordgrass and little bluestem stands that are both naturally and artificially established in a variety of locations, soil types, and microclimes. Once a list of potential organisms is compiled, studies will then need to be conducted on individual species so that the research community can better understand the biology of each organism. Better-informed decisions can then be made concerning breeding strategies and planting, crop management, and biomass harvest timing. For example, if one can determine that the development of physiological races is likely to occur in *P. sparganioides* on prairie cordgrass (i.e., virulence on specific resistance genes), then one will know that breeding efforts should focus on durable rust resistance, gene pyramiding, and/or the development of multilines which vary significantly in the gene(s) responsible for rust resistance (McDonald and Linde 2002; Mundt 2002; Parker and Gilbert 2004). Such research has historically been conducted using traditional genetic analysis; however, the use of modern genomics tools in this field could accelerate our efforts and allow us to be one step ahead of the potential pathogens and pests of prairie cordgrass and little bluestem.

Summary

It is clear that meeting the new expectations from bioenergy crops while concurrently increasing food production for an increasing global human population will demand not only an increased productivity of food crops in current cropland, but the use of class V soils for biofuel production of prairie cordgrass and little bluestem. The literature suggests that these species have received little attention even though they can be more productive in particular marginal lands than other species being considered, such as switchgrass. The work needed to develop these other native species into competitive dedicated biomass crops should focus around a rapid development of highly productive cultivars coupled with the development of sustainable agronomic practices. Studies focusing on the biology, genetics, and natural variability of traits determining biomass yield are urgently needed.

Recent agricultural history reminds us of the importance of plant pathogens and pests. We know that sooner rather than later all crops, when grown in large areas, will become susceptible to the attack of pathogens of different nature and to changes in insect populations. To prevent epidemics, it is imperative that the scientific community takes a proactive approach.

Although the development of transgenic cultivars is feasible in switchgrass and *Miscanthus*, transformation in these and other potential species has not been extensively attempted. In addition, the widespread release of transgenic cultivars of native perennial grasses may face a great deal of opposition due to potential environmental and regulatory issues.

References

Ainouche M. L.; Baumel A.; Salmon A. *Spartina angelica* c. E. Hubbard: A natural model system for analysing early evolutionary changes that affect allopolyploid genomes. *Biol. J. Linn. Soc.* 82: 475–484; 2004. doi:10.1111/j.1095-8312.2004.00334.x.

Ainouche M. L.; Baumel A.; Salmon A.; Yannic G. Hybridization, polyploidy and speciation in *Spartina* (Poaceae). *New Phytol.* 161: 165–172; 2004. doi:10.1046/j.1469-8137.2003.00926.x.

Alderman S. C.; Halse R. R.; White J. F. A reevaluation of the host range and geographical distribution of Claviceps species in the United States. *Plant Dis.* 88: 63–81; 2004. doi:10.1094/PDIS.2004.88.1.63.

Alderson J.; Sharp W. C. Grass varieties in the United States. USDA-SCS, Washington, DC 1994.

Al-Kaisi M. M.; Grote J. B. Cropping system effects on improving soil carbon stocks of exposed subsoil. *Soil Sci. Soc. Am. J.* 71: 1381–1388; 2007. doi:10.2136/sssaj2006.0200.

Anderson W. F.; Dien B. S.; Brandon S. K.; Peterson J. D. Assessment of bermudagrass and bunch grasses as feedstock for conversion to ethanol. *Appl. Biochem. Biotechnol.* 145: 13–21; 2008. doi:10.1007/s12010-007-8041-y.

Anderson K. L.; Aldous A. Improvement of *Andropogon scoparius* Michx. By breeding and selection. *J. Am. Soc. Agronom.* 30: 862–869; 1938.

Baisakh N.; Subudhi P. K.; Parami N. P. cDNA-AFLP analysis reveals differential gene expression in response to salt stress in a halophyte *Spartina alterniflora* Loisel. *Plant Sci.* 170: 1141–1149; 2006. doi:10.1016/j.plantsci.2006.02.001.

Baisakh N.; Subudhi P.; Varadwaj P. Primary responses to salt stress in a halophyte, smooth cordgrass (*Spartina alterniflora* Loisel.). *Funct. Integr. Genom.* 8: 287–300; 2008. doi:10.1007/s10142-008-0075-x.

Balachandran S.; Hurry V. M.; Kelley S. E.; Osmond C. B.; Robinson S. A.; Rohozinski J.; Seaton G. G. R.; Sims D. A. Concepts of plant biotic stress. Some insights into the stress physiology of virus-infected plants, from the perspective of photosynthesis. *Physiol Plant* 100: 203–213; 1997. doi:10.1111/j.1399-3054.1997.tb04776.x.

Barney J. N.; Ditomaso J. M. Nonnative species and bioenergy: are we cultivating the next invader? *Bioscience* 58: 64–70; 2008. doi:10.1641/B580111.

Baumel A.; Ainouche M. L.; Bayer R. J.; Ainouche A. K.; Misset M. T. Molecular phylogeny of hybridizing species from the genus *Spartina* Schreb. (Poaceae). *Mol. Phylogenet. Evol.* 22: 303–314; 2002. doi:10.1006/mpev.2001.1064.

Bennett, M. D.; Leitch, I. J. Plant DNA c-values database (release 4.0, Oct. 2005); 2005.

Blum M. J.; Sloop C. M.; Ayres D. R.; Strong D. R. Characterization of microsatellite loci in *Spartina* species (Poaceae). *Mol. Ecol. Notes* 4: 39–42; 2004. doi:10.1046/j.1471-8286.2003.00556.x.

Boe A.; Bortnem R. Morphology and genetics of biomass in little bluestem. *Crop Sci.* 49: 411–418; 2009.

Boe A.; Keeler K. H.; Normann G. A.; Hatch S. L. The indigenous bluestems of the western hemisphere and gambagrass. In: Moser L.; Burson B.; Sollenberger L. (eds) Warm-season (c_4) grasses. ASA, Madison, WI, pp 873–908; 2004.

Boe A. Lee D. K. Genetic variation for biomass production in prairie cordgrass and switchgrass. *Crop Sci.* 47: 929–934; 2007.

Bouton J. H. Molecular breeding of switchgrass for use as a biofuel crop. *Curr. Opin. Genet Dev.* 17: 553–558; 2007. doi:10.1016/j.gde.2007.08.012.

Bruner, J. Systematics of the *Schizachyrium scoparium* group (Poaceae) in North America. *Am. J. Bot.* 70 Part 2. (5): 108; 1983.

Brunt A.; Crabtree K.; Dallwitz M.; Gibbs A.; Watson L. Viruses of plants: descriptions and lists from the VIDE database. CAB International, Wallingford, UK; 1996.

Cao D.; Craig B.; Doerge R. A model selection-based interval mapping method for autotetraploids. *Genetics* 169: 2371–2382; 2005. doi:10.1534/genetics.104.035410.

Cardona Ca; Sanchez O. J. Fuel ethanol production: Process design trends and integration opportunities. *Bioresour. Technol.* 98: 2415–2457; 2007. doi:10.1016/j.biortech.2007.01.002.

Cassman K. G.; Liska A. J. Food and fuel for all: Realistic or foolish? *Biofuels Bioprod Bioref* 1: 18–23; 2007. doi:10.1002/bbb.3.

Ceccato P.; Cressman K.; Giannini A.; Trzaska S. The desert locust upsurge in West Africa (2003–2005): Information on the desert locust early warning system and the prospects for seasonal climate forecasting. *Int. J. Pest Manag.* 53: 7–13; 2007. doi:10.1080/09670870600968826.

Ceotto, E. Grasslands for bioenergy production: A review. *Agron. Sustain. Dev.* 28: 47–55; 2008.

Chen F.; Dixon R. A. Lignin modification improves fermentable sugar yields for biofuel production. *Nat. Biotechnol.* 25: 759–761; 2007. doi:10.1038/nbt1316.

Christian D. G.; Riche A. B.; Yates N. E. Growth, yield and mineral content of *Miscanthus x giganteus* grown as a biofuel for 14 successive harvests. *Ind. Crop Prod.* 28: 320–327; 2008. doi:10.1016/j.indcrop.2008.02.009.

Cornelius D. R. The effect of source of little bluestem grass seed on growth, adaptation, and use in revegetation seeding. *J. Agric. Res.* 74: 133–143; 1947.

Da Silva J.; Sorrells M.; Burnquist W.; Tanskley S. RFLP linkage map and genome analysis of *Saccharum spontaneum*. *Genome* 36: 182–791; 1993.

Demirbas A. Recent progress in biorenewable feedstocks. *Energy Educ. Sci. Technol.* 22: 69–95; 2008.

Dien B. S.; Jung H. J. G.; Vogel K. P.; Casler M. D.; Lamb J. F. S.; Iten L.; Mitchell R. B.; Sarath G. Chemical composition and response to dilute-acid pretreatment and enzymatic saccharification of alfalfa, reed canarygrass, and switchgrass. *Biomass Bioenergy* 30: 880–891; 2006. doi:10.1016/j.biombioe.2006.02.004.

Diener T. Physiology of virus-infected plants. *Annu. Rev. Phytopathol.* 1: 197–218; 1963. doi:10.1146/annurev.py.01.090163.001213.

Diwan N.; Bouton J.; Kochert G.; Creagan P. Mapping of simple sequence repeat (SSR) DNA markers in diploid and tetraploid alfalfa. *Theor. Appl. Genet.* 101: 165–172; 2000. doi:10.1007/s001220051465.

Doolittle J. J.; Malo, D. D., Kunze, B. O.; Winter, S. D.; Schaefer Jr., W. T.; Millar, J. B.; Shurtliff, D. R. Land judging in South Dakota. ABS 8-01; South Dakota State Univ.; 2002.

Evanylo G. K.; Abaye A. O.; Dundas C.; Zipper C. E.; Lemus R.; Sukkariyah B.; Rockett J. Herbaceous vegetation productivity, persistence, and metals uptake on a biosolids-amended mine soil. *J. Environ. Qual.* 34: 1811–1819; 2005. doi:10.2134/jeq2004.0329.

Fargione J.; Hill J.; Tilman D.; Polasky S.; Hawthorne P. Land clearing and the biofuel carbon debt. *Science* 319: 1235–1238; 2008. doi:10.1126/science.1152747.

Farr D. F.; Rossman A. Y. Fungal databases. Systematic Mycology and Microbiology Laboratory, ARS, USDA, Baltimore; 2009.

Freyre R.; Warnke S.; Sosinski B.; Douches D. Quantitative trait locus analysis of tuber dormancy in diploid potato (*Solanun spp*). *Theor. Appl. Genet.* 89: 474–480; 1994. doi:10.1007/BF00225383.

Fu Y. B.; Phan A. T.; Coulman B.; Richards W. Genetic diversity in natural populations and corresponding seed collections of little bluestem as revealed by AFLP markers. *Crop Sci.* 44: 2254–2260; 2004.

Garrett Ka; Mundt C. C. Epidemiology in mixed host populations. *Phytopathology* 89: 984–990; 1999. doi:10.1094/PHYTO.1999.89.11.984.

Gaunt J. L.; Lehmann J. Energy balance and emissions associated with biochar sequestration and pyrolysis bioenergy production. *Environ. Sci. Tech.* 42: 4152–4158; 2008. doi:10.1021/es071361i.

Gilbert W. L.; Perry L. J.; Stubbiendieck J. Dry matter accumulation of four warm season grasses in the Nebraska sandhills. *J. Range Manag.* 52–54; 1979. doi:10.2307/3897385.

Gomez L. D.; Steele-King C. G.; McQueen-Mason S. J. Sustainable liquid biofuels from biomass: The writing's on the walls. *New Phytol.* 178: 473–485; 2008. doi:10.1111/j.1469-8137.2008.02422.x.

Gravert C.; Tiffany L.; Munkvold G. Outbreak of smut caused by *Tilletia maclaganii* on cultivated switchgrass in Iowa. *Plant Dis.* 84: 596; 2000. doi:10.1094/PDIS.2000.84.5.596A.

Heaton E.; Clifton-Brown J. C.; Voigt T.; Jones M.; Long S. P. Miscanthus for renewable energy generation: European Union experience and projections for Illinois. *Mitig. Adapt. Strategies Glob. Chang.* 9: 433–451; 2004. doi:10.1023/B:MITI.0000038848.94134.be.

Heaton E. A.; Dohleman F. G.; Long S. P. Meeting US biofuel goals with less land: The potential of Miscanthus. *Glob. Chang. Biol.* 14: 2000–2014; 2008. doi:10.1111/j.1365-2486.2008.01662.x.

Hitchcock A. S. Manual of the grasses of the United States. USDA Misc. Pub. No. 200. U.S. Gov. Print. Office, Washington, D.C. 1935.

Hockensmith R. D.; Steele J. G. Recent trends in the use of the land-capability classification. *Soil Sci. Soc. Amer. J.* 14: 383–388; 1950.

Huff D. R.; Quinn J. A.; Higgins B.; Palazzo A. J. Random amplified polymorphic DNA (RAPD) variation among native little bluestem [*Schizachyrium scoparium* (Michx.) Nash] populations from sites of high and low fertility in forest and grassland biomes. *Mol. Ecol.* 7: 1592–1597; 1998. doi:10.1046/j.1365-294x.1998.00473.x.

Hultquist S. J.; Vogel K. P.; Lee D. J.; Arumuganathan K.; Kaeppler S. Chloroplast DNA and nuclear DNA content variations among cultivars of switchgrass, *Panicum virgatum* L. *Crop Sci.* 36: 1049–1052; 1996.

Ingham R. E.; Detling J. K. Effects of root-feeding nematodes on aboveground net primary production in a North American grassland. *Plant Soil* 121: 279–281; 1990. doi:10.1007/BF00012321.

Jacobson E. T.; Tober D. A.; Haas R. J.; Darris D. C. The performance of selected cultivars of warm season grasses in the northern prairie and plains states. In: Clambey G. K.; Pemble R. H. (eds) The 9th North American Prairie Conference. Tri-College University Centre for Environmental Studies, Moorhead, MN, pp 215–221; 1984.

James C. Global status of commercialized biotech/GM crops. ISAAA Brief Number 39. ISAAA, Ithaca, NY2008.

Johnson S. R.; Knapp A. K. Impact of *Ischnodemus falicus* (Hemiptera: Lygaeidae) on photosynthesis and production of *Spartina pectinata* wetlands. *Environ. Entomol.* 25: 1122–1127; 1996.

Johnson J. R.; Larsen G. E. Grassland plants of South Dakota and the northern great plains. South Dakota Agricultural Experiment Station, South Dakota State University, Brookings, SD1; 999.

Kalo P.; Endre G.; Zimanyi L.; Csnadi G.; Kiss G. Construction of an improved linkage map of diploid alfalfa (*Medicago sativa*)*Theor. Appl. Genet.* 100: 641–657; 2000.

Keshwani D. R.; Cheng J. J. Switchgrass for bioethanol and other value-added applications: A review. *Bioresour. Technol.* 100: 1515–1523; 2009. doi:10.1016/j.biortech.2008.09.035.

Kittelson P. M.; Handler S. D. Genetic diversity in isolated patches of the tallgrass prairie forb, *Lithospermum canescens* (Boraginaceae). *J. Torrey Bot. Soc.* 133: 513–518; 2006. doi:10.3159/1095-5674(2006)133[513:GDIIPO]2.0.CO;2.

Knapp A. K.; Briggs J. M.; Blair J. M.; Turner C. L. Patterns and controls of aboveground net primary production in tallgrass prairie. In: Knapp A. K. (ed) Grassland dynamics: Long-term ecological research in tallgrass prairie. Oxford University Press, Oxford, UK; 1998.

Leslie J. F.; Zeller K. A.; Logrieco A.; Mule G.; Moretti A.; Ritieni A. Species diversity of and toxin production by *Gibberella fujikuroi* species complex strains isolated from native prairie grasses in Kansas. *Appl. Environ. Microbiol.* 70: 2254–2262; 2004. doi:10.1128/AEM.70.4.2254-2262.2004.

Linsmaier E. M.; Skoog K. Organic growth factor requirement for tobacco tissue cultures. *Physiol. Plant* 18: 100–127; 1965. doi:10.1111/j.1399-3054.1965.tb06874.x.

Luo Z. W.; Hackett C. A.; Bradshaw J. E.; McNichol J. W.; Milbourne D. Construction of a genetic linkage map in tetraploid species using molecular markers. *Genetics* 157: 1369–1385; 2001.

Luo Z. W.; Zhang R. M.; Kearsey M. J. Theoretical basis for genetic linkage analysis in autotetraploid species. *Proc. Natl. Acad. Sci. U. S. A.* 101: 7040–7045; 2004. doi:10.1073/pnas.0304482101.

Luo Z. W.; Zhang Z.; Leach L.; Zhang R. M.; Bradshaw J. E.; Kearsey M. J. Constructing genetic linkage maps under a tetrasomic model. *Genetics* 172: 2635–2645; 2006. doi:10.1534/genetics.105.052449.

Madakadze I. C.; Coulman B. E.; Mcelroy A. R.; Stewart K. A.; Smith D. L. Evaluation of selected warm-season grasses for biomass production in areas with a short growing season. *Bioresour. Technol.* 65: 1–12; 1998. doi:10.1016/S0960-8524(98)00039-X.

Mankin C. J. Diseases of grasses and cereals in South Dakota: a check list. Agricultural Experiment Station, South Dakota State University, Brookings, SD, p 28; 1969.

Marchant C. J. Corrected chromosome numbers for *Spartina x townsendii* and its parents. *Nature* 199: 929; 1963. doi:10.1038/199929a0.

Marchant C. J. Evolution in *Spartina* (Gramineane). III. Species chromosome numbers and their taxonomic significance. *Bot. J. Linn. Soc.* 60: 411–417; 1968. doi:10.1111/j.1095-8339.1968.tb00097.x.

Martinez-Reyna J. M.; Vogel K. P. Heterosis in switchgrass: spaced plants. *Crop Sci.* 48: 1312–1320; 2008. doi:10.2135/cropsci2007.12.0695.

Masters R. A.; Mislevy P.; Moser L. E.; Rivas-Pantoja F. Stand establishment. In: Moser L. E.; Burson B. L.; Sollenberger L. E. (eds) Warm-season (C_4) grasses. Agronomy Monograph No. 45. ASA, CSSA, SSSA, Madison, WI, pp 145–178; 2004.

McDonald B. A.; Linde C. Pathogen population genetics, evolutionary potential, and durable resistance. *Annu. Rev. Phytopathol.* 40: 349–379; 2002. doi:10.1146/annurev.phyto.40.120501.101443.

McLaughlin S. B.; Kszos L. A. Development of switchgrass (*Panicum virgatum*) as a bioenergy feedstock in the United States. *Biomass Bioenergy* 28: 515–535; 2005. doi:10.1016/j.biombioe.2004.05.006.

Missaoui A. M.; Paterson A. H.; Bouton J. H. Investigation of genomic organization in switchgrass (*Panicum virgatum* L.) using DNA markers. *Theor. Appl. Genet.* 110: 1372–1383; 2005. doi:10.1007/s00122-005-1935-6.

Mitchell C. E.; Tilman D.; Groth J. V. Effects of grassland plant species diversity, abundance, and composition on foliar fungal disease. *Ecology* 83: 1713–1726; 2002.

Mobberley D. G. Taxonomy and distribution of the genus *Spartina*. *Iowa State Coll. J. Sci.* 30: 471–574; 1956.

Montemayor M. B.; Price J. S.; Rochefort L.; Boudreau S. Temporal variations and spatial patterns in saline and waterlogged peat fields. *Environ. Exp. Bot.* 62: 333–342; 2008.

Mulkey V. R.; Owens V. N.; Lee D. K. Management of switchgrass-dominated conservation reserve program lands for biomass production in South Dakota. *Crop Sci.* 46: 712–720; 2006. doi:10.2135/cropsci2005.04-0007.

Mulkey V. R.; Owens V. N.; Lee D. K. Management of warm-season grass mixtures for biomass production in South Dakota USA. *Bioresour. Technol.* 99: 609–617; 2008. doi:10.1016/j.biortech.2006.12.035.

Mundt C. C. Use of multiline cultivars and cultivar mixtures for disease management. *Annu. Rev. Phytopathol.* 40: 381–410; 2002. doi:10.1146/annurev.phyto.40.011402.113723.

Nutter F. W.; Guan J.; Gotlieb A. R.; Rhodes L. H.; Grau C. R.; Sulc R. M. Quantifying alfalfa yield losses caused by foliar diseases in Iowa, Ohio, Wisconsin, and Vermont. *Plant Dis.* 86: 269–277; 2002. doi:10.1094/PDIS.2002.86.3.269.

Orolaza N. P.; Lamari L.; Balance G. M. Evidence of a host-specific chlorosis toxin from *Pyrenophora-tritici-repentis*, the causal agent of tan spot of wheat. *Phytopathology* 85: 1282–1287; 1995. doi:10.1094/Phyto-85-1282.

Parker I. M.; Gilbert G. S. The evolutionary ecology of novel plant–pathogen interactions. *Annu. Rev. Ecol. Evol. Syst.* 35: 675–700; 2004. doi:10.1146/annurev.ecolsys.34.011802.132339.

Parrish D. J.; Fike J. H. The biology and agronomy of switchgrass for biofuels. *Crit. Rev. Plant Sci.* 24: 423–459; 2005. doi:10.1080/07352680500316433.

Perlack R.; Wright L.; Turhollow A.; Graham R.; Stokes B.; Erbach D. Biomass as feedstock for a bioenergy and bioproducts industry: the technical feasibility of a billion-ton annual supply. U.S. Dept. of Commerce, Springfield, VA; 2005.

Phan A. T.; Smith Jr. S. R. Seed yield variation in blue grama and little bluestem plant collections in southern Manitoba, Canada. *Crop Sci.* 40:555-561; 2000.

Plantard O.; Valette S.; Gross M. F. The root-knot nematode producing galls on *Spartina altemiflora* belongs to the genus *Meloidogyne*: Rejection of *Hypsoperine* and *Spartonema* spp. *J. Nematol.* 39: 127–132; 2007.

Polley H. W.; Wilsey B. J.; Derner J. D. Dominant species constrain effects of species diversity on temporal variability in biomass production of tallgrass prairie. *Oikos* 116: 2044–2052; 2007. doi:10.1111/j.2007.0030-1299.16080.x.

Potter L.; Bingham M. J.; Bajer M. G.; Long S. P. The potential of two perennial C4 grasses and a perennial C4 sedge as ligno-cellulosic fuel crop in N.W. Europe crop establishment and yields in E. England. *Ann. Bot.* 76: 520; 1995. doi:10.1006/anbo.1995.1127.

Reed C. List of insect species which may be tallgrass prairie specialists. University of Minnesota, St. Paul, MN; 1996.

Reeder J. R. Chromosome numbers in western grasses. *Am. J. Bot.* 64: 102–110; 1977. doi:10.2307/2441882.

Riedell W. E.; Gustin R. D.; Beck D. L. Effect of irrigation on root-growth and yield of plants damaged by western corn-rootworm larvae. *Maydica* 37: 143–148; 1992.

Russelle M. P.; Morey R. V.; Baker J. M.; Porter P. M.; Jung H.-J. Comment on "carbon-negative biofuels from low-input high-diversity grassland biomass". *Science* 316: 1567b; 2007. doi:10.1126/science.1139388.

Ryan, A. B. Agronomic and molecular characterization of Louisiana native *Spartina alterniflora* accessions. Agronomy. Dissertation, Louisiana State University and Agricultural and Mechanical College; 2003.

Sanderson M. A.; Adler P. R. Perennial forages as second generation bioenergy crops. *Int. J. Mol. Sci.* 9: 768–788; 2008. doi:10.3390/ijms9050768.

Sanderson M. A.; Adler P. R.; Boateng A. A.; Casler M. D.; Sarath G. Switchgrass as a biofuels feedstock in the USA. *Can. J. Plant Sci.* 86: 1315–1325; 2006.

Sanderson M. A.; Brink G. E.; Higgins K. F.; Naugle D. E. Alternative uses of warm-season forage grasses. In: Moser L. E.; Burson B. L.; Sollenberger L. E. (eds) Warm-season (C_4) grasses. Agronomy Monograph No. 45. ASA, CSSA, SSSA, Madison, WI, pp 389–416; 2004.

Sarath G.; Mitchell R. B.; Sattler S. E.; Funnell D.; Pedersen J. F.; Graybosch R. A.; Vogel K. P. Opportunities and roadblocks in utilizing forages and small grains for liquid fuels. *J. Ind. Microbiol. Biotechnol.* 35: 343–354; 2008. doi:10.1007/s10295-007-0296-3.

Schmer M. R.; Vogel K. P.; Mitchell R. B.; Perrin R. K. Net energy of cellulosic ethanol from switchgrass. *Proc. Natl. Acad. Sci. U. S. A.* 105: 464–469; 2008. doi:10.1073/pnas.0704767105.

Schnoor J. L. Biofuels and the environment. *Environ. Sci. Tech.* 40: 4042–4042; 2006. doi:10.1021/es0627141.

Searchinger T.; Heimlich R.; Houghton R. A.; Dong F. X.; Elobeid A.; Fabiosa J.; Tokgoz S.; Hayes D.; Yu T. H. Use of us croplands for biofuels increases greenhouse gases through emissions from land-use change. *Science* 319: 1238–1240; 2008. doi:10.1126/science.1151861.

Simpson T. W.; Sharpley A. N.; Howarth R. W.; Paerl H. W.; Mankin K. R. The new gold rush: Fueling ethanol production while protecting water quality. *J. Environ. Qual.* 37: 318–324; 2008. doi:10.2134/jeq2007.0599.

Sims P. L.; Risser P. G. Grasslands. In: Barbour M. G.; Billings W. D. (eds) North American terrestrial vegetation. Cambridge Univ. Press, Cambridge, MS, pp 323–356; 2000.

Sloop C. M.; McGray H. G.; Blum M. J.; Strong D. R. Characterization of 24 additional microsatellite loci in *Spartina* species(Poaceae). *Conservat. Genet.* 6: 1049–1052; 2005. doi:10.1007/s10592-005-9084-7.

Smiley R. W.; Easley S. A.; Gourlie J. A. Annual spring wheat yields are suppressed by root-lesion nematodes in Oregon. *Phytopathology* 96: S171–S171; 2006. doi:10.1094/PHYTO-96-0171.

Smolik J. D. The role of nematodes in a South Dakota grassland ecosystem. Plant Science Department, South Dakota State University, Brookings, SD, p 74; 1973.

Somleva M. N.; Snell K. D.; Beaulieu J. J.; Peoples O. P.; Garrison B. R.; Patterson N. A. Production of polyhydroxybutyrate in switchgrass, a value-added co-product in an important lignocellulosic biomass crop. *Plant Biotechnol.* 6: 663–678; 2008. doi:10.1111/j.1467-7652.2008.00350.x.

Song K.; Lu P.; Tang K.; Osborn T. C. Rapid genome change in synthetic polyploids of Brassica and its implications for polyploid evolution. *Proc. Natl. Acad. Sci. U. S. A.* 92: 7719–7723; 1995. doi:10.1073/pnas.92.17.7719.

Songstad D. D.; Chen C. H.; Boe A. A. Plant regeneration in callus cultures derived from young inflorescences of little bluestem. *Crop Sci.* 26: 827–829; 1986.

Stewart D. A. B. Economic losses in cereal crops following damage by the African migratory locust, *Locusta migratoria migratorioides* (Reiche & Fairmaire) (Orthoptera: Acrididae), in the northern province of South Africa. *Afr. Entomol.* 5: 167–170; 1997.

Stubbiendieck J.; Hatch S. L.; Kjar K. J. North American range plants. University of Nebraska Press, Lincoln, NE; 1982.

Tilman D.; Hill J.; Lehman C. Carbon-negative biofuels from low-input high-diversity grassland biomass. *Science* 314: 1598–1600; 2006. doi:10.1126/science.1133306.

Tober D. A.; Duckwitz W.; Knudson W. Big bluestem trials in North Dakota, South Dakota, and Minnesota. USDA NRCS Plant Materials Center, Bismark, ND; 2008.

Tobias C. M.; Hayden D. M.; Twigg P.; Sarath G. Genic microsatellite markers derived from EST sequences of switchgrass (*Panicum virgatum* L.). *Mol. Ecol. Notes* 6: 185–187; 2006. doi:10.1111/j.1471-8286.2006.01187.x.

Tobias C. M.; Sarath G.; Twigg P.; Lindquist E.; Pangilinan J.; Penning B. J.; Barry K.; McCann M. C.; Carpita N. C.; Lazo G. R. Comparison of switchgrass ESTs with the Sorghum genome and development of EST-SSR markers. *Plant Genome* 2: 111–124; 2008. doi:10.3835/plantgenome2008.08.0003.

Tobias C. M.; Twigg P.; Hayden D. M.; Vogel K. P.; Mitchell R. M.; Lazo G. R.; Chow E. K.; Sarath G. Analysis of expressed sequence tags and the identification of associated short tandem repeats in switchgrass. *Theor. Appl. Genet* 111: 956–964; 2005. doi:10.1007/s00122-005-0030-3.

Todd M. C.; Washington R.; Cheke R. A.; Kniveton D. Brown locust outbreaks and climate variability in southern Africa. *J. Appl. Ecol.* 39: 31–42; 2002. doi:10.1046/j.1365-2664.2002.00691.x.

Travis S. E.; Hester M. W. A space-for-time substitution reveals the long-term decline in genotypic diversity of a widespread salt marsh plant, *Spartina alterniflora*, over a span of 1500 years. *J. Ecol.* 93: 417–430; 2005. doi:10.1111/j.0022-0477.2005.00985.x.

U.S. DOE. Breaking the biological barriers to cellulosic ethanol: A joint research agenda, DOE/SC-0095 U.S. Department of Energy Office of Science and Office of Energy and Renewable Energy (http://www.doegenomestolife.org/biofuels/); 2006.

USDA-NRCS. Summary report, 1997 National Resources Inventory. Revised Dec. 2000 - Table 4 (http://www.nrcs.usda.gov/technical/NRI/1997/summary_reports/table4.html website verified 5 Feb. 2009); 2000.

USDA-SCS. Land resource regions and major land resource areas of the United States. In: USDA-SCS (ed) Superintendent of documents. U.S. Government Print Office, Washington, DC; 1981.

Varvel G. E.; Vogel K. P.; Mitchell R. B.; Follett R. F.; Kimble J. M. Comparison of corn and switchgrass on marginal soils for bioenergy. *Biomass Bioenergy* 32: 18–21; 2008. doi:10.1016/j.biombioe.2007.07.003.

Vogel K. P.; Brejda J. J.; Walters D. T.; Buxton D. R. Switchgrass biomass production in the Midwest USA: Harvest and nitrogen management. *Agron. J.* 94: 413–420; 2002.

Vogel K. P.; Jung H. G. Genetic modification of herbaceous plants for feed and fuel. *Crit. Rev. Plant Sci.* 20: 15–49; 2001. doi:10.1016/S0735-2689(01)80011-3.

Weaver J. E. Extent of communities and abundance of the most common grasses in prairie. *Bot. Gaz.* 122: 25–33; 1960. doi:10.1086/336082.

Weaver J. E. North American prairie. Johnsen, Lincoln, NE; 1954.

Weaver J. E.; Fitzpatrick T. J. Ecology and relative importance of the dominants of the tall-grass prairie. *Bot. Gaz.* 93: 113–150; 1932. doi:10.1086/334244.

Wells G. R.; Fribourg H. A.; Schlarbaum S. E.; Ammons J. T.; Hodges D. G. Alternate land uses for marginal soils. *J. Soil Water Conserv.* 58: 73–81; 2003.

Willis W. O.; Bauer A.; Black A. L. Water conservation: Northern great plains. In: Dregne H. E.; Willis W. O. (eds) Dryland agriculture. Agronomy Monograph 23. ASA, CSSA, SSSA, Madison, WI; 1983.

Wipff J. K. Nomenclature combinations of *Schizachyrium* (Poaceae: Andropogoneae). *Phytologia* 80: 35–39; 1996.

Wu R.; Gallo-Meagher M.; Littell R. C.; Zeng Z. A general polyploid model for analyzing gene segregation in outcrossing tetraploid species. *Genetics* 159: 869–882; 2001.

Wu R.; Ma C. A general framework for statistical linkage analysis in multivalent tetraploids. *Genetics* 170: 899–907; 2005. doi:10.1534/genetics.104.035816.

Young D. K. Remnant dependence among prairie-inhabiting and savanna-inhabiting insects. *Nat. Area J.* 15: 290–292; 1995.

Young Sa; Park S. K.; Rodgers C.; Mitchell R. E.; Bender C. L. Physical and functional-characterization of the gene-cluster encoding the polyketide phytotoxin coronatine in *Pseudomonas-syringae pv-glycinea. J. Bacteriol. Virol.* 174: 1837–1843; 1992.

Chapter 6
Development and Status of Dedicated Energy Crops in the United States

Russell W. Jessup

Abstract The biofuel industry is rapidly growing because of increasing energy demand and diminishing petroleum reserves on a global scale. A multitude of biomass resources have been investigated, with high-yielding, perennial feedstocks showing the greatest potential for utilization as advanced biofuels. Government policy and economic drivers have promoted the development and commercialization of biofuel feedstocks, conversion technologies, and supply chain logistics. Research and regulations have focused on the environmental consequences of biofuels, greatly promoting systems that reduce greenhouse gas emissions and life-cycle impacts. Numerous biofuel refineries using lignocellulosic feedstocks and biomass-based triglycerides are either in production or pre-commercial development phases. Leading candidate energy crops have been identified, yet require additional efforts to realize their full potential. Advanced biofuels, complementing conventional biofuels and other renewable energy sources such as wind and solar, provide the means to substantially displace humanity's reliance on petroleum-based energy.

Keywords Biofuels • Cellulosic ethanol • Biodiesel • Energy crops

Introduction

Current world projections include a 57% increase in energy consumption between 2004 and 2030 (EIA 2007), while petroleum industry forecasts speculate the maximum output of economically extractable oil is either approaching or already passed (Campbell 2002; Almeida and Silva 2009; Bardi 2009). In this context, alternative sources of energy, fuels, and feedstocks are increasingly valuable to humankind.

R.W. Jessup (✉)
Department of Soil & Crop Sciences, Texas A&M University, College Station, TX 77843-2474, USA
e-mail: rjessup@tamu.edu

Biofuels—solid, liquid, or gaseous energy sources derived from renewable biomass sources—are currently the only direct substitute for fossil fuels available on a significant scale. Biomass can be burned directly for thermal energy or converted to high-value fuels (such as ethanol, diesel, methanol, methane, or hydrogen) for electricity, heating, and transportation. Solid biofuels displace mostly coal in electric power generation, liquid biofuels substitute primarily for petroleum fuels in transportation, and biogas best serves to displace natural gas.

Substantive biofuel production in the United States began in 1978 and grew at a steady pace of roughly 150 mgy (million gallons per yr) through 2004. Mandates driven by rising petroleum costs and desire for renewable energy sources have recently led to dramatic increases in biofuel industry growth, including an average rate of 2.4 bgy (billion gallons per yr) between 2005 and 2008 and projected average rate of 1.9 bgy between 2009 and 2022. Public and private funding sources have similarly increased the number of biofuel refineries and bioenergy research centers (Hoekman 2009). While not providing a complete solution to energy demand, the sustainable use of biomass resources would be sufficient to displace up to 30% of current petroleum consumption in the U.S. (Perlack et al. 2005).

First-generation 'conventional' biofuels were primarily artifacts of cereal and oilseed food crops, such as ethanol derived from corn (*Zea mays* L.) starch, and sugarcane (*Saccharum officinarum* L.) and biodiesel derived from soybean (*Glycine max* L.) and rapeseed (*Brassica napus* L.). Early 'advanced' biofuels have recently included non-food annual crop alternatives, such as sorghum (*Sorghum bicolor* [L.] Moench) starch and sugars for ethanol and camelina (*Camelina sativa* [L.] Crantz) for biodiesel. Crops having the largest potential as biofuel feedstocks, however, are biomass derived, perennial, and high yielding. Leading candidates include perennial grasses (switchgrass, *Panicum virgatum* L.; miscanthus, *Miscanthus* spp.; energy cane, *Saccharum* spp.), short-rotation forests (willow, *Salix* spp.; poplar, *Populus* spp.) for lignocellulosic ethanol, and perennial oilseed species (jatropha, *Jatropha curcas* L.) and algae for biomass-derived biodiesel.

Policy and Economic Drivers

The renewable fuel standard (RFS), established by the Energy Policy Act (EPAct) of 2005 and expanded by the Energy Independence and Security Act (EISA) of 2007, has been a major contributor to the recent growth of the U.S. biofuel industry. As summarized in Table 6.1, the RFS mandates increased production of conventional biofuels (such as corn ethanol) from 3.9 to more than 9.0 bgy between 2005 and 2008 (Renewable Fuels Association 2009a). The EPAct of 2005 also established the Volumetric Ethanol Excise Tax Credit (VEETC) of $0.51 per gallon, with an additional subsidy of $0.10 per gallon for refineries producing less than 15 mgy (Parcell and Westhoff 2005). Extension of the VEETC by the EPAct of 2005 further spurred growth of the U.S. biodiesel market, providing $1.00 per gallon for biodiesel producers using agricultural feedstocks and $0.50 per gallon for

Table 6.1 RFS renewable fuel volumes based on EIA (2007)

Year	Total RFS	Conventional biofuel[y]	Advanced biofuel		
			Cellulosic biofuel	Biomass-based diesel[y]	Undifferentiated advanced biofuel[y]
2006	4.00	4.00			
2007	4.70	4.70			
2008	9.00	9.00			
2009	11.10	10.50	0.60	0.50	0.10
2010	12.95	12.00	0.95	0.10	0.65	0.20
2011	13.95	12.60	1.35	0.25	0.80	0.30
2012	15.20	13.20	2.00	0.50	1.00	0.50
2013	16.55	13.80	2.75	1.00	[z]	1.75
2014	18.15	14.40	3.75	1.75	[z]	2.00
2015	20.50	15.00	5.50	3.00	[z]	2.50
2016	22.25	15.00	7.25	4.25	[z]	3.00
2017	24.00	15.00	9.00	5.50	[z]	3.50
2018	26.00	15.00	11.00	7.00	[z]	4.00
2019	28.00	15.00	13.00	8.50	[z]	4.50
2020	30.00	15.00	15.00	10.50	[z]	4.50
2021	33.00	15.00	18.00	13.50	[z]	4.50
2022	36.00	15.00	21.00	16.00	[z]	5.00

[z] EPA administrator determination
[y] May include biodiesel

biodiesel producers utilizing waste grease and oils. U.S. biodiesel production concomitantly increased from 75 to 700 million gallons per yr between 2005 and 2008 (National Biodiesel Board 2009).

Acknowledging that conversion of all the corn and soybeans in the U.S. into liquid biofuels would meet only 12% of US energy needs (Hill et al. 2006), the RFS caps conventional biofuel levels at 15 billion gallons per year as of 2015 and mandates an additional 21 bgy of advanced biofuels (renewable fuels other than corn ethanol) by 2022. Renewable biomass sources for advanced biofuels include biomass crops, grains other than corn, sugar crops, waste materials, crop residues, and algae, with cellulosic biofuels expected to supply the majority (16 bgy) of the mandate (Perlack et al. 2005). The 2007 Farm Bill, enacted by Congress in May 2008, reduced the tax credit for corn-based ethanol from $0.51 to $0.45 per gallon and specifically introduced a tax credit of $1.01 per gallon for cellulose-based ethanol.

Additional supports for the biofuel industry entail tariffs, state policies, and research funding. Federal import barriers include a $0.54 per gallon and 2.5% *ad valorem* tariff on foreign-produced ethanol (Steenblik 2007). Biofuel incentives with renewable portfolio standards and renewable energy goals have been enacted in 28 and five states, respectively (DSIRE 2009). Several states also offer some form of excise tax exemption. The EPAct of 2005 funded biofuel research $632 million in 2007, $743 million in 2008, and $852 million in 2009.

Industry Development and Projections

The number of operating ethanol refineries rose from 81 to 170 between 2005 and 2008 (Renewable Fuels Association 2009b) and the number of biodiesel refineries using conventional biofuels increased from 45 to 176 (National Biodiesel Board 2005, 2008) in the same timeframe. Bioethanol produced from corn held a 99% market share and 5-yr annual growth rate of 18% (Scaff and Reca 2005), with a majority of U.S. production during 2008 used in gasoline blends (E85, 85% ethanol and 15% gasoline; E10, 10% ethanol and 90% gasoline) or fuel additives in place of methyl tertiary butyl ether (MTBE). RFS mandates for biomass-derived diesel began in 2009 and for cellulosic ethanol in 2010. Despite the aggressive timelines within the RFS for advanced biofuel production, only a few pilot plants are in operation. However, many commercial refineries are under development (Table 6.2).

Cumulatively, a vision for U.S. renewable energy mandates calls for 20% of transportation fuels, 55% of bioproducts, and 7% of the nation's power to be produced from biomass by 2030 (BRDI 2006). Achieving these goals will require roughly 907 million Mg (1 billion tons) of biomass annually (Perlack et al. 2005). An estimated 22.3 million ha of the 181.4 million ha of USA cropland would be needed, primarily from cropland pasture (9.1 million ha), hayland (4.2 million ha), CRP (4 million ha), and reallocation of existing cropland (5.7 million ha). Roughly 7% of energy consumption during 2006 in the USA (~99 EJ, total) was supplied by renewable sources, 48% of which was generated from biomass (EIA 2008). This is only a fraction of production potential, and the U.S. land base could therefore supply both biomass and existing food, feed, and export demands. As much as 40% of biomass feedstocks are forecast to be produced in the southern U.S. (English et al. 2006).

Waste streams, such as oil and animal fats for biodiesel production, will be an important but minor component of future energy supplies (Tashtoush et al. 2004). Several waste streams may be used to produce methane, including processed fruit and vegetable solids (Gunaseelan 2004), captured landfill gases (Weeks 2005), and manure emissions (Schwart et al. 2005). Cost savings associated with production windows will prove advantageous for biofuels in comparison to other renewable energy forms. Biomass generation, not subject to the random availability inherent of wind or solar energy (Smith et al. 2004), will lower risk and increase capital recovery (Neuhoff 2005).

Greenhouse gas (GHG) and carbon emission policies will become more prevalent, strengthening the impetus for development of sustainable biofuel systems. Agriculture produces approximately 7% of total GHG emissions in the U.S. (USDA 2004), of which effectively deployed biofuels offer great potential to reduce (Adler et al. 2007). Optimized agronomic practices through reduced tillage, fertilization, and inputs further impart significant reductions in net GHG emissions (West and Marland 2002). While converting forest or productive grassland to bioenergy crop production may incur a large carbon debt, producing biomass on marginal or degraded lands will be advantageous and likely not incur such a carbon debt (Fargione et al. 2008). Perennial crops would further improve degraded lands by increasing soil organic matter and structure (Sartori et al. 2006).

Table 6.2 Operating or planned advanced biofuel refineries in the US

Company	Plant location	Biofuel	Status	Feedstock	Capacity (mgy)
AE Biofuels	MT	Cellulosic ethanol	Operating	Corn, corn stover	0.15
KL Energy Corp.	WY	Cellulosic ethanol	Operating	Wood	1.5
Poet	SD	Cellulosic ethanol	Operating	Corn cobs	0.02
Verenium	LA	Cellulosic ethanol	Operating	Sugarcane, energy cane	1.4
Abengoa	NB	Cellulosic ethanol	Planned	Corn stover, residual starch	10
Abengoa Bioenergy	KS	Cellulosic ethanol	Planned	Herbaceous biomass	30
BlueFire	CA	Cellulosic ethanol	Planned	Green landfill waste	3.2
BlueFire Mecca LLC	CA	Cellulosic ethanol	Planned	Green waste	17
Citrus Energy LLC	FL	Cellulosic ethanol	Planned	Citrus waste	4
Clemson University	SC	Cellulosic ethanol	Planned	Wood waste, algae	10
Colusa Biomass	CA	Cellulosic ethanol	Planned	Rice hulls	12.5
Coskata	PA	Cellulosic ethanol	Planned	Woody biomass, waste	0.04
Dupont Danisco	TN	Cellulosic ethanol	Planned	Switchgrass, stover	0.25
Ecofin LLC	KY	Cellulosic ethanol	Planned	Corn cobs	1
Flambeau River	WI	Cellulosic ethanol	Planned	Forest, paper waste	6
Fulcrum	NV	Cellulosic ethanol	Planned	Municipal waste	10.5
Gulf Coast Energy	FL	Cellulosic ethanol	Planned	Woody biomass	25
Gulf Coast Energy	AL	Cellulosic ethanol	Planned	Wood waste	0.4
ICM Inc	MO	Cellulosic ethanol	Planned	Switchgrass, sorghum	1.5
KL Energy Corp.	CO	Cellulosic ethanol	Planned	Wood pellets	5
Liberty Industries	FL	Cellulosic ethanol	Planned	Forest waste	7
Mascoma Corp.	MI	Cellulosic ethanol	Planned	Woody biomass and residues	40
NewPage	WI	Cellulosic ethanol	Planned	Woody biomass	5.5
Pacific Ethanol	OR	Cellulosic ethanol	Planned	Wheat straw, poplar	2.7
PetroSun	AZ	Biomass-based diesel	Planned	Algae	30
PetroSun	TX	Biomass-based diesel	Planned	Algae	4.4
Poet	IA	Cellulosic ethanol	Planned	Corn stover and cobs	25

(continued)

Table 6.2 (continued)

Company	Plant location	Biofuel	Status	Feedstock	Capacity (mgy)
PureVision	CO	Cellulosic ethanol	Planned	Corn stalks	2
Range Fuels	GA	Cellulosic ethanol	Planned	Woody biomass and residues	20
RSE Pulp & Chem	ME	Cellulosic ethanol	Planned	Woody biomass	2.2
Smiling Earth Energy	VA	Biomass-based diesel	Planned	Jatropha	320
SunOpta	MN	Cellulosic ethanol	Planned	Corn stover, waste	10
Sustainable Oils	MT	Biomass-based diesel	Planned	Camelina	100
University of Florida	FL	Cellulosic ethanol	Planned	Bagasse	2
US Envirofuels LLC	FL	Cellulosic ethanol	Planned	Sorghum, sugar cane	20
Verenium-BP	FL	Cellulosic ethanol	Planned	Energy cane, sorghum	36
West Biofuels	CA	Cellulosic ethanol	Planned	Wood chips	0.18

Biofuel Feedstocks

A 'dedicated energy crop' (DEC), as used in this text, is a crop grown exclusively for the purpose of converting its harvested biomass into bioenergy. These dedicated bioenergy systems generally refer to herbaceous and/or woody plant species, but have more recently included algae. Bioenergy derived from wastes and residues are technically byproducts from other enterprises and not dedicated systems.

It is unlikely that any single species will be a universal feedstock for the biofuel industry. Climatic zones will likely be best suited for different DECs. Further, productivity may be maximized by introducing or maintaining species diversity in energy cropping systems (Florine et al. 2006). Finally, it would be difficult to identify today those species that will become the feedstocks for future biorefineries because processing plants and conversion technologies are in early stages of development. The species that will eventually serve as feedstocks will likely be highly engineered from existing species (DOE 2007).

Some desired attributes of DECs are implicit—perennial, high-yielding, non-food/feed crop, readily adopted by producers, minimal input requirements, and adapted to the region where biorefineries are located. The following points further address some critical features of viable DECs:

1. *Bioengineering amenability.* Traits that do not exist within a species or its accessible gene pools, as well as traits with added value upon genetic improvement beyond the potential of conventional breeding, must readily afford transgenic manipulations that can be subsequently transmitted via seed.
2. *Seed resources.* Seed or seed stock (propagules) should be readily propagated and easily planted. This largely implicates seed propagation in order to achieve a significant portion of the 17 million ha proposed for perennial energy crop culture in the billion-ton study (Perlack et al. 2005).
3. *Vigorous establishment.* Energy crops should have fast growth rates in order to out-compete weeds and produce sufficient biomass for a full harvest in the establishment year. Species that require 3 yr or more to reach full productive potential would have less economic potential as DECs.
4. *Non-invasive.* Some candidate biofuel species can spread aggressively when introduced into ecosystems (Raghu et al. 2006). Thus, native species occurring naturally within an ecoregion are more desirable DECs.
5. *Harvest flexibility.* Species affording multiple harvests per growing season provide greater flexibility to growers and benefit biorefineries with a year-round feedstock. Nutrient cycling and fertilization requirements, however, will impact such cropping systems. Post-senescence harvests allow nitrogen and other minerals to be translocated from the shoots into the crown and roots, and other nutrients such as potassium to be leached out of dead biomass into soil, in order to support the next season's growth (Parrish and Fike 2005; Adler et al. 2006). Both nitrogen translocation and nutrient leaching have been shown to occur in *Miscanthus* (Lewandowski et al. 2003), switchgrass (Parrish et al. 2003), and

perennial grass systems (McKendrick et al. 1975; Clark 1977) when managed with delayed harvests for bioenergy production.

6. *Standability*. Energy crops that could be left as standing fodder in the field after the growing season without major biomass losses would provide flexibility in use of harvest equipment and labor, allowing the dried material to be baled or otherwise densified at optimal times for refinery schedules.

7. *Stress tolerance*. DECs to be grown on millions of hectares must be relatively free of pests and diseases while minimizing the use of chemical controls for both economic and ecological reasons. Deployment of multi-species energy crop systems may reduce the threat of major outbreaks (Tilman et al. 2006b) through increased biodiversity within the plant community. Simplified weed suppression options are in particular critical during the establishment phase in order to prevent crop failures or large economic costs associated with chemical controls. DECs with increased heat and drought tolerance would be preferred, including considerations for potential effects of global climate change on the bioenergy industry (Tuck et al. 2006).

8. *Wildlife habitat*. DEC plantations should provide wildlife habitat and improve biodiversity, as has been demonstrated for switchgrass (Roth et al. 2005) and *Miscanthus* spp. (Semere and Slater 2007).

9. *Carbon sequestration*. Perennial species sequester carbon to varying extents in below-ground biomass, which is beneficial towards achieving carbon negative cropping systems (Tilman et al. 2006a). Comparatively, perennial biomass crops use fewer inputs, produce more energy, and reduce GHG emissions to a greater extent than annual cropping systems (NRCS 2006; Adler et al. 2007; DOE 2007). No-till methods will further minimize erosive potential and soil carbon losses. The DOE/USDA biomass feasibility study (Perlack et al. 2005) suggested that corn and other annual biofuel crops should also use no-till methods, but that even this technology might be insufficient to attain acceptable soil loss rates.

10. *Agronomically integrated*. Hybrid DECs are feasible in the mid- to long-term and will undoubtedly enhance biomass potential. Large-seeded crops with vigorous establishment will simplify biofuel production systems. Delayed flowering through photoperiodism will afford greater biomass accumulation and potentially prevent seed-borne weed risks. Sterility (cytoplasmic-, genetic-, or wide hybrid-based) will enable hybrid energy crop production, greater commodity control for seed companies, and reduced invasiveness potential.

Leading Candidate Dedicated Energy Crops

Forest resources. Woody biomass supplies roughly 7% of world energy supplies (Mead 2005). Forestlands are the largest renewable biomass resources in the U.S. in terms of standing biomass (18 billion Mg) and land area (303 million ha; Perlack et al. 2005). Willow (*Salix* spp.) and poplar (*Populus* spp.) have been studied most extensively for biomass because of their rapid growth and establishment rates

(Dickmann 2006), and additional species are under evaluation (Geyer 2006). Sustainability is a potential limitation for large scale wood-to-fuel systems, especially in regions of the developing world where reliance on forest lands for heating and cooking has resulted in deforestation-related environmental consequences (Schulte-Bisping et al. 1999). Forest removal on infertile soils in the U.S. would also negatively impact site productivity (Scott and Dean 2006), sediment losses (Malik et al. 2000), and potentially the thermodynamics of using wood-for-fuel plantations (Patzek and Pimentel 2005). While forests have the largest land-based inventory, agricultural lands have greater biomass production potential via greater management intensity (Perlack et al. 2005). Whether woody crops can become ideal feedstocks in the long-term is yet undetermined; however, examples such as sustainable forest thinning in the western United States combines biofuel harvest with reduced fire hazard (Fecko 2008).

Switchgrass. Switchgrass (*Panicum virgatum* L.), a C4 native perennial grass, has been a focal point in bioenergy crop research in the U.S. for the past two decades (McLaughlin and Kszos 2005). The majority of the resulting switchgrass research is in the public domain, serving as a reference to extend agronomic, breeding, genomic, ecological, and economic resources towards improving additional native species (big bluestem, *Andropogon gerardii* Vitman; Indian grass, *Sorghastrum nutans* [L.] Nash; little bluestem, *Schizachyrium scoparium* [Michx.] Nash; sideoats grama, *Bouteloua curtipendula* [Michx.] Torr.; prairie cordgrass, *Spartina pectinata* Bosc ex Link; eastern gamagrass, *Tripsacum dactyloides* [L.] L.) and non-native species(miscanthus, *Miscanthus* spp.; bermudagrass, *Cynodon* spp.; napiergrass, *Pennisetum purpureum* L.) with potential as biofeedstocks. Positive attributes of switchgrass include high biomass potential across diverse environments, adaptation to marginal lands, relatively low input requirements, and environmental benefits (Bouton 2007). Potential disadvantages of switchgrass include slow establishment, small seed size, and recently identified risks to yield-reducing nematode (Cassida et al. 2005) and fungal smut (Gravert et al. 2000) infestations. Well-adapted cultivars for the southern (Alamo) and northern (Cave-in-Rock) U.S. are publicly available. Newer cultivars with improved biomass yield and chemical composition have been more recently released (Vogel et al. 1996; Burns et al. 2008a; Burns et al. 2008b), including those intended for commercial use as DEC (Delta Farm Press 2009). Comparatively, switchgrass has received limited plant breeding and most cultivars are not far removed from native germplasm (Bouton 2007; Casler et al. 2007). There is therefore tremendous genetic variability and potential for germplasm improvement (Rose et al. 2008), including successful genetic transformation (Somleva et al. 2002) and demonstrated heterosis in F_1 hybrids (Taliaferro 2002).

Miscanthus. *Miscanthus*, a C4 grass genus native to Asia, includes intraspecific, interspecific, and intergeneric derivatives with potential as energy crops (Jezowski 2008; Miguez et al. 2008). A sterile hybrid, *M.×giganteus* Greef et Deu. (*M. sacchariflorus×M. sinensis*) with substantial biomass potential has been widely researched and used as a biofuel in Europe (Lewandowski et al. 2003). *Miscanthus* has superior cold tolerance for a C4 species (Farage et al. 2006) and low input requirements due to efficient nutrient recycling (Lewandowski and Schmidt 2006) and late winter harvests (El-Bassam 1998). *Miscanthus* is used as a forage crop in Japan (Ogura et al. 1999) but largely known in the U.S. as an ornamental plant. *M.×giganteus*' sterility is a benefit towards non-invasiveness, and *in vitro* techniques have greatly improved its propagation potential (Holme et al. 1997; Holme 1998). Comparisons between *Miscanthus* yields in Europe and switchgrass yields in North America suggest *Miscanthus* has capacity to produce greater biomass (Heaton et al. 2004a, 2004b), and *Miscanthus* yielded 33% more biomass than 'Kanlow' switchgrass in a recent European trial (Boehmel et al. 2008). *M. sinensis* has been shown to be suitable for seed propagation (Christian et al. 2005) and have yield potential equal to or greater than that of *M.×giganteus* (Clifton-Brown et al. 2001). Successful *Miscanthus–Saccharum* hybridizations (Li et al. 1948; Chen and Lo 1989) further provide opportunities to combine and optimize value-added biofuel traits.

Energy Cane. Energy cane refers to high biomass sugarcane (*Saccharum* spp.) hybrids with fiber content at least 20% higher and juice content at least 10% lower than that of conventional cane. Energy cane can potentially yield 10 to more than 30 t dry matter per acre in diverse environments, and public cultivars exist (St. John et al. 2007). The lack of cold tolerance characteristic of sugarcane has begun to be addressed by introgression efforts involving *S. spontaneum* (Wang et al. 2008). As part of the *Saccharum* Complex (Hodkinson et al. 2002), energy cane improvement programs will likely use *Miscanthus* and *Erianthus* intergeneric hybridizations. Sorghum is also within the *Saccharum* Complex, and demonstrated *Saccharum–Sorghum* hybrids (Gupta et al. 1978) suggest great potential to introgress drought tolerance and large seed size from sorghum into high-biomass, perennial energy canes.

Jatropha. Jatropha (*Jatropha curcas* L.) is a drought-tolerant, perennial shrub in the Euphorbiaceae family, native to Mexico and Central America. Yield is variable, but seed production up to 5 t/ha and 37% oil content has been reported (Achten et al. 2008). The oil can be combusted directly as fuel without being refined, and byproducts make suitable organic fertilizers and insecticides. Jatropha is productive on marginal soils but lacks cold tolerance requirements for cultivation in temperate latitudes (Openshaw 2000). Jatropha oil quality is suitable for biodiesel production, but additional research is required to improve its crude oil quality (de Oliveira et al. 2009) and yield-limiting asynchronous seed maturation (Achten et al. 2008).

Algae. While a direct comparison with higher plants is not possible, algae have potential to be engineered as productive and sustainable DECs. Algae are good candidates for biodiesel production, based on high photosynthetic efficiency, biomass production, and growth rates. Microalgae convert carbon dioxide to numerous biofuels, including methane, biodiesel (Banerjee et al. 2002), and hydrogen (Melis 2002). Algae reproduce rapidly and can double biomass in fewer than 24 h. Algae are also exceedingly high in oil content, with average lipid contents up to 90% of dry weight under ideal conditions (Spolaore et al. 2006). Algae oil composition differs from that of animal and vegetable sources and has been the focus of improvement efforts (Certik and Shimizu 1999). Genetic transformation of algae is also more straightforward than in higher plants, allowing greater potential to engineer biofuels (Roessler et al. 1994). The first commercial endeavor in single cell oils was unsuccessful (Ratledge 2004), and infrastructure requirements and cost competitiveness remain largely unanswered for algal-derived biodiesel. Open-pond systems are estimated to be at least twice as expensive as petroleum diesel, face contamination from plant and microbial weeds, and require abundant land and water in warm geographies. Closed bioreactors avoid many of these problems but are even more cost prohibitive due to additional infrastructure requirements. Despite these challenges, high fuel yield potentials, RFS qualification, and the likelihood of large advances through biotechnology will likely drive development of algal biodiesel.

Conclusions

Rapid growth is occurring in the U.S. biofuels industry, driven by concerns for alternative energy sources in the face of growing demand and finite supplies. Current emphasis on corn-based ethanol is transitioning to biofuels produced from lignocellulosic materials and biomass-derived triglycerides. No single approach is likely, with regional variation in feedstocks and infrastructure probable. Biofuels likewise have potential to displace substantial amounts of petroleum-based fuels but will largely serve as one of many components advancing renewable energy systems. U.S. natural resources are ultimately sufficient to develop and implement biofuels, if handled with sustainability and stewardship in mind.

References

Achten W.; Verchot L.; Franken Y.; Mathijs E.; Singh V.; Aerts R.; Muys B. Jatropha bio-diesel production and use. *Biomass Bioenerg.* 3212: 1063–1084; 2008 doi:10.1016/j.biombioe.2008.03.003.
Adler P.R.; Sanderson M.A.; Boateng A.A.; Weimer P.J.; Jung H.J.G. Biomass yield and biofuel quality of switchgrass harvested in fall or spring. *Agron. J.* 98: 1518–1525; 2006 doi:10.2134/agronj2005.0351.
Adler P.R.; Del Grosso S.J.; Parton W.J. Life cycle assessment of net greenhouse gas flux for bioenergy cropping systems. *Ecol. Appl.* 17: 675–691; 2007 doi:10.1890/05-2018.
Almeida P.; Silva P.D. The peak of oil production—timings and market recognition. *Energy Policy* 37: 1267–1276; 2009 doi:10.1016/j.enpol.2008.11.016.
Banerjee A.; Sharma R.; Chisti Y.; Banerjee U.C. *Botryococcus braunii*: a renewable source of hydrocarbons and other chemicals. *Crit. Rev. Biotechnol.* 22: 245–279; 2002 doi:10.1080/07388550290789513.
Bardi U. Peak oil: The four stages of a new idea. *Energy* 34: 323–326; 2009 doi:10.1016/j.energy.2008.08.015.
Boehmel C.; Lewandowski I.; Claupein W. Comparing annual and perennial energy cropping systems with different management intensities. *Agr Syst* 96: 224–236; 2008 doi:10.1016/j.agsy.2007.08.004.
BRDI (Biomass Research and Development Initiative) (2006) Vision for bioenergy and biobased products in the United States; Available via: http://www.brdisolutions.com/Site%20Docs/Final%202006%20Vision.pdf Cited 12 March 2009

Bouton J.H. Molecular breeding of switchgrass for use as a biofuel crop. *Curr. Opin. Genet. Dev.* 17: 553–558; 2007 doi:10.1016/j.gde.2007.08.012.

Burns J.C.; Godshalk E.B.; Timothy D.H. Registration of 'Performer' switchgrass. *J. Plant Registrations* 2: 29–30; 2008a doi:10.3198/jpr2007.02.0093crc.

Burns J.C.; Godshalk E.B.; Timothy D.H. Registration of 'BoMaster' switchgrass. *J. Plant Registrations* 2: 31–32; 2008b doi:10.3198/jpr2007.02.0094crc.

Campbell C.J. Petroleum and people. *Pop. Envir.* 24: 193–207; 2002 doi:10.1023/A:1020752205672.

Casler M.D.; Stendal C.A.; Kapich L.; Vogel K.P. Genetic diversity, plant adaptation regions, and gene pools for switchgrass. *Crop Sci.* 47: 2261–2273; 2007 doi:10.2135/cropsci2006.12.0797.

Cassida K.A.; Kirkpatrick T.L.; Robbins R.T.; Muir J.P.; Venuto B.C.; Hussey M.A. Plant-parasitic nematodes associated with switchgrass (*Panicum virgatum* L.) grown for biofuel in South Central United States. *Nematropica* 35: 1–10; 2005.

Certik M.; Shimizu S. Biosynthesis and regulation of microbial polyunsaturated fatty acid production. *J. Biosci. Bioeng.* 87: 1–14; 1999 doi:10.1016/S1389-1723(99)80001-2.

Chen Y.H.; Lo C.C. Disease resistance and sugar content in *Saccharum–Miscanthus* hybrids. *Taiwan Sugar* 363: 9–12; 1989.

Christian D.G.; Yates N.E.; Riche A.B. Establishing *Miscanthus sinensis* from seed using conventional sowing methods. *Ind. Crops Prod.* 21: 109–111; 2005 doi:10.1016/j.indcrop.2004.01.004.

Clark F.L. Internal cycling of nitrogen in shortgrass prairie. *Ecology* 58: 1322–1333; 1977 doi:10.2307/1935084.

Clifton-Brown J.C.; Lewandowski I.; Andersson B.; Basch G.; Christian D.G.; Bonderup-Kjeldsen J.; Jørgensen U.; Mortensen J.; Riche A.B.; Schwarz K.U.; Tayebi K.; Teixeira F. Performance of 15 *Miscanthus* genotypes at five sites in Europe. *Agron. J.* 93: 1013–1019; 2001.

DSIRE (Database of State Initiatives for Renewables and Efficiency) (2009). Renewables portfolio standards. Available via: http://www.dsireusa.org/documents/SummaryMaps/RPS_Map.ppt Cited 12 March 2009

de Oliveira J.S.; Leite P.M.; de Souza L.B.; Mello V.M.; Silva E.C.; Rubim J.C.; Meneghetti S.M.P.; Suarez P.A.Z. Characteristics and composition of *Jatropha gossypiifolia* and *Jatropha curcas* L. oils and application for biodiesel production. *Biomass Bioenerg.* 33: 449–453; 2009 doi:10.1016/j.biombioe.2008.08.006.

Delta Farm Press (2009) Ceres: first seed for dedicated energy crops. Available via: http://deltafarmpress.com/biofuels/ceres-energy-0213/ Cited 12 March 2009

Dickmann D.I. Silviculture and biology of short-rotation woody crops in temperate regions: then and now. *Biomass Bioenerg.* 30: 696–705; 2006 doi:10.1016/j.biombioe.2005.02.008.

DOE (US Department of Energy) (2007) Breaking the Biological Barriers for Cellulosic Ethanol: A Joint Research Agenda. DOE/SC-0095, US Department of Energy Office of Science and Office of Energy Efficiency and Renewable Energy. Available via: www.doegenomestolife.org/biofuels Cited 12 March 2009

EIA (US Energy Information Administration) (2007) International Energy Outlook 2007 with Projections to 2030. Available via: http://www.eia.doe.gov///oiaf/aeo/index.html Cited 12 March 2009

EIA (US Energy Information Administration) (2008) U.S. Energy Consumption by Energy Source. Available via: http://www.eia.doe.gov/cneaf/solar.renewables/page/trends/table1.html Cited 12 March 2009

El-Bassam N. Energy plant species: their use and impact on environment and development. James and James, London 1998.

English B.C.; De La Torre U.D.G.; Walsh M.E.; Hellwinkel C.; Menard J. Economic competitiveness of bioenergy production and effects on agriculture of the southern region. *J. Agric. Appl. Econ.* 38: 389–402; 2006.

Farage P.K.; Blowers D.; Long S.P.; Baker N.R. Low growth temperatures modify the efficiency of light use by photosystem II for CO_2 assimilation in leaves of two chilling-tolerant C4 species,

Cyperus longus L. and *Miscanthus* × *giganteus*. *Plant Cell Environ.* 29: 720–728; 2006 doi:10.1111/j.1365-3040.2005.01460.x.

Fargione J.; Hill J.; Tilman D.; Polasky S.; Hawthorne P. Land clearing and the biofuel carbon debt. *Science* 319: 1235–1236; 2008 doi: 10.1126/science.1152747.

Fecko R.M. Effects of mechanized thinning and prescription fire on stand structure, live crown, and mortality in Jeffrey pine. *J. Sust. Forestry* 264: 241–283; 2008 doi:10.1080/10549810701879743.

Florine S.E.; Moore K.J.; Fales S.L.; White T.A.; Burras C.L. Yield and composition of herbaceous biomass harvested from naturalized grassland in southern Iowa. *Biomass Bioenerg.* 30: 522–528; 2006 doi: 10.1016/j.biombioe.2005.12.007.

Geyer W.A. Biomass production in the Central Great Plains USA under various coppice regimes. *Biomass Bioenerg.* 30: 778–783; 2006 doi:10.1016/j.biombioe.2005.08.002.

Gravert C.E.; Tiffany L.H.; Munkvold G.P. Outbreak of smut caused by *Tilletia maclaganii* on cultivated switchgrass in Iowa. *Plant Dis.* 84: 596; 2000 doi: 10.1094/PDIS.2000.84.5.596A.

Gunaseelan V.N. Biochemical methane potential of fruits and vegetable solid waste feedstocks. *Biomass Bioenerg.* 26: 389–399; 2004 doi:10.1016/j.biombioe.2003.08.006.

Gupta S.C.; de Wet J.M.J.; Harlan J.R. Morphology of *Saccharum–Sorghum* hybrid derivatives. *Am. J. Bot.* 65: 936–942; 1978 doi: 10.2307/2442680.

Heaton E.; Voigt T.; Long S.P.A. quantitative review comparing the yields of two candidate C4 perennial biomass crops in relation to nitrogen, temperature, and water. *Biomass Bioenerg.* 27: 21–30; 2004a doi:10.1016/j.biombioe.2003.10.005.

Heaton E.A.; Clifton-Brown J.; Voigt T.B.; Jones M.B.; Long S.P. *Miscanthus* for renewable energy generation: European Union experience and projections for Illinois. *Mitig. Adapt. Strat. Glob. Change* 9: 433–451; 2004b doi: 10.1023/B:MITI.0000038848.94134.be.

Hill J.; Nelson E.; Tilman D.; Polasky S.; Tiffany D. Environmental, economic, and energetic costs and benefits of biodiesel and ethanol biofuels. *Proc. Nat. Acad. Sci. U S A* 103: 11206–11210; 2006 doi:10.1073/pnas.0604600103.

Hodkinson T.R.; Chase M.W.; Lledó M.D.; Salamin N.; Renvoize S.A. Phylogenetics of *Miscanthus*, *Saccharum* and related genera (Saccharinae, Andropogoneae, Poaceae) based on DNA sequences from ITS nuclear ribosomal DNA and plastid trnL intron and trnL-F intergenic spacers. *J. Plant. Res.* 115: 381–392; 2002 doi: 10.1007/s10265-002-0049-3.

Hoekman S.K. Biofuels in the U.S.—challenges and opportunities. *Renew. Energ.* 34: 14–22; 2009 doi:10.1016/j.renene.2008.04.030.

Holme I.B. Growth characteristics and nutrient depletion of *Miscanthus x ogiformis* Honda 'Giganteus' suspension cultures. *Plant Cell Tiss. Org. Cult.* 53: 143–151; 1998 doi: 10.1023/A:1006001419989.

Holme I.B.; Krogstrup P.; Hansen J. Embryogenic callus formation, growth and regeneration in callus and suspension cultures of *Miscanthus×ogiformis* Honda 'Giganteus' as affected by proline. *Plant Cell Tiss. Organ. Cult.* 50: 203–210; 1997 doi: 10.1023/A:1005981300847.

Jezowski S. Yield traits of six clones of *Miscanthus* in the first 3 yr following planting in Poland. *Ind. Crop. Prod.* 27: 65–68; 2008 doi:10.1016/j.indcrop.2007.07.013.

Lewandowski I.; Clifton-Brown J.C.; Andersson B.; Basch G.; Christian D.G.; Jorgensen U. Environment and harvest time affects the combustion qualities of *Miscanthus* genotypes. *Agron. J.* 95: 1274–1280; 2003.

Lewandowski I.; Schmidt U. Nitrogen, energy, and land use efficiencies of *Miscanthus*, reed canary grass and triticale as determined by the boundary line approach. *Agric. Ecosyst. Environ.* 112: 335–346; 2006 doi:10.1016/j.agee.2005.08.003.

Li H.W.; Loh C.S.; Lee C.L. Cytological studies on sugarcane and its relatives I. Hybrids between *Saccharum officinarum*, *Miscanthus japonicus* and *Saccharum spontaneum*. *Bot. Bull. Acad. Sinica* 2: 147–160; 1948.

Malik R.K.; Green T.H.; Brown G.F.; Mays D. Use of cover crops in short rotation hardwood plantations to control erosion. *Biomass Bioenerg.* 18: 479–487; 2000.

McKendrick J.D.; Owensby C.E.; Hyde R.M. Big bluestem and Indian-grass vegetative reproduction and annual reserve carbohydrate and nitrogen cycle. *Agro-Ecosystems* 2: 75–93; 1975 doi: 10.1016/0304-3746(75)90007-4.

McLaughlin S.B.; Kszos L.A. Development of switchgrass (*Panicum virgatum*) as a bioenergy feedstock in the United States. *Biomass Bioenerg.* 28: 515–535; 2005 doi:10.1016/j.biombioe.2004.05.006.

Mead D.J. Forests for energy and the role of planted trees. *Crit. Rev. Plant Sci.* 24: 407–421; 2005 doi: 10.1080/07352680500316391.

Melis A. Green alga hydrogen production: progress, challenges and prospects. *Int. J. Hydrogen Energy* 27: 1217–1228; 2002 doi: 10.1016/S0360-3199(02)00110-6.

Miguez F.E.; Villamil M.B.; Long S.P.; Bollero G.A. Meta-analysis of the effects of management factors on *Miscanthus x giganteus* growth and biomass production. *Agric. Forest. Meteorol.* 148: 1280–1292; 2008 doi:10.1016/j.agrformet.2008.03.010.

National Biodiesel Board (2005) Biodiesel production soars: 2005 production expected to triple last year's figures. Available via: http://www.biodiesel.org/resources/pressreleases/gen/20051108_productionvolumes05nr.pdf Cited 12 March 2009

National Biodiesel Board (2008) Commercial biodiesel production plants. Available via: http://www.biodiesel.org/buyingbiodiesel/producers_marketers/ProducersMap-Existing.pdf Cited 12 March 2009

National Biodiesel Board (2009) Production estimate graph. Available via: http://www.biodiesel.org/pdf_files/fuelfactsheets/Production_Graph_Slide.pdf Cited 12 March 2009

Neuhoff K. Large-scale deployment of renewables for electricity generation. *Oxford Rev. Econ. Policy* 21: 88–110; 2005 doi:10.1093/oxrep/gri005.

NRCS (Natural Resources Conservation Service) (2006) Crop residue removal for biomass energy production: effects on soils and recommendations. Soil Quality Technical Note No. 19. Available via: http://www.soils.usda.gov/sqi Cited 12 March 2009

Ogura S.; Kosako T.; Hayashi Y.; Dohi H. Effect of eating mastication on *in-vitro* ruminal degradability of *Zoysia japonica*, *Miscanthus sinensis* and *Dactylis glomerata*. *Grassland Sci.* 451: 92–94; 1999.

Openshaw K.A. Review of *Jatropha curcas*: an oil plant of unfulfilled promise. *Biomass Bioenerg.* 19: 1–15; 2000 doi: 10.1016/S0961-9534(00)00019-2.

Parcell J.L.; Westhoff P. Economic effects of biofuel production on states and rural communities. *J. Agric. Appl. Econ.* 38: 377–387; 2005.

Parrish D.J.; Fike J.H. The biology and agronomy of switchgrass for biofuels. *Crit. Rev. Plant Sci.* 24: 423–459; 2005 doi:10.1080/07352680500316433.

Parrish D.J.; Wolf D.D.; Fike J.H.; Daniels W.L. Switchgrass as a Biofuels Crop for the Upper Southeast: Variety Trials and Cultural Improvements. Final Report for 1997 to 2001. ORNL/SUB-03-19XSY163/01; 2003

Patzek T.W.; Pimentel D. Thermodynamics of energy production from biomass. *Crit. Rev. Plant Sci.* 24: 327–364; 2005 doi:10.1080/07352680500316029.

Perlack R.D.; Wright L.L.; Turhollow A.F.; Graham R.L.; Stokes B.J.; Erbach D.C. (2005) Biomass as feedstock for a bioenergy and bioproducts industry: the technical feasibility of a billion ton annual supply. Oak Ridge National Laboratory, Oak Ridge, TN. Available via: http://www1.eere.energy.gov/biomass/publications.html. Cited 12 March 2009

Raghu S.; Anderson R.C.; Daehler C.C.; Davis A.S.; Wiedenmann R.N.; Simberloff D. Adding biofuels to the invasive species fire. *Science* 313: 1742; 2006 doi:10.1126/science.1129313.

Ratledge C. Fatty acid biosynthesis in microorganisms being used for single cell oil production. *Biochimie* 4: 1–9; 2004.

Renewable Fuels Association. Renewable fuels standard. Available via: http://www.ethanolrfa.org/resource/standard/ Cited 12 March 2009; 2009a.

Renewable Fuels Association. 2009 ethanol industry outlook. Available via: http://www.ethanolrfa.org/objects/pdf/outlook/RFA_Outlook_2009.pdf Cited 12 March 2009; 2009b.

Roessler P.G.; Brown L.M.; Dunabay T.G.; Heacox D.A.; Jarvis E.E.; Schneider J.C. Genetic-engineering approaches for enhanced production of biodiesel fuel from microalgae. *ACS Symp. Ser.* 566: 255–270; 1994.

Rose L.W.; Das M.K.; Taliaferro C.M. Estimation of genetic variability and heritability for biofuel feedstock yield in several populations of switchgrass. *Ann. Appl. Biol.* 152: 11–17; 2008 doi:10.1111/j.1744-7348.2007.00186.x.

Roth A.M.; Sample D.W.; Ribic C.A.; Paine L.; Undersander D.J.; Bartelt G.A. Grassland bird response to harvesting switchgrass as a biomass energy crop. *Biomass Bioenerg.* 28: 490–498; 2005 doi: 10.1016/j.biombioe.2004.11.001.

Sartori F.; Lal R.; Ebinger M.H.; Parrish D.J. Potential soil carbon sequestration and CO_2 offset by dedicated energy crops in the USA. *Crit. Rev. Plant Sci.* 25: 441–472; 2006 doi:10.1080/07352680600961021.

Scaff R.; Reca A. Ethanol in the US: changing corn's market dynamics. *Int. Sugar J.* 107: 148–150; 2005.

Schulte-Bisping H.; Bredemeier M.; Beese F.O. Global availability of wood and energy supply from fuelwood and charcoal. *Ambio* 28: 592–594; 1999.

Schwart R.; Jackson R.; Herbst B.; Whitney R.; Lacewell R.; Mjelde J. (2005) Methane Generation. Final Report to the State Energy Conservation Office under CAFO Methane Digester Project Contract CM433 with the Texas Agricultural Experiment Station. p. 47.

Scott D.A.; Dean T.J. Energy trade-offs between intensive biomass utilization, site productivity loss, and ameliorative treatments in loblolly pine plantations. *Biomass Bioenerg.* 30: 1001–1010; 2006 doi:10.1016/j.biombioe.2005.12.014.

Ground flora, small mammal and bird species diversity in *Miscanthus* (*Miscanthus x giganteus*) and reed canary-grass (*Phalaris arundinacea*) fields. *Biomass Bioenerg.* 31: 20–29; 2007 doi: 10.1016/j.biombioe.2006.07.001.

Smith J.C.; DeMeo E.A.; Parson B.; Milligan M. (2004) Wind Power Impacts on Electric Power System Operating Costs: A Summary and Perspective on Work to Date. National Renewable Energy Lab. NREL/CP-500-35946. Available via: http://www.nrel.gov/docs/fy04osti/35946.pdf Cited 12 March 2009

Somleva M.N.; Tomaszewski Z.; Conger B.V. Agrobacterium mediated genetic transformation of switchgrass. *Crop Sci.* 42: 2080–2087; 2002.

Spolaore P.; Joannis-Cassan C.; Duran E.; Isambert A. Commercial applications of microalgae. *J. Biosci. Bioeng.* 101: 87–96; 2006 doi:10.1263/jbb.101.87.

St. John J.B., Boethel D., Breaux J. Notice of release of high fiber sugarcane variety Ho 00-961. Agricultural Center, Louisiana State University;2007.

Steenblik R. (2007) Biofuels at what cost: government support for ethanol and biodiesel in selected OECD countries, IISSD, Global Subsidies Initiative. Available via: http://www.iisd.org/pdf/2007/biofuels_oecd_synthesis_report.pdf Cited 12 March 2009

Taliaferro C.M. (2002) Breeding and selection of new switchgrass varieties for increased biomass production. Oak Ridge National Laboratory. Available via: http://www.ornl.gov/~Ewebworks/cppn/y2001/rpt/116285.pdf Cited 12 March 2009

Tashtoush G.M.; Al-Widyan M.I.; Al-Jarrah M.M. Experimental study on evaluation and optimization of conversion of waste animal fat into biodiesel. *Energ. Convers. Manag.* 45: 2697–2711; 2004 doi:10.1016/j.enconman.2003.12.009.

Tilman D.; Hill J.; Lehman C. Carbon-negative biofuels from low-input high-diversity grassland biomass. *Science* 314: 1598–1600; 2006a doi: 10.1126/science.1133306.

Tilman D.; Reich P.B.; Knops J.M.H. Biodiversity and ecosystem stability in a decade-long grassland experiment. *Nature* 441: 629–632; 2006b doi:10.1038/nature04742.

Tuck G.; Glendining M.J.; Smith P.; House J.I.; Wattenbach M. The potential distribution of bioenergy crops in Europe under present and future climate. *Biomass Bioenerg.* 30: 183–197; 2006 doi: 10.1016/j.biombioe.2005.11.019.

USDA (United States Department of Agriculture) (2004) U.S. forestry and agriculture greenhouse gas inventory: 1990–2001. Global Change Program Office, Office of the Chief Economist, U.S. Department of Agriculture. Technical Bulletin No. 1907. 163 p. March 2004; Available via: http://www.usda.gov/oce/global_change/gg_inventory.htm. Cited 12 March 2009

Vogel K.P.; Hopkins A.A.; Moore K.J.; Johnson K.D.; Carlson I.T. Registration of 'Shawnee' switchgrass. *Crop Sci.* 36: 1713; 1996.

Wang L.P.; Jackson P.A.; Lu X.; Fan Y.-H.; Foreman J.W.; Chen X.-K.; Deng H.-H.; Fu C.; Ma L.; Aitken K.S. Evaluation of sugarcane x *Saccharum spontaneum* progeny for biomass composition and yield components. *Crop Sci.* 48: 951–961; 2008 doi:10.2135/cropsci2007.10.0555.

Weeks J. Landfills expand energy output. *BioCycle* 48: 50–54; 2005.

West T.O.; Marland G. A. synthesis of carbon sequestration, carbon emissions, and net carbon flux in agriculture: comparing tillage practice in the United States. Agric. Ecosyst. Environ. 91: 261–287; 2002 doi:10.1016/S0167-8809(01)00233-X.

Chapter 7
Genetic Improvement of C4 Grasses as Cellulosic Biofuel Feedstocks

Katrin Jakob, Fasong Zhou, and Andrew H. Paterson

Abstract C4 grasses are among the most productive plants and most promising cellulosic biofuel feedstocks. Successful implementation of cellulosic biofuel feedstocks will depend on the improvement of critical crop characteristics and subsequent conversion technologies. The content and composition of lignin, cellulose, and hemicellulose, their biomass yields, and their biotic and abiotic stress tolerances are critical factors which can be enhanced by molecular breeding methods, including marker-assisted selection and transgenic approaches. To maximize biomass yield, no flowering or late flowering and no grain set would be ideal for cellulosic biofuel crops. Reducing fecundity also reduces the risk of undesired gene transfer and invasiveness, thus accelerating deregulation processes and permitting faster implementation of highly improved genotypes in cellulosic feedstock production.

Keywords Biomass yield • Lignin • Germplasm • *Miscanthus* • Switchgrass • *sorghum* • SSRs • SNPs • QTL • Marker-assisted selection • Transcription factor • Transgenic

Why Do We Need to Develop Cellulosic Biofuels?

The hunt for carbon neutral energy sources has become one of the primary challenges of the twenty-first century. The use of ethanol is a proven concept for replacing gasoline and reducing CO_2 emission from fossil oil. Currently, ethanol is produced from sugar- and starch-rich crops, which are grown by labor- and machine-intensive agriculture and require high-nitrogen fertilization. These practices in turn negatively impact the overall energy and CO_2 balance for this particular production chain. Cellulosic biofuels are a promising component in a future mix of

K. Jakob (✉)
Mendel Biotechnology, Inc., 3935 Point Eden Way, Hayward, CA 94545, USA
e-mail: kjakob@mendelbio.com

alternative renewable energy solutions. There are challenges which exist in the use of cellulosic biofuel crops; but with continuously developing plant breeding, crop development, and farming as well as conversion technologies, cellulosic biofuel crops will emerge as strong contenders in the race for sustainable energy sources. Among the many choices for cellulosic biofuels are trees and short rotation coppice, agricultural waste material, and by-products from crops and biomass grasses. This review will focus on the genetic improvement of selected C4 grasses including *Sorghum bicolor*, *Miscanthus* species, hybrids between these species, and hybrids between *Miscanthus* × sugarcane (Miscane), and *Panicum virgatum* (switchgrass) for the development of dedicated cellulosic biofuel crops.

Characteristics of Sustainable Cellulosic Biofuel Crops

The transition from the Mesozoic to Neolithic period (around 10,000 BCE) is defined by the development of nascent agricultural practices and the selection and genetic manipulation of cereal grasses to meet the qualitative and quantitative requirements of food and feedstock. In this respect, the development of cellulosic biofuel feedstock will be similar. Tailoring crops for biofuel production is critical in improving the overall economy of ethanol production from cellulosic feedstock (Wyman 2007). Lignin content and composition of cell walls determine, among other factors, sugar release and thus the efficiency of the ethanol extraction process (Davison et al. 2006; Chen and Dixon 2007; Vermerris et al. 2007; Yoshida et al. 2008). Hence, biomass composition, in particular cell wall composition, plays an important role and is considered a major trait for improvement of cellulosic biofuels (Sticklen 2006; McCann and Carpita 2008). Lignin, cellulose, and hemicellulose are main components of cell walls and their biosynthesis is facilitated by complex pathways. While the lignin content and biosynthesis pathway have been extensively studied for digestibility purposes for animal feedstock like corn, other grasses, and alfalfa (Jung and Vogel 1986; Baucher et al. 1999; Barrière et al. 2004; Ralph et al. 2004; Reddy et al. 2005) and for improving pulping qualities in wood (Hu et al. 1999; Pilate et al. 2002; Baucher et al. 2003; Li et al. 2003), we are only beginning to unravel the details of cellulose and hemicellulose biosynthesis (Paredez et al. 2006; DeBolt et al. 2007; Desprez et al. 2007; Lindeboom et al. 2008; Paredez et al. 2008; Wang et al. 2008) and the regulation of these pathways (Mitsuda et al. 2005; Zhong et al. 2006; Mitsuda et al. 2007; Zhong et al. 2007a, b; Zhong and Ye 2007; Xu et al. 2008; Zhong et al. 2008). Reliable and accurate measurements for cell wall components are labor intensive and standards are not consistently established (Kelley et al. 2004; Labbé et al. 2008) which complicate the exact definition of the most favorable ratio of lignin, cellulose, and hemicellulose in the cell wall for ethanol production. Given the complexity, cell type specificity, and lack of efficient screening and precise analysis methods, engineering of cell wall composition will undoubtedly be the most challenging task of cellulosic feedstock improvement.

Biomass yield is critical for the successful implementation of cellulosic biofuels. C4 plants have the potential to produce exceptionally high grain yield as well as stem and leaf biomass yield (Piedade et al. 1991) and, compared with C3 plants, appear to be more promising as biomass-producing plants for cellulosic biofuels.

C4 plants convert energy more efficiently into biomass than C3 plants and have up to 60% higher water and nutrient use efficiency (Beadle and Long 1985; Beale et al. 1999; Heaton et al. 2008b). By selecting for high-biomass-yielding species, combined with high nutrient and water use efficiency, economically efficient production of biofuel feedstock may be realized on less optimal land without pressuring prime grain crop territories. To thrive in these proposed suboptimal environments, biofuel crops need to be adapted to various stresses such as drought, salt, cold, and low nutrient availability. Criticism of the development of biofuels for ethanol production is often based on a fear of continued displacement of food crops with that of biofuel crops. The expansion of agriculture to provide plant biomass for production of fuels or chemical feedstocks will require greater use of marginal lands. This will make the production of low per-unit value biomass economical. Perennial crops are essential to bringing marginal lands into sustainable biomass production (Wagoner 1990; Scheinost et al. 2001; Cox et al. 2002), maximizing ecosystem productivity (Field 2001) and minimizing losses of topsoil (Gantzer et al. 1990; Pimentel et al. 1995), water, and nutrients (Randall and Mulla 2001).

Flowering time significantly affects biomass production as the transition from vegetative to reproductive phase negatively impacts biomass accumulation. For the maximum biomass yield of cellulosic biofuel crops, flowering is undesired and, in the case of perennials, plants should senesce only shortly before the end of the vegetation period to relocate nutrients into the rhizome. Manipulating flowering time has long been a breeding goal to enable or prevent flowering at desired times and to increase yield in photoperiod-sensitive crops independent of geographical region (e.g., sorghum) and for other applications. In one biomass crop, sugarcane, flowering alone reduces yields enough to wipe out profit margins in some years (Julien and Soopramanien 1976; Long 1976; Julien et al. 1978; Heinz 1987) and increases disease susceptibility (Ricaud et al. 1980).

Biomass yield is a complex trait and influenced by many plant characteristics. Plant height, tiller number per plant, tiller density, and stem thickness are important biomass yield determinants with complex genetics underlying these traits. Tiller number and density will eventually define the number of plants needed per unit area for optimal biomass production and hence influence the establishment costs for plantations. Depending on the species to be developed, propagation capability (seed, *in vitro* culture, or rhizome), rapid field establishment, and pest and disease resistance are also important traits in cellulosic biofuel crops.

Invasiveness remains a concern for nonnative biofuel crops as well as for native biofuel crops grown outside their normal ecosystems (Rhagu et al. 2006). Many plant species such as corn, rice, wheat, soybean, barley, and sorghum are spectacularly successful as crops grown outside of their natural ranges but show varying degrees of invasiveness ranging from none to high. Invasive characteristics like nodal shoot and aggressive rhizome growth or seed dispersal need to be minimized or eliminated in biofuel crops through the breeding process.

Miscanthus x *giganteus*, a sterile triploid hybrid of a cross between *Miscanthus sinensis* and *Miscanthus sacchariflorus* (Hodkinson et al. 2002b), has been successfully grown for several years as a biofuel crop in Europe (Clifton-Brown et al. 2001). This particular genotype already combines many of the traits desirable for a biofuel crop. As a perennial C4 plant, it produces consistently high biomass yields over many years with little or no nitrogen application. Rhizome growth is locally contained and sterility provides protection from outcrossing. However, the yield potential might not be fully used when this variety is cultivated under varying climatic conditions. Furthermore, growing only one single clone holds risk with regard to disease susceptibility. Additionally, nutrient content as well as cell wall composition need to be better amenable to conversion. Early flowering at lower latitudes like the Southern USA currently prevents full realization of its yield potential. Breeding *Miscanthus* genotypes adapted to a wide range of growing conditions with lower lignin and ash content, resulting in more efficient biomass conversion, will further improve its suitability as a biofuel crop. Genetic diversity for *Miscanthus* breeding can be drawn from a large gene pool of different *Miscanthus* species like *M. sinensis*, *M. sacchariflorus*, *M. floridulus*, *Triarrhena lutarioriparia*, and others. Switchgrass (*P. virgatum*) has been utilized as a forage grass and has been tested as a biofuel crop in recent years. However, its present yield is lower than *Miscanthus* (Heaton et al. 2008a) but improved varieties could be a useful addition to the biofuel crop repertoire for sites less suitable for *Miscanthus*. *S. bicolor* is known for its adaptation to several stress conditions, in particular drought, and could complement perennial biofuel crops in establishment years or in arid climates. Breeding perennial sorghum would be of interest as well as interspecific crosses between sorghum and *Miscanthus*. Similarly, Miscane, a hybrid between sugarcane and *Miscanthus*, could potentially combine the high productivity of both species with the perenniality and adaptation of *Miscanthus* to colder climates. Of the main breeding goals, improving cell wall composition and its digestibility is presently deemed most critical for enhancing sugar extraction efficiency and overall economy of cellulosic ethanol production.

Genetic Manipulation to Improve Biofuel Crops

Breeding cellulosic feedstock can initially take advantage of the knowledge and technologies developed for food and forage crops in their long breeding history. This ranges from germplasm collection and characterization to breeding strategy development including marker-assisted selection (MAS) and transgenic modifications. The starting point of a biofuel crop breeding program can vary significantly depending on the extent of genetic improvement. For sorghum and sugarcane, well-established breeding programs have existed for many years. While the breeding goal has not been for biomass production (e.g., high sugar content versus high biomass for sugarcane), attention to issues such as harvestability and disease resistance may indirectly benefit biofuel breeding. Others, like *Miscanthus* and switchgrass, are essentially "wild" undomesticated plants. The genetic improvement of *Miscanthus* and switchgrass has to be initiated with the collection of germplasm but at the same time holds much potential for realizing the genetic gain in the first few generations of variety improvement.

Germplasm Collection

Germplasm collections provide breeders with genetic resources for trait improvement in energy crop breeding programs. By cross-pollination and subsequent selection in segregating progenies, plant breeders can reassemble targeted traits in one variety for its optimized performance. Sufficient germplasm variations and genetic information relating to interesting traits are essential for the success of a breeding program.

Looking closely at the available germplasm collections in biofuel crops, such as *S. bicolor*, *Miscanthus*, switchgrass, and Miscane, different pictures emerge for each crop. A large sorghum germplasm collection of >36,000 accessions exists in the USA (Reddy et al. 2006) and at the International Crops Research Institute for the Semiarid Tropics (Saballos 2008). In contrast, there is practically no public germplasm collection for *Miscanthus*, despite a few collection activities by research institutes, botanical gardens, and private companies who have compiled a small number of landraces and ornamental accessions. Nevertheless, there is rich genetic variation in wild populations of both switchgrass and *Miscanthus*. Switchgrass is native to North America and has been bred as a forage crop. *Miscanthus*, on the other hand, originated in East Asia and nearby Pacific islands. Eleven to 20 species have been identified in *Miscanthus* and ongoing molecular analysis is still being applied to refine our understanding of the phylogenetic relationships between *Miscanthus* species (Hodkinson et al. 2002; Clifton-Brown et al. 2008). The center of diversity for *Miscanthus* is in East China, South Korea, Taiwan, and Japan. In China, limited breeding work has been done in *Miscanthus* to improve its fiber quality and yield as a feedstock for the paper industry (He et al. 1998). Several species of *Saccharum* can be explored for Miscane/energy cane breeding. Sugarcane's genetic disposition includes several species of the *Saccharum* complex (Dillon et al. 2007) and a large number of collections of *Saccharum* have been maintained in Brazil, India, the USA, and China.

Access to genetic resources is vital to breeders. Because of restricted biomaterial exchange between countries, it is a huge challenge for breeders to access genetic resources if they want to establish a breeding program in a species outside of its natural geographic locations. For example, several obstacles have to be overcome for a breeder in the US to access *Miscanthus* germplasm. While collections of several *Miscanthus* accessions from different locations in China, Japan, Taiwan, and Korea exist, export and import restrictions and ambiguous regulations under the convention of biological diversity protocol (CBD) complicate the acquisition and transfer of *Miscanthus* germplasm outside these countries for commercial product development. The limited capacity for *Miscanthus* quarantine further delays the germplasm import into the USA. It can take several years before a significant number of *Miscanthus* accessions are available for breeders outside Asia. Similar restrictions have applied to sorghum and sugarcane germplasm exchange since the CBD protocol became effective in 1993.

Germplasm Characterization

Genetic and phenotypic analyses need to be conducted once the germplasm is available to the breeder. Core germplasm accessions should be selected as parents to be used in crossbreeding programs. Of the potential C4 biofuel species, most (except sorghum) are either polyploid or have large complex genomes, complicating genetic analysis. *Miscanthus*, switchgrass, sugarcane, energy cane, big bluestem, Bermuda grass, napier grass, and others have varying ploidy levels from diploid up to octoploid, and their chromosome numbers can be up to 100 or more. *S. bicolor* provides a much needed model for biofuel crop genomics and breeding research. *S. bicolor* is diploid and has a relatively small genome of about 730 Mb that has been completely sequenced (Paterson et al. 2009). The sorghum sequence will be a valuable tool for comparative genetic studies. With the sorghum sequence as a reference, syntenic regions or candidate gene sequences can be exploited to identify the locations of genes that govern critical trait inheritance. Subsequently, these syntenic regions or candidate gene sequences can be used to develop markers linked to the traits of interest for marker-assisted selection or cloning of the underlying genes.

Germplasm characterization at the molecular level can shed light onto the genetic diversity and trait inheritance information that are needed by breeders in planning their crosses and subsequent selections. Keeping a degree of genetic diversity in the crossing parents is a common measure to enhance the adaptability of a variety to a broad range of environments. Molecular markers, such as restriction fragment length polymorphism (RFLP), amplified fragment length polymorphism (AFLP), simple sequence repeat (SSR), Diversity Arrays Technology (DArT), and single-nucleotide polymorphism (SNP), are frequently used in germplasm characterizations. Each type of marker has somewhat different properties in terms of the nature of polymorphism assayed, the assay methods, equipment requirements, costs, and assay throughput. RFLP, AFLP, and DArT markers do not require sequence information; they can be used for initial genetic characterization of germplasm collections that have little or no genomic sequences available. DArT markers have been applied in sorghum (Mace et al. 2008), rice (Jaccaud et al. 2001), barley (Wenzl et al. 2004), wheat (Akbari et al. 2006), and others (James et al. 2008). The method has demonstrated good reproducibility and has been applied to higher ploidy levels and consequently large genome size [as demonstrated in hexaploid wheat (Akbari et al. 2006)] and could ultimately be applied for polymorphism analysis in *Miscanthus* and other biofuel candidate crops with higher ploidy levels.

SSR and SNP markers rely on the availability of sequence information for the design of primers or allele-specific oligos. Currently, neither the *Miscanthus* nor the switchgrass genome has been sequenced. Nevertheless, SSRs have become the marker of choice for genetic diversity study and genotype identifications in biofuel crops for the following reasons: (1) large numbers of SSRs have been identified in related species, such as sorghum, sugarcane, corn, wheat, barley, and rice (http://www.gramene.org/), and can be used in *Miscanthus* and switchgrass (Hernández et al. 2001). Sugarcane and sorghum are more closely related to *Miscanthus* than corn (Hodkinson et al. 2002) and as such a much higher success rate is expected for sorghum and sugarcane SSRs being validated in *Miscanthus*. (2) SSR polymorphism is derived from numbers of repeats in the flanking sequence and has a high power to differentiate many alleles. (3) SSRs are polymerase-chain-reaction-based markers and thus relatively easy to use. Multiplexing and automation of SSR assays can be performed using capillary-based DNA analyzers in combination with various color florescence-labeled primers.

SNPs are powerful for genetic mapping due to their abundance in the genome. Mining of expressed sequence tag libraries is commonly used for SNP discovery (Grivet et al. 2003; Cordeiro et al. 2006; Novaes et al. 2008) and a database for SNP search is available for rice, barley, and brassica (Duran et al. 2009). Conserved intron scanning primers have also been employed for SNP discovery by using conserved locations of introns to design primers followed by identifying sequence variation within the intron (Feltus et al. 2006b). The absence of sufficient genome sequences used to be a bottleneck for SNP discovery application to biofuel crops. However, the new generation of high-throughput DNA sequencing technology such as Illumina (Pavy et al. 2008), 454 (Barbazuk et al. 2007), and SOLiD (Smith et al. 2008) has made SNP discovery rapid, more affordable, and widely available.

Identification of Trait-Linked Markers

Identification of trait-linked molecular markers by genetic mapping is a first step in implementing marker-assisted selection in breeding improved biomass crops. Draft linkage maps have been published for *M. sinensis* (Atienza et al. 2002) and switchgrass (Missaoui et al. 2005). However, higher densities of molecular markers are needed. Biomass yield and chemical composition are two major goals in biofuel crop breeding. Biotic and abiotic stress tolerance are also important for broad adaptability and yield stability.

Biomass yield of energy crops positively correlates with plant height, stem density, and stem thickness as demonstrated in *Miscanthus* and switchgrass (Atienza et al. 2003; Das et al. 2004; Boe and Beck 2008; Jezowski 2008). In different *Miscanthus* species, plant height can vary from 1.5 to over 5.0 m. Thus, there is large potential for the genetic improvement of plant height in *Miscanthus*. *T. lutarioriparia* is generally tall with thick stems, but stem density is usually low. In contrast, *M. sinensis* is usually short and with stems clumped together at very high density. The stem thickness can vary within a large range, with *M. floridulus* and *M. sacchariflorus* being intermediate types for these traits. In addition to the enormous variations between *Miscanthus* species, the difference between ecotypes of the same species can also be large. Study of the genetic control of the three yield components and subsequently optimizing their value in genetically improved varieties would help breeders achieve their breeding goals for enhanced biomass yield.

Many traits that are high priority for genetic improvement in *Miscanthus* and other biofuel crops are "domestication traits" for which there exists substantial knowledge of genetic control in sorghum and/or other crops and for which the locations of controlling genes/quantitative trait loci (QTLs) often correspond across divergent grasses (Lin et al. 1995; Paterson et al. 1995a, b; Ming et al. 2002; Hu et al. 2003). For example, QTLs controlling tiller number and culm height are known in sorghum (Quinby and Karper 1954; Lin et al. 1995; Hart et al. 2001) and the synteny relationships of plant height QTLs between sorghum and sugarcane have been studied (Ming et al. 2002). Stem diameter itself is a biomass component and also enhances lodging tolerance. A stem diameter QTL which accounted for thicker stems and reduced lodging was identified in rice (Kashiwagi et al. 2008). However, biomass composition and lignin content have to be kept favorable for ethanol extraction with the modification of these traits. Flowering time and photoperiod sensitivity play a critical role in biomass yield through determination of growth period. Delayed or no flowering is desirable for biomass production because an extended vegetative growth period helps produce more biomass and sterility would either reduce or eliminate the invasiveness potential of *Miscanthus* and other biofuel grasses. Several genes and QTLs involved in photoperiod and flowering control have been recognized in sorghum and sugarcane (Quinby and Karper 1945; Rooney and Aydin 1999; Ming et al. 2002), corn, and rice. Similar to plant height QTLs, QTLs controlling flowering often correspond in sugarcane and sorghum and may be conserved in other Poaceae (Lin et al. 1995; Ming et al. 2002).

The optimal chemical composition of energy crops is determined by the subsequent conversion technologies. Two major conversion technologies are being tested to produce liquid biofuel from plant biomass. One is thermal conversion and the other is bioconversion. These approaches have different requirements for biomass compositions in order to obtain high conversion rates, but both require low ash content. Nutrient recycling of elements such as N, P, and K from the shoots back to roots before harvesting would aid in reducing the ash content. This in turn requires less nutrient uptake by the plant as the nutrients are available from the rhizomes in the following spring. Therefore, reducing ash content not only increases biomass to fuel conversion rate but also reduces overall nutrient uptake. The natural senescence of the plant is essential for nutrient recycling and, in an approach to align maps of two sorghum populations with one common parent, QTLs for leaf senescence could be found (Feltus et al. 2006a). Cell wall composition is cell specific and highly variable depending on the cell type (Nakashima et al. 2008).

Parenchyma and collenchyma cells have only primary cell walls whereas sclerenchyma cells have both primary and secondary cell walls. Lignin content is usually higher in secondary cell walls compared to primary cell walls. Studies of altered gene expression in the monolignol biosynthesis pathway in *Arabidopsis* and other dicots as well as in trees revealed changes in both lignin content and composition (Vanholme et al. 2008). Lignin content was negatively correlated with sugar yield in *Miscanthus* (Yoshida et al. 2008). In fact, lignin content could be reduced to a level where untreated transgenic alfalfa yielded higher saccharification efficiency than pretreated control plants (Chen and Dixon 2007). Less information is available on lignin manipulation for grasses; however, the brown midrib, originally a naturally occurring mutant identified in corn, sorghum, and millet, is perhaps the most widely studied trait. While reduced lignin improves saccharification efficiency, it can also reduce biomass yield and survival and also cause lodging (Casler et al. 2002; Vogel et al. 2002; Pedersen et al. 2005; Colemann et al. 2008). However, this seems to be dependent on the genetic background. There are also examples of reduced lignin without obvious phenotypic changes. QTLs for lignin and cell wall components in corn have been successfully identified (Cardinal et al. 2003; Krakowsky et al. 2006; Barrière et al. 2008).

Invasiveness is a concern for novel biofuel crop and can often be contributed to aggressively spreading rhizomes, seed dispersal, or nodal growth. Rhizome growth is generally desired to fill in space between plants but should not be so aggressive that it takes over beyond the field location. Seed set in biofuel crops is only desired in case of seed production but has not yet been excluded in grasses grown in their native habitats. QTLs for rhizomatousness in rice (Hu et al. 2003) and studies of rhizome growth behavior in *Sorghum propinquum* and *S. bicolor* (Paterson et al. 1995; Jang et al. 2006, 2008) have begun to uncover the genes involved, which are the basis for selection and effective manipulation. Mutations in two genes for seed shattering in rice and a single QTL in sorghum have also been identified (Paterson et al. 1995b; Konishi et al. 2006; Li et al. 2006). The nonshattering gene or QTL-linked markers should be applicable for molecular marker screens in *Miscanthus* or other germplasm. Seed size, seedling vigor, and early cold germination are important traits for fast and successful crop establishment but could also affect invasiveness. These traits need to be balanced and managed in combination with other measures for bioconfinement.

Given the wealth of genetic information for biomass, chemical compositions, and stress tolerance obtained from the study in model plant species such as *Arabidopsis*, rice, sorghum, and corn, a candidate gene approach to identify genes or synteny regions of interest for equivalent biofuel traits could be effective. This approach relies on the conservation of trait-related synteny/collinearity and gene function between model plants and the biofuel crop of interest. Candidate gene approaches have been successfully demonstrated in plant disease resistance gene mapping and the subsequent cloning (McIntyre et al. 2005). As shown above, many QTLs important for biofuel traits have been identified in sorghum, sugarcane, and rice. The use of conserved markers in genetic mapping should identify related genes or their locations in *Miscanthus* and other biofuel crops.

Traditionally, trait mapping was performed in segregating populations that were derived from biparental crosses. The mapped trait may be restricted by the choice of available genetic resources. Due to the development of high-density SNP markers with a genome-wide coverage, trait-associating mapping becomes feasible using a broad panel of selected genetic resources, such as germplasm accessions, breeding lines, and commercial varieties. The association mapping approach eliminates the necessity of mapping population construction and allows breeders to focus on their most valuable genetic resources. Furthermore, recently high-throughput deep sequencing may yield more closely linked markers or directly identify the gene for the targeted traits without going through tedious genetic mapping procedures (Lister et al. 2009).

Marker-Assisted Selection

Once the genes controlling traits of interest or phenotype-associated molecular markers have been identified, direct selection for desired genotypes becomes possible through molecular marker analysis. MAS is especially valuable for the improvement of biomass yield, which is complex partially because gene–environment interactions reduce the accuracy of direct phenotype selection. Likewise, diagnostic markers for cell wall composition would help to mitigate the lack of efficient assays for breeders to do rapid assessment (ideally in the field). The implementation of MAS in biofuel crop breeding programs should expedite development of novel cultivars with high biomass and desired chemical compositions.

MAS can significantly improve breeding efficiency by the accurate selection of desired genotypes in early generations or at early growth stages. MAS can also reduce the impact of nongenetic factors that can interfere with phenotype selection and help to quantify the stability of particular genes (or QTLs) across environments. Distinguishing genes that are stable from those that are environment specific may help to tailor genotypes to particular production systems (Saranga et al. 2001; Paterson et al. 2003; Saranga et al. 2004).

Genotype selection by MAS in early generations or at early growth stages is particularly beneficial in biofuel crop breeding programs because most potential biomass crops are perennials and their phenotype expression can be delayed for multiple years. As an extreme example, application of a marker-assisted backcrossing selection in perennial oil palm would theoretically reduce the breeding cycle from 19 to 13 years compared to a conventional approach (Wong and Bernardo 2008). Furthermore, biofuel crops tend to be large plants and each plant requires space in greenhouses and breeding nurseries. Plant breeders need to screen thousands of breeding lines or individuals in each generation. A *Miscanthus* plantation reaches its full yield potential in 2 to 3 years after planting. Any selection that can be practiced by applying diagnostic DNA markers in the first year or even at the seedling stage would significantly save space and time in the breeding process. MAS can be performed in different stages of a breeding program. Marker-assisted germplasm fingerprinting can be used for parent selections. In fact, parental selection based on molecular marker screening has been deemed critical in starting a successful MAS wheat breeding program (Anderson et al. 2007).

MAS has been implemented for both qualitative and quantitative traits in crop breeding programs (Castro et al. 2003; Concibido et al. 2004; Joseph et al. 2004; Schmierer et al. 2005; Zhou et al. 2005; Collard and Mackill 2008; Buerstmayr et al. 2009; Zhou et al. 2009). Major gene-controlled traits can be directly identified by marker analysis and applied for targeted improvement. For example, *Xa21*-mediated disease resistance (Chen et al. 2000) and *Bph1* and *Bph2*-mediated insect resistance (Sharma et al. 2004) have been identified in elite germplasm collections and introduced into rice cultivars through marker-assisted backcrossing. In the latter case, an additional advantage of MAS has been employed, pyramiding of several traits to maximize phenotype expression.

Marker-assisted gene pyramiding for the development of sustainable disease resistance will become critical in biofuel crop breeding. Large acreage plantings of a vegetative propagated biofuel crop could lead to monoculture of a few selected genotypes, with potentially high vulnerability to disease epidemics. Pyramiding different disease resistance genes against various pathogen strains in elite breeding lines would help reduce genetic vulnerability in released cultivars (Castro et al. 2003). Huang et al. (1997) reported pyramiding of four bacterial blight resistance genes in rice, achieving broad-spectrum disease resistance.

Beyond single-gene traits such as disease resistance, more complex traits such as aroma and root growth have also employed MAS for variety improvement (Steele et al. 2006). Rates of gain in complex (quantitative) trait improvement through MAS depend on the number of QTLs involved and their genetic effects. A range of results have been reported for the improvement of quantitative traits, such as grain yield, by MAS. Bouchez et al. (2002) reported on an extensive MAS program for the improvement of corn elite lines based on QTLs for earliness and grain yield. Testing of the improved varieties showed the expected earliness, but grain yield was negatively impacted. A MAS approach for corn yield under drought stress produced lines that generated higher corn yields under heavy drought stress but these lines lost their superiority over the control lines when the drought stress was less severe (Ribaut and Ragot 2007) and conventional breeding produced the same results (Ribaut and Ragot 2007). Phenotypic screening in both programs was executed only at the very end of the breeding process. As a result, the authors recommend including phenotypic screens during the MAS process. On the other hand, breeding for the quantitatively inherited *Fusarium* head blight resistance could be successfully accomplished by MAS (Anderson et al. 2007). The major QTL (*Fhb1*) produced a consistent phenotype over different genotypes and environments and closely linked markers were available (Anderson et al. 2007; Buerstmayr et al. 2009).

One of the most successful MAS breeding programs to date has been implemented by Monsanto (Eathington et al. 2007; Edgerton 2009). Researchers compared 248 unique soybean populations in a conventional selection with MAS. MAS-derived lines in the Monsanto program were higher performing for a combination of traits including grain yield, but the amount of gains for both methods varied between years. Similar results were reported for sunflower for the improvement of grain yield, oil content, and other traits and in corn for grain yield and other traits (Eathington et al. 2007). A high genetic gain was observed in early populations. The genetic gain of each selection cycle is very much dependent on the frequency increase of favorable genes. Therefore, marker-assisted genotype evaluation and subsequent selection can apparently help to accumulate desired genotypes in the selected populations.

Although the application of MAS for quantitatively inherited traits is presently more challenging, further development of genetic maps and marker technology will accelerate implementation. Biomass yield, an important target trait in biofuels, is a complex trait and improvement by MAS will pose challenges. Parental selection in a large number of populations seems crucial for selecting the best potential parents as well as the integration of phenotypic selections steps between the marker selection stages (Anderson et al. 2007; Eathington et al. 2007). MAS can be readily applied to biofuels crops once candidate genes have been identified for biomass yield and cell wall composition. MAS will play a significant role in future enhancement of biofuel crops as well as food and feed crops (Edgerton 2009). An even more intriguing approach is the combination of MAS with the transgene technology.

Transgenic Cellulosic Biofuel Crops

Genetically modified crops have successfully been grown for over 10 years since their debut in 1996. The majority of present transgenic crops comprise just two traits, herbicide resistance and insect resistance, but these have been widely applied to soybean, corn, cotton, and canola. The land cultivated with these crops covered a remarkable 125 million hectares in 2008 (ISAAA 2008: Executive Summary). Since their introduction, transgenic technology and crops have been under scrutiny relative to human health and the environment although transgenic crops which are commercialized are considered generally recognized as safe and approved by various government agencies throughout the world for use in food, feed, and fuel. Gene flow and escape into wild relatives and populations, unintended admixture with nontransgenic crops, and development of herbicide-resistant weeds and plants are concerns that are manageable based on current accepted practices (Council for Agricultural Science and Technology 2007).

The development of cellulosic biofuels feedstock will undoubtedly benefit from transgenic approaches. *Miscanthus*, switchgrass, and other novel crops are obligate outcrossers; hence, gene flow is a potential problem. However, cellulosic nonfood crops have a key advantage over transgenic food crops. Flowering and seed development are undesired. In fact, a cellulosic biofuel crop produces biomass more efficiently if resources are not directed into flowering. The risk of transgene escape may be dramatically reduced if nonflowering biofuel crops can be produced in the field. The sorghum photoperiod-sensitive gene, e.g. *Ma5* and *Ma6*, involved in flowering of sorghum have been shown to delay flowering significantly when dominantly expressed (Rooney and Aydin 1999). Tobacco plants produced higher biomass although at reduced plant height when an *Arabidopsis* flower repressor (FLC) was expressed causing delayed flowering (Salehi et al. 2005). Additionally, *CENTRORADIALIS* homologs were shown to delay flowering and vastly increased the height of tobacco plants (Amaya et al. 1999). Such findings ultimately provide opportunities to manipulate flowering time in grasses by a transgenic approach. On the other hand, flowering of biofuel crops is desired for recombination in breeding nurseries and crossing fields. Given the restricted areas for breeding purpose, gene escape should be easily manageable by growing crops afar from potential undesired crossing partners. Switchgrass is likely to be grown in its native habitat and a major concern is that transgenic switchgrass could easily hybridize with local population. *S. bicolor*, although self-fertile, can also hybridize with species of the *Saccharum* complex. Ideally, only sterile genotypes would be grown in the field as currently practiced for the sterile *M.* x *giganteus*. Sterile *Miscanthus* hybrid production is one of the breeding goals which entails the remake of the cross of high-yielding *M. sinensis* with *M. sacchariflorus* to produce a hybrid that outyields *M.* x *giganteus*. However, due to high propagation costs, the development of high-yielding locally adapted *M. sinensis* and *M. sacchariflorus* which could be sown and produce seeds is an opportunity to reduce establishment costs. Several options for genetic manipulation are available to prevent gene escape (Daniell 2002). Transgenic approaches like male sterility which have been deployed to produce hybrids (Hartley 1988, 1989; Mariani et al. 1991; Li et al. 2007; Gils et al. 2008) or transgene expression in only maternally inherited chloroplasts can reduce risks associated with cross-pollination (Ruf et al. 2001; Wurbs et al. 2007) and are examples for strategies to prevent outcrossing.

Genetically modified food crops are subject to a rigorous screening, testing, and deregulation process. Since biofuel crops are not intended for human or animal consumption, regulatory processes for biofuel crops could become less elaborative and cheaper and potentially accelerate the development and release of transgenic biofuel varieties and reduce costs in the development of novel biofuel crops.

The transfer of foreign DNA into plants has commonly been facilitated by *Agrobacterium*-mediated transformation or particle bombardment for major crops. Of the novel biofuel plants, switchgrass has been transformed with *Agrobacterium* (Somleva et al. 2002) and *M. sacchariflorus* was reportedly transformed via particle bombardment (Zili et al. 2004). High transformation efficiency protocols for *Agrobacterium*-mediated transformation need to be developed for *Miscanthus* to enable the effective employment of promising genes for crop improvement in further plant generations.

As described above, lignin composition as well as lignin, cellulose, and hemicellulose content most critically determines the efficiency of lignin degradation and saccharification and ultimately affects ethanol yield. Hence, these components are considered main targets of genetic improvement of biofuel feedstock. Transcriptional profiling has only just begun to unravel the complexity of genes regulating the lignin biosynthesis and the effect on genes beyond this pathway. Different levels of lignin and gene expression in various tissues, as well as changes in pathways beyond the lignin pathway, are beginning to illustrate the complexity of lignin biosynthesis. Interestingly, but not surprisingly, an array of transcription factors (TFs) are surfacing as key regulators of secondary cell wall biosynthesis in *Arabidopsis* and trees (Patzlaff et al. 2003; Goicoechea et al. 2005; Zhong et al. 2006; Zhong and Ye 2007; Zhong et al. 2008). TFs of the NAC family appear to be cell-type-specific which activate a cascade of downstream MYB and KNAT TFs which then activate genes of the biosynthetic pathways for cellulose and hemicellulose (Kubo et al. 2005; Mitsuda et al. 2007; Zhong et al. 2008). MYB and LIM TFs have also been suggested to be specifically involved in the lignin pathway (Kawaoka et al. 2000; Patzlaff et al. 2003a, b; Karpinska et al. 2004; Goicoechea et al. 2005; Yang et al. 2007). More studies should identify specific TFs regulating either the lignin or cellulose biosynthesis pathway. Such knowledge can be applied to manipulate particular TFs to optimize secondary cell wall composition. Increased cellulose and hemicellulose content in combination with reduced lignin for enhanced sugar and ethanol yield would produce a particularly valuable cellulosic feedstock. Cell- and tissue-specific gene expression analysis and their interaction in networks will be a trademark of further identification of genes and TFs involved in cell wall biosynthesis.

In addition to lignin and cellulose modification, direct expression of cellulases in plants has been discussed as an alternative way to improve decomposition of plant cell walls and reduce recalcitrance to processing. The complexity of cellulose biosynthesis is expected to require a complex, concerted, and tissue-specific use of enzyme action for cellulose degradation (Taylor et al. 2008). Heterologous expression of glycosyl hydrolases has been used to enhance glucose extraction in rice (Oraby et al. 2007) and corn (Ransom et al. 2007) and could be a strategy for improving overall sugar extraction. The correct expression of these enzymes would be critical to initiate the cellulose degradation shortly before, at, or immediately after harvest. Suitable promoters or chemical induction methods would be needed to accomplish this step. Another interesting aspect of improving the economic prospects of ethanol production could be to extract higher-value products from the lignin biosynthetic pathway by manipulating this pathway for different by-products and at the same time accomplish to simplify lignin structure for improved degradation (Anderson et al. 2005; Somleva et al. 2008).

Overexpression of transcription factors has been widely demonstrated to enhance resistance to biotic and abiotic stress in model plants as well as in crop plants (Century et al. 2008). Cellulosic biofuel feedstocks will likely be grown on less fertile soil while more fertile soils remain cultivated with food crops. Hence, tolerance to stresses such as drought, salt, cold, and heat will be preferred traits for biofuels. TFs are particularly promising transgenic tools as they regulate not only one gene but complex and often interacting pathways directed toward numerous stress responses. Therefore, changes in gene expression caused by overexpression of a particular TF often provide increased resistance to more than one abiotic stress factor. For example, the NAC family, which also comprises TFs involved in cell wall biosynthesis, has provided cold and salt tolerance when overexpressed in rice and tested under field conditions (Hu et al. 2006). CBF transcription factors of the

AP2/ERF family have been demonstrated to provide cold tolerance in *Arabidopsis* (Jaglo-Ottosen et al. 1998; Novillo et al. 2007), as well as orthologs in rice (Ito et al. 2006), corn (Qin et al. 2004), and birch (Welling and Palva 2008) or when AtCBF1 was overexpressed in poplar (Benedict et al. 2006). Moreover, drought and salt tolerance has also been observed in combination with cold tolerance when CBF rice orthologs were constitutively overexpressed in rice (Dubouzet et al. 2003).

Similar to abiotic stress resistance, TFs can also regulate defense responses to pests and diseases. TFs of the ERF subfamily have been demonstrated to enhance disease resistance in *Arabidopsis* and crop plants (Berrocal-Lobo et al. 2002; Guo et al. 2004; Zuo et al. 2007). WRKY TFs have been shown to act at the junction between the jasmonic and the salicylic acid pathways, thereby modifying the cross-talk between these pathways (Li et al. 2004; Mao et al. 2007; Higashi et al. 2008) and defining the plant defense response to pathogen attack.

Breeding for disease resistance in biofuel crops will become imperative with increasing cultivation. Pests and diseases will have a more profound effect on the economy of the biofuel crop if significant damage or total plant loss occurs. *Miscanthus*, switchgrass, and other novel biofuel crops have not yet been grown extensively, but bacterial, viral, and fungal pathogens as well as nematodes have already been reported to attack *Miscanthus* (Christian et al. 1994; O'Neill and Farr 1996; Thinggard 1997; Gams et al. 1999; Halbert and Remaudiere 2000) and switchgrass (Gravert and Munkvold 2002; Gustafson et al. 2003; Krupinski et al. 2004; Carris et al. 2008). Interestingly, *Miscanthus* has been used in sugarcane breeding for improving disease resistance (Chen and Lo 1989; Miller et al. 2005), implicating that there is useful variation for disease resistance in *Miscanthus* germplasm.

A database is now available integrating gene regulatory information for maize, rice, sorghum, and sugarcane (http://www.grassius.org Yilmaz et al. 2009) which could further advance the knowledge of regulatory genes and networks in biofuel grasses and their incorporation into breeding programs for improved genotypes. Gene stacking and the use of tissue-specific or inducible promoters for precise expression will be needed for the genetic engineering of biofuel crops because of the interaction networks of transcription factors and the potential need to add multiple traits. Transgenic corn with more than one introduced trait (one for glyphosate herbicide resistance and the other for resistance for the European corn borer and corn rootworm) is currently in commercial use (Dill et al. 2008) and shows no adverse effects on forage quality (McCann et al. 2007; Drury et al. 2008). New strategies for pyramiding several genes or integrating gene regions into plants via minichromosomes in corn are also being developed (Carlson et al. 2007) and could be beneficial for biofuel crop advancement.

Cell wall degradability, biomass yield, and stress resistance are the major traits which will ultimately define the suitability and use of cellulosic biofuels. Optimizing these traits through transgene technology can be considered a promising tool for effectively developing cellulosic biofuel crops. Given that flowering is an undesired trait in biofuel crops, gene flow could much more easily be avoided by eliminating or altering flowering. In addition, deregulation of transgenic biofuel crops should be more straightforward because biofuels are not used for human or animal consumption. In summary, applying transgenic technology to biofuel crops has fewer hurdles than transgenic food crops and should enable much faster variety development.

Conclusion

The potential for significant genetic improvement of C4 grasses as biofuel crops is good. Full exploration of natural genetic resources through plant breeding with the aid of molecular tools could dramatically increase biomass yield of dedicated biofuel crops and thus meet the demand of feedstocks for biofuel production without a significant impact on our food supply and natural environment.

Acknowledgements The authors thank Oliver Ratcliffe, Bob Creelman, Erik Sacks, Timothy Smith, and several anonymous reviewers for helpful discussions and critical comments on the manuscript.

References

Akbari M.; Wenzl P.; Caig V.; Carling J.; Xia L.; Yang S.; Uszynski G.; Mohler V.; Lehmensiek A.; Kuchel H.; Hayden M. J.; Howes N.; Sharp P.; Vaughan P.; Rathmell B.; Huttner H.; Kilian A. Diversity arrays technology (DArT) for high-throughput profiling of the hexaploid wheat genome. *Theor. Appl. Genet* 113: 1409–1420; 2006. doi:10.1007/s00122-006-0365-4.

Amaya I.; Ratcliffe O. J.; Bradley D. J. Expression of *CENTRORADIALIS (CEN)* and *CEN*-like genes in tobacco reveals a conserved mechanism controlling phase change in diverse species. *Plant. Cell* 11: 1405–1417; 1999.

Anderson J. A.; Chao S.; Liu S. Molecular breeding using a major QTL for *Fusarium* head blight resistance in wheat. *Crop Sci* 47: S112–S119; 2007. doi:10.2135/cropsci2006.05.0359.

Anderson W. F.; Peterson J.; Akin D. E.; Morrison W. H. Enzyme pretreatment of grass lignocellulose for potential high-value co-products and an improved fermentable substrate. *Appl. Biochem. Biotechnol* 121: 303–310; 2005 doi:10.1385/ABAB:121:1-3:0303.

Atienza S. G.; Satovic Z.; Petersen K. K.; Dolstra O.; Martín A. Preliminary genetic linkage map of *Miscanthus sinensis* with RAPD markers. *Theor. Appl. Genet* 105: 946–952; 2002 doi:10.1007/s00122-002-0956-7.

Atienza S. G.; Satovic Z.; Petersen K. K.; Dolstra O.; Martín A. Identification of QTLs influencing agronomic traits in *Miscanthus sinensis* Anderss. I. Total height, flag-leaf height and stem diameter. *Theor. Appl. Genet* 107: 123–129; 2003 doi:10.1007/s00122-003-1218-z.

Barbazuk W. B.; Emrich S. J.; Chen H. D.; Li L.; Schnable P. S. SNP discovery via 454 transcriptome sequencing. *Plant. J* 51: 910–918; 2007 doi:10.1111/j.1365-313X.2007.03193.x.

Barrière Y.; Ralph J.; Méchin V.; Guillaumie S.; Grabber J. H.; Argillier O.; Chabbert B.; Lapierre C. Genetic and molecular basis of grass cell wall biosynthesis and degradability. II. Lessons from brown-midrib mutants. *C. R. Biol* 327: 847–860; 2004 doi:10.1016/j.crvi.2004.05.010.

Barrière Y.; Thomas J.; Denoue D. QTL mapping for lignin content, lignin monomeric composition, *p*-hydroxycinnamate content, and cell wall digestibility in the maize recombinant inbred line progeny F838 × F286. *Plant Sci* 175: 585–595; 2008 doi:10.1016/j.plantsci.2008.06.009.

Baucher M.; Bernard-Vailhé M. A.; Chabbert B.; Besle J. M.; Opsomer C.; Van Montagu M.; Botterman J. Down-regulation of cinnamyl alcohol dehydrogenase in transgenic alfalfa

(*Medicago sativa* L.) and the effect on lignin composition and digestibility. *Plant Mol. Biol* 39: 437–447; 1999 doi:10.1023/A:1006182925584.

Baucher M.; Halpin C.; Petit-Conil M.; Boerjan W. Lignin: genetic engineering and impact on pulping. *Crit. Rev. Biochem. Mol. Biol* 38: 305–350; 2003 doi:10.1080/10409230391036757.

Beadle C. L.; Long S. P. Photosynthesis—is it limiting to biomass production? *Biomass* 8: 119–168; 1985 doi:10.1016/0144-4565(85)90022-8.

Beale C. V.; Morison J. I. L.; Long S. P. Water use efficiency of C-4 perennial grasses in a temperate climate. *Agric. Forest Meteorol* 96: 103–115; 1999 doi:10.1016/S0168-1923(99)00042-8.

Benedict C.; Skinner J. S.; Meng R.; Chang Y.; Bhalerao R.; Huner N. P.; Finn C. E.; Chen T. H.; Hurry V. The CBF1-dependent low temperature signaling pathway, regulon and increase in freeze tolerance are conserved in *Populus* spp. *Plant Cell Environ* 29: 1259–1272; 2006 doi:10.1111/j.1365-3040.2006.01505.x.

Berrocal-Lobo M.; Molina A.; Solano R. Constitutive expression of ETHYLENE-RESPONSE-FACTOR1 in *Arabidopsis* confers resistance to several necrotrophic fungi. *Plant J* 29: 23–32; 2002 doi:10.1046/j.1365-313x.2002.01191.x.

Boe A.; Beck D. L. Yield components of biomass in switchgrass. *Crop. Sci* 48: 1306–1311; 2008 doi:10.2135/cropsci2007.08.0482.

Bouchez A.; Hospital F.; Causse M.; Gallais A.; Charcosset A. Marker-assisted introgression of favorable alleles at quantitative trait loci between maize elite lines. *Genetics* 162: 1945–1959; 2002.

Buerstmayr H.; Ban T.; Anderson J. A. QTL mapping and marker-assisted selection for Fusarium head blight resistance in wheat: a review. *Plant Breed* 128: 1–26; 2009 doi:10.1111/j.1439-0523.2008.01550.x.

Cardinal A. J.; Lee M.; Moore K. J. Genetic mapping and analysis of quantitative trait loci affecting fiber and lignin content in maize. *Theor. Appl. Genet* 106: 866–874; 2003.

Carlson S. R.; Rudgers G. W.; Zieler H.; Mach J. M.; Luo S.; Grunden E.; Krol C.; Copenhaver G. P.; Preuss D. Meiotic transmission of an *in vitro*-assembled autonomous maize minichromosome. *Publ. Libr. Sci. Genetics* 310: e179; 2007.

Carris L. M.; Castlebury L. A.; Zale J. First report of *Tilletia pulcherrima* bunt on switchgrass (*Panicum virgatum*) in Texas. *Plant Dis* 92: 1707; 2008 doi:10.1094/PDIS-92-12-1707C.

Casler M. D.; Buxton D. R.; Vogel K. P. Genetic modification of lignin concentration affects fitness of perennial herbaceous plants. *Theor. Appl. Genet* 104: 127–131; 2002 doi:10.1007/s001220200015.

Castro A. J.; Capettini F.; Corey A. E.; Filichkina T.; Hayes P. M.; Kleinhofs A.; Kudrna D.; Richardson K.; Sandoval-Islas S.; Rossi C.; Vivar H. Mapping and pyramiding of qualitative and quantitative resistance to stripe rust in barley. *Theor. Appl. Genet* 107: 922–930; 2003 doi:10.1007/s00122-003-1329-6.

Century K.; Reuber T. L.; Ratcliffe O. J. Regulating the regulators: the future prospects for transcription-factor-based agricultural biotechnology products. *Plant Physiol* 147: 20–29; 2008 doi:10.1104/pp.108.117887.

Chen F.; Dixon R. A. Lignin modification improves fermentable sugar yields for biofuel production. Nat. Biotechnol 25: 759–761; 2007 doi:10.1038/nbt1316.

Chen S.; Lin X. H.; Xu C. G.; Zhang Q. Improvement of bacterial blight resistance of 'Minghui 63', an elite restorer line of hybrid rice, by molecular marker-assisted selection. *Crop Sci* 40: 239–244; 2000.

Chen Y. H.; Lo C. C. Disease resistance and sugar content in *Saccharum–Miscanthus* hybrids. *Taiwan Sugar* 36: 9–12; 1989.

Christian D. G.; Lamptey J. N. L.; Forde S. M. D.; Plumb R. T. First report of barley yellow dwarf luteovirus on *Miscanthus* in the United Kingdom. *Eur. J. Plant Pathol* 100: 167–170; 1994 doi:10.1007/BF01876249.

Clifton-Brown J.; Chiang Y-C.; Hodkinson T. *Miscanthus*: genetic resources and breeding potential to enhance bioenergy production. In: Vermerris W. (ed) Genetic improvement of bioenergy crops. Springer, New York, pp 273–294; 2008.

Clifton-Brown J. C.; Lewandowski I.; Andersson B.; Basch G.; Christian D. G.; Bonderup-Kjeldsen J.; Jørgensen U.; Mortensen V.; Riche A. B.; Schwarz K. U.; Tayebi K.; Teixeira F. Performance of 15 *Miscanthus* genotypes at five sites in Europe. *Agron. J* 93: 1013–1019; 2001.

Coleman H. D.; Samuels A. L.; Guy R. D.; Mansfield S. D. Perturbed lignification impacts tree growth in hybrid poplar—a function of sink strength; vascular integrity; and photosynthetic assimilation. *Plant Physiol* 148: 1229–1237; 2008 doi:10.1104/pp.108.125500.

Collard B. C. Y.; Mackill D. J. Marker-assisted selection: an approach for precision plant breeding in the twenty-first century. *Phil. Trans. R. Soc. B* 363: 557–572; 2008 doi:10.1098/rstb.2007.2170.

Concibido V. C.; Diers B. W.; Arelli P. R. A decade of QTL mapping for cyst nematode resistance in soybean. *Crop Sci* 44: 1121–1131; 2004.

Council for Agricultural Science and Technology (CAST) Implications of gene flow in the scale-up and commercial use of biotechnology-derived crops: economic and policy considerations. issue paper 37. CAST, Ames, Iowa; 2007.

Cordeiro G. M.; Eliott F.; McIntyre C. L.; Casu R. E.; Henry R. J. Characterization of single nucleotide polymorphisms in sugarcane ESTs. *Theor. Appl. Genet* 113: 331–343; 2006 doi:10.1007/s00122-006-0300-8.

Cox T. S.; Bender M.; Picone C.; Van Tassel D. L.; Holland J. B.; Brummer E. C.; Zoeller B. E.; Paterson A. H.; Jackson W. Breeding perennial grain crops. *Crit. Rev. Plant Sci* 21: 59–91; 2002 doi:10.1080/0735-260291044188.

Daniell H. Molecular strategies for gene containment in transgenic crops. *Nature Biotechnol*. 20: 581–586; 2002 doi:10.1038/nbt0602-581.

Das M. K.; Fuentes R. G.; Taliaferro C. M. Genetic variability and trait relationships in switchgrass. *Crop Sci* 44: 443–448; 2004.

Davison B. H.; Drescher S. R.; Tuskan G. A.; Davis M. F.; Nghiem N. P. Variation of S/G ratio and lignin content in a *Populus* family influences the release of xylose by dilute acid hydrolysis. *Appl. Biochem. Biotechnol* 130: 427–435; 2006 doi:10.1385/ABAB:130:1:427.

DeBolt S.; Gutierrez R.; Ehrhardt D. W.; Melo C. V.; Ross L.; Cutler S. R.; Somerville C.; Bonetta D. Morlin; an inhibitor of cortical microtubule dynamics and cellulose synthase movement. *Proc. Natl. Acad. Sci. U. S. A.* 104: 5854–5859; 2007 doi:10.1073/pnas.0700789104.

Desprez T.; Juraniec M.; Crowell E. F.; Jouy H.; Pochylova Z.; Parcy F.; Höfte H.; Gonneau M.; Vernhettes S. Organization of cellulose synthase complexes involved in primary cell wall synthesis in *Arabidopsis thaliana*. *Proc. Natl. Acad. Sci. U. S. A.* 104: 15572–15577; 2007 doi:10.1073/pnas.0706569104.

Dill G. M.; Cajacob C. A.; Padgette S. R. Glyphosate-resistant crops: adoption; use and future considerations. *Pest. Manag. Sci* 64: 326–331; 2008 doi:10.1002/ps.1501.

Dillon S. L.; Shapter F. M.; Henry R. J.; Cordeiro G.; Izquierdo L.; Lee L. S. Domestication to crop improvement: genetic resources for *Sorghum* and *Saccharum* (Andropogoneae). *Ann. Bot (Lond)* 100: 975–989; 2007 doi:10.1093/aob/mcm192.

Drury S. M.; Reynolds T. L.; Ridley W. P.; Bogdanova N.; Riordan S.; Nemeth M. A.; Sorbet R.; Trujillo W. A.; Breeze M. L. Composition of forage and grain from second-generation insect-protected corn MON 89034 is equivalent to that of conventional corn (*Zea mays* L.). *J. Agric. Food Chem* 56: 4623–4630; 2008 doi:10.1021/jf800011u.

Dubouzet J. G.; Sakuma Y.; Ito Y.; Kasuga M.; Dubouzet E. G.; Miura S.; Seki M.; Shinozaki K.; Yamaguchi-Shinozaki K. OsDREB genes in rice; *Oryza sativa* L.; encode transcription activators that function in drought-; high-salt- and cold-responsive gene expression. *Plant J* 33: 751–763; 2003 doi:10.1046/j.1365-313X.2003.01661.x.

Duran C.; Appleby N.; Clark T.; Wood D.; Imelfort M.; Batley J.; Edwards D. AutoSNPdb: an annotated single nucleotide polymorphism database for crop plants. *Nucleic Acids Res.* 37: D951–D953; 2009.

Eathington S. R.; Crosbie T. M.; Edwards M. D.; Reiter R. S.; Bull J. K. Molecular markers in a commercial breeding program. *Crop Sci* 47: S154–S163; 2007 doi:10.2135/cropsci2007.04.0015IPBS.

Edgerton M. D. Increasing crop productivity to meet global needs for feed; food; and fuel. *Plant Physiol* 149: 7–13; 2009 doi:10.1104/pp.108.130195.

Feltus F. A.; Hart G. E.; Schertz K. F.; Casa A. M.; Kresovich S.; Abraham S.; Klein P. E.; Brown P. J.; Paterson A. H. Alignment of genetic maps and QTLs between inter- and intra-specific sorghum populations. *Theor. Appl. Genet* 112: 1295–305; 2006a doi:10.1007/s00122-006-0232-3.

Feltus F. A.; Singh H. P.; Lohithaswa H. C.; Schulze S. R.; Silva T. D.; Paterson A. H. A comparative genomics strategy for targeted discovery of single-nucleotide polymorphisms and conserved-noncoding sequences in orphan crops. *Plant Physiol* 140: 1183–1191; 2006b doi:10.1104/pp.105.074203.

Field C. B. Sharing the garden. *Science* 294: 2490–2491; 2001 doi:10.1126/science.1066317.

Gams W.; Klamer M.; O'Donnell K. *Fusarium miscanthi* sp. nov. from *Miscanthus* litter. *Mycol* 91: 263–268; 1999 doi:10.2307/3761371.

Gantzer C. J.; Anderson S. H.; Thompson A. L.; Brown J. R. Estimating soil erosion after 100 years of cropping on Sanborn Field. *J. Soil Water Conserv* 45: 641–644; 1990.

Gils M.; Marillonnet S.; Werner S.; Grützner R.; Giritch A.; Engler C.; Schachschneider R.; Klimyuk V.; Gleba Y. A novel hybrid seed system for plants. *Plant Biotechnol. J* 6: 226–235; 2008 doi:10.1111/j.1467-7652.2007.00318.x.

Goicoechea M.; Lacombe E.; Legay S.; Mihaljevic S.; Rech P.; Jauneau A.; Lapierre C.; Pollet B.; Verhaegen D.; Chaubet-Gigot N.; Grima-Pettenati J. *Eg*MYB2; a new transcriptional activator from *Eucalyptus* xylem; regulates secondary cell wall formation and lignin biosynthesis. *Plant J* 43: 553–567; 2005 doi:10.1111/j.1365-313X.2005.02480.x.

Gravert C. E.; Munkvold G. P. Fungi and diseases associated with cultivated switchgrass in Iowa. *J. Iowa Acad Sci* 109: 30–34; 2002.

Grivet L.; Glaszmann J. C.; Vincentz M.; da Silva F.; Arruda P. ESTs as a source for sequence polymorphism discovery in sugarcane: example of the *Adh* genes. *Theor. Appl. Genet* 106: 190–197; 2003.

Guo Z. J.; Chen X. J.; Wu X. L.; Ling J. Q.; Xu P. Overexpression of the AP2/ EREBP transcription factor OPBP1 enhances disease resistance and salt tolerance in tobacco. *Plant Mol. Biol* 55: 607–618; 2004 doi:10.1007/s11103-004-1521-3.

Gustafson D. M.; Boe A.; Jin Y. Genetic variation for *Puccinia emaculata* infection in switchgrass. *Crop Sci* 43: 755–759; 2003.

Halbert S. E.; Remaudiere G. A new oriental *Melanaphis* species recently introduced in North America [*Hemiptera*; *Aphididae*]. *Rev. Fr. Entomol.* 22: 109–117; 2000.

Hart G. E.; Schertz K. F.; Peng Y.; Syed N. H. Genetic mapping of *Sorghum bicolor* (L.) Moench QTLs that control variation in tillering and other morphological characters. *Theor. Appl. Genet* 103: 1232–1242; 2001 doi:10.1007/s001220100582.

Hartley R. W. Bamase and barstar: expression of its cloned inhibitor permits expression of a cloned ribonuclease. *J. Mol. Biol* 202: 913–915; 1988 doi:10.1016/0022-2836(88)90568-2.

Hartley R. W. Bamase and barstar: two small proteins to fold and fit together. Trends Biochem. Sci 14: 450–454; 1989 doi:10.1016/0968-0004(89)90104-7.

He L. Z.; Zhou P. H.; Liu X. M.; Cao X. J.; Cao M. D.; Liu Y. S. Studies on the autotetraploid of *Triarrhena lutarioriparia* L. Liou sp. *nov*. *Acta. Genetica. Sinica* 25: 49–55; 1998.

Heaton E. A.; Dohleman F. G.; Long S. P. Meeting US biofuel goals with less land: the potential of Miscanthus. *Global Change Biol* 14: 2000–2014; 2008a doi:10.1111/j.1365-2486.2008.01662.x.

Heaton E. A.; Mascia P.; Flavell R.; Thomas S.; Long P. S.; Dohleman F. G. Energy crop development: current progress and future prospects. *Curr. Opin. Biotechnol* 19: 202–209; 2008b doi:10.1016/j.copbio.2008.05.001.

Heinz D. Sugarcane improvement through breeding. Elsevier, Amsterdam 1987.

Hernández P.; Dorado G.; Laurie D. A.; Martín A.; Snape J. W. Microsatellites and RFLP probes from maize are efficient sources of molecular markers for the biomass energy crop Miscanthus. *Theor. Appl. Genet.* 102:616–622; 2001.

Higashi K.; Ishiga Y.; Inagaki Y.; Toyoda K.; Shiraishi T.; Ichinose Y. Modulation of defense signal transduction by flagellin-induced WRKY41 transcription factor in *Arabidopsis thaliana*. *Mol. Genet. Genom* 279: 303–312; 2008 doi:10.1007/s00438-007-0315-0.

Hodkinson T. R.; Chase M. W.; Lledó M. D.; Salamin N.; Renvoize S. A. Phylogenetics of *Miscanthus*; *Saccharum* and related genera (Saccharinae; Andropogoneae; Poaceae) based on DNA sequences from ITS nuclear ribosomal DNA and plastid trnL intron and trnL-F intergenic spacers. *J. Plant Res* 115: 381–392; 2002a doi:10.1007/s10265-002-0049-3.

Hodkinson T. R.; Chase M. W.; Takahashi C.; Leitch I. J.; Bennett M. D.; Renvoize S. A. The use of DNA sequencing (ITS and *trnL-F*); AFLP; and fluorescent *in situ* hybridization to study allopolyploid *Miscanthus* (Poaceae). *Amer. J. Botan* 89: 279–286; 2002b doi:10.3732/ajb.89.2.279.

Hu F. Y.; Tao D. Y.; Sacks E.; Fu B. Y.; Xu P.; Li J.; Yang Y.; McNally K.; Khush K. S.; Paterson A. H.; Li Z-K. Convergent evolution of perenniality in rice and sorghum. *Proc. Natl. Acad. Sci. U. S. A.* 100: 4050–4054; 2003 doi:10.1073/pnas.0630531100.

Hu H.; Dai M.; Yao J.; Xiao B.; Li X.; Zhang Q.; Xiong L. Overexpressing a NAM, ATAF, and CUC (NAC) transcription factor enhances drought resistance and salt tolerance in rice. *Proc. Natl. Acad. Sci. U. S. A.* 103: 12987–12992; 2006.

Hu H.; You J.; Fang Y.; Zhu X.; Qi Z.; Xiong L. Characterization of transcription factor gene SNAC2 conferring cold and salt tolerance in rice. *Plant Mol. Biol* 67: 169–181; 2008 doi:10.1007/s11103-008-9309-5.

Hu W-J.; Harding S. A.; Lung J.; Popko J. L.; Ralph J.; Stokke D. D.; Tsai C-J.; Chiang V. L. Repression of lignin biosynthesis promotes cellulose accumulation and growth in transgenic trees. *Nature Biotech* 17: 808–812; 1999 doi:10.1038/11758.

Huang N.; Angeles E. R.; Domingo J.; Magpantay G.; Singh S.; Zhang G.; Kumaravadivel N.; Bennett J.; Khush G. S. Pyramiding of bacterial blight resistance genes in rice: marker-assisted selection using RFLP and PCR. *Theor. Appl. Genet* 95: 313–320; 1997 doi:10.1007/s001220050565.

ISAAA. Brief 38-2008: Executive summary. Global status of commercialized biotech/GM crops: 2008 the first thirteen years; 1996 to 2008 http://www.isaaa.org/resources/publications/briefs/39/executivesummary/default.html; 2008.

Ito Y.; Katsura K.; Maruyama K.; Taji T.; Kobayashi M.; Seki M.; Shinozaki K.; Yamaguchi-Shinozaki K. Functional analysis of rice DREB1/CBF-type transcription factors involved in cold-responsive gene expression in transgenic rice. *Plant Cell Physiol.* 47: 141–153; 2006 doi:10.1093/pcp/pci230.

Jaccoud D.; Peng K.; Feinstein D.; Kilian A. Diversity arrays: a solid state technology for sequence information independent genotyping. *Nucleic Acids Res.* 29: E25; 2001 doi:10.1093/nar/29.4.e25.

Jaglo-Ottosen K. R.; Gilmour S. J.; Zarka D. G.; Schabenberger O.; Thomashow M. F. Arabidopsis CBF1 overexpression induces cor genes and enhances freezing tolerance. *Science* 280: 104–106; 1998 doi:10.1126/science.280.5360.104.

James K. E.; Schneider H.; Ansell S. W.; Evers M.; Robba L.; Uszynski G.; Pedersen N.; Newton A. E.; Russell S. J.; Vogel J. C.; Kilian A. Diversity arrays technology (DArT) for pan-genomic evolutionary studies of non-model organisms. *Publ. Libr. Sci. ONE* 32: e1682; 2008.

Jang C. S.; Kamps T. L.; Skinner D. N.; Schulze S. R.; Vencill W. K.; Paterson A. H. Functional classification; genomic organization; putatively cis-acting regulatory elements; and relationship to quantitative trait loci; of sorghum genes with rhizome-enriched expression. *Plant Physiol* 142: 1148–1159; 2006 doi:10.1104/pp.106.082891.

Jang C. S.; Kamps T. L.; Tang H.; Bowers J. E.; Lemke C.; Paterson A. H. Evolutionary fate of rhizome-specific genes in a non-rhizomatous *Sorghum* genotype. *Heredity* 102: 266–273; 2008 doi:10.1038/hdy.2008.119.

Je owski S. Yield traits of six clones of *Miscanthus* in the first 3 years following planting in Poland. *Ind. Crop Prod* 27: 65–68; 2008 doi:10.1016/j.indcrop.2007.07.013.

Joseph M.; Gopalakrishnan S.; Sharma R. K.; Singh V. P.; Singh A. K.; Singh N. K.; Mohapatra T. Combining bacterial blight resistance and Basmati quality characteristics by phenotypic and molecular marker-assisted selection in rice. *Mol. Breed* 13: 377–387; 2004 doi:10.1023/B:MOLB.0000034093.63593.4c.

Julien M.; Delaveau P.; Soopramanien G.; Martine J. Age; time of harvest; and environment as factors influencing differences in yield between flowering and vegetative canes. *Proc. Int. Soc. Sugarcane Technol* 16: 1771–1789; 1978.

Julien M.; Soopramanien G. The effect of flowering on yield in sugarcane. *Rev. Agric. Sucr. Ile Maurice* 55: 151–158; 1976.

Jung H. G.; Vogel K. P. Influence of lignin on digestibility of forage cell wall material. *J. Anim. Sci.* 62: 1703–1712; 1986.

Karpinska B.; Karlsson M.; Srivastava M.; Stenberg A.; Schrader J.; Sterky F.; Bhalerao R.; Wingsle G. MYB transcription factors are differentially expressed and regulated during secondary vascular tissue development in hybrid aspen. *Plant Mol. Biol* 56: 255–270; 2004 doi:10.1007/s11103-004-3354-5.

Kashiwagi T.; Togawa E.; Hirotsu N.; Ishimaru K. Improvement of lodging resistance with QTLs for stem diameter in rice (*Oryza sativa* L.). *Theor. Appl. Genet* 117: 749–757; 2008 doi:10.1007/s00122-008-0816-1.

Kawaoka A.; Kaothien P.; Yoshida K.; Endo S.; Yamada K.; Ebinuma H. Functional analysis of tobacco LIM protein Ntlim1 involved in lignin biosynthesis. *Plant J* 22: 289–301; 2000 doi:10.1046/j.1365-313x.2000.00737.x.

Kelley S. S.; Rowell R. M.; Davis M.; Jurich C. K.; Ibach R. Rapid analysis of the chemical composition of agricultural fibers using near infrared spectroscopy and pyrolysis molecular beam mass spectrometry. *Biomass Bioenergy* 27: 77–88; 2004 doi:10.1016/j.biombioe.2003.11.005.

Konishi S.; Izawa T.; Lin S. Y.; Ebana K.; Fukuta Y.; Sasaki T.; Yano M. An SNP caused loss of seed shattering during rice domestication. *Science* 312: 1392–1396; 2006 doi:10.1126/science.1126410.

Krakowsky M. D.; Lee M.; Coors J. G. Quantitative trait loci for cell wall components in recombinant inbred lines of maize (*Zea mays* L.) II: leaf sheath tissue. *Theor. Appl. Genet* 112: 717–726; 2006 doi:10.1007/s00122-005-0175-0.

Krupinsky J. M.; Berdahl J. D.; Schoch C. L.; Rossman A. Y. A new leaf spot disease on switchgrass (*Panicum virgatum*) caused by *Bipolaris oryzae*. *Can. J. Plant Pathol* 26: 371–378; 2004.

Kubo M.; Udagawa M.; Nishikubo N.; Horiguchi G.; Yamaguchi M.; Ito J.; Mimura T.; Fukuda H.; Demura T. Transcription switches for protoxylem and metaxylem vessel formation. *Genes Dev* 19: 1855–1860; 2005 doi:10.1101/gad.1331305.

Labbé N.; Ye P. X.; Franklin J. A.; Womac A. R.; Tyler D. D.; Rials T. G. Analysis of switchgrass characteristics using near infrared techniques. *BioRes* 3: 1329–1348; 2008.

Li C.; Zhou A.; Sang T. Rice domestication by reducing shattering. *Science* 311: 1936–1939; 2006 doi:10.1126/science.1123604.

Li J.; Brader G.; Palva E. T. The WRKY70 transcription factor: a node of convergence for jasmonate-mediated and salicylate-mediated signals in plant defense. *Plant Cell* 16: 319–331; 2004 doi:10.1105/tpc.016980.

Li L.; Zhou Y.; Cheng X.; Sun J.; Marita J. M.; Ralph J.; Chiang V. L. Combinatorial modification of multiple lignin traits in trees through multigene cotransformation. *Proc. Natl. Acad. Sci. U. S. A.* 100: 4939–4944; 2003 doi:10.1073/pnas.0831166100.

Li S. F.; Iacuone S.; Parish R. W. Suppression and restoration of male fertility using a transcription factor. *Plant Biotechnol. J* 5: 297–312; 2007 doi:10.1111/j.1467-7652.2007.00242.x.

Lin Y. R.; Schertz K. F.; Paterson A. H. Comparative analysis of QTLs affecting plant height and maturity across the Poaceae; in reference to an interspecific sorghum population. *Genetics* 141: 391–411; 1995.

Lindeboom J.; Mulder B. M.; Vos J. W.; Ketelaar T.; Emons A. M. Cellulose microfibril deposition: coordinated activity at the plant plasma membrane. *J. Microsc* 231: 192–200; 2008 doi:10.1111/j.1365-2818.2008.02035.x.

Lister R.; Gregory B. D.; Ecker J. R. Next is now: new technologies for sequencing of genomes; transcriptomes; and beyond. *Curr. Opinion Plant Biol* 12: 1–12; 2009 doi:10.1016/j.pbi.2008.12.005.

Long A. A large varietal difference in cane deterioration due to flowering. *Pro. South Afr. Sugar Technol. Assoc* 50: 78–81; 1976.

Mace E. S.; Xia L.; Jordan D. R.; Halloran K.; Parh D. K.; Huttner E.; Wenzl P.; Kilian A. DArT markers: diversity analyses and mapping in *Sorghum bicolor*. *BMC Genomics* 9: 26; 2008 doi:10.1186/1471-2164-9-26.

Mao P.; Duan M.; Wei C.; Li Y. WRKY62 transcription factor acts downstream of cytosolic NPR1 and negatively regulates jasmonate-responsive gene expression. *Plant Cell Physiol* 48: 833–842; 2007 doi:10.1093/pcp/pcm058.

Mariani C.; Goldberg R. B.; Leemans J. Engineered male sterility in plants. *Symp. Soc. Exp. Biol* 45: 271–279; 1991.

McCann M. C.; Carpita N. C. Designing the deconstruction of plant cell walls. *Curr. Opin. Plant Biol* 11: 314–320; 2008 doi:10.1016/j.pbi.2008.04.001.

McCann M. C.; Trujillo W. A.; Riordan S. G.; Sorbet R.; Bogdanova N. N.; Sidhu R. S. Comparison of the forage and grain composition from insect-protected and glyphosate-tolerant MON 88017 corn to conventional corn (*Zea mays* L.). *J. Agric. Food Chem* 16: 4034–4042; 2007 doi:10.1021/jf063499a.

McIntyre C. L.; Casu R. E.; Drenth J.; Knight D.; Whan V. A.; Croft B. J.; Jordan D. R.; Manners J. M. Resistance gene analogues in sugarcane and sorghum and their association with quantitative trait loci for rust resistance. *Genome* 48: 391–400; 2005 doi:10.1139/g05-006.

Miller J. D.; Tai P. Y.; Edme S. J.; Comstock J. C.; Glaz B. S.; Gilbert R. A. Basic germplasm utilization in the sugarcane development program at Canal Point; FL; USA. *Int. Soc. Sugar Cane Technol. Proc.* 2: 532–536; 2005.

Ming R.; Del Monte T. A.; Hernandez E.; Moore P. H.; Irvine J. E.; Paterson A. H. Comparative analysis of QTLs affecting plant height and flowering among closely-related diploid and polyploid genomes. *Genome* 45: 794–803; 2002 doi:10.1139/g02-042.

Missaoui A. M.; Paterson A. H.; Bouton J. H. Investigation of genomic organization in switchgrass (*Panicum virgatum* L.) using DNA markers. *Theor. Appl. Genet* 110: 1372–1383; 2005 doi:10.1007/s00122-005-1935-6.

Mitsuda N.; Iwase A.; Yamamoto H.; Yoshida M.; Seki M.; Shinozaki K.; Ohme-Takagi M. NAC transcription factors; NST1 and NST3; are key regulators of the formation of secondary walls in woody tissues of Arabidopsis. *Plant Cell* 19: 270–280; 2007 doi:10.1105/tpc.106.047043.

Mitsuda N.; Seki M.; Shinozaki K.; Ohme-Takagi M. The NAC transcription factors NST1 and NST2 of *Arabidopsis* regulate secondary wall thickening and are required for anther dehiscence. *Plant Cell* 17: 2993–3006; 2005 doi:10.1105/tpc.105.036004.

Nakashima J.; Chen F.; Jackson L.; Shadle G.; Dixon R. A. Multi-site genetic modification of monolignol biosynthesis in alfalfa (*Medicago sativa*): effects on lignin composition in specific cell types. *New Phytol* 179: 738–750; 2008 doi:10.1111/j.1469-8137.2008.02502.x.

Novaes E.; Drost D. R.; Farmerie W. G.; Pappas G. J. Jr; Grattapaglia D.; Sederoff R. R.; Kirst M. High-throughput gene and SNP discovery in *Eucalyptus grandis*; an uncharacterized genome. *BMC Genomics* 309: 312; 2008.

Novillo F.; Medina J.; Salinas J. Arabidopsis CBF1 and CBF3 have a different function than CBF2 in cold acclimation and define different gene classes in the CBF regulon. *Proc. Natl. Acad. Sci. U. S. A.* 104: 21002–21007; 2007 doi:10.1073/pnas.0705639105.

O'Neill N. R.; Farr D. F. *Miscanthus* blight; a new foliar disease of ornamental grasses and sugarcane incited by *Leptosphaeria* sp. and its anamorphic state *Stagonospora* sp. *Plant Dis* 80: 980–987; 1996.

Oraby H.; Venkatesh B.; Dale B.; Ahmad R.; Ransom C.; Oehmke J.; Sticklen M. Enhanced conversion of plant biomass into glucose using transgenic rice-produced endoglucanase for cellulosic ethanol. *Transgenic. Res* 16: 739–749; 2007 doi:10.1007/s11248-006-9064-9.

Paredez A. R.; Persson S.; Ehrhardt D. W.; Somerville C. R. Genetic evidence that cellulose synthase activity influences microtubule cortical array organization. *Plant Physiol* 147: 1723–1734; 2008 doi:10.1104/pp.108.120196.

Paredez A. R.; Somerville C. R.; Ehrhardt D. W. Visualization of cellulose synthase demonstrates functional association with microtubules. *Science* 312: 1491–1495; 2006 doi:10.1126/science.1126551.

Paterson A. H.; Bowers J. E.; Bruggmann R.; Dubchak I.; Grimwood J.; Gundlach H.; Haberer G.; Hellsten U.; Mitros T.; Poliakov A.; Schmutz J.; Spannagl M.; Tang H.; Wang X.; Wicker T.; Bharti A. K.; Chapman J.; Feltus F. A.; Gowik U.; Grigoriev I. V.; Lyons E.; Maher C. A.; Martis M.; Narechania A.; Otillar R. P.; Penning B. W.; Salamov A. A.; Wang Y.; Zhang L.; Carpita N. C.; Freeling M.; Gingle A. R.; Hash C. T.; Keller B.; Klein P.; Kresovich S.; McCann M. C.; Ming R.; Peterson D. G.; Rahman M.; Ware D.; Westhoff P.; Mayer K. F. X.; Messing J.; Daniel S.; Rokhsar D. S. The *Sorghum bicolor* genome and the diversification of grasses. *Nature* 457: 551–556; 2009 doi:10.1038/nature07723.

Paterson A. H.; Lin Y. R.; Li Z.; Schertz K. F.; Doebley J. F.; Pinson S. R.; Liu S. C.; Stansel J. W.; Irvine J. E. Convergent domestication of cereal crops by independent mutations at corresponding genetic-loci. *Science* 269: 1714–1718; 1995a doi:10.1126/science.269.5231.1714.

Paterson A. H.; Saranga Y.; Menz M.; Jiang C. X.; Wright R. J. QTL analysis of genotype x environment interactions affecting cotton fiber quality. *Theor. Appl. Genet* 106: 384–396; 2003.

Paterson A. H.; Schertz K. F.; Lin Y-R.; Liu S-C.; Chang Y-L. The weediness of wild plants: Molecular analysis of genes influencing dispersal and persistence of johnsongrass; *Sorghum halepense* (L.). Pers. *Proc. Natl. Acad. Sci. U. S. A.* 92: 6127–6131; 1995b doi:10.1073/pnas.92.13.6127.

Patzlaff A.; McInnis S.; Courtenay A.; Surman C.; Newman L. J.; Smith C.; Bevan M. W.; Mansfield S.; Whetten R. W.; Sederoff R. R.; Campbell M. M. Characterization of a pine MYB that regulates lignification. *Plant J* 36: 743–754; 2003a doi:10.1046/j.1365-313X.2003.01916.x.

Patzlaff A.; Newman L. J.; Dubos C.; Whetten R. W.; Smith C.; McInnis S.; Bevan M. W.; Sederoff R. R.; Campbell M. M. Characterization of PtMYB1; an R2R3-MYB from pine xylem. *Plant Mol. Biol* 53: 597–608; 2003b doi:10.1023/B:PLAN.0000019066.07933.d6.

Pavy N.; Pelgas B.; Beauseigle S.; Blais S.; Gagnon F.; Gosselin I.; Lamothe M.; Isabel N.; Bousquet J. Enhancing genetic mapping of complex genomes through the design of highly-multiplexed SNP arrays: application to the large and unsequenced genomes of white spruce and black spruce. *BMC Genomics* 9: 21; 2008 doi:10.1186/1471-2164-9-21.

Pedersen J. F.; Vogel K. P.; Funnell D. L. Impact of reduced lignin on plant fitness. *Crop Sci.* 45: 812–819; 2005 doi:10.2135/cropsci2004.0155.

Piedade M. T. F.; Junk W. J.; Long S. P. The productivity of the C4 grass *Echinochloa polystachya* on the Amazon floodplain. *Ecology* 72: 1456–1463; 1991 doi:10.2307/1941118.

Pilate G.; Guiney E.; Holt K.; Petit-Conil M.; Lapierre C.; Leplé J-C.; Pollet B.; Mila I.; Webster E. A.; Marstorp H. G.; Hopkins D. W.; Jouanin L.; Boerjan W.; Schuch W.; Cornu D.; Halpin C. Field and pulping performances of transgenic trees with altered lignification. *Nature Biotechnol* 20: 607–612; 2002 doi:10.1038/nbt0602-607.

Pimentel D.; Harvey C.; Resosudarmo P.; Sinclair K.; Kurz D.; McNair M.; Crist S.; Shpritz L.; Fitton L.; Saffouri R.; Blair R. Environmental and economic costs of soil erosion and conservation benefits. *Science* 267: 1117–1123; 1995 doi:10.1126/science.267.5201.1117.

Qin F.; Sakuma Y.; Li J.; Liu Q.; Li Y. Q.; Shinozaki K.; Yamaguchi-Shinozaki K. Cloning and functional analysis of a novel DREB1/CBF transcription factor involved in cold-responsive gene expression in *Zea mays* L. *Plant Cell Physiol* 45: 1042–1052; 2004 doi:10.1093/pcp/pch118.

Quinby J. R.; Karper R. The inheritance of three genes that influence time of floral initiation and maturity date in milo. *J. Am. Soc. Agron* 37: 916–936; 1945.

Quinby J. R.; Karper R. E. Inheritance of height in sorghum. *Agron. J* 46: 211–216; 1954.

Raghu S.; Anderson R. C.; Daehler C. C.; Davis A. S.; Wiedenmann R. N.; Simberloff D.; Mack R. N. Ecology. Adding biofuels to the invasive species fire? *Science* 313: 1742; 2006 doi:10.1126/science.1129313.

Ralph J.; Guillaumie S.; Grabber J. H.; Lapierre C.; Barrière Y. Genetic and molecular basis of grass cell-wall biosynthesis and degradability. III. Towards a forage grass ideotype. *C. R. Biol* 327: 467–479; 2004 doi:10.1016/j.crvi.2004.03.004.

Randall G. W.; Mulla D. Nitrate nitrogen in surface waters as influenced by climatic conditions and agricultural practices. *J. Environm. Qual* 30: 337–344; 2001.

Ransom C.; Balan V.; Biswas G.; Dale B.; Crockett E.; Sticklen M. Heterologous *Acidothermus cellulolyticus* 1,4-beta-endoglucanase E1 produced within the corn biomass converts corn stover into glucose. *Appl. Biochem. Biotechnol* 137–140: 207–219; 2007 doi:10.1007/s12010-007-9053-3.

Reddy G.; Upadhyaya H. D.; Gowda C. C. L. Current status of sorghum genetic resources at ICRISAT: their sharing and impacts. *J. SAT Agric. Res.* 2: 5; 2006 http://www.icrisat.cgiar.org/Journal/archives.htm.

Reddy M. S.; Chen F.; Shadle G.; Jackson L.; Aljoe H.; Dixon R. A. Targeted down-regulation of cytochrome P450 enzymes for forage quality improvement in alfalfa (*Medicago sativa* L.). *Proc. Natl. Acad. Sci. U. S. A.* 102: 16573–16578; 2005 doi:10.1073/pnas.0505749102.

Ribaut J. M.; Ragot M. Marker-assisted selection to improve drought adaptation in maize: the backcross approach; perspectives; limitations; and alternatives. *J. Exp. Bot* 58: 351–360; 2007 doi:10.1093/jxb/erl214.

Ricaud C.; Land A.; Sullivan S. Losses from the recurrence of yellow spot epiphytotics in Mauritius. *Sugar Azucar* 75: 28–29; 1980.

Rooney W. L.; Aydin S. Genetic control of a photoperiod-sensitive response in *Sorghum bicolor* (L.) Moench. *Crop Sci* 39: 397–400; 1999.

Ruf S.; Hermann M.; Berger I. J.; Carrer H.; Bock R. Stable genetic transformation of tomato plastids and expression of a foreign protein in fruit. *Nat. Biotechnol* 19: 870–875; 2001 doi:10.1038/nbt0901-870.

Saballos A. Development and utilization of sorghum as a bioenergy crop. In: Vermerris W. (ed) Genetic improvement of bioenergy crops. Springer, New York, pp 211–248; 2008.

Salehi H.; Ransom C. B.; Oraby H. F.; Seddighi Z.; Sticklen M. B. Delay in flowering and increase in biomass of transgenic tobacco expressing the *Arabidopsis* floral repressor gene *FLOWERING LOCUS C*. *J. Plant Physiol* 162: 711–717; 2005 doi:10.1016/j.jplph.2004.12.002.

Saranga Y.; Jiang C. X.; Wright R. J.; Yakir D.; Paterson A. H. Genetic dissection of cotton physiological responses to arid conditions and their inter-relationships with productivity. *Plant Cell Environ* 27: 263–277; 2004 doi:10.1111/j.1365-3040.2003.01134.x.

Saranga Y.; Menz M.; Jiang C. X.; Wright R. J.; Yakir D.; Paterson A. H. Genomic dissection of genotype x environment interactions conferring adaptation of cotton to arid conditions. *Gen. Res* 11: 1988–1995; 2001 doi:doi:10.1101/gr.157201.

Scheinost P. L.; Lammer D. L.; Cai X.; Murray T. D.; Jones S. S. Perennial wheat: a sustainable cropping system for the Pacific Northwest. *Am. J. Alternative Agric* 16: 147–151; 2001.

Schmierer D. A.; Kandemir N.; Kudrna D. A.; Jones B. L.; Ullrich S. E.; Kleinhofs A. Molecular marker-assisted selection for enhanced yield in malting barley. *Mol. Breed* 14: 463–473; 2005 doi:10.1007/s11032-005-0903-9.

Sharma P. N.; Torii A.; Takumi S.; Mori N.; Nakamura C. Marker-assisted pyramiding of brown planthopper (*Nilaparvata lugens* Stål) resistance genes *Bph1* and *Bph2* on rice chromosome 12. *Hereditas* 140: 61–69; 2004 doi:10.1111/j.1601-5223.2004.01726.x.

Smith D. R.; Quinlan A. R.; Peckham H. E.; Makowsky K.; Tao W.; Woolf B.; Shen L.; Donahue W. F.; Tusneem N.; Stromberg M. P.; Stewart D. A.; Zhang L.; Ranade S. S.; Warner J. B.; Lee C. C.; Coleman B. E.; Zhang Z.; McLaughlin S. F.; Malek J. A.; Sorenson J. M.; Blanchard A. P.; Chapman J.; Hillman D.; Chen F.; Rokhsar D. S.; McKernan K. J.; Jeffries T. W.; Marth G. T.; Richardson P. M. Rapid whole-genome mutational profiling using next-generation sequencing technologies. *Genome. Res* 18: 1638–1642; 2008 doi:10.1101/gr.077776.108.

Somleva M. N.; Snell K. D.; Beaulieu J. J.; Peoples O. P.; Garrison B. R.; Patterson N. A. Production of polyhydroxybutyrate in switchgrass; a value-added co-product in an important lignocellulosic biomass crop. *Plant Biotechnol. J* 6: 663–678; 2008 doi:10.1111/j.1467-7652.2008.00350.x.

Somleva M. N.; Tomaszewski Z.; Conger B. V. *Agrobacterium*-mediated genetic transformation of switchgrass. *Crop Sci* 42: 2080–2087; 2002.

Steele K. A.; Price A. H.; Shashidhar H. E.; Witcombe J. R. Marker-assisted selection to introgress rice QTLs controlling root traits into an Indian upland rice variety. *Theor. Appl. Genet* 112: 208–221; 2006 doi:10.1007/s00122-005-0110-4.

Sticklen M. Plant genetic engineering to improve biomass characteristics for biofuels. *Curr. Opin. Biotechnol* 17: 315–319; 2006 doi:10.1016/j.copbio.2006.05.003.

Taylor L. E. II; Dai Z.; Decker S. R.; Brunecky R.; Adney W. S.; Ding S. Y.; Himmel M. E. Heterologous expression of glycosyl hydrolases in planta: a new departure for biofuels. *Trends Biotechnol* 26: 413–424; 2008 doi:10.1016/j.tibtech.2008.05.002.

Thinggaard K. Study of the role of *Fusarium* in the field establishment problem of *Miscanthus*. *Acta Agric. Scand. B. Plant Soil Sci.* 47: 238–241; 1997.

Vanholme R.; Morreel K.; Ralph J.; Boerjan W. Lignin engineering. *Curr. Opin. Plant Biol* 11: 278–285; 2008. doi:10.1016/j.pbi.2008.03.005.

Vermerris W.; Saballos A.; Ejeta G.; Mosier N. S.; Ladisch M. R.; Carpita N. C. Molecular breeding to enhance ethanol production from corn and sorghum. *Stover Crop Sci* 47: 142–153; 2007.

Vogel K. P.; Hopkins A. A.; Moore K. J.; Johnson K. D.; Carlson I. T. Winter survival in switchgrass populations bred for high- IVDMD. *Crop Sci* 42: 1857–1862; 2002.

Wagoner P. Perennial grain development: past efforts and potential for the future. *Crit. Rev. Plant Sci* 9: 381–408; 1990 doi:10.1080/07352689009382298.

Wang J.; Elliott J. E.; Williamson R. E. Features of the primary wall CESA complex in wild type and cellulose-deficient mutants of *Arabidopsis thaliana*. *J. Exp. Bot* 59: 2627–2637; 2008 doi:10.1093/jxb/ern125.

Welling A.; Palva E. T. Involvement of CBF transcription factors in winter hardiness in birch. *Plant Physiol* 147: 1199–1211; 2008 doi:10.1104/pp.108.117812.

Wenzl P.; Carling J.; Kudrna D.; Jaccoud D.; Huttner E.; Kleinhofs A.; Kilian A. Diversity arrays technology (DArT) for whole-genome profiling of barley. *Proc. Natl. Acad. Sci. U.S.A.* 101: 9915–9920; 2004 doi:10.1073/pnas.0401076101.

Wong C. K.; Bernardo R. Genomewide selection in oil palm: Increasing selection gain per unit time and cost with small populations. *Theor. Appl. Genet* 116: 815–824; 2008 doi:10.1007/s00122-008-0715-5.

Wurbs D.; Ruf S.; Bock R. Contained metabolic engineering in tomatoes by expression of carotenoid biosynthesis genes from the plastid genome. *Plant J* 49: 276–288; 2007 doi:10.1111/j.1365-313X.2006.02960.x.

Wyman C. E. What is (and is not) vital to advancing cellulosic ethanol. *Trends Biotechnol* 25: 153–157; 2007 doi:10.1016/j.tibtech.2007.02.009.

Xu S. L.; Rahman A.; Baskin T. I.; Kieber J. J. Two leucine-rich repeat receptor kinases mediate signaling linking cell wall biosynthesis and ACC synthase in Arabidopsis. *Plant Cell* 20: 3065–3079; 2008 doi:10.1105/tpc.108.063354.

Yang C.; Xu Z.; Song J.; Conner K.; Barrena G. V.; Wilson Z. A. Arabidopsis MYB26/MALE STERILE35 regulates secondary thickening in the endothecium and is essential for anther dehiscence. *Plant Cell* 19: 534–548; 2007 doi:10.1105/tpc.106.046391.

Yilmaz A.; Nishiyama M. Y. Jr; Garcia-Fuentes B.; Souza G. M.; Janies D.; Gray J.; Grotewold E. GRASSIUS: a platform for comparative regulatory genomics across the grasses. *Plant Physiol* 149: 171–180; 2009 doi:10.1104/pp.108.128579.

Yoshida M.; Liu Y.; Uchida S.; Kawarada K.; Ukagami Y.; Ichinose H.; Kaneko S.; Fukuda K. Effects of cellulose crystallinity; hemicellulose; and lignin on the enzymatic hydrolysis of *Miscanthus sinensis* to monosaccharides. *Biosci. Biotechnol. Biochem* 72: 805–810; 2008 doi:10.1271/bbb.70689.

Zhong R.; Demura T.; Ye Z-H. SND1; a NAC domain transcription factor; is a key regulator of secondary wall synthesis in fibers of Arabidopsis. *Plant Cell* 18: 3158–3170; 2006 doi:10.1105/tpc.106.047399.

Zhong R.; Lee C.; Zhou J.; McCarthy R. L.; Ye Z. H. A battery of transcription factors involved in the regulation of secondary cell wall biosynthesis in *Arabidopsis*. *Plant Cell* 20: 2763–2782; 2008 doi:10.1105/tpc.108.061325.

Zhong R.; Richardson E. A.; Ye Z. H. The MYB46 transcription factor is a direct target of SND1 and regulates secondary wall biosynthesis in *Arabidopsis*. *Plant Cell* 19: 2776–2792; 2007a doi:10.1105/tpc.107.053678.

Zhong R.; Richardson E. A.; Ye Z. H. Two NAC domain transcription factors; SND1 and NST1; function redundantly in regulation of secondary wall synthesis in fibers of Arabidopsis. *Planta* 225: 1603–1611; 2007b doi:10.1007/s00425-007-0498-y.

Zhong R.; Ye Z-H. Regulation of cell wall biosynthesis. *Curr. Opin. Plant Biol* 10: 564–572; 2007 doi:10.1016/j.pbi.2007.09.001.

Zhou R.; Zhu Z.; Kong X.; Huo N.; Tian Q.; Li P.; Jin C.; Dong Y.; Jia J. Development of wheat near-isogenic lines for powdery mildew resistance. *Theor. Appl. Gen* 110: 640–648; 2005 doi:10.1007/s00122-004-1889-0.

Zhou Y. L.; Xu J. L.; Zhou S. C.; Yu J.; Xie X. W.; Xu M. R.; Sun Y.; Zhu L. H.; Fu B. Y.; Gao Y. M.; Li Z. K. Pyramiding *Xa23* and *Rxo1* for resistance to two bacterial diseases into an elite indica rice variety using molecular approaches. *Mol. Breed* 23: 279–287; 2009 doi:10.1007/s11032-008-9232-0.

Zili Y.; Puhua Z.; Chengcai C.; Xiang L.; Wenzhong T.; Li W.; Shouyun C.; Zuoshun T. Establishment of genetic transformation system for *Miscanthus sacchariflorus* and obtaining of its transgenic plants. *High Tech. Lett* 10: 27–31; 2004.

Zuo K. J.; Qin J.; Zhao J. Y.; Ling H.; Zhang L. D.; Cao Y. F.; Tang K. X. Overexpression GbERF2 transcription factor in tobacco enhances brown spots disease resistance by activating expression of downstream genes. *Gene* 391: 80–90; 2007 doi:10.1016/j.gene.2006.12.019.

Chapter 8
Short-Rotation Woody Crops for Bioenergy and Biofuels Applications

Maud Hinchee, William Rottmann, Lauren Mullinax,
Chunsheng Zhang, Shujun Chang, Michael Cunningham,
Les Pearson, and Narender Nehra

Abstract Purpose-grown trees will be part of the bioenergy solution in the United States, especially in the Southeast where plantation forestry is prevalent and economically important. Trees provide a "living biomass inventory" with existing end-use markets and associated infrastructure, unlike other biomass species such as perennial grasses. The economic feasibility of utilizing tree biomass is improved by increasing productivity through alternative silvicultural systems, improved breeding and biotechnology. Traditional breeding and selection, as well as the introduction of genes for improved growth and stress tolerance, have enabled high growth rates and improved site adaptability in trees grown for industrial applications. An example is the biotechnology-aided improvement of a highly productive tropical *Eucalyptus* hybrid, *Eucalyptus grandis* × *Eucalyptus urophylla*. This tree has acquired freeze tolerance by the introduction of a plant transcription factor that up-regulates the cold-response pathways and makes possible commercial plantings in the Southeastern United States. Transgenic trees with reduced lignin, modified lignin, or increased cellulose and hemicellulose will improve the efficiency of feedstock conversion into biofuels. Reduced lignin trees have been shown to improve efficiency in the pre-treatment step utilized in fermentation systems for biofuels production from lignocellulosics. For systems in which thermochemical or gasification approaches are utilized, increased density will be an important trait, while increased lignin might be a desired trait for direct firing or co-firing of wood for energy. Trees developed through biotechnology, like all transgenic plants, need to go through the regulatory process, which involves biosafety and risk assessment analyses prior to commercialization.

Keywords Bioenergy • Biofuels • Biomass • Biotechnology • *Eucalyptus* • Flowering control • Lignin modification • Purpose-grown trees • *Populus* • Pine • *Pinus* • Productivity • Silviculture

M. Hinchee (✉)
ArborGen, Inc, Summerville, SC 29484, USA
e-mail: mahinch@arborgen.com

Introduction

Woody biomass represents a renewable resource with multiple industrial applications. It serves as feedstock for the pulp and paper industry but also can be planted specifically to address the feedstock needs for the energy or biofuels industry. Trees and wood have been identified as part of the bioenergy solution in the "Billion Ton Report" (Perlack et al. 2005). This report investigated the feasibility of producing the estimated 1 billion dry tons of lignocellulosic biomass needed annually to meet the "'30×'30" goal for a 30% replacement of United States petroleum consumption with biofuels by 2030. In this report, trees grown for bioenergy applications were included under the heading of agricultural resources as part of the broadly defined "perennial energy crops". Purpose-grown trees are expected to account for 377 million dry tons of the 1.37 billion dry ton total biomass resource potential at projected yields of 8 dry tons/acre/yr (Perlack et al. 2005).

It is expected that short-rotation woody crops, such as fast growing species *Populus*, *Salix*, and *Eucalyptus* and their respective hybrids, will be planted as purpose-grown wood on sites that enable high productivity and proximity to the processing plant. Short-rotation, purpose-grown trees have a variety of inherent logistical benefits and economic advantages relative to other lignocellulosic energy crops. Many of these advantages are driven by the fact that trees can typically be harvested year-round and continue growing year after year providing a "living inventory" of available biomass. Due to the flexibility associated with harvest time, trees have reduced storage and inventory holding costs and can minimize shrinkage or degradation losses typically associated with storage of annually-harvested biomass. Since trees can be harvested after several years and at different times, tree biomass mitigates the risk of annual yield fluctuations due to drought, disease and pest pressures, as well as other biotic or abiotic stresses. This allows a better matching of biomass supply with demand. An excess supply of an annually-harvested crop is necessary to hedge against years in which low yields are experienced in order to ensure full capacity utilization at a processing plant. Year-round harvest of trees enables the harvest and transport of wood to be distributed throughout the year, reducing infrastructure needs relative to annually-harvest crops (Sims and Venturi 2004).

Purpose-grown trees would minimize environmental impacts associated with biomass production since multi-year rotations of trees allow for extended periods between harvests with limited disturbance to the land. The multi-year rotation of trees also offers deployment and logistical benefits by reducing the land footprint that must be planted and harvested each year. While the acreage to feed a bioenergy plant may be similar between trees and other bioenergy crops with similar productivity, only a fraction of that total footprint would need to be planted or harvested in any given year for trees (Table 8.1). Purpose-grown trees for biomass also provide feedstock growers greater economic flexibility relative to other energy crops. The grower is provided a choice in harvest time as well as multiple end uses: traditional forest products and energy products such as cellulosic ethanol and power generation through direct firing, co-firing, or wood pellet systems.

Table 8.1 Total and annual acreage needs for trees relative to annually-harvested energy crops, data (generated from an ArborGen financial model by L. Mullinax)

Feedstock	Biomass needed (MM green tons/year)	Productivity (green tons/acre/year)	Rotation length (years)	Total acres needed (MM)	Acres planted annually (MM)
Trees	100	20	6	5	0.83
Annually-harvested crop	100	20	1	5	5

Current Limitations to the Use of Woody Feedstock for Biofuel Production

It has been projected by the U.S. Department of Energy and others that a productivity rate of 8 to 10 dry tons/acre/yr will be required for the long-term feasibility of renewable energy production (English et al. 2006). Research has been conducted on growing short-rotation trees for bioenergy in the United States and other countries (Short Rotation Forestry Handbook 1995). Short-rotation coppicing of hardwoods offers the promise to produce biomass for bioenergy. Coppicing is the process by which new shoots and trees are regenerated from a cut stump following harvest. The use of coppiced hardwoods for this purpose is not novel, although it is the subject of renewed interest and focused research (Andersson et al. 2002; Dickmann 2006). There are currently 12,000 acres of intensively managed short-rotation hardwoods. There are two main silvicultural systems: (1) moderately dense stands of cottonwood, and (2) dense stands with 1- to 4-yr rotations usually using willows (*Salix* species) or sycamore (*Plantanus occidentalis*). Loblolly pine (*Pinus taeda*) and sweetgum (*Liquidambar styraciflua*) plantations are also being considered for bioenergy applications (Davis and Trettin 2006; Dickmann 2006). However, the typical biomass productivity for these species is most likely not cost effectively adequate to meet the demand. For example, short-rotation willow crop yields range from 3 to 7 oven-dry tons/acre/yr (Mead 2005). Sweetgum and sycamore plantations grown for 7 yr on old agricultural land had, respectively, productivities of 1 and 2.3 oven-dry tons/yr, although this is expected to increase later in the rotation (Davis and Trettin 2006). *Populus deltoides* (eastern cottonwood) planted on good sites can produce an average yield of 5 dry tons/acre/yr. Loblolly pine grown to a 20-yr rotation can produce an average 4 dry tons/acre/yr (Mercker 2007).

It is clear that to achieve the productivity gains required for tree biofuel and bioenergy applications, a significant research effort is needed for improving tree genetics and silvicultural practices. The economics of a purpose-grown tree feedstock for energy may not be feasible unless there is significant genetic improvement in the base growth rate. There are two basic biotechnological strategies to achieve dramatic improvements in growth in plantation trees. The first is to genetically improve productivity of indigenous trees. The second strategy is to genetically improve the adaptability of exceptionally productive introduced trees so they can grow in the United States.

Genetics, Silviculture, and Biotechnology Enable Short-Rotation Trees

Productivity improvements, such as provided through advanced planting stock, optimal siliviculture and biotech traits, will be important for the cost-effective deployment of short-rotation trees for biomass applications. The improvement mechanism will be dependent on the biological and genetic limitations of the species utilized, in addition to the climate, soil, and moisture conditions of the tree plantation. Both native and exotic tree species can be considered potential feedstock sources. The approaches for successful short-rotation forestry for both native and exotic trees will be discussed.

Increased Biomass Productivity through Genetic Improvement of Native Species

Populus Various *Populus* species and their hybrids are among the most rapidly growing trees adapted to temperate climates. However, the high inherent growth potential of trees in this genus is often manifested only at the most favorable sites. The challenge is to determine if specific *Populus* genotypes can demonstrate wide adapatability across a range of sites and environmental conditions. Unless strategies to increase productivity are employed together with tolerance to abiotic and biotic stresses, plantations of *Populus* species and hybrids will remain limited.

Several strategies offer potential to overcome these limitations and allow *Populus* to play an increasingly important role in bioenergy initiatives. The first and most straightforward of these strategies is through traditional breeding to generate hybrids and varieties that grow fast, have high volume increments, and can grow across a wide range of sites. For example, advanced *Populus* clones are being developed by companies such as ArborGen, Inc and Greenwood Resources (http://www.greenwoodresources.com) to have greater productivity and adaptability. These programs typically consist of breeding among selected genotypes within a species or between species and then testing the seedling progeny in a series of field trials. The first tests are in a nursery at close spacing to evaluate the genotypes for broad adaptability and resistance to various pests; this is followed by one or more series of vegetatively propagated field trials in which the varieties are further screened for suitability to diverse planting sites. Commercial candidates are typically selected based on projected yields and wood properties after as many as 10 yr of field testing.

In addition, the Oak Ridge National Laboratory, in conjunction with the Bioenergy Feedstock Development Program and Boise Cascade Corp. is developing methods to identify drought-tolerant genotypes based on the presence of certain leaf metabolites (Oak Ridge National Laboratory, http://bioenergy.ornl.gov/papers/

misc/drotpopl.html). These techniques could reduce the cost and improve the efficiency of breeding and selection of *Populus* varieties adapted to upland sites.

The second approach involves direct genetic modification to add or modify genes that increase growth, increase stress tolerance and improve adaptability. It has been suggested that suboptimal nutrient and water availability limit *Populus* adaptability and productivity on many sites. Genes and gene families have been identified that have the ability to alter plant responses to water and nutrient limitations (Tuskan et al. 2006). Introduced genes being tested in *Populus* include the *Populus tremula* and *Arabidopsis* stable protein 1 (SP-1) gene, as well as genes involved in metabolic processes responsive to drought, redox proteins, transporter proteins, signal transduction proteins, and transcription factors (Polle et al. 2006).

A third approach would be to add genes to already widely adapted genotypes to improve their growth rate and productivity. Improved growth has been achieved in *Populus* through gene insertion technology. Kirby and co-workers at Rutgers University, who studied over-expression of a conifer cytosolic glutamine synthetase (GS1) in *Populus* (Fu et al. 2003; Man Hui-min et al. 2005), showed that greenhouse-grown GS1 transgenic trees had a greater than 100% increase in leaf biomass relative to controls when grown under low nitrogen conditions (Man Hui-min et al. 2005). The effect was less marked when more nitrogen was available. Glutamine synthetase (GS), found in either cytosolic or plastid located isoforms, is responsible for NH_4^+ assimilation in an ATP-requiring reaction that produces the amino acid glutamine from glutamic acid (reviewed in Good et al. 2004). The effect of the transgene was confirmed by measurement of increased glutamine synthetase enzyme activity, along with decreased foliar NH_4^+ and increased amino acids. Transgenic *Populus* characterized by over-expression of pine cytosolic glutamine synthetase gene exhibits other beneficial phenotypes including enhanced tolerance to water stress (El-Khatib et al. 2004) and enhanced nitrogen use efficiency (Man Hui-min et al. 2005).

Another enzyme in nitrogen assimilation is Fdx-GOGAT, which is predominant in leaves and believed to act to recycle NH_4^+ released in photorespiration. NADH-GOGAT is expressed more in roots and cotyledons, so it is believed to recycle NH_4^+ from catabolism of amino acids, including senescence. A strategy of over-expressing Fdx-GOGAT using a strong constitutive promoter should increase NH_4^+ recycling. When NADH-GOGAT was over-expressed in tobacco tissues, there was a 30% increase in foliar biomass but the nitrogen to carbon ratio remained unchanged (Chichkova et al. 2001). It is not known whether the level of GS becomes limiting in these transgenic plants, but it is conceivable that a stacked GS plus GOGAT construct would further improve nitrogen utilization further.

Genes involved in cell wall development have also been shown to affect tree growth. A β-1,4-endoglucanase (*cel*1) involved in cell wall modification during cell growth has improved growth in *Populus* (Shani et al. 2004). Genes involved in leaf size and structure, stem development, timing of bud flush and leaf senescence all may influence biomass accumulation, rotation time and growth rate. For example, it has been shown that a gibberellin catabolism gene, GA 2-oxidase can affect tree height (Busov et al. 2003). It is also possible that genes that confer stress tolerance or delay senescence could also improve growth under environmental stress conditions or could extend the growing season, and it has been shown that suppression of a gene, deoxyhypusine synthase (DHS) which is part of the stress-response pathway, increases vegetative and reproductive growth in the model plant *Arabidopsis* (Duguay et al. 2007).

Pinus Loblolly pine, as a native North American species, has the advantage of wide adapatability across sites at less than 2,000 ft in elevation. Currently, it is the most widely planted forestry species in the world, with an average 900 million seedlings planted annually in the southeastern United States alone (McKeand et al. 2003). Its wood, because of its lignin chemistry, is currently best suited for bioenergy applications that utilize direct firing or gasification technologies, although scientists believe that enzymatic processes might also be utilized in the future (Frederick et al. 2008). However, the economic limitation for loblolly pine in biofuel and bioenergy applications is its relatively long rotation time (15 yr for pulp wood applications and 23 yr for sawtimber applications).

To address this limitation, ArborGen and other tree-breeding organizations are employing advanced breeding and crossing methods to develop high-performing traditional seedlings that have improved growth, disease resistance, and form. ArborGen and two other companies (CellFor and Weyerhaeuser) have also utilized a tissue culture process called somatic embryogenesis to mass-propagate selected elite loblolly genotypes. Improvements in traditional breeding and selection are predicted to achieve 35% volume gains and sawtimber rotation times of approximately 20 yr. Biotech gene insertion methods will be necessary to develop loblolly pine with the productivity levels most desirable for bioenergy applications. Early research results indicate that rotation times of 15 yr may be possible. ArborGen has introduced genes into loblolly pine that demonstrated nearly double the normal biomass production in the first 3 yr of field trials (unpublished results).

Increased Biomass Productivity through Genetic Improvement of Introduced Species

Eucalyptus is an ideal energy crop with certain species and hybrids having excellent biomass productivity, relatively low lignin content and a short rotation time. In Brazil, commonly planted *Eucalyptus* hybrids such as *Eucalyptus grandis* × *Eucalyptus urophylla*, routinely yield 10–12 dry tons/acre/yr. A study with *E. grandis* in Florida indicated that this species could achieve total biomass productivity values exceeding 30 green tons (~15 dry tons) /acre/yr, with the potential to reach 55 green tons/acre/yr (Stricker et al. 2000). This scale of productivity addresses the biomass requirements for cost-effective generation of biofuels and bioenergy from lignocellulosic feedstocks. *Eucalyptus* species and hybrids with this level of productivity are adapted to growth in the tropics and are highly sensitive to freezing temperatures.

Through a significant advance in the understanding of freezing tolerance in *Arabidopsis*, a freeze-tolerant *Eucalyptus* has been developed. This advance was the discovery of the C-repeat/dehydration-responsive element binding factor (*CBF/DREB*) cold-response pathway (Jaglo-Ottosen et al. 1998; Liu et al. 1998). The functions of *Arabidopsis* CBF genes and CBF homologs identified in many species of plants have been studied extensively using transgenic plants. Over-expression of *CBF* genes conferred freezing tolerance and drought tolerance, as well as salt tolerance in *Arabidopsis* (Liu et al. 1998; Kasuga et al. 1999). Over-expression of *Arabidopsis CBF* genes in *Brassica napus* and tobacco induced the expression of orthologs of *Arabidopsis CBF*-targeted genes and increased freezing and drought tolerance of transgenic plants (Jaglo-Ottosen et al. 2001; Kasuga et al. 2004). Heterologous expression of *Arabidopsis CBF1* resulted in enhanced chilling tolerance but not freezing tolerance in transgenic tomato (Hsieh et al. 2002; Zhang et al. 2004). Over-expression of a *CBF* homolog obtained from pepper, *CaPF1*, in tobacco and *Arabidopsis* improved freezing tolerance of the transgenic plants (Yi et al. 2004). Ectopic expression of *Arabidopsis CBF1* in *Populus* increased freezing tolerance of the transgenic *Populus* (Benedict et al. 2006). In rice, four *CBF* homologs have been identified, and over-expression of one of the homologs (*OsDREB1A*) in *Arabidopsis* conferred enhanced freezing tolerance and high-salinity tolerance (Dubouzet et al. 2003; Ito et al. 2006). A *CBF* homolog, *ZmDREB1A*, was found in maize, and over-expression of this gene in *Arabidopsis* resulted in improved freezing and drought tolerance (Qin et al. 2004). Several *CBF* homologs were identified in wheat, and it was found that these genes clustered in a chromosome locus. It has been shown that this locus mediated *Cor/Lea* gene expression and freezing tolerance in common wheat (Jaglo-Ottosen et al. 2001; Kobayashi et al. 2005; Vagujfalvi et al. 2005). All of the above research strongly suggests that CBF genes are playing an important role in stress tolerance, especially the freezing tolerance, in plants.

The *CBF2* gene is part of the C-repeat/dehydration-responsive element binding factor (*CBF/DREB*) cold-response pathway (Jaglo-Ottosen et al. 1998; Zhang et al. 2004). It is known from the literature that over-expression of CBF genes under control of a constitutive promoter can increase cold tolerance but can also promote dwarfing (Zhang et al. 2004). To overcome this problem, stress-inducible plant promoters with a low background expression level have been used in conjunction with the cold tolerance genes (Yamaguchi-Shinozaki and Shinozaki 1994).

ArborGen has introduced the *Arabidopsis* CBF2 transcription factor driven by the *Arabidopsis* rd29a stress-inducible promoter (Yamaguchi-Shinozaki and

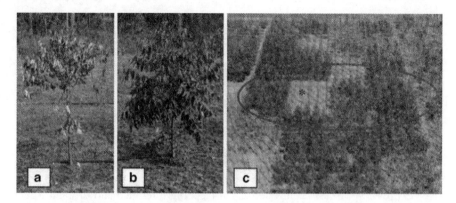

Fig. 8.1 A Eucalyptus hybrid, with or without the addition of a freeze tolerance gene, after a typical winter in the Southeast United States. (**a**) photograph of control Eucalyptus after winter temperatures of 16°F in South Carolina, (**b**): rd29a::CBF2 transgenic EH1, photograph taken from the same field trial and time as the tree in photograph (**a**). Photograph (**c**) an aerial photograph of block plots of different lines of rd29a::CBF2 *Eucalyptus* in a field trial in Alabama after winter temperatures of 19°F. A control tree block is marked with a "The * marking the control plot is missing. Is it possible to add that over the leafless block in the center of the circle in photo "c"?", and all other similar blocks are also control blocks

Shinozaki 1993) into a highly productive tropical *Eucalyptus E. grandis* × *E. urophylla* genotype. The new variety, Freeze-tolerant Eucalyptus, has demonstrated tolerance to 16°F across multiple years and multiple field trial locations while essentially maintaining its exceptional productivity (ArborGen unpublished data; Fig 8.1).

The yields achievable with Freeze-tolerant Eucalyptus are predicted to meet or exceed those that have been defined by DOE and others for the long-term-feasibility of renewable energy production (i.e., 8 to 10 dry tons/acre/yr; English et al. 2006). The application of total biomass-driven management systems could further increase yields and reduce delivered costs. As with many other hardwood species, an added benefit of Freeze-tolerant Eucalyptus is its ability to coppice when managed appropriately. Coppicing allows for subsequent crops without the added costs of establishment (site preparation, seedling, and planting costs), which can provide a higher return to landowners. Coppice crops can show increases in productivity relative to the initial single-stem harvest (Sims 2001), but coppice yields will decline over time. Re-planting will then become economically attractive as new varieties become available.

Improvements like those discussed above will be necessary to make forest trees a sustainable and economical feedstock option for the production of cellulosic ethanol and other forms of bioenergy. Table 8.2 summarizes the theoretical acreage needed to meet the "advanced biofuels" target in the 2007 Renewable Fuel Standard (RFS) in the southeastern United States based on current productivity assumptions for loblolly pine and *Eucalyptus* under pulpwood and high-density coppicing scenarios.

Table 8.2 Approximate productivity and total planted acreage needed to meet the Renewable Fuel Standard (RFS) in the southeastern United States using purpose-grown pine or *Eucalyptus*

	Pinus taeda	*Eucalyptus urograndis*	
		Pulpwood management	Total biomass management
Productivity (green tons/acre/yr)	10^z	20^y	30^x
Planted acres (million) needed to meet target 118 million green tons/yr^a	17	6	4

On December 19, 2007, the Energy Independence and Security Act of 2007 (H.R. 6) was signed into law. This comprehensive energy legislation amends the Renewable Fuel Standard (RFS) signed into law in 2005, growing to 36 billion gallons of biofuels available in 2022

^{z}ArborGen, unpublished data, assumes as 10-yr rotation with a planting density of 1,000 trees per acre

^{y}ArborGen, unpublished data, assumes as 7-yr rotation, with 450 trees/acre

^{x}ArborGen unpublished data, assumes an average product of an initial harvest at 3 yr, followed by three coppice rotations of approximately 3 yr coppice rotation, using a similar coppicing regime as described in Sims 2001)

Altering Wood Quality to Improve Feedstock Conversion Efficiency

Once woody biomass can be grown productively with a cost-effective delivered cost at the conversion plant, improvements in the wood itself will have great value for biofuels or bioenergy conversion. If wood is used as a feedstock for ethanol production, there are two general target areas for modifying the wood to increase ethanol yield per ton of wood. Foremost, the polysaccharides should be made more easily degradable and accessible to the enzymes and/or microorganisms used to break them down. Another avenue is to reduce the concentrations of compounds that inhibit the fermentation of the sugars once they are released. These factors contribute to the "recalcitrance" of cellulosic feedstocks to saccharification relative to sucrose and starch. The DOE has funded several bioenergy science centers to address biological and technological barriers to cost-effective production of biofuels from lignocellusic feedstocks. One such center, the BioEnergy Science Center managed by Oak Ridge National Laboratory, has focused its research on the reducing factors that contribute to the recalcitrance of wood in biofuels conversion (see http://genomicsgtl.energy.gov/centers/center_ORNL.shtml).

One approach that addresses both areas outlined above would be to reduce the lignin content of the wood or at least make it easier to remove. Reduction of lignin has long been a target of interest in crop and forestry species, because lignin interferes with digestion of plant materials by farm animals and removal of lignin is a costly step in the production of paper. This has motivated researchers to identify and isolate from a variety of species many of the genes coding for enzymes in the lignin pathway. A large body of research involving manipulation of the lignin biosynthetic pathway in transgenic plants has accumulated over the past 15 yr; this has been exhaustively reviewed (Anterola and Lewis 2002; Boerjan et al. 2003; Li et al. 2008; Vanholme et al. 2008; Weng et al. 2008). In broad strokes, experiments have shown that lignin content can be significantly reduced in trees and herbaceous plants. For example, several of the transgenic plants with down-regulated genes for lignin biosynthetic enzymes in the early steps of the pathway showed that syringyl lignin (S-lignin) content was more strongly affected than guaiacyl lignin (G-lignin) content (Vanholme et al. 2008).

Optimal reduction of lignin with negligible negative effects will require use of carefully selected promoters and target genes, as lignin reduction is often seen with other associated effects. Examples of pleiotropic effects from lignin reduction in *Populus* are: (1) the accumulation of sugars in leaves and concomitant reduction of photosynthetic capacity following down-regulation of *p*-coumaroyl shikimate 3'-hydroxylase (Coleman et al. 2008) and (2) the reduction of hemicellulose that occurred with down-regulation of cinnamoyl-coenzyme A reductase (Leplé et al. 2007). Increased cavitation and vessel collapse have also been observed in some plants with reduced lignin (Coleman et al. 2008). Another means of modifying lignin production that may address these negative pleiotropic effects is to alter the expression of transcription factors or other regulators which in turn modify the expression of a whole suite of genes that participate in the lignin biosynthesis pathway. Masaru Ohme-Takagi and colleagues showed that two plant-specific transcription factors, designated NAC secondary wall thickenings promoting factor 1 (NST1) and NST3, regulate the formation of secondary walls in woody tissues of *Arabidopsis*. (Mitsuda et al. 2007).

An increased syringyl lignin to guaiacyl lignin (S/G) ratio is thought to be desirable because the more oxygen-rich S-lignin is easier to remove through chemical treatment during pulping (Chiang and Funaoka 1990). This can be achieved by over-expression of ferulate 5-hydroxylase (F5H), also known as coniferaldehyde 5-hydroxylase (Cald5H; Huntley et al. 2003). Interestingly, it has been reported that the composition of lignin does not have a strong effect on biological degradation of cell walls (Grabber et al. 1997).

The polysaccharides themselves contribute a large part to recalcitrance because of their insolubility. In both conifer and hardwood (angiosperm) secondary cell walls, cellulose is approximately 45% of the dry weight of the wood while hemicellulose is approximately 20%. Hemicellulose composition varies strongly depending upon the wood source. In the hemicellulose of hardwoods such as *Populus*, xylans comprise about 80% with the remainder being mannans (10%), galactans (5%), and arabinans (<5%). In pine, mannans comprise about 50%, xylans 30%, galactans 10%, and arabinans 5% of the total hemicellulose. Although these composition percentages reflect data obtained by the authors in their own research, these closely match other published numbers for hardwoods and conifers (Rowell 2005).

If wood is to be used simply as a fuel for burning, the two obvious targets for modification are increasing density and increasing lignin content. Increasing wood density has the potential of increasing the yield of fuel per acre and decreasing transportation and storage costs (increasing the energy yield per truckload of wood or chips), although the amount of energy per unit weight of wood would remain the same. Production of biofuels via gasification would also benefit from these aspects of increased wood density. It seems unlikely that it would be possible to insert significantly more material into the cell wall matrix, but decreasing the ratio of lumen volume to cell wall volume is a plausible route to increasing wood density. The methods by which the lumen-to-wall ratio can be decreased are rather speculative because the molecular biology of xylem development is only beginning to be understood. Goicoechea et al. (2005) describes how strong expression of a *Eucalyptus gunnii* MYB transcription factor (EgMYB2) in tobacco led to significant thickening of xylem fiber cell walls. Because the transcription factor was first isolated based on its ability to bind the promoters of lignin biosynthetic genes, and these genes are up-regulated in the EgMYB2 over-expressing lines (Goicoechea et al. 2005), it is possible that the phenotype is due to an overall rate increase in deposition of the S2 layer of the secondary cell wall.

Increasing lignin content would increase the thermal energy of wood. Pure cellulose has a calorific value of ~8,000 BTU/lb; pure lignin is ~11,000 BTU/lb (White 1987). Increasing lignin content from 25% to 35% would increase the calorific value of wood by approximately 450 BTU/lb. This might be accomplished by achieving the opposite of the lignin reduction strategies mentioned above. The over-expression of an enzyme that is a kinetic barrier in the pathway is one possible approach, as is up-regulation of the whole set of biosynthetic genes with a regulatory protein. There have been several reports of increased lignin deposition due to over-expression of MYB transcription factors (Patzlaff et al. 2003; Goicoechea et al. 2005).

As the DOE has funded 3 collaborative centers for research focused on improving lignocellulosic feedstock and related downstream processes for improved conversion efficiency, it is anticipated that within the next 5 yr much more will be known about genes that can be used to address the recalcitrance of plant cell walls to enzymatic biofuels conversion methods.

Regulatory Requirements and Associated Risk Assessment for Biotech Trees

It is likely that the use of woody feedstock for liquid biofuels production will require biotech traits in order to provide an economically feasible process. Currently, no biotech trees are planted for industrial forestry in the United States. Biotech trees for biofuels production will require regulatory oversight and de-regulation prior to being commercialized.

Domestication, breeding, and selection of plants, including forest tree species, has resulted in direct and indirect change in the genetic makeup of plants grown for food and industrial applications. Forest tree domestication accelerated during the latter half of the twentieth century with conventional breeding methods applied to forest tree populations to improve growth, volume, and wood quality traits (Burdon and Libby 2006). The application of biotechnology to forest tree species is expected to further accelerate such improvement (Sedjo 2001), including new developments for bioenergy applications. Commercialization of improved planting stocks, based on new varieties generated through clonal propagation and advanced breeding programs, as well as further improved biotech trees with high value traits, will occur in the near future. These trees will further enhance the quality and productivity of plantation forests (Nehra et al. 2005) and will provide a renewable resource for industrial applications.

In 1986, the United States developed the Coordinated Framework for Regulation of Biotechnology. A key aspect of the framework was the understanding that the characteristics, composition, and intended use of the genetically modified (GM) product were the important considerations in their regulation, not the methods by which they were developed. Under this framework, the United States Department of Agriculture, Animal and Plant Health Inspection Service (USDA-APHIS), Food and Drug Administration (FDA), and Environmental Protection Agency (EPA) coordinate the regulation of GM crops. The agencies' roles in assessing the safety and approval of GE products depend on the intended use of the product; for example, FDA oversees food and feed uses while EPA oversees insect pest management applications (Re et al. 1996; Nehra et al. 2005; Just et al. 2006). APHIS plays a major role in overseeing field testing, risk assessment and approval of genetically modified plants for planting. In 2002, APHIS created the Biotechnology Regulatory Services (BRS) unit that manages all activities with respect to risk assessment of genetically modified organisms, including tree species (www.aphis.usda.gov/brs/). Developers must obtain permits from BRS prior to any release into the environment in field trials. BRS conducts in-depth analyses as part of the permitting process for field trials and their assessment of petitions for non-regulated status. All genetically modified plants are considered as regulated articles by BRS, and prior to commercialization developers provide data from field trials and other analyses to BRS in the form of a petition requesting non-regulated status. APHIS-BRS oversight of genetically modified plants is provided by the Plant Protection Act of 2000. This Act gives them the authority to assess any potential noxious weed or plant pest risk. BRS' assessment also involves analyses in fulfillment of its obligations under the National Environmental Policy Act (NEPA) and Endangered Species Act. The US system for oversight and regulation of transgenic plants has worked effectively for more than 23 yr, ensuring the safety of biotech crops and protection of the environment. BRS is now considering a number of proposals for further improvement to make regulation more streamlined and proportional to product novelty and risk (http://federalregister.gov/).

Since the commercialization of the first biotech crops in 1996, the adoption rate of biotech crops has increased rapidly worldwide, with more than 2 billion acres planted in 25 countries (ISAAA 2008). Field tests of genetically modified trees are being conducted in several countries, with the majority of these field tests occurring in the US (van Frankenhuyzen and Beardmore 2004; http://www.isb.vt.edu/). To date, only two tree species, papaya and plum, have been granted non-regulated status for planting in the United States; these fruit trees are resistant to viruses that can have devastating impacts on fruit production and quality. There are currently no large-scale plantings of transgenic forest trees other than an insect resistant poplar that is being grown in China (Hu et al. 2001). More recently, a petition was submitted to BRS requesting non-regulated status for the Freeze-tolerant Eucalyptus hybrid described above.

It is recognized that the use of biotechnology for tree improvement can bring significant economic, social and environmental benefits, but some concerns must also be addressed (van Frankenhuyzen and Beardmore 2004), particularly those associated with the potential dispersal of pollen, seeds or vegetative propagules. Gene flow via pollen and seed dispersal is an important natural phenomenon for genetic improvement and evolution in plant species. The potential for gene flow from crops, be they from traditional breeding or developed through biotechnology, is considered to be less for non-native self-pollinated crops compared to native and wind pollinated species. Perennial wind pollinated species models predict that a small proportion of pollen and seed can travel long distance (Nathan et al. 2002; Williams 2005; Williams and Davis 2005). However, for there to be any consequences of such dispersal these models assume that: (1) viable pollen is able to fertilize receptive ovules of a related species resulting in viable seed production, followed by establishment of this seed in the environment and (2) adverse consequences can occur only if the inserted genes are considered a significant risk to other organisms or can cause unintended effects on the fitness of the species. Traits including improved growth, wood quality, and abiotic or biotic stress tolerance that are also being altered via traditional breeding might be considered as being inherently low risk, particularly when using genes from the tree itself, or other plant genes that are homologous to genes already present in the tree, where no unintended phenotypes have been observed after extensive field testing. In some cases, especially for biomass production for bioenergy uses, short-rotation trees may not even produce abundant pollen or seed prior to harvest. In addition, there are proven and well-tested technologies that selectively prevent or reduce pollen formation without affecting other functions of the plant species (Gomez Jimenez et al. 2006; Nasrallah et al. 1999; Yanofsky 2006). ArborGen has adapted this technology for use in tree species and demonstrated high levels of efficacy (Fig 8.2). Therefore, a number of tools exist that can minimize the potential risk of gene flow via pollen or seed dispersal from plantations. It is important, however, that any risk assessments

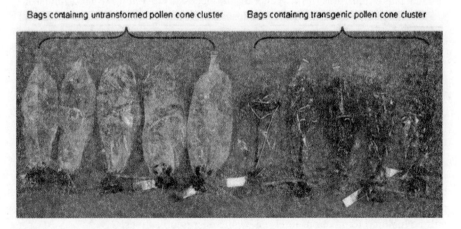

Fig. 8.2 Bags containing male cone clusters of untransformed control and transgenic lines of *P. taeda* (ArborGen unpublished data). *Yellow-colored* pollen is clearly visible inside bags containing untransformed male cones (*left*), while no pollen was found inside the bags containing male cones of lines transformed with genes for pollen ablation (*right*)

for trees take into account scientifically informed arguments for incorporating any such gene flow control mechanisms on a case-by-case basis.

Conclusion

The high productivity of purpose-grown, short-rotation trees, such as Freeze-tolerant Eucalyptus, is expected to improve the economic feasibility of bioenergy and biofuels production from woody biomass plantations. Bioenergy is already becoming a substantial market outlet for wood. According to TimberMart-South (2008), 16 new bioenergy projects were announced for the U.S. south within the last 2 yr, with an anticipated increase in wood consumption of 9 million green tons. The development of a bioenergy sector in the southeastern United States holds great economic promise, and it is anticipated that purpose-grown, short-rotation trees will be planted to address the 120 million green tons of biomass that will be needed annually as a feedstock for advanced biofuels and bioenergy. At an estimated price of $20 to $30 per green ton, this represents 2 to 4 billion dollars in economic opportunity associated with biomass production for the southeastern United States. Upside demand exists because of the suitability of wood for other bioenergy applications, such as the production of electricity through direct burning of wood or co-firing with coal.

The inherent logistical benefits of trees in combination with the high productivity of new varieties of short-rotation trees, such as Freeze-tolerant Eucalyptus, make it an ideal biomass for traditional industrial end uses such as pulp and paper, as well as for energy products such as cellulosic ethanol and electric power generation. Short-rotation trees will generate more wood on less land, requiring a smaller plantation footprint to generate the necessary dry tons to feed industrial processing plants. This, in turn, will lessen pressure to harvest from native and old-growth forests in order to meet society's demand for pulp, paper and energy. The addition of biotech traits to elite varieties of purpose-grown trees will achieve the required rotation times and productivity to make lignocellulosic feedstocks cost effective.

The choice of which energy crops to plant must take into consideration regional conditions and needs, both in minimizing transportation costs as well as in avoiding the current long-distance distribution limitations of ethanol. In the southeastern United States, where accessible inventory and harvesting infrastructure for forestry operations are already well established, trees provide a clear advantage for biomass production compared to annual crops. Although trees will play a significant role in helping to meet renewable energy standards, it is recognized that multiple, integrated approaches with a variety of different crop species and production systems will be required to meet our total renewable energy objectives.

Open Access This article is distributed under the terms of the Creative Commons Attribution Noncommercial License which permits any noncommercial use, distribution, and reproduction in any medium, provided the original author(s) and source are credited.

References

Andersson G.; Asikainen A.; Bjorheden R.; Hall P. W.; Hudson J. B.; Jirjis R.; Mead D. J.; Nurmi J.; Weetman G. F. Production of forest energy. In: Richardson J.; Bjorheden R.; Hakkila P.; Lowe A. T.; Smith C. T. (eds) Bioenergy from sustainable forestry: guiding principles and practice. Kluwer, The Netherlands; 2002. Dordrecht 49-123.

Anterola A. M.; Lewis N. G. Trends in lignin modification: a comprehensive analysis of the effects of genetic manipulations/mutations on lignification and vascular integrity. *Phytochemistry* 61: 221–294; 2002.

Benedict C.; Skinner J. S.; Meng R.; Chang Y.; Bhalerao R.; Huner N. P.; Finn C. E.; Chen T. H.; Hurry V. The CBF1-dependent low temperature signaling pathway, regulon and increase in freeze tolerance are conserved in *Populus* spp. *Plant Cell Environ.* 29: 1259–1272; 2006.

Boerjan W.; Ralph J.; Baucher M. Lignin biosynthesis. *Ann. Rev. Plant Biol.* 54: 519–546; 2003.

Burdon R. W.; Libby W. J. Genetically modified forests: from Stone age to modern biotechnology. Forest History Society, Durham, North Carolina, USA; 2006.

Busov V. B.; Meilan R.; Pearce, D. W.; Ma C.; Rood S. B.; Strauss S. H. Activation tagging of a dominant gibberellin catabolism gene (*GA 2-oxidase*) from poplar that regulates tree stature. *Plant Phys.* 132: 1–9; 2003.

Chiang V. L.; Funaoka M. The difference between guaiacyl and guaiacyl-syringyl lignins in their responses to Kraft delignification. *Holzforschung* 44: 309–313; 1990.

Chichkova S.; Arellano J.; Vance C. P.; Hernandex G. Transgenic tobacco plants that overexpress alfalfa NADH-glutamate synthase have higher carbon and nitrogen content. *J. Exp. Bot.* 52: 2079–2084; 2001.

Coleman H. D.; Samuels A. L.; Guy R. D.; Mansfield S. D. Perturbed lignification impacts tree growth in hybrid poplar—a function of sink strength, vascular integrity, and photosynthetic assimilation. *Plant Phys.* 148: 1229–1237; 2008.

Davis A. A.; Trettin C. C. Sycamore and sweetgum plantation productivity on former agricultural land in South Carolina. *Biomass and Bioenergy.* 30: 769–777; 2006.

Dickmann D. L. Silviculture and biology of short rotation woody crops in temperate regions: then and now. *Biomass Bioenergy.* 30: 696–705; 2006.

Dubouzet J. G.; Sakuma Y.; Ito Y.; Kasuga M.; Dubouzet E. G.; Miura S.; Seki M.; Shinozaki K.; Yamaguchi-Shinozaki K. OsDREB genes in rice, *Oryza sativa* L., encode transcription activators that function in drought-, high-salt- and cold-responsive gene expression. *Plant J* 33: 751–763; 2003.

Duguay J.; Jamal S.; Wang T. W.; Thompson J. E. Leaf-specific suppression of deoxyhypusine synthase in *Arabidopsis thaliana* enhances growth without negative pleiotropic effects. *J. Plant Phys.* 164: 408–420; 2007.

El-Khatib R.; Hamerlynck E. P.; Gallardo F.; Kirby E. G. Transgenic poplar characterized by ectopic expression of a pine cytosolic glutamine synthetase gene exhibits enhanced tolerance to water stress. *Tree Phys.* 24: 729–736; 2004.

English B. C.; De La Torre, Ugarte D. G.; Jensen K.; Hellwinckel C.; Menard J.; Wilson B.; Roberts R.; Walsh M. (2006) 25% Renewable energy for the United States by 2025: Agricultural and Economic Impacts. University of Tennessee Agricultural Economics, http://www.25x25.org/storage/25x25/documents/RANDandUT/UT-EXECsummary25X25FINALFF.pdf 2006.

Frederick Jr. W. J.; Lien S. J.; Courchene C. E.; DeMartini N. A.; Ragauskas A. J.; Iisa K. Production of ethanol from carbohydrates from loblolly pine: a technical and economic assessment. *Bioresour Technol* 99: 5051–5057; 2008.

Fu J.; Sampalo R.; Gallardo F.; Canavos F. M.; Kirby E. G. Assembly of a cytosolic pine glutamine synthetase holoenzyme in leaves of transgenic poplar leads to enhanced vegetative growth in young plants. *Plant Cell Envir.* 26: 411–418; 2003.

Goicoechea M.; Lacombe E.; Legay S.; Mihaljevic S.; Rech P.; Jauneau A.; Lapierre C.; Pollet B.; Verhaegen D.; Chaubet-Gigot N.; Grima-Pettenati J. EgMYB2, a new transcriptional activator from *Eucalytpus* xylem, regulates secondary cell wall formation and lignin biosynthesis. *Plant J.* 43: 553–567; 2005.

Gomez Jimenez M. D.; Canas Clemente L. A.; Madueno Albi F.; Beltran Porter J. P. Sequence regulating the anther-specific expression of a gene and its use in the production of androsterile plants and hybrid seeds. U.S. Patent No. 7078593; 2006.

Grabber J. H.; Ralph J.; Hatfield R. D.; Quideau S. p-hydroxyphenyl, guaiacyl, and syringyl lignins have similar inhibitory effects on cell wall degradation. *J. Agric. Food Chem.* 45: 2530–2532; 1997.

Good A. G.; Swarat A. K.; Muench D. G. Can less yield more? Is reducing nutrient input into the environment compatible with maintaining crop production? *Trends Plant Sci.* 9(597–6): 05; 2004.

Hsieh T. H.; Lee J. T.; Yang P. T.; Chiu L. H.; Charng Y. Y.; Wang Y. C.; Chan M. T. Heterologous expression of the *Arabidopsis* C-repeat/dehydration response element binding factor 1 gene confers elevated tolerance to chilling and oxidative stresses in transgenic tomato. *Plant Physiol.* 129: 1086–1094; 2002.

Hu J. J.; Tian Y. C.; Han Y. F.; Li L.; Zhang B. E. Field evaluation of insect resistant transgenic *Populus nigra* trees. *Euphytica* 121: 123–127; 2001.

Huntley S. K.; Ellis D.; Gilbert M.; Chapple C.; Mansfield S. D. Significant increases in pulping efficiency in C4H–F5H-transformed poplars: improved chemical savings and reduced environmental toxins. *J. Agric. Food Chem.* 51: 6178–6183; 2003.

ISAAA. The International service for the acquisition of Agri-Biotech Applications (ISAAA) report, Global Status of Commercialized Biotech/GM Crops: 2008, http://www.isaaa.org 2008.

Ito Y.; Katsura K.; Maruyama K.; Taji T.; Kobayashi M.; Seki M.; Shinozaki K.; Yamaguchi-Shinozaki K. Functional analysis of rice DREB1/CBF-type transcription factors involved in cold-responsive gene expression in transgenic rice. *Plant Cell Phys.* 47: 141–153; 2006.

Jaglo-Ottosen K. R.; Gilmour S. J.; Zarka D. G.; Schabenberger O.; Thomashow M. F. *Arabidopsis* CBF1 overexpression induces COR genes and enhances freezing tolerance. *Science* 280: 104–106; 1998.

Jaglo-Ottosen K. R.; Kleff S.; Amundsen K. L.; Zhang X.; Haake V.; Zhang J. Z.; Deits T.; Thomashow M. F. Components of the *Arabidopsis* C-repeat/dehydration-responsive element binding factor cold-response pathway are conserved in *Brassica napus* and other plant species. *Plant Phys.* 127: 910–917; 2001.

Just R. E., Alston J. M., Zilberman D. (eds.). Regulating Agricultural Biotechnology: Economics and Policy. Springer, New York; 2006.

Kasuga M.; Liu Q.; Miura S.; Yamaguchi-Shinozaki K.; Shinozaki K. Improving plant drought, salt, and freezing tolerance by gene transfer of a single stress-inducible transcription factor. *Nature Biotech* 17: 287–291; 1999.

Kasuga M.; Miura S.; Shinozaki K.; Yamaguchi-Shinozaki K. A combination of the *Arabidopsis* DREB1A gene and stress-inducible rd29A promoter improved drought- and low-temperature stress tolerance in tobacco by gene transfer. *Plant Cell Physiol.* 45: 346–350; 2004.

Kobayashi F.; Takumi S.; Kume S.; Ishibashi M.; Ohno R.; Murai K.; Nakamura C. Regulation by Vrn-1/Fr-1 chromosomal intervals of CBF-mediated Cor/Lea gene expression and freezing tolerance in common wheat. *J Exp Bot.* 56: 887–895; 2005.

Leplé J.-C.; Dauwe R.; Morreel K.; Storme V.; Lapierre C.; Pollet B.; Naumann A.; Kang K.-Y.; Kim H.; Ruel K.; Lefèbvre A.; Joseleau J.-P.; Grima-Pettenati J.; De Rycke R.; Andersson-Gunneräs S.; Erban A.; Fehrle I.; Petit-Conil M.; Kopka J.; Polle A.; Messens E.; Sundberg B.; Mansfield S. D.; Ralph J.; Pilate G.; Boerjan W. Down regulation of cinnamoyl-coenzyme A reductase in poplar: multiple-level phenotyping reveals effects on cell wall polymer metabolism and structure. *Plant Cell* 19: 3669–3691; 2007.

Li X.; Jing-Ke Weng J.-K.; Chapple C. Improvement of biomass through lignin modification. *Plant J.* 54: 569–581; 2008.

Liu Q.; Kasuga M.; Sakuma Y.; Abe H.; Miura S.; Yamaguchi-Shinozaki K.; Shinozaki K. Two transcription factors, DREB1 and DREB2, with an EREBP/AP2 DNA binding domain separate two cellular signal transduction pathways in drought- and low-temperature-responsive gene expression, respectively, in *Arabidopsis*. *Plant Cell* 10: 1391–406; 1998.

Man Hui-min R.; Boriel R.; El-Khatib R.; Kirby E. G. Characterization of transgenic poplar with ectopic expression of pine cyotsolic glutamine synthetase under conditions of varying nitrogen availability. New Phytol (167): 31–39; 2005.

McKeand S.; Mullin T.; Byram T.; White T. Deployment of genetically improved loblolly and slash pine in the South. *J. Forestry.* 101(3): 32–37; 2003.

Mead D. J. Forests for energy and the role of planted trees. *Crit. Rev. Plant sci* 24: 407–421; 2005.

Mercker D. Short rotation woody crops for biofuels. University of Tennessee Agricultural Experiment Station. http://www.utextension.utk.edu/publications/spfiles/SP702-C.pdf 2007

Mitsuda N.; Iwase, A.; Yamamoto H.; Yoshida, M.; Seki, M.; Shinozaki K.; Ohme-Takagi M. NAC transcription factors, NST1 and NST3, are key regulators of the formation of secondary walls in woody tissues of *Arabiodopsis*. *Plant Cell* 19: 270–280; 2007.

Nasrallah M. E.; Nasrallah J. B.; Thorsness M. K. Isolated DNA elements that direct pistil-specific and anther-specific gene expression and methods of using same. United States Patent No. 5,859,328; 1999.

Nathan R.; Katul G. G.; Horn H. S.; Thomas S. M.; Oren R.; Avissar R.; Pacala S. W.; Levin S. A. Mechanisms of long-distance dispersal of seeds by wind. *Nature* 418: 409–413; 2002.

Nehra N. S.; Becwar M. R.; Rottmann W. H.; Pearson L.; Chowdhury K.; Chang S.; Wilde H. D.; Kodrzycki R. J.; Zhang C.; Gause K. C.; Parks D. W.; Hinchee M. A. Forest biotechnology: innovative methods, emerging opportunities. *In Vitro Cell. Dev. Biol. Plant.* 41: 701–717; 2005.

Patzlaff A.; McInnis S.; Courtenay A.; Surman C.; Newman L. J.; Smith C.; Bevan M. W.; Mansfield S.; Whetten R. W.; Sederoff R. R.; Campbell M. M. Characterisation of a pine MYB that regulates lignification. *Plant J.* 46: 743–754; 2003.

Perlack R. D.; Turhollow W. L. L.; AF G. R. L.; Stokes B. J.; Erbach D. C. Biomass as a feedstock for a bioenergy and bioproducts industry: the technical feasibility of a billion-ton annual supply. US Department of Energy, Oak Ridge National Laboratory, Oak Ridge, TN; 2005.

Polle A.; Altman A.; Jiang X. Towards genetic engineering for drought tolerance in trees. In: Fladung M.; Ewald D. (eds) Tree transgenesis recent developments. Springer, Berlin; 2006.

Qin F.; Sakuma Y.; Li J.; Liu Q.; Li Y. Q.; Shinozaki K.; Yamaguchi-Shinozaki K. Cloning and functional analysis of a novel DREB1/CBF transcription factor involved in cold-responsive gene expression in *Zea mays* L. *Plant Cell Phys.* 45: 1042–1052; 2004.

Re D. B.; Rogers S. G.; Stone T. B.; Serdy F. S. Herbicide tolerant plants developed through biotechnology: regulatory considerations in the United States. In: Duke S. O. (ed) Herbicide resistant crops. CRC, New York, pp 341–347; 1996.

Rowell R. M. Handbook of Wood Chemistry and wood composites. Taylor & Francis, a CRC Press Book. ISBN 0849315883, 9780849315886; 2005.

Sedjo R. A. Biotechnology in forestry: considering the costs and benefits. *Resour. Future* 145: 10–12; 2001.

Shani Z.; Dekel M.; Tsabary G.; Goren R.; Shoseyov O. Growth enhancement of transgenic poplar plants by over expression of *Arabidopsis thaliana* endo-1, 4-β–glucanase (cel1). *Mol Breed* 14: 321–330; 2004.

Short Rotation Forestry Handbook. University of Aberdeen. http://www.abdn.ac.uk/wsrg/srfh-book 1995.

Sims R. H. Short rotation coppice tree species selection for woody biomass production in New Zealand. *Biomass Bioenergy* 20: 329–335; 2001.

Sims R. H.; Venturi P. All year-round harvesting of short rotation coppice *Eucalyptus* compared with the delivered costs of biomass from more conventional short season, harvesting systems. *Biomass Bioenergy* 26: 27–37; 2004.

Stricker J. A.; Rockwood D. L.; Segrest S. A.; Alker G. R; Prine G. M.; Carter D. R. Short Rotation Woody Crops For Florida. University of Florida http://www.treepower.org/papers/strickerny.doc 2000.

TimberMart-South Market News Quarterly 2008 13:1 pg. 28. http://www.tmart-south.com/tmart/pdf/Qtr_01Q08news.pdf

Tuskan G.; DiFazio S.; Hellsten U.; Jansson S.; Rombauts S.; Putnam N.; Sterck L.; Bohlmann J.; Schein J.; Bhalerao R. R.; Bhalerao R. P.; Blaudez D.; Boerjan W.; Brun A.; Brunner A.; Busov V.; Campbell M.; Carlson J.; Chalot M.; Chapman J.; Chen G.; Cooper D.; Coutinho P. M.; Couturier J.; Covert S.; Cunningham R.; Davis J.; Degroeve S.; dePamphilis C.; Detter J.; Dirks B.; Dubchak I.; Duplessis S.; Ehlting J.; Ellis B.; Gendler K.; Goodstein D.; Gribskov M.; Grigoriev I.; Groover A.; Gunter L.; Hamberger B.; Heinze B.; Helariutta Y.; Henrissat B.;

Holligan D.; Islam-Faridi N.; Jones-Rhoades M.; Jorgensen R.; Joshi C.; Kangasjärvi J.; Karlsson J.; Kelleher C.; Kirkpatrick R.; Kirst M.; Kohler A.; Kalluri U.; Larimer F.; Leebens-Mack J.; Leplé J. C.; Déjardin A.; Pilate G.; Locascio P.; Lucas S.; Martin F.; Montanini B.; Napoli C.; Nelson D. R.; Nelson C. D.; Nieminen K. M.; Nilsson O.; Peter G.; Philippe R.; Poliakov A.; Ralph S.; Richardson P.; Rinaldi C.; Ritland K.; Rouzé P.; Ryaboy D.; Salamov A.; Schrader J.; Segerman B.; Sterky F.; Souza C.; Tsai C.; Unneberg P.; Wall K. The genome of black cottonwood, populus trichocarpa (Torr. & Gray). *Science* 313(5793): 1596–1604; 2006.

Vagujfalvi A.; Aprile A.; Miller A.; Dubcovsky J.; Delugu G.; Galiba G.; Cattivelli L. The expression of several Cbf genes at the Fr-A2 locus is linked to frost resistance in wheat. *Mol Genet. Genomics.* 274: 506–514; 2005.

van Frankenhuyzen K.; Beardmore T. Current status and environmental impact of transgenic forest trees. *Can. J. For. Res.* 34: 1163–1180; 2004.

Vanholme R.; Morreel K.; Ralph J.; Boerjan W. Lignin engineering. *Curr. Opinion Plant Biol.* 11: 278–285; 2008.

Weng J.-K.; Li X.; Bonawitz N.; Chapple C. Emerging strategies of lignin engineering and degradation for cellulosic biofuel production. *Curr. Opinion Biotech.* 19: 166–172; 2008.

White R. H. Effect of lignin content and extractives on the higher heating value of wood. *Wood Fiber Sci.* 19: 446–452; 1987.

Williams C. G. Framing the issues on transgenic forests. *Nat. Biotechnol* 23: 530–532; 2005.

Williams C. G.; Davis B. Rate of transgene spread via long-distance seed dispersal in *Pinus taeda*. *For Ecol Manag.* 217: 95–102; 2005.

Yamaguchi-Shinozaki K.; Shinozaki K. Characterization of the expression of a desiccation-responsive *rd29* gene of *Arabidopsis thaliana* and analysis of its promoter in transgenic plants. *Mol. Gen. Genet.* 236: 331–340; 1993.

Yamaguchi-Shinozaki K.; Shinozaki K. A novel *cis*-acting element in an *Arabidopsis* gene is involved in responsiveness to drought, low temperature, or high salt-stress. *Plant Cell.* 6: 251–264; 1994.

Yanofsky M. F. Methods of Suppressing Flowering in Transgenic Plants. United States Patent No. 6,987,214 B1; 2006.

Yi S. Y.; Kim J. H.; Joung Y. H.; Lee S.; Kim W. T.; Yu S. H.; Choi D. The pepper transcription factor CaPF1 confers pathogen and freezing tolerance in *Arabidopsis*. *Plant Physiol.* 136: 2862–2874; 2004.

Zhang X.; Fowler S. G.; Cheng H.; Lou Y.; Rhee S. Y.; Stockinger E. J.; Thomashow M. F. Freezing-sensitive tomato has a functional CBF cold response pathway, but a CBF regulon that differs from that of freezing-tolerant *Arabidopsis*. *Plant J.* 39: 905–919; 2004.

Chapter 9
The Brazilian Experience of Sugarcane Ethanol Industry

Sizuo Matsuoka, Jesus Ferro, and Paulo Arruda

Abstract Biomass has gained prominence in the last few years as one of the most important renewable energy sources. In Brazil, a sugarcane ethanol program called ProAlcohol was designed to supply the liquid gasoline substitution and has been running for the last 30 yr. The federal government's establishment of ProAlcohol in 1975 created the grounds for the development of a sugarcane industry that currently is one of the most efficient systems for the conversion of photosynthate into different forms of energy. Improvement of industrial processes along with strong sugarcane breeding programs brought technologies that currently support a cropland of 7 million hectares of sugarcane with an average yield of 75 tons/ha. From the beginning of ProAlcohol to the present time, ethanol yield has grown from 2,500 to around 7,000 l/ha. New technologies for energy production from crushed sugarcane stalk are currently supplying 15% of the electricity needs of the country. Projections show that sugarcane could supply over 30% of Brazil's energy needs by 2020. In this review, we briefly describe some historic facts of the ethanol industry, the role of sugarcane breeding, and the prospects of sugarcane biotechnology.

Keywords Sugarcane • Brazilian ethanol • Biofuel • Biotechnology

Introduction

Since the introduction of the Kyoto Protocol in 1997, a worldwide concern about climate change and its impact on global warming has motivated unprecedented discussions on energy sustainability (Cox et al. 2000; Hansen et al. 2005; Wigley

---

P. Arruda (✉)
Alellyx S.A., Rua James Clerk Maxwell, 360, Condomínio Techno Park, Via Anhanguera, Km 104, 13069-380, Campinas, São Paulo, Brazil
and
Departamento de Genética e Evolução, Instituto de Biologia, Universidade Estadual de Campinas (UNICAMP), CP 6109, 13083-970 Campinas, São Paulo, Brazil
e-mail: parruda@alellyx.com.br

2005). It is generally agreed that the current energy resources, largely based on fossil fuels, are not sustainable for the long term (Chu et al. 2007; FAO 2008). A global effort to develop sustainable energy sources is urgent in order to both preserve the natural resources and mitigate the effects of CO_2 emissions (Fischer et al. 2008). Currently, fossil fuels, represented by oil, coal, and natural gas, meet more than 80% of global primary energy demand, while renewable forms represent around 13%, of which biomass contributes 10% (FAO 2008). Modern forms of renewable energy sources represented by wind, solar, geothermal, and ocean tide still represent an insignificant portion of the total current energy use (Chu et al. 2007). The use of biomass as an energy source is a practice as old burning wood, which still has a significant contribution in less-developed countries (FAO 2008). However, the Brazilian example of producing sugarcane ethanol as a liquid fuel has shown that dedicated renewable biomass crops will make a significant contribution to the world's energy needs and, at the same time, contribute to reducing the CO_2 and other greenhouse gas emissions (Nemir 1983; Rosillo-Calle 1984; Boddey 1993; FAO 2008).

The initial Brazilian ethanol experience based on sugarcane started during the 1970s oil crisis with the rapid rise of the cost of petroleum. The risk of a diminishing supply caused the federal government to launch ProAlcohol, an alternative liquid fuel program based on the ethanol obtained from sugarcane (Nemir 1983; Rosillo-Calle 1984; Natale Netto 2005; Andrietta et al. 2007; Xavier 2007). At first, a mandate was created to blend the gasoline with 5% ethanol. This stimulated the sugarcane industry to rapidly expand its activities, backed by research and development programs to both increase sugarcane yield and optimize the industrial processes (Martines-Filho et al. 2006). In addition, the logistics for collection, distribution, and commercialization of ethanol all over the country, together with the establishment of a price control system, played an essential role. Later, the car industry was supported in producing engines fueled with 100% ethanol which firmly increased the demand for ethanol (Rosillo-Calle and Cortez 1998; Moreira and Goldemberg 1999; Xavier 2007). The ProAlcohol program that originally was devised to counteract the increasing petroleum price and to strategically alleviate the country's dependence on external liquid fuel, incidentally came to be one of the best renewable liquid fuel models in the world (Goldemberg 2007; Amaral et al. 2008; FAO 2008; Goldemberg et al. 2008). In this review, we present the historic steps that contributed to the success of the program and the benefit brought by sugarcane breeding in Brazil. The future benefits from this technology implemented in Brazil over the last few years are also briefly described.

Ethanol as Car Fuel, a More Than 100-yr-Long History

Even when motorized vehicles were first a substitute for draught animals at the beginning of nineteenth century, ethanol was a fuel choice in Europe. In 1861, an engine developed by the Otto brothers used ethanol as the preferred fuel (Natale Netto 2005). The first vehicles arriving in Brazil at the end of the nineteenth and beginning of twentieth century were preferentially fueled by ethanol, a novelty that excited Brazilians with a nationalist idea of replacing the imported kerosene, used as lamp fuel and motor propulsion, with sugarcane ethanol produced in Brazil (Rosillo-Calle and Cortez 1998; Natale Netto 2005; Andrietta et al. 2007).

During the First World War, the scarcity of vehicle fuels associated with the economic crisis, the difficulty in paying for oil importation coupled with the overproduction of sugar made it difficult to sell the surplus sugar at reasonable prices. This influenced politicians and industry leaders to lobby the government to consider the production of ethanol as a pendulum to balance the excess sugar production and at the same time to have an alternative vehicle fuel (Natale Netto 2005). The idea received public support through the newspapers, and special meetings were organized to discuss the plan with the stakeholders. In addition, famous car racers adopted the idea and encouraged the development of race cars fueled by ethanol (Natalle Neto 2005). Soon some state governments created mandates to blend the gasoline with 5–10% of sugarcane ethanol (Xavier 2007). In 1931, the federal government established a mandate to blend gasoline with 5% ethanol at the national level. Fuel ethanol brands such as Usina Serra Grande-AL and Azulina (the product produced by a group of sugarcane mills from Pernambuco) were the first brands of fuel ethanol which appeared in Pernambuco and Alagoas states, the most important northeast sugarcane producers at that time. During the Second World War, the blending of gasoline with sugarcane ethanol reached 42% but afterward, it decreased to 3–7% in 1970 in the State of São Paulo (Natale Netto 2005).

In 1933, the federal government created the Instituto do Açúcar e do Alcool (IAA), with the aim of establishing the regulatory rules to control the sugarcane industry in Brazil. One of the IAA mandates was to support the sugarcane producers and promote the installation of large scale sugarcane mills to produce dehydrated ethanol in order to balance the sugar production and supply all the country's needs for liquid fuels (Nemir 1983; Natale Netto 2005). However, due to the subsequent economic crises, the creation of the Brazilian oil company Petrobras in 1943 along with pressures coming from both the oil cartel and the car industry, the plan of building up the fuel ethanol industry in Brazil was dropped and the mandate to nationally blend gasoline with 5% ethanol failed (Natale Netto 2005).

ProAlcohol, the Program that Boosted Sugarcane Ethanol in Brazil

In November 1975, the federal government created the national alcohol program ProAlcohol to face the threat of the sharp oil price increase and to counteract the collapse of sugar prices resulting from world overproduction (Natale Netto 2005; Andrietta et al. 2007; Xavier 2007). The government was convinced that sugarcane ethanol would be essential to minimize the country's high expenditure for oil importation as well as a strong strategic need to curtail the high dependence on imported oil. This political decision created the grounds for successful technological development focused on breeding for increased sugarcane yield together with the amelioration of agriculture management procedures and the modernization of the industrial processes in the sugarcane mills.

The ProAlcohol program has experienced several phases in popularity and successfulness (Andrietta et al. 2007; Xavier 2007). In the first phase, from 1975 to 1979, the program created a mandate to blend gasoline with 5% of anhydrous ethanol (Walter and Cortez 1999). In the second phase, from 1979 to 1985, due to the second oil crisis, the government decided to subsidize the car industry to produce cars fueled with 100% hydrated ethanol. The decision was based on the work carried out at the Instituto Tecnológico da Aeronáutica that proved the viability of engines fueled exclusively with ethanol (Natale Netto 2005; Andrietta et al. 2007). In the third phase, from 1985 to 1990, due to the decrease in the international oil price, ethanol production cost became higher than that of gasoline, and the ProAlcohol program faced some drawbacks, with ethanol disappearing from the gas stations throughout Brazil for a certain period (Andrietta et al. 2007). This fact led to a deterioration of consumer confidence in the program and a questioning of the real advantage of having ethanol as a gasoline substitute. The sales of ethanol fueled new cars which reached 76% of total cars sold in 1986 had dropped to less than 1% by 1997 (Fig. 9.1). In the fourth phase, from 1990 to 2002, economic uncertainties remained and consumers continued to have little confidence in the ProAlcohol program so the sales of ethanol fueled new cars continued at a very low level (Kheshgi et al. 2000; Fig.9.1). Also during this period, the sugarcane sector initiated privatization which provoked uncertainties and disturbances in the production and commercialization of both sugar and ethanol (Natale Netto 2005). The fifth phase, from 2003 to present, was characterized by a renewed enthusiasm in the ethanol program. The flex-fuel technology launched in 2003 completely changed the consumer's belief in fuel ethanol (Martines-Filho et al. 2006; Xavier 2007).

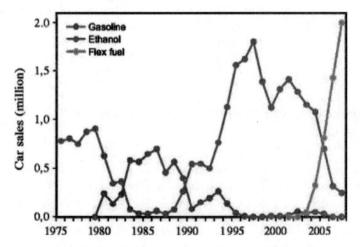

Fig. 9.1 Evolution of sales of flex-fuel cars that can be fueled with 100% gasoline, 100% ethanol, or any mixture proportion between the two fuels. Source: ANFAVEA (www.anfavea.com.br)

This technology offered consumers the possibility to run their cars with 100% gasoline, 100% hydrated ethanol, or any proportional mixture between the two fuels. The sales of flex-fuel new cars increased rapidly and reached the actual level of about 86% of new car sales (ANFAVEA 2009; IBGE 2009; Fig.9.1). The increasing adoption of flex-fuel cars together with the old fleet of ethanol vehicles resulted in an excess of ethanol sold at the pump in relation to gasoline in 2008.

This new scenario attracted a surge of investments in the ethanol industry, resulting in a significant increase in production and, consequently, in a favorable price of ethanol at the pump. In the recent years, the price of ethanol fluctuated from 70% to nearly half the price of gasoline. This, associated with the concern about renewable energy, sustainability, and global warming, stimulated investments in the sugarcane ethanol industry, resulting in over 100 new large-scale sugarcane mills being projected for construction over the next 5 yr (Goldemberg 2008). During the ProAlcohol fifth phase period, the sugarcane growing area increased from 3.5 million to ~7 million hectares with an increase rate of 10%/yr in the last 3 yr (Fig.9.2a ;

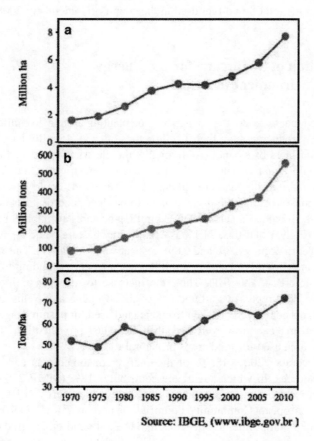

Fig. 9.2 Evolution of (**a**) cultivated area, (**b**) total production and (**c**) yield of sugarcane in Brazil since 1970. Source: IBGE (www.ibge.gov.br)

IBGE 2009). Half of the sugarcane grown is currently used to produce the 22 billion liters of ethanol, 90% of which is consumed as liquid fuel in the country (IBGE 2009).

Another important technological achievement of the sugarcane industry was the development of cogeneration technologies to produce electricity from the crushed sugarcane stalk (bagasse; Jank 2008). The modern sugarcane mills are self-sufficient in energy and produce an excess of electricity that is sold into the grid (Scaramucci et al. 2006; Jank 2008). This "bio-electricity" is increasing at a very rapid rate and it is estimated that today, bagasse contributes 15% of the total electricity consumed in the country (BNDES 2008; MME 2008). All of these achievements made sugarcane a premium crop compared to other crops to produce renewable energy (Jank 2008).

The ProAlcohol program was pivotal in establishing the basis of the Brazilian ethanol program that today can be considered well-consolidated. Pure gasoline is no longer sold in the gas stations in Brazil; it is nationally blended with 20% to 25% of ethanol (Goldemberg 2008). It is estimated that in the near future, with the increasing proportion of flex-fuel car sales (Fig.9.1), sugarcane ethanol will soon reach 30% of the total liquid fuel used in the country (Goldemberg 2008).

Contribution of Sugarcane for the Energy and the Environment in Brazil

It is predicted that by 2020, the country will be planting around 14 million hectares of sugarcane, producing more than 1 billion tons of cane, 45 million tons of sugar, and 65 billion liters of ethanol. Additionally, the electricity produced by burning bagasse should equal or surpass the hydropower electricity (Jank 2008). The projected increase in the land area planted with sugarcane could have little or no impact in terms of occupation of areas used to produce food or areas occupied by natural forest. Brazil has an estimated 232 million hectares available for agriculture and pasture (Fig.9.3). Of this, 74.1% is occupied by pasture while the major crops soybean and corn occupy 8.8% and 6.0%, respectively (Table 9.1). The area planted with sugarcane currently occupies 7 million hectares corresponding to 3% of the total available area (Table 9.1). Thus, the increase to reach 14 million hectares estimated by 2020 represents around 6% of the available area. This additional 7 million hectares could well be taken from the land used for pasture (Fig.9.3). Thus, the sugarcane area can grow even more than the values projected for 2020 without affecting the area used for food crops or natural forest.

As of December 2007, 48.7% of the energy consumed was generated from renewable sources, in which sugarcane contributes 18% (MME 2008; Fig.9.4). As a comparison, the average world use of renewable energy is less than 10% and, in the European Community countries, less than 7% (FAO 2008). It has been estimated that pollutant emissions in São Paulo and other large cities have been reduced to a quarter of that before the ethanol program (Goldemberg et al. 2008). A 2003 estimate revealed an emission reduction in the atmosphere of 27.5 million tons of CO_2 (Goldemberg et al. 2008). Thus, a program that initially was

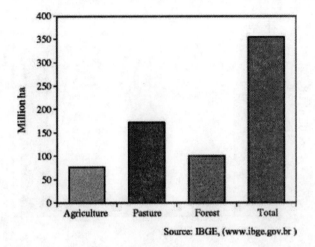

Source: IBGE, (www.ibge.gov.br)

Fig. 9.3 Distribution of land area in Brazil among different activities. *Agriculture* represents all land cultivated with different annual and perennial crops. *Pasture* represents natural and planted pasture. *Forest* represents native forest for permanent preservation, legal reserves, natural forest, and planted forest. Source: IBGE (www.ibge.gov.br)

Table 9.1 Distribution of different crops among available agricultural land in Brazil as of 2007

Crop	Cultivated land (1,000 ha)	% of cultivated land	% of territory
Pasture	172,333	74.1	46.5
Soybean	20,571	8.8	5.6
Corn	14,010	6.0	3.8
Sugarcane	7,086	3.0	1.9
Beans	3,975	1.7	1.1
Rice	2,915	1.3	0.8
Coffee	2,280	1.0	0.6
Cassava	1,941	0.8	0.5
Wheat	1,855	0.8	0.5
Cotton	1,131	0.5	0.3
Orange	821	0.3	0.2
Cocoa	685	0.3	0.2
Sorghum	671	0.3	0.2
Others	4,393	1.9	1.2
Total	232,492	100	62.8

Source: IBGE (www.ibge.gov.br)

mainly motivated by economic and strategic reasons later turned out to be one of the most successful cases of sustainable energy production in the world. Today, sugarcane ethanol is recognized as the most commercially successful biomass fuel worldwide (FAO 2008; Zuurbier and van de Vooren 2008) with the highest positive energy balance (Macedo 1998; Goldemberg 2008; Macedo and Seabra 2008).

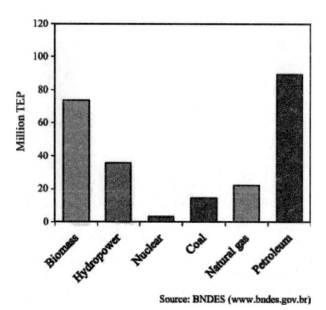

Fig. 9.4 Profile of energy use in Brazil as of 2007. Source: BNDES (www.bndes.gov.br)

Sugarcane Breeding in Brazil in the Last 40 yr

When ProAlcohol was created, there were two major sugarcane breeding programs in Brazil: one supported by the cooperative of sugar mills of the State of São Paulo, now called Centro de Tecnologia Canavieira (CTC) and another supported by IAA. The former one was initiated in 1969 and the second in 1970 (Matsuoka et al. 2005; Landell and Bressiani 2008). Both programs started with the motivation to help improve the technological level of the Brazilian sugar industry. A survey on the status of the Brazilian sugarcane industry in 1966 revealed that great attention should be given to establish sugarcane breeding programs as a basis to sustain the industry (Mangelsdorf 1967). CTC established its breeding program focusing mainly in the sugarcane production areas of the State of São Paulo (Machado et al. 1987) while IAA established its breeding program focused in all the sugarcane producing areas of the country (Matsuoka et al. 2005; Natale Netto 2005). When ProAlcohol was created in 1975, the two breeding programs were well-established and running their research programs at a scale large enough to support the growing industry demand. After the first 30 yr of the ProAlcohol program, it was recognized that the success of the Brazilian sugarcane ethanol experience was due to the research developed almost entirely inside the country by local scientists (Natale Netto 2005; Martines-Filho et al. 2006; Goldemberg et al. 2008).

The development of high-yielding sugarcane varieties adapted to different environments along the country remains a major challenge to breeders. This is compounded by the fact that sugarcane has a complex genome and its genetics are not well-understood. Sugarcane has been bred since the end of the nineteenth century, when the breeders crossed *Saccharum officinarum* to *Saccharum spontaneum*, a wild and vigorous relative (Berding and Roach 1987). Since then, the interspecific hybrids have passed a few cycles of intercrossing and selection. *S. officinarum* has a basic chromosome number of $x=10$ while *S. spontaneum* has a basic chromosome number of $x=8$ (D'Hont et al. 1998; Ha et al. 1999), and both species have a high degree of ploidy with *S. officinarum* presenting $2n=80$ chromosomes and *S. spontaneum* presenting $2n$ complement varying from 40 to 128 (Stevenson 1965; Sreenivasan et al. 1987). The modern sugarcane cultivars have a complex mixture of both species genomes with ploidy varying between $2n=100$ and $2n=130$ chromosomes. The difference in the basic chromosome numbers of both species results in the coexistence in the hybrids of 15–25% of chromosomes from *S. spontaneum* and a variable proportion of recombinants between homologous chromosomes (D'Hont et al. 1996). The complex chromosome organization makes each cross a unique unpredictable event because of the random sorting of chromosomes from both species and the formation of recombinants affecting the distribution of favorable and nonfavorable alleles (Grivet and Arruda 2001). This, associated with the presence of several distinct alleles at each locus (Lu et al. 1994; Jannoo et al. 1999), makes the breeding process in sugarcane a very unpredictable and difficult task. To overcome the difficulty imposed by the genome complexity of sugarcane, the breeding programs had to evaluate hundreds of thousands of progenies obtained from a large number of crosses between hundreds of parental clones.

The major focus of the early sugarcane hybridizations was to control the diseases that had been threatening the plantations as well as to develop varieties for intensive cultivation (Stevenson 1965). Genetic materials resulting from the early hybridizations in Indonesia and India were shared with other important sugarcane growing regions that initiated their own breeding programs (Tew 1987). The main focus of the programs was breeding for better varieties with a broad range of commercially important traits. These include adaptability to distinct local environments, sucrose yield, resistance to important pathogenic viral, fungal and bacterial diseases, agronomic manageability, and good milling characteristics (Matsuoka et al. 1994; Matsuoka and Meneghin 1999). Those aims have been successful achieved by CTC and IAA in the development of Sao Paulo (SP) and Republica do Brasil (RB) varieties, respectively. Those varieties have significantly contributed to the steady increase of productivity registered in the last 30 yr (Fig.9.2 C; Matsuoka 1999; Natale Netto 2005; Martines-Filho et al. 2006).

From the beginning of ProAlcohol in 1975 until 1985, there were approximately ten varieties cultivated in over 80% of the planting area, but two varieties, namely CB41-76 and NA56-79, were preferred and covered over 50% of the plantings (Fig. 9.5). Those few genotypes planted in a large area imposed several constraints in yield because of specific biotic and abiotic stresses associated with particular environments. During this period, the sugarcane yield averaged around 55 tons/ha (Fig.9.2 C). Since then, new varieties were released mainly by the CTC program, and from 1985 to 1995, three of them, SP70-1143, SP71-6163, and SP71-1406, occupied 40% of the planted area (Fig.9.5). It took 10 yr for the release of these new varieties, which

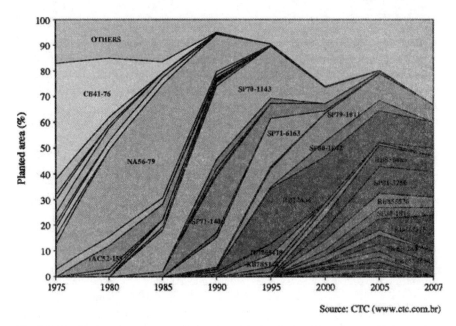

Fig. 9.5 Distribution of sugarcane varieties planted during the last 30 yr. Source: CTC (www.ctc.com.br)

corresponds to the usual 9 to 12 yr required for breeding new sugarcane varieties (Matsuoka et al. 2005). It is also important to note that from 1975 to 1990, the sugarcane production was more than doubled, and most of this resulted from the area expansion (Fig.9.2 A, B). From 1990 to the present, the sugarcane planted area continually expanded, reaching the current 7 million hectares. During this time, the newly bred varieties contributed to a continuous yield increase, reaching ~75 tons/ha (Fig. 2 C). Thus, the overall increase of sugarcane in the last 18 yr greatly benefited from the new varieties developed by the SP and RB breeding programs. The two programs released over 60 varieties over the past 30 yr, with more than 25 varieties being effectively adopted by the sugarcane growers. At present, more than 30 varieties are being grown throughout the country (Fig.9.5). Contrasting with the 1970s and 1980s, a period in which few varieties were planted, in the majority of the planted area today, most of the sugarcane varieties occupy not more than 2–5% each, although top varieties such as RB72454 and SP701011 comprise around 15–20% of the planting area (Fig.9.5). It is important to mention that in the last 10 yr, the released sugarcane varieties have also incorporated the capability to extend the harvest period, so that early season varieties, midseason varieties, and late season varieties can be planned to adequately supply the sugarcane mills from beginning of April until the end of November (Matsuoka 1999).

The sugarcane breeding history reveals that diseases were the main cause of yield losses. As can be seen in Fig. 9.5, predominant varieties dropped out rapidly because of susceptibility to smut and rust in the case of NA56-79 (Bergamin Filho and Amorim 1996; Matsuoka et al. 1994), rust in the case of SP70-1143 (Matsuoka et al. 1994), and poor agronomic performance due to complex biotic and abiotic stresses in the case of SP71-6163 (Matsuoka and Meneghin 1999). The release of an increasing number of new varieties over the years reduced the risk of crop losses by sudden epidemic diseases. This contributed to the 50% increase in the average crop productivity from 1975 to the present (Fig. 2 C) with a dramatic impact in the ethanol yield from ~2,500 l/ha in 1975 to ~7,000 l/ha at the present (Goldemberg 2008). It is estimated that sugarcane breeding can contribute significantly to realize the predicted 9,000 l/ha alcohol production for the next decade.

Ethanol from sugar or starch is referred to as the first-generation biofuel (Yuan et al. 2008). A second-generation ethanol based on the saccharification of cellulosic feedstock has received strong support in developed countries (Somerville 2006). The sugarcane bagasse can also serve as cellulosic feedstock and its potential use could double the present contribution of the crop for the bioethanol industry (Hailing and Simms-Borre 2008; Sticklen 2008). In this scenario, there is great potential for further gain with the development of what has been called "energy cane", a cane that instead of producing high-sugar stalk produces stalk with less sugar but higher fiber content. As a result, energy cane has increased biomass yield and also increased resistance to biotic and abiotic stresses (Alexander 1985). The energy cane offers several advantages as it produces more biomass per unit area. The net energy ratio of the energy cane is estimated to be higher than that of the current elite sugarcane varieties.

Sugarcane Biotechnology

Biotechnology has been viewed to be of special importance for sugarcane for the introduction of new genes and traits through molecular marker-assisted breeding and genetic engineering. Evaluation of transgenic sugarcane in Brazil has been continuing since 1993. Studies carried out by CTC and various collaborators has produced transgenic sugarcane (Falco et al. 2000), and field trials have been conducted evaluating genes responsible for herbicide tolerance, virus resistance, insect resistance, flower inhibition, and increased sucrose yield (Table 9.2). Herbicide resistant transgenic sugarcane has also been tested in field trials conducted by BASF (Table 9.2). In 1998, sugarcane biotechnology received great attention in Brazil after the formation of a network to sequence and analyze the sugarcane transcriptome (Arruda 2001). The network, called SUCEST (for sugarcane expressed sequence tags (ESTs)), produced a database of around 300,000 ESTs from a collection of cDNA libraries from different organs and tissues sampled at different developmental stages or conditions (Vettore et al. 2001). The SUCEST database allowed the tagging of over 80% of the sugarcane transcriptome (Vettore et al. 2003) and has served as a tool for the identification of genes involved in sugar accumulation and response to different biotic and abiotic stresses (Nogueira et al. 2003; Rossi et al. 2003; Nogueira et al. 2005; Calsa and Figueira 2007; Borges et al. 2007;

Table 9.2 Field trials of transgenic sugarcane approved in Brazil by the National Biosafety Commission (CTNBio)

Institution	Trait	No. of field trials	Year of approval
CTC	Herbicide tolerance	9	1997, 1998, 1999, 2000
	Virus resistance	3	1999, 2000
	Insect resistance	1	1999
	Flowering inhibition	1	2002
	Sucrose yield	5	2005, 2006
Alellyx SA	Virus resistance	2	2005, 2006
	Sucrose yield	6	2006, 2007, 2008
	Drought tolerance	3	2007, 2008
	Herbicide tolerance + Insect resistance	4	2007, 2008
BASF S.A.	Herbicide tolerance	4	1999, 2000, 2001, 2002

Source: CTNBIO (www.ctnbio.gov.br)

Rocha et al. 2007). Functional analysis of the sugarcane transcriptome has mapped the signal transduction network of sugarcane linking important regulatory cascades associated with abiotic stresses (Papini-Terzi et al. 2005; Rocha et al. 2007). The SUCEST database has also been used for the development of molecular markers and the identification of microsatellites in sugarcane (da Silva and Bressiani 2005). Molecular maps have been generated that allowed the identification of loci associated with the variation of phenotypic traits, as well as loci associated with important traits, such as sugar yield and disease resistance (Pinto et al. 2004; Garcia et al. 2006; Pinto et al. 2006; Oliveira et al. 2007). A large-scale experiment conducted in two interspecific S. *offinarum* x S. *spontaneum* crosses was designed to investigate variation in the sucrose content of the stalk (Lima et al. 2002). Many independently segregating alleles were identified and could be assigned to eight distinct loci. When compared with maps of other grasses, particularly maize, some of these alleles were given a candidate identity (Lima et al. 2002; Oliveira et al. 2007).

The strong impact of genomics in Brazil attracted the attention of investors interested in the creation of startup biotech companies. In 2002, a biotech company called Alellyx was founded with investments from Votorantim New Business with the aim of developing technologies directed to the yield improvement of sugarcane and other crops. Soon a sugarcane breeding company called Canavialis was also founded by Votorantim New Business. Canavialis and Alellyx joined the efforts to bring to the market new sugarcane varieties with, besides the superior agronomic performance, biotech traits. Both companies, recently acquired by Monsanto, are currently engaged in a strong sugarcane breeding and biotechnology programs to meet the demand of the growing sugarcane industry in Brazil. The two companies are focused on the development of herbicide and insect resistance as well as increased biomass, drought resistance, and sugar yield. They are currently conducting several field trials for those traits (Table 9.2). It is expected that in the next few years, the first commercial transgenic sugarcane variety will be available for the sugarcane growers in Brazil. Canavialis is also engaged in a strong program of molecular markers to accelerate breeding advances. Thousands of microsatellite markers were identified. By using few markers from this collection, it was possible to discriminate all hybrid varieties and *Saccharum* species and related genera from major germplasm collections (Maccheroni et al. 2009). These results support the projection that using ten of those markers it will be possible to discriminate all varieties and related genotypes grown in the world, qualifying this system to become an important tool to all those involved in sugarcane genetic improvement (Maccheroni et al. 2009).

Conclusions

The world needs renewable energy sources to counteract the increasing threat of global warming and greenhouse gas emissions, and the use of biomass remains an important option to address this global issue. The current technologies together with those in development for conversion of cellulosic biomass into energy have the potential to contribute to the world energy needs and alleviate the impact of increasing CO_2 emissions.

Brazilian sugarcane ethanol has been in place since 1975 and successfully proved that a dedicated crop can help supply a significant proportion of the energy needs of the country and contribute to amelioration of environmental impact. The development of such a large-scale program benefited from several key factors. First is the economics driven by the need for alternative fuel combined with the political will to encourage success through incentives and guidelines. Second is the existence of a local industrial base to provide research and development support for the technology needs. Third is the existence of an innovative local automobile industry eager to engage in technological development of engines suited for ethanol fuels. Fourth is the existence of extensive agricultural land in a tropical condition and a favorable socioeconomical land-use policy. Fifth is the establishment of a fuel distribution system with logistical support to deliver ethanol to the customer efficiently.

The use of sugarcane to produce sugar, ethanol, and electricity is not accomplishing the full potential of this crop. The full development of the process for the commercial production of second-generation ethanol—the cellulosic ethanol—in the years to come can be a transformational change in the sugarcane industry. However, the projected increase in the sugarcane cultivated area will require development of new varieties adapted to poor soils, less water, and which is resistant to attacks by pests, fungi, and bacterial diseases. To this end, sugarcane breeding that efficiently incorporates new technologies such as molecular markers will be essential. Biotechnology should be increasingly important in the near future, as the current elite sugarcane varieties will express additional traits incorporated through genetic engineering.

References

Alexander A. G. The energy cane alternative. Elsevier, Amsterdam; 1985.

Amaral W. A. N.; Marinho J. P.; Tarasantchi R.; Beber A.; Giuliani E. Environmental sustainability of sugarcane ethanol in Brazil. In: Zuurbier P.; van de Vooren J. (eds) Sugarcane ethanol: Contribution to climate change mitigation and the environment. Wageningen Academic, Wageningen, pp 113–138; 2008.

Andrietta M. G. S.; Andrietta S. R.; Steckelberg C.; Stupiello E. N. A. Bioethanol—Brazil, 30 years of Proalcool. *Int. Sugar J* 109: 195–200; 2007.

ANFAVEA. Produção de autoveículos por tipo e combustível, 2008. http://www.anfavea.com.br/tabelas.html Cited Feb. 16, 2009.

Arruda P. Sugarcane transcriptome. A landmark in plant genomics in the tropics. *Genet. Mol. Biol.* 24: 1; 2001.

Berding N.; Roach B. T. Germplasm collection, maintenance, and use. In: Heinz D. J. (ed) Sugarcane improvement through breeding. Elsevier, Amsterdam, pp 143–210; 1987.

Bergamin Filho A.; Amorim L. Doenças de Plantas Tropicais: Epidemiologia e Controle Econômico. Ed. Agronômica Ceres, São Paulo; 1996.

BNDES (2008) Bioetanol de Cana-de-açúcar. Energia para o Desenvolvimento Sustentável. BNDES, Rio de Janeiro. http://www.bioetanoldecana.org Cited Nov. 11, 2008

Boddey R. Green energy from sugar cane. *Chem. Ind* 10: 355–358; 1993.

Borges J. C.; Cagliari T. C.; Ramos C. H. I. Expression and variability of molecular chaperones in the sugarcane expressome. *J. Plant Physiol* 164: 505–513; 2007.

Calsa T.; Figueira A. Serial analysis of gene expression in sugarcane (Saccharum spp.) leaves revealed alternative C-4 metabolism and putative antisense transcripts. *Plant Mol. Biol* 63: 745–762; 2007.

Cox P. M.; Betts R. A.; Jones C. D.; Spall S. A.; Totterdell I. J. Acceleration of global warming due to carbon-cycle feedbacks in a coupled climate model. *Nature* 408: 184–187; 2000.

Chu S.; Goldemberg J.; Arungu Olende S.; El-Ashry M.; Davis G.; Johansson T.; Keith D.; Jinghai L.; Nakicenovic N.; Pachauri R.; Shafie-Pour M.; Shpilrain E.; Socolow R.; Yamaji J.; Luguang Y. Lighting the way: Toward a sustainable energy future. Inter Academy Council, Amsterdam; 2007.

D'Hont A.; Grivet L.; Feldmann P.; Rao S.; Berding N.; Glaszmann J. C. Characterization of the double genome structure of modern sugarcane cultivars (Saccharum spp) by molecular cytogenetics. *Mol. Gen. Genet* 250: 405–413; 1996.

D'Hont A.; Ison D.; Alix K.; Roux C.; Glaszmann J. C. Determination of basic chromosome numbers in the genus Saccharum by physical mapping of ribosomal RNA genes. *Genome* 41: 221–225; 1998.

da Silva J. A. G.; Bressiani J. A. Sucrose synthase molecular marker associated with sugar content in elite sugarcane progeny. *Genet. Mol. Biol* 28: 294–298; 2005.

Falco M. C.; Tullman Neto A.; Ulian E. C. Transformation and expression of a gene for herbicide resistance in a Brazilian sugarcane. *Plant Cell Rep* 19: 1188–1194; 2000.

FAO The State of Food and Agriculture. Part I. Biofuels: Prospects, risks and opportunities. FAO Agriculture Series no. 39. FAO, Rome; 2008.

Fischer G.; Teixeira E.; Hizsnyik E. T.; Velthuizen H. Land use dynamics and sugarcane production. In: Zuurbier P.; van de Vooren J. (eds) Sugarcane ethanol: Contribution to climate change mitigation and the environment. Wageningen Academic, Wageningen, pp 29–62; 2008.

Garcia A. A. F.; Kido E. A.; Meza A. N.; Souza H. M. B.; Pinto L. R.; Pastina M. M.; Leite C. S.; da Silva J. A. G.; Ulian E. C.; Figueira A.; Souza A. P. Development of an integrated genetic map of a sugarcane (Saccharum spp.) commercial cross, based on a maximum-likelihood approach for estimation of linkage and linkage phases. *Theor. Appl. Genet* 112: 298–314; 2006.

Goldemberg J. Ethanol for a sustainable energy future. *Science* 315: 808–810; 2007.

Goldemberg J. The Brazilian biofuels industry. *Biotechnol Biofuels* 1: 6; 2008. doi:10.1186/1754-6834-1-6.

Goldemberg J.; Coelho S. T.; Guardabassi P. The sustainability of ethanol production from sugarcane. *Energy Policy* 36: 2086–2097; 2008.

Grivet L.; Arruda P. Sugarcane genomics: depicting the complex genome of an important tropical crop. *Curr. Opin. Plant Biol* 5: 122–127; 2001.

Ha S.; Moore P. H.; Heinz D.; Kato S.; Ohmido N.; Fukui K. Quantitative chromosome map of the polyploid Saccharum spontaneum by multicolor fluorescence in situ hybridization and imaging methods. *Plant Mol. Biol* 39: 1165–1173; 1999.

Hailing P.; Simms-Borre P. Overview of lignocellulosic feedstock conversion into ethanol—focus on sugarcane bagasse. *Int. Sugar J* 110: 191–194; 2008.

Hansen J.; Nazarenko L.; Ruedy R.; Sato M.; Willis J.; Del Genio A.; Koch D.; Lacis A.; Lo K.; Menon S.; Novakov T.; Perlwitz J.; Russell G.; Schmidt G. A.; Tausnev N. Earth's energy imbalance: Confirmation and implications. *Science* 308: 1431–1435; 2005.

IBGE. Instituto Brasileiro de Geografia e Estatística. Censo Agropecuario, 2006. www.ibge.gov.br/home/estatistica/economia/agropecuaria/censoagro/2006/default htm. Cited Feb 5 2009.

Jank, M. S. Cane for sugar, ethanol and bioelectricity: a global economy. UNICA, the Brazilian Sugarcane Industry. http://www.unica.com.br Cited; 2008.

Jannoo N.; Grivet L.; Seguin M.; Paulet F.; Domaingue R.; Rao P. S.; Dookun A.; D'Hont A.; Glaszmann J. C. Molecular investigation of the genetic base of sugarcane cultivars. *Theor. Appl. Genet* 99: 171–184; 1999.

Kheshgi H. S.; Prince R. C.; Marland G. The potential of biomass fuels in the context of global climate change: Focus on transportation fuels. *Ann. Rev. Energy Environ* 25: 199–244; 2000.

Landell M. G. A.; Bressiani J. A. Melhoramento genético, caracterização e manejo varietal. In: Dinardo-Miranda L. L. et al. (eds) Cana-de-açúcar. Instituto Agronômico, Campinas, pp 101–155; 2008.

Lima M. L. A.; Garcia A. A. F.; Oliveira K. M.; Matsuoka S.; Arizono H.; de Souza C. L.; de Souza A. P. Analysis of genetic similarity detected by AFLP and coefficient of parentage among genotypes of sugar cane (Saccharum spp.). *Theor. Appl. Genet* 104: 30–38; 2002.

Lu Y. H.; D'Hont A.; Paulet F.; Grivet L.; Arnaud M.; Glaszmann J. C. Molecular diversity and genome structure in modern sugarcane varieties. *Euphytica* 78: 217–226; 1994.

Maccheroni, W.; Jordão, H.; Degaspari, R.; Moura, G. L.; Matsuoka, S. Development of a dependable microsatellite-based fingerprinting system for sugarcane. *Sugar Cane Int.* 27: 47–52; 2009.

Macedo I. C. Greenhouse gas emissions and energy balance in bioethanol production and utilization in Brazil. *Biomass Bioenergy* 14: 77–81; 1998.

Macedo I. C.; Seabra E. A. Mitigation of GHG emissions using sugarcane bioethanol. In: Zuurbier P.; van de Vooren J. (eds) Sugarcane ethanol: Contribution to climate change mitigation and the environment. Wageningen Academic, Wageningen, pp 95–111; 2008.

Machado, Jr. G. R.; Silva, W. M.; Irvine, J. E. Sugarcane breeding in Brazil: The Copersucar program. In: Copersucar International Sugarcane Breeding Workshop São Paulo, Copersucar, pp 217–232; 1987.

Mangelsdorf A. J. Um programa de melhoramento da cana-de-açúcar para a agroindústria canavieira do Brasil. Brasil Açucar 69: 208–223; 1967.

Martines-Filho, J.; Burnquist, H. L.; Vian, C. E. F. Bioenergy and the rise of sugarcane-based ethanol in Brazil. Choices, AAEA, 2nd Quarter, http://www.choicesmagazine.org 2006.

Matsuoka S. The recent evolution of sugarcane varieties in Brazil. *STAB* 17: 37; 1999.

Matsuoka S.; Bassinello A. I.; Martins S.; Arizono H. A retrospective analysis of crop damage caused by sugarcane rust in Brazil. II. Losses in spring planted cane. In: Rao G. P. et al. (ed) Current trends in sugarcane pathology. International Books and Periodicals Supply Service, New Delhi, pp 27–35; 1994.

Matsuoka S.; Garcia A. A. F.; Arizono H. Melhoramento da cana-de-açúcar. In: Borém A (ed) Melhoramento de Espécies Cultivadas. Editora UFV, Viçosa, Minas Gerais, 2nd ed, pp 225–274; 2005.

Matsuoka S.; Meneghin S. P. Yellow leaf syndrome and alleged pathogen: casual and not causal relationship. *Proc. ISSCT Congress* 23: 382–389; 1999.

MME – Ministério das Minas e Energia. Matriz energética 2007 Brasil. Cited Jan. 12, 2009; 2008.

Moreira J. R.; Goldemberg J. The alcohol program. *Energy Policy* 27: 227–2291; 1999.

Natale Netto J. A Saga do Álcool. Novo Século Editora, Osasco; 2005.

Nemir A. S. Alcohol fuels—the Brazilian experience and its implications for the United States. *Sugar J* 45: 10–13; 1983.

Nogueira F. T. S.; de Rosa V. E.; Menossi M.; Ulian E. C.; Arruda P. RNA expression profiles and data mining of sugarcane response to low temperature. *Plant Physiol* 132: 1811–1824; 2003.

Nogueira F. T. S.; Schlogl P. S.; Camargo S. R. et al. SsNAC23, a member of the NAC domain protein family, is associated with cold, herbivory and water stress in sugarcane. *Plant Sci* 169: 93–106; 2005.

Oliveira K. M.; Pinto L. R.; Marconi T. G.; Margarido G. R. A.; Pastina M. M.; Teixeira L. H. M.; Figueira A. V.; Ulian E. C.; Garcia A. A. F.; Souza A. P. Functional integrated genetic linkage map based on EST-markers for a sugarcane (Saccharum spp.) commercial cross. *Mol. Breed* 20: 189–208; 2007.

Papini-Terzi F. S.; Rocha F. R.; Vencio R. Z. N.; Oliveira K. C.; Felix J. D.; Vicentini R.; Rocha C. D.; Simoes A. C. Q.; Ulian E. C.; Di Mauro S. M. Z.; Da Silva A. M.; Pereira C. A. D.; Menossi M.; Souza G. M. Transcription profiling of signal transduction-related genes in sugarcane tissues. *DNA Research* 12: 27–38; 2005.

Pinto L. R.; Oliveira K. M.; Marconi T.; Garcia A. A. F.; Ulian E. C.; de Souza A. P. Characterization of novel sugarcane expressed sequence tag microsatellites and their comparison with genomic SSRs. *Plant Breed* 125: 378–384; 2006.

Pinto L. R.; Oliveira K. M.; Ulian E. C.; Garcia A. A. F.; de Souza A. P. Survey in the sugarcane expressed sequence tag database (SUCEST) for simple sequence repeats. *Genome* 47: 795–804; 2004.

Rocha F. R.; Papini-Terzi F. S.; Nishiyama M. Y.; Vencio R. Z. N.; Vicentini R.; Duarte R. D. C.; de Rosa V. E.; Vinagre F.; Barsalobres C.; Medeiros A. H.; Rodrigues F. A.; Ulian E. C.; Zingaretti S. M.; Galbiatti J. A.; Almeida R. S.; Figueira A. V. O.; Hemerly A. S.; Silva-Filho M. C.; Menossi M.; Souza G. M. Signal transduction-related responses to phytohormones and environmental challenges in sugarcane. *BMC Genomics* 8: 71; 2007.

Rosillo-Calle F. A re-assessment of the Brazilian National Alcohol Programme (PNA). *Ind. Biotech* 3: 11–16; 1984.

Rosillo-Calle F.; Cortez L. A. B. Towards ProAlcool II—a review of the Brazilian bioethanol programme. *Biomass Bioenergy* 14: 115–124; 1998.

Rossi M.; Araujo P. G.; Paulet F.; Garsmeur O.; Dias V. M.; Chen H.; Van Sluys M. A.; D'Hont A. Genomic distribution and characterization of EST-derived resistance gene analogs (RGAs) in sugarcane. *Mol. Genet. Genom* 269: 406–419; 2003.

Scaramucci J. A.; Perin C.; Pulino P. et al. Energy from sugarcane bagasse under electricity rationing in Brazil: a computable general equilibrium model. *Energy Policy* 34: 986–992; 2006.

Somerville C. The billion-ton biofuels vision. *Science* 312: 1277; 2006.

Sreenivasan T. V.; Ahloowalia B. S.; Heinz D. J. Cytogenetics. In: Heinz D. J. (ed) Sugarcane improvement through breeding. Elsevier, Amsterdam, pp 211–253; 1987.

Stevenson G. C. Genetics and breeding of sugarcane. Longmans, London. 1965.

Sticklen M. B. Plant genetic engineering for biofuel production: towards affordable cellulosic ethanol. *Nat. Rev. Genet* 9: 433–443; 2008.

Tew T. L. New varieties. In: Heinz D. J. (ed) Sugarcane improvement through breeding. Elsevier, Amsterdam, pp 559–594; 1987.

Vettore A. L.; da Silva F. R.; Kemper E. L. et al. The libraries that made SUCEST. *Genet. Mol. Biol.* 24: 1–7; 2001.

Vettore A. L.; da Silva F. R.; Kemper E. L. et al. Analysis and functional annotation of an expressed sequence tag collection for tropical crop sugarcane. *Genome Res.* 13: 2725–2735; 2003.

Xavier M.R. The Brazilian sugarcane ethanol experience. Competitive Enterprise Institute, Washington, DC200714p. http://www.cei.org.

Yuan J. S.; Tiller K. H.; Al-Ahmad H.; Stewart N. R.; Stewart C. N. Plants to power: Bioenergy to fuel the future. *Trend Plant Sci.* 13: 421–429; 2008.

Walter A. Cortez, L. An historical overview of the Brazilian bioethanol program. *Renew. Energy Dev.* 11no. 1: 1–4; 1999.

Wigley T. M. L. The climate change commitment. *Science* 307: 1766–1769; 2005.

Zuurbier P.; van de Vooren J. Sugarcane ethanol: Contribution to climate change mitigation and the environment. Wageningen Academic, Wageningen; 2008.

Chapter 10
Biofuels: Opportunities and Challenges in India

Mambully Chandrasekharan Gopinathan and Rajasekaran Sudhakaran

Abstract Energy plays a vital role in the economic growth of any country. Current energy supplies in the world are unsustainable from environmental, economic, and societal standpoints. All over the world, governments have initiated the use of alternative sources of energy for ensuring energy security, generating employment, and mitigating CO_2 emissions. Biofuels have emerged as an ideal choice to meet these requirements. Huge investments in research and subsidies for production are the rule in most of the developed countries. India started its biofuel initiative in 2003. This initiative differs from other nations' in its choice of raw material for biofuel production—molasses for bioethanol and nonedible oil for biodiesel. Cyclicality of sugar, molasses, and ethanol production resulted in a fuel ethanol program which suffered from inconsistent production and supply. The restrictive policies, availability of molasses, and cost hampered the fuel ethanol program. Inconsistent policies, availability of land, choice of nonnative crops, yield, and market price have been major impediments for biodiesel implementation. However, a coherent, consistent, and committed policy with long-term vision can sustain India's biofuel effort. This will provide energy security, economic growth, and prosperity and ensure a higher quality of life for India.

Keywords Biofuels • Biodiesel • Fuel ethanol • India

Global Energy Overview

Ensuring an adequate and reliable energy supply at competitive prices to support economic growth and meet essential population needs is vital for any country. The volatility of the market and of energy prices, declining production rates, and

M.C. Gopinathan (✉)
Research and Development Centre, EID PARRY (India) Ltd, 145, Devanahalli Road,
Off. Old Madras Road, Bangalore, 560049, India
e-mail: gopinathanmc@parry.murugappa.com

recent geopolitical acts of war and terrorism has underscored the vulnerability of the current global energy system to supply disruptions. According to World Energy Outlook (2008), current energy supplies are unsustainable from environmental, economic, and societal standpoints. In addition, it is projected that world energy demands will continue to expand by 45% from 2008 to 2030, an average rate of increase in 1.6%/yr. In 2007, the intergovernmental panel on climate change (IPCC 2007) released its fourth assessment report confirming that climate change is accelerating and if current trends continue, energy-related emissions of carbon dioxide (CO_2) and other greenhouse gases will rise inexorably, pushing up average global temperature by as much as 6°C in the long term. Recent floods, cyclones, tsunamis, sea rise, droughts, and famines throughout the world were implicated as a part of climate change resulting from unabated burning of fossil fuels (IPCC 2008). Climate change threatens water, food production, human health, and the quality of land on a global scale (OCC 2006; IPCC 2008). Preventing catastrophic and irreversible damage to the global climate ultimately requires a major decarbonization drive. Globally, 80% of total primary energy supply depends on the fossil fuels coal, gas, and petroleum-based oils. Renewable energy sources represent only 13% of total primary energy supply currently, with biomass (the material derived from living organisms) dominating with 10% in renewable sector (IEA 2007a). Traditional biomass, including fuel wood, charcoal, and animal dung, continues to provide important sources of bioenergy for most of the world population who live in extreme poverty and who use this energy mainly for cooking. More advanced and efficient conversion technologies now allow the extraction of biofuels in solid, liquid, and gaseous forms from a wide range of biomass sources such as woods crops and biodegradable plant and animal wastes. Biofuels can be classified according to source, type, and technological process of conversion under the categories of first, second, third, and fourth generation biofuels. First generation biofuels are biofuels made from biomass consisting of sugars, starch, vegetable oils, animal fats, or biodegradable output wastes from industry, agriculture, forestry, and households using conventional technologies. Second generation biofuels are derived from lignocellulosic biomass to liquid technology, including cellulosic biofuels from nonfood crops such as the stalks of wheat, corn, wood, and energy-dedicated biomass crops, such as miscanthus. Many second generation biofuels are under development such as biohydrogen, biomethanol, dimethyl furan, dimethyl ether, Fischer–Tropsch diesel, biohydrogen diesel, mixed alcohols, and wood diesel. Third generation biofuels are in the nascent stage of development and are derived from low input/high output production organisms such as algal biomass. Fourth generation biofuels are derived from the bioconversion of living organisms (microorganisms and plants) using biotechnological tools (Rutz and Janseen 2007; FAO 2008).

National governments are setting targets and developing strategies, policies, and investment plans in biofuels to enhance energy security and exploit alternative energy to mitigate CO_2 emission. The recent increase of oil prices, energy security fears, and the domestic reform of agricultural policies (in the context of international negotiation for agricultural trade liberation) give cause for a more serious consideration of biofuel in most of countries. USA, Europe, and Brazil are leading proponents of these initiatives. Mandates for blending biofuel into vehicle fuels have been enacted in at least 37 countries (Martinot 2008). Most mandates require blending of 5–10% ethanol with gasoline and 2–5% biodiesel with diesel fuel. In developed countries, government support for the domestic production of energy crops for biofuel seems to be the rule (Dufey 2006). In the USA, estimated subsidies to the biofuel industry may reach US $13 billion in 2008 and federal tax credit could cost US $19 billion/yr by 2022 (Koplow 2007). In the European Union (EU), biofuel support of €0.52/l will end up costing its tax payers €34 billion/yr (Kutas et al. 2007; Steenblik 2007; Bailey 2008). These initiatives contributed to the rapid growth of liquid biofuels in terms of volume and share of transport fuels. Since 2001, biofuel production has increased almost six fold to 6 billion liters in 2006 and is projected to grow to 3.0–3.5% of total global transport energy by 2030 from the present 1.9% (IEA 2007b; Worldwatch Institute 2007).

However, environmental groups have been raising concerns about the trade-off in food *vs.* fuel and effectiveness of biofuels in mitigating green house gas emissions. Recent rise in food prices, shortage of food, conflicting demands of arable land, heavy use of fertilizers for biofuel production, and deforestation of rain forests escalated the debate to a global scale (Worldwatch Institute 2007; Bailey 2008; FAO 2008; Mitchell 2008; Searchinger et al. 2008; World Bank 2008). On the other hand, several studies show that biofuel production can be significantly increased without affecting food crops. Further reports suggest that Brazil's sugar-based ethanol production has not contributed to the food crisis (Dufey et al. 2007; Banse et al. 2008; DEFRA 2008; UNICA 2008a, b). Many reports suggest that the success of second and third generation technologies dealing with nonfood biomass will play much bigger role than expected in coming years (IEA 2007a, b; FAO 2008). However, the investment, trade, and subsidy policies around these technologies continue to play critical role for successful exploitation of biofuels.

Indian Energy Challenges

India is a rapidly expanding large economy and faces a formidable challenge to meet its energy needs in a responsible and sustainable manner. To sustain India's 8% average annual economic growth and to support its growing population, India needs to generate two- to threefold more energy than the present (IEA 2007b). This means an increase in energy supply from 542 million tons of oil equivalent in 2006 to 1,516 million tons of oil equivalent in 2031–2032 (GOI 2006a). The nature, dimensions, and complexities of achieving this challenge are analyzed based on the present energy capacity, context, and potential. The country is rich in coal and abundantly endowed with renewable energy in the form of solar, wind, and hydrogenerated energy, bioenergy, and large reserves of thorium. Unfortunately, reserves of hydrocarbon, gas, and uranium are meager. At the current level of production and consumption, India's coal reserves are estimated to last more than 200 yr. India is currently the third largest coal-producing country in the world (behind China and the USA) and accounts for about 7.5% of the world's annual coal production (IEO 2008). India is also currently the third largest coal-consuming country (behind the China and the USA) and accounts for nearly 9% of the world's total annual coal consumption (MoC 2009). More than half of India's energy needs are met by coal, and about 80% of India's electricity generation is now fueled by coal. The annual demand for coal has been steadily increasing over the past decade. Despite a production increase from 70 million tons in early 1970s to 456 million tons in 2007–2008 (CIL 2009), India continues to face shortages of high quality coal for steel manufacturing (44 million tons in 2007–2008) which is imported. Over the last 7 yr, imports have doubled from 20 million tons in 2000–2001 to 44 million tons 2007–2008 and are expected to triple in 2030 (EIA 2008).

The country has made significant progress toward the augmentation of power infrastructure with an installed capacity of 147,457 MW as of January 2009. Of this, 93,392 MW is accounted for by thermal power plants (coal, gas, diesel), 36,762 MW by large hydroelectric plants, 4,120 MW by nuclear, and the remainder from renewable sources (CEA 2009). Despite the significant growth in electricity generation, significant problems persist, such as poor quality, power shortages, load shedding, fluctuating voltage, erratic frequency, and frequent power cuts. On top of this, currently 400 million Indians are reported to have no access to electricity (IEA 2007b). Even after the signing of a nuclear cooperation treaty with USA, India's nuclear contribution to the energy mix is at best expected to be 3–4% unless vast thorium resources are exploited.

It is estimated that India has only 0.4% of the world's proven reserves of crude oil. The production of crude oil in the country has increased from 6.82 million tons in 1970–1971 to 34.12 million tons in 2007–2008 (MoP 2009). However India's oil consumption increased by 5.7% per annum from 1980 to 2001 periods to 11.9% from 2001 to 2006, and it now stands at 156 million tons, or 3% of global oil consumption (IEA 2007b; MoP 2009). In India, oil provides energy for 95% of transportation needs and the demand for diesel is fivefold higher than the demand for petrol. Over 80% of passengers and about 60% of freight are transported by road. With the increased economic growth and expendable income over the last two decades, demand has also increased for all transport services by road, rail, and air. Vehicle ownership has increased, with the number of private motor cars growing by 16%, two wheelers by 20%, and goods vehicles by 13%/yr from 1991 to 2003. The latest available statistics indicate that the total number of vehicles has increased more than threefold, from 1991 to 2007–2008 and projected to grow by 12–15% reaching 373 million in 2035 (Fig. 10.1). This growth is expected to fuel 5–8% in the demand for petroleum-based energy in India (GOI 2006b; MoP 2009)

In India, natural gas is currently a minor fuel in the overall energy mix, representing 10% of total primary energy consumption in 2008. Natural gas demand has been growing at the rate of about 6.5% during the last 10 yr. Industries such as power generation, fertilizer, and petrochemical production are shifting toward natural gas. Although recent discoveries are expected to boost gas production to bridge the gap, a growing share of gas requirements need to be met by imports.

Today, India has one of the highest potentials for the effective use of renewable energy (Table 10.1). India is the world's fourth largest producer of wind power after

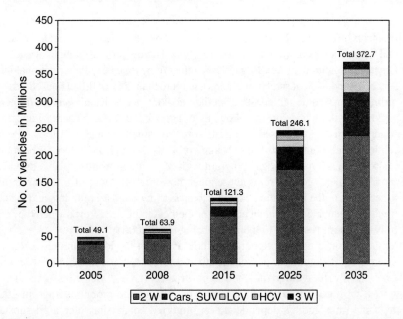

Fig. 10.1 Projected growth of automobiles in India. *2W* motorcycles, *3W* 3 wheeler, *HCV* heavy duty commercial vehicle, *LCV* light duty commercial vehicle, *SUV* sport utility vehicle. Source: http://www.adb.org/Documents/Reports/Energy-Efficiency-Transport/chap01.pdf

Table 10.1 Renewable energy resources (Mtoe/yr)

Resources	Present	Potential
Hydropower (MW)	32,326	150,000
Biomass		
Wood	140	620
Biogas	0.6	4
	0.1	15
Biofuels		
Biodiesel		20
Ethanol	<1	10
Solar		
Photovoltaic		1,200
Thermal		1,200
Wind energy	<1	10
Small hydropower	<1	5

Denmark, Germany, and Spain. There is a significant potential in India for generation of power from renewable energy sources such as small hydro (less than 25 MW), biomass, and solar energy. The country has an estimated small hydropower potential of about 150,000 MW. India produces 13,242 MW renewable energy excluding large hydropower (MNRE 2009) representing 9% of total electricity production. Other renewable energy technologies, including biomass, wind, solar, small hydro (less than 25 MW), bagasse and waste to energy are also growing.

Despite increasing dependence on commercial fuels, a sizeable quantum of energy requirements (40% of total), especially in the rural household sector, is met by noncommercial energy sources, which include fuel wood, crop residue, animal waste, and human and draft animal power. Regardless of the progress achieved after national independence, around 86% of rural households and more than 20% of urban households still rely primarily on traditional fuels to meet their cooking needs. Biomass is the domestic fuel used for cooking and consists of mainly of agricultural waste, gathered woods, and cow dung. Biomass is also used as industrial fuel by small cottage industries. The use of traditional fuels continues to cause health problems arising from indoor air pollution. India also has a 40-yr-old biogas program with 3.7 million installed plants providing energy requirements for the rural households; however, only half of these are in use. Large segments of the population continue to have a low standard of living, and the task of providing clean and convenient energy for their essential needs, even when they cannot fully pay for it, is critical to their well-being (GOI 2006a).

Per capita consumption of energy in India is one of the lowest in the world, 439 kg oil equivalent (kgoe) per person, compared to 1,090 kgoe in China, 7,835 kgoe in USA, and world average of 1,688 kgoe in 2003 (GOI 2006a). It is expected to grow to 1,250 kgoe in 2032 which would be 74% of the global average in 2003. At the same time, India's dependence on imported energy has increased substantially over the years. Up from 17.85% of Total Primary Commercial Energy Supply (TPCES) in 1991, imports accounted for 30% of our TPCES in 2004–2005 (GOI 2006a). Due to limited domestic crude oil reserves, India meets about 70% of crude oil and petroleum products (diesel and aviation fuel) requirement through imports, which are expected to expand in coming years. The quantity of crude oil imported increased nine fold from (11.66 million tons) during 1970–1971 to (121 million tons) by 2007–2008. During the last 7 yr, India's oil import expenditure has increased fivefold because of the escalation of global oil prices (MoP 2009; Fig. 10.2).

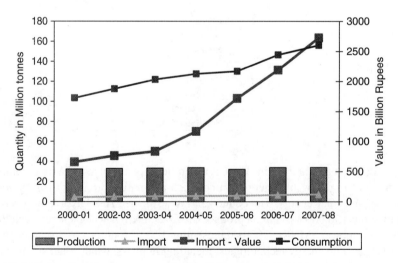

Fig. 10.2 Crude oil production, consumption, and import trends in India

In addition, the economic cost of oil dependence is even greater because the government of India spends US $57.8 billion in subsidies: an amount more than 3% of country's GDP (Ringwald 2008). It is estimated that at a growth rate in demand of 2.9%/yr, India needs to import 6 million barrels/d crude oil in 2030. This would make India the world's third largest oil importer after the USA and China (GOI 2006a; IEA 2007b). Coal imports are likely to grow substantially over time. Hence, energy security has become a growing concern for India's energy policy.

Policy Initiatives

To address these issues in an integrated manner during 2000–2006 period, the Planning Commission constituted a series of committees such as Hydrocarbon vision-2025, India vision-2020, and Integrated Energy Policy-2006 (GOI 2000a, b, 2006a) and prepared an integrated energy policy linked with sustainable development addressing all aspects of energy use and supply. The broad vision behind the energy policy was to reliably meet the demand for energy services of all sectors at competitive prices. In addition, essential energy needs of all households must be met even if that entails subsidies to vulnerable households. The demand must be met through safe, clean, and convenient forms of energy at the least cost in a technically efficient, economically viable, and environmentally sustainable manner.

Based on the committee reports, the Planning Commission projected the country's energy requirements until 2031–2032 based on various growth rates of GDP. To meet these requirements, India needs, at the very least, to increase its primary energy supply by three to four times and its electricity generation capacity/supply by five to six times of their 2003–2004 levels in 2031–2032 (Table 10.2). By 2031–2032, power generation capacity must increase to nearly 800,000 MW from the current capacity of around 160,000 MW inclusive of all plants in production. Similarly supply of coal, the dominant fuel in India's energy mix will need to expand to over 2 billion tons/annum based on domestic quality of coal (GOI 2006a).

Table 10.2 Projected primary commercial energy requirements at 8% GDP growth rate

Year	Hydro	Nuclear	Coal	Oil	Natural gas	TPCES
2011–2012	12	17	257	166	44	496
2016–2017	18	31	338	214	64	665
2021–2022	23	45	464	278	97	907
2026–2027	29	71	622	365	135	1,222
2031–2032	35	98	835	486	197	1,651
CAGR—%	5.9	11.2	5.9	5.1	7.2	6
Per capita consumption 2032 (kgoe)	24	67	569	331	134	1,124
In 2004 (kgoe)	6.5	4.6	157	111	27	306
Ratio 2032/2004	3.7	14.6	3.6	2.9	5.2	3.7

Table 10.3 Range of utilization of different fuels in 2031–2032

Resource	Supply sources (Mtoe) in 2031–2032	Utilization (Mtoe) in 2004
Oil	463–493	116.00
Natural gas	114–224	27.65
Coal	573–1,082	184.35
Hydro	5–50	<1
Nuclear	3–89	<1
Solar	1–4	<1
Wind	0–12	<1
Fuel wood	0–69	115.44
Ethanol	0–4	<1
Biodiesel	0–8	<1

Source: GOI 2003, Planning Commission report.

Meeting this vision requires that India pursues all available fuel options and forms of energy, both conventional and nonconventional. IEA (2007b) estimated that from 2006 to 2030, India will need to invest the massive amount of US $1.25 trillion in the energy infrastructure, three fourths of which will be in the power sector: a huge challenge for meeting sustainable economic growth.

A disturbing fact that emerges from these various scenarios is that even if India somehow succeeds in raising the contribution of renewable energy more than 40-fold by 2031–2032, the contribution of renewable energy to the overall energy mix will not go beyond 5.6% of total energy required in 2031–2032 (Table 10.3). The actual share of modern renewable energy in India's energy mix is significantly lower (about 2% of the total). This is based on the premise that India is neither a significant contributor to greenhouse gas emissions nor will it be so in the foreseeable future. Also, India has made environmental impact reports, called "green clearance", mandatory for most development projects.

However, the current growth in transport activity is a significant environmental concern given the fact that India's carbon emissions are growing at an average of 3.2% per annum, making it one of the top five global contributors to carbon emissions. Furthermore, at the present economic growth rate, India is set to become the third largest carbon dioxide emitter by 2015 (IEA 2007b). It is also estimated that the annual carbon dioxide emission could increase to 1 billion tons to 5.5 billion tons/yr by 2031–2032. There has been a per capita increase of carbon dioxide emissions in India from 2.6 to 3.6 tons, compared to 2004 levels of 20 tons in the USA and a global average of 4.5 tons. The Planning Commission in its integrated energy policy also indicated the carbon emission scenarios are significant (GOI 2006a). In addition, according to the calculation of Carbon Disclosed Project (CDP), the impact of climate change will be greater than in other countries and the cost of climate change in India could even be as high as 9–13% loss in GDP by 2100 (CDP 2007). These impacts will be experienced by a majority of the rural Indian population (60% or 700 million), who directly depend upon on climate sensitive sectors such as agriculture, forestry, and fisheries for their livelihood.

Although India's per capita emission of air pollutants remains low, the population size and the high density of automobiles in the urban areas produce some of the cities with the worst air quality. Hence, the Government of India's transport policy targets Euro III and Euro IV norms (GOI 2003) for the vehicles, which will require clean quality fuel. With the current planning of energy mix, it is not possible to mitigate projected carbon emissions because of heavy dependence on coal. This trend is not sustainable and India is bound to face serious international pressures to reduce carbon emissions, despite politically strong arguments such as the need for economic growth, and exceptions on account of a dense population and poverty. To create opportunities for growth and sustainable livelihood for its citizens, balancing the economic growth and environmental demands requires a paradigm shift in the energy policy.

Fuel Ethanol Overview

Ethanol is a biofuel produced from sugar and starch raw materials by fermentation and has been found to be an excellent substitute for petrol. In a large number of countries, ethanol obtained a predominant position among biofuels as a blending agent with petrol because of its oxygenation properties, energy balance, environmentally friendly nature, possible employment benefits in the rural sector, and contribution to energy security at the national level (GTZ-TERI 2007; Faaji et al. 2008; Zuurbier and van de Vooran 2008). Global production of fuel ethanol increased by 18% over 2006 to 46 billion liters in 2007, marking the sixth consecutive year of double-digit growth (Worldwatch Institute 2009). The USA became the leading fuel ethanol producer in 2007, producing over 24.5 billion liters and jumping ahead of longstanding leader Brazil. Brazil and the USA accounted for 95% of all ethanol production in 2007. Several important political, technological, and federal policies and incentives led to both countries becoming world leaders in the use of bioethanol. Other countries implementing fuel ethanol programs are Australia, Canada, China, Colombia, the Dominican Republic, France, Germany, India, Jamaica, Malawi, Poland, South Africa, Spain, Sweden, Thailand, and Zambia (Dufey 2006; DEFRA 2008; IEA 2008; Faaji et al. 2008).

Biodiesel Overview

Biodiesel is technically defined as a fuel comprised of mono-alkyl esters of long chain fatty acids derived from vegetable oils or animal fats. It is produced by modification of oil through a chemical process of transesterification, neutralizing the free fatty acids, removing the glycerin, and creating an alcohol ester. There are several methods for carrying out this reaction including the common chemical batch process, supercritical processes, ultrasonic methods, and even microwave methods (Bruce et al. 2004; Janulis 2004). After this processing, biodiesel has combustion properties very similar to those of petroleum diesel and can be used as a direct motor fuel or supplement depending on the type and model of vehicle. A by-product of the transesterification process is the production of glycerol. For every 1 tonne of biodiesel that is manufactured, 100 kg of glycerol are produced (Gonsalves 2006; GTZ-TERI 2007; glycerol is presently used in cosmetics, soaps, pharmaceuticals, alkaline resins, and polyglycerols). A variety of oils can be used to produce biodiesel. These include oils from main crops such as rapeseed, soybean, mustard, flax, sunflower, palm oil, waste oils, animal fats, and nonedible crops such as *Jatropha* and hemp. Sunflower and rapeseed are the raw materials used in Europe whereas soybean is used in USA. Thailand uses palm oil, and Ireland uses frying oil and animal fats (FAO 2008).

The world market for biodiesel has expanded rapidly in recent years. Large numbers of countries have implemented a broad range of laws that support biodiesel usage. Currently, a biodiesel mandate for use motor fuel has been set in 28 countries with various incentives and support (FAO 2008). Hence, biodiesel has steadily emerged from pilot plants to commercial production and marketing products with wide acceptance as a fuel for the diesel vehicle industry. Around 10 billion liters of biodiesel were produced in 2007, an 11-fold increase since 2000. Most biodiesel was produced in the EU (6 billion liters) followed by USA (2 billion liters), Indonesia (0.4 billion liters), and Malaysia (0.3 billion liters; FAO 2008). Various research studies, evaluations, tests, and certifications from a large number of countries confirmed biodiesel as clean alternative fuel having the potential to reduce carbon emission from transport vehicles (Gonsalves 2006). Biodiesel is considered a clean fuel since it has no aromatics and almost no sulfur and has about 10% to 11% built-in oxygen, which helps it to burn fully (GTZ-TERI 2007). Its higher octane number improves the ignition quality even when blended with petroleum diesel. Energy content of biodiesel is close to that of diesel. Fuel efficiency is the same as diesel. Fuel economy, power, and torque are proportional to the heating value of biodiesel or biodiesel blend. Due to these favorable properties, biodiesel can be used as fuel for diesel engines (as either, B5-a blend of 5% Bio-diesel in petrodiesel fuel or B20 or B100). USA uses B20 and B100 biodiesel; France uses B5 as mandatory in all diesel fuel (Martinot 2008).

In India, food security is a national priority and therefore, India cannot afford to use (or promote) either cereal grains for ethanol production or edible oil for biodiesel production as is done in other biofuel promoting regions (EU and USA). India is one of the leading importers of vegetable oil in the world as demand outstrips domestic production. Production of food grains like wheat, corn, and coarse cereals has been relatively stagnant in recent years, forcing India to import wheat in 2006 after being an exporter for several years. A recent spurt in global prices for cereals and vegetable oils have been an additional concern for the government, which does not want to aggravate the crisis by promoting the use of food commodities for biofuel. Hence, India's biofuel program is centered on bioalcohol from sugarcane molasses and biodiesel from nonedible oil crops such oil-bearing trees.

Ethanol in India: Conflicting Interests

The processes by which ethanol can be produced are diverse as from sugarcane, molasses, sweet sorghum, wheat, corn, sugar beet, sweet sorghum, rice, cassava, and potato. Unlike Brazil, where ethanol is produced directly from sugar cane juice, and the USA, which uses corn for production, India produces ethanol from molasses, a by-product of sugar manufacturing. Alcohol is also a raw material for industrial use in the production of potable alcohol and chemicals. Hence, ethanol production in India has an intrinsic relationship and dependence with industry structure, government policies, and controls followed in sugar and other related industries.

The sugar industry in India is the second largest agricultural industry after cotton textiles and is located mainly in rural India. The sugar industry has a turnover of US $14 billion per annum and it contributes almost US $700 million to the central and state exchequer in additional taxes every year (KPMG 2007; MoCFA 2007). With more than 516 sugar mills operating in more than 18 states of the country, the Indian sugar industry has been a focal point for socioeconomic development in the rural areas. Around 249 sugar mills are in the cooperative sector and balance are in the private or public sector. Out of 516 operating units, the majority have small capacities (below 5,000 tonnes crushed/d): 64 are of medium size (above 5,000 tons crushes/d) and only eight units have large capacities (above 1,000 tons crushes/d; MoCFA 2009). About 50 million sugarcane farmers and a large number of agricultural laborers are involved in sugarcane cultivation and ancillary activities, constituting 12% of the rural population. In addition, the industry provides employment to about two million skilled or semiskilled workers and others mostly from the rural areas.

The sugar industry is a primary source of raw material for the alcohol industry in India, and sugarcane is the key raw material for the manufacture of sugar and alcohol. The sugarcane growing areas of India may be broadly classified into three regions based on climatic conditions, yield of cane, and sugar content. These regions are (a) the subtropical northern belts representing Uttar Pradesh, Uttaranchal, Bihar, Punjab, and Haryana; (b) the subtropical peninsular region representing Maharashtra, Gujarat, and Karnataka; and (c) the tropical Tamil Nadu, Andhra Pradesh, and Orissa (Table 10.4).

The sugarcane supply to mill is dependent on the cane production from a large number of small farmers because the mills cannot own land according to the Indian land ceiling act. The average size of farm holdings are less than 1 ha and only 25% are more than 4 ha. This means that a mill of 3,000 tons crushes/d must procure cane from 18,000 farmers (KPMG 2007). Cane cultivation and harvesting in India is manual and mechanization is mainly limited to plowing and transport. Availability of labor is becoming critical and mechanization of small farms will be a challenging task in the coming years. Another difficulty is that the crop cycle is limited to 2 or 3 yr because of extreme climatic conditions in most parts of India, compared to 6 to 7 yr cycles in other countries. This requires high flexibility of farmers to shift to other crops in the absence of profitability. The value chain of the sugar industry has significant variations from regions to regions in its profitability to farmers and

Table 10.4 Classification of sugarcane belts of India

Region	State	Average yield tones/ha	Average recovery (%)	Average crushing days	Temperature Min (°C)	Max (°C)
Subtropical—north	Bihar	42.91	9.13	93.00	7.7	41.5
	Uttar Pradesh	57.57	9.62	134.42	3.6	42.6
	Uttaranchal	57.95	9.54	131.00	2.1	42.1
	Punjab	60.21	9.60	107.85	4.6	43.6
	Haryana	60.54	10.00	136.28	4.1	43.3
Subtropical—central	Gujarat	72.08	10.70	154.14	11.1	40.9
	Maharashtra	72.77	11.46	116.71	10.9	42.8
	Karnataka	83.74	10.56	141.85	14.4	41.5
Tropical—south	Orissa	59.01	9.33	72.42	11.5	41.2
	Andhra Pradesh	76.64	10.16	123.85	13.6	41.0
	Tamil Nadu	100.25	9.59	185.85	18.5	37.5

Source: Sugar data from Cooperative Sugar Journal, published by Indian Sugar Mills Association. Temperature data from www.indiawaterportal.org. Values are average of 2001–2002 to 2006–2007.

millers, due to different levels of productivity of cane, cost structure, sugar recovery, and multiple and complex taxes and levies on sugar and its by-products (KPMG 2007; ISMA 2008).

In India, the sugar industry is beginning to diversify to an integrated complex with cogeneration of power, alcohol for industrial and fuel uses. The sugar industry is a green industry and largely self-sufficient in its energy needs because of the use of bagasse for power and steam. In fact, the sugar industry generates exportable surplus power through cogeneration and contributes to the reduction of the energy deficit. The realization from exportable power is dependent on long-term power purchase agreements with governments and power companies. The cogeneration also has proven revenue potential from CDM.

Sugar Policy

Sugar is a controlled commodity in India under Essential Commodities Act 1955 and regulated across the value chain. The heavy regulations in the sector artificially impact the demand–supply forces resulting in market imbalance. Since 1993, the regulations have been progressively eased out. These include delicensing of the industry in 1998 and the removal of control on storage and distribution in 2002. However, central and state governments still have control over the sugar value chain from mandatory and state advisory cane price (statutory minimum price (SMP) and state advisory price (SAP)), mill capacity expansion, distance of operation, by-product use and transportation, levy and free sale (10:90) release mechanism and exports, and various forms of taxes at central and state level (KPMG 2007).

Cyclical Sugarcane and Sugar Production

Since independence, the land area under cane cultivation, cane production, productivity, and sugar production have increased dramatically (Figs. 10.3 and 10.4). Regardless of increased growth in area and productivity, the production of sugarcane and sugar fluctuates considerably from year to year in India (Fig. 10.3). Natural factors for this volatility include distribution of rainfall, climatic conditions of flood and droughts, pests and diseases, fluctuations, in prices of gur and khandasari (traditional Indian sugar products), and changes in returns from competing crops. Man-made factors are primarily government policies regarding sugarcane price, release mechanism, taxes, and export and import controls. Sugarcane prices are determined independently of sugar market prices and have been increasing year after year. The uptrend in the sugar cycle starts with timely cane payments by millers from the increased profits for sugar produced and sold in the markets. This will result in increased cane planting by farmers, bumper production of cane and factories to crush it, and produce more sugar, over supply in the market, decline in sugar prices, lower profitability for mills, and delayed payment to farmers. When there is a wide disparity due to high sugarcane price and low sugar prices in markets, millers are not able to make payments on time and arrears to farmers start mounting. The farmers

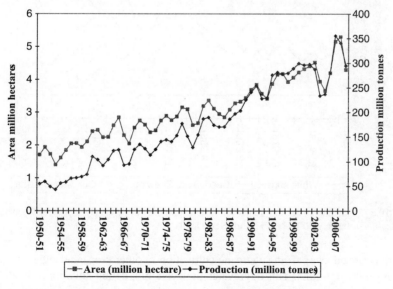

Fig. 10.3 Indian sugarcane area and production—a cyclical trend. *Source*: Cooperative Sugar 40 (5), January 2009, NFCSF Ltd., New Delhi

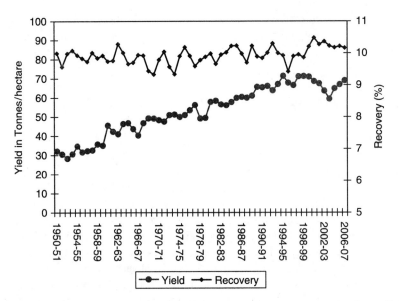

Fig. 10.4 Indian cane yield and sugar recovery percentage. *Source*: Cooperative Sugar 40 (5), January 2009, NFCSF Ltd., New Delhi

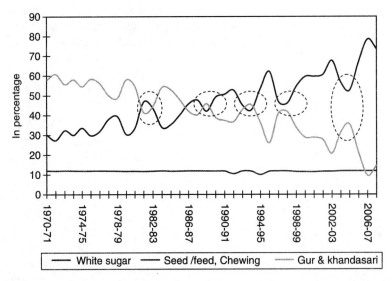

Fig. 10.5 Sugarcane utilization trends for various purposes in percentage. Source: Cooperative Sugar 40 (5), January 2009, NFCSF Ltd. New Delhi

are thus forced away from sugarcane cultivation to other crops and sugarcane and sugar production falls. This aggravates the cycle in a deficit situation, causing an increased diversion of cane to gur and khandasari and resulting in less availability of cane for white sugar manufacturing (Fig. 10.5). All these factors result

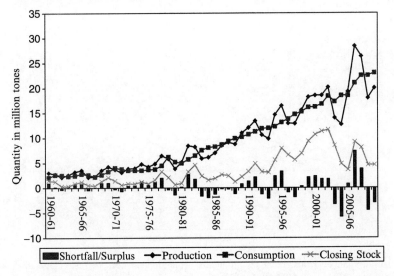

Fig. 10.6 Indian sugar production and consumption—vicious cycle. *Source*: ISMA 2008 and incorporating 2008–2009 data

again in reduced sugar production, higher sugar prices, turnaround of industry and timely cane payments, and then this vicious cycle continues. In the past, these cycles arose every 4 to 5 yr (Fig. 10.6). In recent years, these deficit/surplus gaps are becoming wider irrespective of stock positions, existing various control regimes, and policy interventions. However, consumption of sugar in India has been growing at a steady rate of 3% and is currently at 23.1 million tons, with per capita consumption at 18 kg (lower than world average of 22 kg; Fig. 10.6). Consumption trends continue to shift from household to industrial consumers. A nationwide survey conducted in 2007 estimated that 61% of sugar sold in the free market accounted for industrial and small business segments (KPMG 2007).

Molasses and Alcohol Interdependence

India has about 300 distilleries, with a production capacity of approximately 3.2 billion liters of rectified spirits (alcohol) per year, almost all of which is produced from sugar molasses. There has been a steady increase in the production of alcohol in the country, with the production doubling from 887.2 million liters in 1992–1993 to 1,654 million liters in 1999–2000 and was expected to triple to 2300 million liters by 2006–2007 (GOI 2006a).

The sugar industry is a major supplier of molasses for alcohol and ethanol producing units. There had been ups and downs in production of molasses and ethanol over the years as it is directly linked to molasses production from sugar and its cyclicality (Fig. 10.7). Surplus sugar results in increased production of molasses and depresses

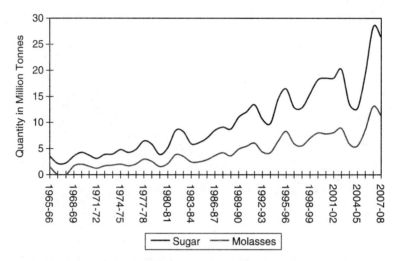

Fig. 10.7 Indian sugar and molasses production—interdependence. *Source*: Cooperative Sugar 40 (5), January 2009, NFCSF Ltd., New Delhi

prices for molasses and alcohol. On the other hand, a shortfall of sugar production results in low molasses production and an increase in the price of molasses and alcohol. Main consumers of alcohol are potable and industrial segments. The industry is tightly controlled by various central and state rules, regulations, and taxes.

Fuel Ethanol: A Turbulent Journey

Affected by the rising oil prices and increased imports of oil for transport, India commenced its bioalcohol transport fuel blending in 2001. In order to ascertain financial and operational aspects of blending 5% ethanol with petrol, government had launched three pilot projects: two in Maharashtra and one in Uttar Pradesh during 2001. Apart from these pilot projects, R and D studies were undertaken simultaneously to evaluate techno-commercial feasibility and identify vehicle modification requirements, if any. Both pilot projects and R and D studies were successful and established blending potential of ethanol up to 5% with petrol and usage of ethanol-doped petrol in vehicles. Discussions were held with stakeholders at the central and the state level; Society for the Indian Automobile Manufacturers, the state governments, and an expert group were set up an for examining various options of blending ethanol with petrol. Considering the logistical and financial advantages, this group has recommended blending of 5% ethanol with petrol at supply locations (terminals/depots) of oil companies. The second phase was aimed to cover the entire country and third phase ethanol blending to be increased to 10%. The availability of molasses and alcohol was estimated to be adequate to meet this requirement, after fully meeting the requirement of the chemical industry and potable sectors. In view of the surplus availability of alcohol, the central government has implemented with effect from January 1, 2003, 5% ethanol-doped petrol supply in the nine states (out of 29) and four contiguous union territories (out of six) as first phase (Gujarat, Andhra Pradesh, Haryana, Karnataka, Maharashtra, Punjab, Tamil Nadu, Uttar Pradesh, Damman and Diu, Goa, Dadra and Nagar Haveli, Chandigarh, Pondicherry).

In addition, during June 2002, the government commissioned a committee under Planning Commission with a mandate of identifying status of biofuels, its commercial use, storage, handling, development of quality standards, identification of prospective sources, cost benefits, R and D requirements, and measures for effective coordination of various ministries. The committee endorsed the initiative of government on the introduction of blending of 5% ethanol with gasoline and projected demand for petrol and diesel and the amount of ethanol and biodiesel required for 5%, 10%, and 20% blending. The committee also estimated demand and supply situation for ethanol until the 12th plan period (Table 10. 5) for blending with 5% taking into consideration of requirements of potable and chemical industry.

Costing of ethanol using sugarcane–molasses ethanol route were also worked out taking into consideration of prevailing prices of molasses at that time and past trends (Rs. 1,000), efficiencies of production (220 l/ton). The ethanol costs were less than Rs. 9/l and quite competitive to that time imported cost of gasoline which was around Rs. 10–12/l (GOI 2003). The committee also took into consideration of past production and consumption trends of molasses, alcohol, and projected surplus alcohol production in the country (Table 10. 6). It is also assessed that the area of 4.36 million hectares under sugar cane may expand to 4.96 million hectares in 2006–2007 yielding additional cane production of 50 million tons. This will provide an adequate base for ethanol for 10% blending even in the tenth plan period. Hence, the committee submitted the findings in April 2003 with following recommendations:

- The country must move toward the use of ethanol as substitute for gasoline.
- Production of molasses and distillery capacity can be expanded to meet 5–10% blend of ethanol.
- Ethanol may be manufactured using molasses or directly from sugar cane juice when sugarcane is surplus.
- Restrictions on movement of molasses and putting up ethanol manufacturing plants may be removed.
- Imported ethanol should be subjected to suitable duties.
- Buyback arrangement with oil companies will be arranged.

Several sugar mills geared up production and supply of ethanol by adding additional capacities (11 factories in Uttar Pradesh, seven units in Tamil Nadu, eight in Karnataka, four units in Andhra Pradesh). Similar steps have also been taken up by the cooperative sector units in Maharashtra, Punjab, and Uttar Pradesh. By the end of the 2004, it is estimated that about 300 million liters capacity would have been created for the production of anhydrous alcohol (Ethanol India 2009).

The 2003–2004 season droughts resulted in a lower sugarcane crop and sugar production (Figs. 10.3, 4, and 5) and consequently a decreased availability of molasses and increased molasses prices. The sugar output dropped to 13 million tons

Table 10.5 Projected demand and supply of ethanol for 5% blending in petrol

Year	Petrol demand (Mt)	Ethanol demand (ML)	Molasses production (Mt)	Ethanol production (ML)			Ethanol utilization (ML)		
				Molasses	Cane	Total	Potable	Industry	Balance
2001–2002	7.07	416.14	8.77	1,775	0	1,775	648	600	527
2006–2007	10.07	592.72	11.36	2,300	1,485	3,785	765	711	2,309
2011–2012	12.85	756.36	11.36	2,300	1,485	3,785	887	844	2,054
2016–2017	16.4	965.30	11.36	2,300	1,485	3,785	1,028	1,003	1,754

Source: GOI 2003, Planning Commission report.

Table 10.6 Molasses and alcohol production consumption trends (in million liters)

Year	Molasses production	Alcohol production	Industrial use	Potable use	Other uses	Surplus availability
1998–1999	7.00	1,411.8	534.4	5,840.0	55.2	238.2
1999–2000	8.02	1,654.0	518.9	622.7	576	455.8
2000–2001	8.33	1,685.9	529.3	635.1	588.0	462.7
2001–2002	8.77	1,775.2	5,398.0	647.8	59.9	527.7
2002–2003	9.23	1,869.7	550.5	660.7	61.0	597.5
2003–2004	9.73	1,969.2	578.0	693.7	70.0	627.5
2004–2005	10.24	2,074.5	606.9	728.3	73.5	665.8
2005–2006	10.79	2,187.0	619.0	746.5	77.2	742.3
2006–2007	11.36	2,300.4	631.4	765.2	81.0	822.8

Source: GOI 2003, Planning Commission report.

(normally 21 million tons), molasses production sunk to 5.9 million tons (normally 9 million tons), and the ethanol manufacturing level decreased to 1,518 million liters (normally 2,000 million liters; Fig. 10.5). The ethanol requirement for 5% blending in the nine states where blending mandatory was 363 million liters in 2003–2004, but the oil companies could only procure 196 million liters. In addition, most of the states have a labyrinth of rules and regulations (interstate movement, high excise duties, and storage charges) to control alcohol for the potable liquor industry. Due to large number of taxes and levies, ethanol blending became commercially unviable in most of the states. The results were that ethanol supplies to the oil companies came to a virtual halt in September 2004. To meet the shortfalls in the year 2004, India imported 447 million liters of ethanol from Brazil. Recognizing the difficulties due to high ethanol prices and low availability, the government of India amended its 5% blending mandate with the notification that 5% ethanol blended petrol shall be supplied in identified areas if (a) the indigenous price of ethanol offered for ethanol blended program is comparable to that offered by the indigenous ethanol industry for alternative uses, (b) the indigenous delivery price of ethanol offered for the ethanol blended petrol program at a particular location is comparable to the import parity price of petrol at that location, and (c) there is adequate supply of ethanol (MoP 2004).

A new government expert committee was commissioned to develop an integrated energy policy to deal with all aspects and forms of energy. The committee reporting on molasses' scarcity differed from the previous committee on the potential of sugarcane ethanol for India. The relative merit of sugarcane ethanol and alternative technologies for ethanol development are still under discussion. In addition, it raises the issues of water scarcity, the lack of area for sugarcane, regional productivity drops, grain shortages, food security, and arguments that the availability of molasses-based alcohol from the sugar industry is unlikely to grow significantly over in coming years. Hence, the committee made the following major recommendations(GOI 2006a):

- Set import tariff on alcohol independent of use and at a level no greater than that for petroleum products.
- Do not mandate blending of ethanol with petrol and prices of ethanol at its economic value vis-à-vis petrol.

- To encourage alternate routes to ethanol, such production may be procured at the full trade parity price of petrol for 5–7 yr instead of being purchased at its true economic value based on calorific content duly adjusted for improved efficiency.
- Create incentives for cellulosic ethanol with investment credits

The bumper monsoon in the year 2005–2006 boosted sugarcane production, availability of molasses, and also increasing prices of petroleum products resulted in a renewed interest in the ethanol program. In August 2005, the government negotiated an agreement between the sugar industry and oil marketing companies to enable the purchase of ethanol, and the ethanol program restarted in a limited number of designated states and union territories. The government of India announced in September 2006 the second phase of the Ethanol Blending Program (EBP) because of the strength of sugar production in that time period. This mandates 5% blending of ethanol with petrol (gasoline) subject to commercial viability in the 20 states and eight union territories with effect from November 2006. Oil marketing companies floated open tenders for ethanol from the domestic producers. Subsequently, bids have been finalized and the EBP has started in about ten states. However, the EBP was not implemented in other states due to high state taxes, excise duties, and levies, which makes the ethanol supply for blending commercially unviable. Consequently, ethanol for blending with petrol in the Indian sugar year 2006–2007 (October–September) is reached at least 250 million liters against the target of 550 million liters.

The sugar industry offered that it could provide ethanol at Rs. 19/l ($0.38/l), which is at a lower cost than the product it would substitute, methyl tertiary butyl ether (MTBE), which costs Rs. 24–26/l ($0.49–0.53/litre) at that time. Petroleum companies purchased fuel grade ethanol from the sugar companies at rates ranging between Rs. 19.0 to 21.5 (47–53 cents)/l during 2006–2007. The cost of production of ethanol depends on the price of molasses, which fluctuates widely during the season. Industry sources estimate the average cost of production of ethanol to range from Rs. 16 to 18 (40–44 cents)/l at 2006 prices of molasses (Rs. 2,000–3,000). The 11th planning commission report (GOI 2007) also states that the economics of sugar production are crucially dependent on the production of by-product ethanol. After stabilization of 5% ethanol blending petrol sales extended to the country as a whole, the content of ethanol in petrol would be considered for increasing up to 10% by the middle of the 11th plan, subject to ethanol availability and commercial viability of blending.

New Biofuel Policy on the Way

Rising prices of petroleum products in 2008 to more than US $100 per barrel and very high import bills of crude oil forced the government to reinitiate a national biofuel policy under the Ministry of New and Renewable Energy. The union cabinet approved the national policy on 11 September 2008 for setting up an empowered national biofuel coordination committee headed by the prime minister and a biofuel steering committee headed by the cabinet secretary (MNRE 2008). Draft policy was circulated for interministerial consultation and deliberations. The policy envisages the following:

- A target of blending bioethanol and biodiesel 20% by 2017
- Biodiesel production from nonedible oils in waste, degraded, or marginal lands
- Community-based biodiesel production
- Minimum support price for biodiesel and minimum purchase price for bioethanol
- Biodiesel and bioethanol may be brought under ambit of "Declared Goods" to ensure unrestricted movements
- No taxes and duties to be levied on biodiesel

The new policy is still in consultation phase and require approvals and parliament clearances. However, as the term of the present government will be ending in May 2009, the future of the biofuel policy is in question.

Biodiesel India: Differing Policy Options

In India, biodiesel research, production, and marketing are in the early stages of development. Oilseeds and edible oils are two of the most sensitive essential commodities. India is one of the largest producers of oilseeds in the world and this sector occupies an important position in the agricultural economy and accounting for the estimated production of 282 lakh tons of oilseeds during the year 2007–2008 (Fig. 10.8). India contributes about 6% to 7% of the world oilseeds production. Climatic conditions enable India to produce wide range of traditional oil seed crops

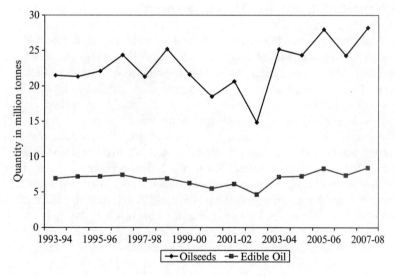

Fig. 10.8 Indian oil seed and edible oil production. Source: Ministry of Agriculture, Production of oil seeds

such as groundnut, mustard and rapeseed, sesame, safflower, linseed, niger seed, and castor. Soya bean and sunflower have also assumed importance in recent years. Coconut is the most important amongst the plantation crops. Efforts are being made to grow oil palm in Andhra Pradesh, Karnataka, Tamil Nadu in addition to Kerala and Andaman and Nicobar Islands. Among the nonconventional oils, rice bran oil and cottonseed oil are the most important. In addition, oilseeds of tree and forest origin are also a significant source of nonedible oils (NOVOD 2008).

Despite the production of diverse and large volume of oils, India is not self-sufficient in edible oils. In the early 1980s, India imported 20–40% of edible oil requirements. Finding this to be a huge drain on foreign exchange resources, the government launched the Oilseeds Technology Mission in 1986 leading to increase in oilseed production and reduction in imports to negligible levels. However, during the last two decades, the edible oil consumption has increased at a compound average growth rate of 4.25% from mere 4.959 million tons in 1986–1987 to 12.191 million tons in 2007–2008. Increased per capita income has also increased per capita consumption of edible oils to 10.23 kg/yr in 2006–2007 from 6.43 kg/yr in 1986–1987. These led to nearly 50% of deficits, which was managed by importing palm oil from Indonesia and Malaysia soybean oil from Argentina and Brazil.

The edible oil consumption in the country is presently growing and likely to remain heavily dependent on imports. According to an estimate by National Council of Applied Economic Research (NCAER 2009), it is predicted that in the year 2015, the demand for edible oil in India would be 20 million tons per annum. Considering the present domestic edible oil supply of 12 million tons per annum, a shortfall of 7 million tons per annum is envisaged in the year 2015. To bridge this gap, an average growth rate of 15% per annum would be required in edible oil production in India. Edible oil production in India is growing at a compound annual growth rate (CAGR) of 4.26%. The situation might worsen if the country fails to maintain the growth in domestic vegetable oil production. Even maintaining the growth rate in production of vegetable oils will not be an easy task, especially when there is increasing competition among the different crops for the cultivable land and irrigation. In the event of failure to achieve the required growth rate, India would continue to spend large sums on the importation of edible oil. In addition, the demand for vegetable oil is becoming linked to the price of petroleum and to the geopolitical complexities of the crude petroleum market.

The edible oil industry of India comprises of 50,000 expellers, 600 solvent extraction plants, 300 vegetable oil refineries, and 175 hydrogenation plants. The edible oil sector occupies a distinct position in Indian economy as it provides job to millions of rural people, achieves on an average a domestic turnover of US $10 billion per annum, and earns foreign exchange of US $90 million per annum from export of by-products of oil. Domestic cost of edible oils is higher than petroleum products. The domestic consumption demand, stagnant growth in production, and foreign exchange requirements of import make edible oil an unviable option for biodiesel in India. Hence, the Indian government decided to explore alternate source for biodiesel development in India.

Biodiesel: A Journey Without Direction

India has a vast untapped potential of nonedible oil-bearing plant species distributed throughout the country: 300 species of trees have been reported to produce oil-bearing seeds (Subramanian et al. 2005). All of them are naturally grown wild species which have not yet been cultivated and harvested systematically for oil production. Seventy-five plant species have been identified (Azam and Nahar 2005) with 30% or more oil content in their seeds or kernels. According to a survey conducted in 2002, 12 species have been selected for its importance of present industrial usage and abundance in distribution. These 12 species of trees identified are *Azadirachta indica, Pongamia glabra, Calophylluum ionophyllum, Hevea braziliensis, Madhuca indica, Shorea robusta, Mesua ferra, Mallotus philippines, Garcinia indica, Ricinus communis, Jatropha curcas and Salvadora*. There are billions of these trees distributed all over India. Collection and processing mechanism of these tree seeds are not yet fully developed. Local people collect small percentage of these seeds (10%) and trade for oil and cake, and the remainder of the seeds go uncollected. Their seeds, oil, and by-products are increasingly being used in modern industry for cosmetics, varnishes and paints, lubricants, resins, adhesive, dyes and inks, explosives, cellophane, pesticides, and pharmaceuticals (GTZ-TERI 2007). It is estimated that the potential availability of tree borne seed oils in India amounts to about 1 million tons/yr; the most abundant oil sources are sal (180,000 tons), mahua (180,000 tons), neem (100,000 tons), and karanja (55,000 tons).

The national and state research laboratories have been investigating the potential of these tree seed oils as a biofuel. Among these, *Jatropha curcas, Pongamia pinnata, Calophyllum inophyllum,* and *Heveia brasiliensis* have been investigated in detail for suitability for Indian conditions and also to meet the automobile blending requirements. There are different advantages to the different species. While *Jatropha* is a native species of South America, pongamia is of Indian origin. Pongamia tree is traditionally planted in several states in road sides, avenues, national highways, and parks and therefore is well known publicly. This tree is not only used for oil but also for animal feed, manure, firewood, and medical purposes. The government of India's biofuels committee submitted a report on biofuels in April 2003, in which it found that *J. curcas* is the most suitable for the biodiesel purpose in India because of following advantages:

- The estimated oil yield per hectare for *Jatropha* is among the highest among tree-borne oil seeds. With an average seed production 3.75 tons/ha, oil content of 30–35% and oil yield of 1,200 kg/ha estimated compared to 375 kg/ha per for soybeans in the USA and 1,000 kg/ha for rapeseed in Europe
- Ability to grow in areas of low rainfall (200 mm/yr), on low fertility, marginal, degraded, fallow, and waste lands
- Relatively easy to collect, plant, and grow without fencing requirements
- Potential use of by-products for manure and biogas generation

- Opportunity to intercrop, integrate with existing social forestry and poverty alleviation programs that deal with land improvement
- Conformation from pilot studies as alternate clean fuel for automotives from India and other parts of world

It is also estimated with a plant density of 2,500 trees/ha can provide an average seed yield of 1.5 kg/tree. A 1-ha plantation can produce an average of 3.75 tons/ha of seed, corresponding yield of 1.2 tons of oil/ha and 2.5 tons of cake. It is estimated that by the end of 11th plan (2011–2012) period that 13.38 million tons of biodiesel for 20% blending will be required, which in turn will require over about 11.2 million hectares of land for *Jatropha*. Cultivation of *Jatropha* was expected to create employment for the rural population.

Potential Availability of Land

India has much underused or unused lands which need to be deforested to prevent degradation. *Jatropha* plantation was intended to rehabilitate these lands, by improving their water retention capacity, stabilizing the soil, and especially for helping the poor. The chain of activities from raising nurseries, planting, maintaining, primary processing, and oil extraction are labor intensive and expected to generate employment opportunities on a large scale, particularly for the rural landless and help them to escape poverty. Hence, 13.4 million hectares of land conversion to *Jatropha* was proposed under various land categories such as forestry management (3 million hectares), hedge plantations (3 million hectares), absentee landlords (2 million hectares), forestry (2.4 million hectares), and public lands park (1 million hectares) and on waste land (2 million hectares). Accordingly, a national mission on biodiesel was proposed with necessary government support to demonstrate the viability of the program in two phases.

Phase I consists of a demonstration project to be implemented by the year 2006–2007 in six "micro-missions" including plantation, procurement of seed and extraction of oil, processing of seed oil into biodiesel (transesterification), blending and marketing, and research and development. This phase was aimed at cultivating 400,000 ha of *Jatropha* installation of a 80,000 tons/yr transesterification plant to produce 0.5 million tons of biodiesel, 10.52 tons compost from the press cake, and the generation of 127.6 million work days in plantation (cumulative basis) for rural poor. The phase II was planned as a self-sustaining expansion of the program leading to the production of biodiesel to meet 20% of the country's diesel requirements by 2011–2012. It plans to accomplish this through accelerating the momentum achieved in the demonstration project, establishing *Jatropha* plantations throughout the country.

The government planned to act as the principle mover ensuring the necessary resources and components, with involvement from all stakeholders. This will involve 400,000 ha of plantations in compact districts, each with an area of 50,000 to 60,000 ha with facilities established for all the activities involved in forward and backward linkages. The Ministry of Forests and Environment and the National Oilseed and Vegetable Oil Development (NOVOD) Board were identified to serve as responsible agencies for the cultivation in the forest and nonforest areas, respectively, by providing the necessary information and financial assistance. Support mechanisms under the National Employment Guarantee Schemes include Comprehensive Land Development Program, Drought Prone Area Program, Watershed Development Fund, and National Food for Work Program. The financial requirement of the demonstration project until 2007–2008 was estimated up to Rs. 14960 million. This includes a government contribution of Rs. 13600 million towards plantation, Rs. 680 million towards administrative expenses and Rs. 680 million towards R & D. In addition to these a mix of entrepreneurs' own contribution of Rs. 160 million (margin money), a subsidy from the government of Rs. 480 million, and a loan of Rs. 960 million from the National Bank of Agricultural and Rural Development in the ratio of 10:30:60. The transesterification unit will be a commercial venture, estimated at Rs. 750 million. The implementation part of the project has divided into four sectors: (1) plantation, production, marketing and trade and research and development, and various stakeholders from concerned central ministries, departments, and research institutes; (2) state departments and universities; (3) petroleum companies and distributors; and (4) private enterprises, nongovernment organizations, and farmers' organizations.

Federal Initiative Progress

Seed development. NOVOD has established national network on *Jatropha* and karanja in 2004 toward development of high-yielding varieties. The network consists of 42 public research institutions and state agricultural universities. Department of biotechnology initiated a "micro-mission" on production and demonstration of quality planting material of *Jatropha*. Work is also in progress on the development of high oil-yielding varieties of *Jatropha* by Department of Biotechnology, Aditya Biotech Research Centre (Raipur), the Indira Gandhi Agriculture University (Raipur), and the Bhabha Atomic Research Centre (Trombay).

Plantation. Agronomic research is under way on *Jatropha* and pongamia in The National Afforestation and Eco-development Board under the guidance of the Ministry of Environment and Forests, The National Oilseed and Vegetable Oil Development Board under the guidance of the Ministry of Agriculture, and The Central Salt and Marine Chemicals Research Institute (Bhavnagar).

Transesterification study. Production of esters from *Madhuca indica, Shorea robusta, Pongamia glabra, Mesua ferra, Mallotus philippines, Garcinia indica, J. curcas,* and *Salvadora* are in progress at Punjab Agricultural University; Indian Institute of Petroleum (IIP), Dehradun; CSMCRI, Bhavnagar; and NBRI Lucknow (CSMCRI, Bhavnagar; NBRI Lucknow; Indian Institute of Chemical Technology; Indian Institute of Technology, Delhi and Madras). Indian Oil Corporation research and development are working on *Jatropha*, karanj oil, mahua oil, and *Salvadora* oil.

Alternate enzymatic esterification on neem, mahua, and linseed is also progressing at Mahindra and Mahindra Ltd.

Pilot plant study. Pilot plants on transesterification with *Jatropha* oil has been carried out by Indian Oil Corporation (research and development), Faridabad; the Indian Institute of Technology, Delhi; the Punjab Harbinsons Biotech, Agricultural University, Ludhiana; the Indian Institute of Chemicals Technology, Hyderabad; the IIP, Dehradun; the Indian Institute of Science, Bangalore; and Southern Railways, Chennai.

NGOs. Organizations *viz.*, Uthan (Allahabad), Sutra (Karnataka); the Institute of Agriculture and Environment (Jind, Haryana); the Bharatiya Agro Industries Foundation Development (Pune, Maharastra); Pan Horti Consultants (Coimbatore); Classic Jatropha Oil (Coimbatore); and Renulakshmi Agro Industries (Coimbatore) are involved in the promotion and creating awareness on biofuel.

Trials. Bharat Petroleum Corporation Limited test marketed through its retail outlets 5% of diesel blends of *Jatropha* and pongamia oil. Railways organize trial runs between Amritsar and Delhi (Shatabdi Express) using blended diesel. Daimler Chrysler, Mahindra and Mahindra Co., and Bombay Electric Supply and Transport tested the blended fuel in Mercedes Cars, Tractors, and Public transport buses, respectively.

State Initiatives

Many states have initiated biodiesel programs based on the central policy directives or their own. Two hundred districts in 19 potential states have been identified on the basis of availability of wasteland, rural poverty ratio, below poverty line census, and agroclimatic conditions suitable for *Jatropha* cultivation over a period of 3 yr. Each district planned to be treated as a block and under each block, a 15,000-ha *Jatropha* plantation is planned to be undertaken through farmers (GOI 2003; DIE 2008). Details of progress are summarized in Table 10.7.

Commercial Initiatives

Large number of small and medium private enterprises was also invested in plantations as well as commercial production of biodiesel; however, the market for biodiesel has not yet emerged on a commercial scale. The current status of their activities is summarized in Table 10.8. In October 2005, the Ministry of Petroleum and Natural Gas initiated a biodiesel purchase policy with effect from January 2006. According to the policy, oil marketing companies are to purchase biodiesel in 20 purchase centers in 12 states (DIE 2008). As per the government notification, biodiesel has been completely exempted from the excise duty.

Table 10.7 Details of progress on biofuel initiatives in various states of India

No.	Name of state	Status of activities
1	Andhra Pradesh	Promotion of pongamia and simaruba with an objective to achieve 100,000 acres of biodiesel plantations in 13 districts was initiated in order to make productive use of degraded land. Forest department is planning to enter into a public private partnership with private company for ensuring buy back agreements. For example, a formal agreement was entered with Reliance Industries for *Jatropha* planting. The company has selected 200 acres of land at Kakinada to grow *Jatropha*. Government has reduced the value added tax for biodiesel to 4% and state road transport corporation was planned to run 10% of its fleet on 5% blending of biodiesel (APGO 2006)
2	Bihar	Plantations have been initiated in districts namely, Araria, Aurangabad, Banka, Betiah (West Champaran), Bhagalpur, Gaya, Jahanabad, Jamui, Kaimur, Latehar, Muzzaffarpur, Munger, and Nawada
3	Chhattisgarh	Chhattisgarh Biofuel Development Authority has been set up for promotion of biofuel. 210 million *Jatropha* saplings were raised for planting in the year 2005 and 2006 and planted on 84,000 ha of farmer's and government fallow land. Pilot demonstration plantation was established on 100 ha in government fallow land in each district. A small transesterification plant was installed for biodiesel production at Raipur. Biodiesel-based power generators for rural electrification in a cluster of 50 remote villages were also installed. As a part of the government plan to electrify all villages by 2012, 400 villages are planned to electrify through *Jatropha* based biodiesel funded by Village Energy Security Program of MNRE. State-of-art laboratory was set up in association with a local NGO, for testing of oils and biodiesel, etc. As a demonstration, chief minister continued to use biodiesel-blended fuel in his official vehicle. Government notification issued for allotting government revenue fallow land on lease to private investors to undertake Jatropha/Karanj plantation and also to setup biodiesel plant
4	Jharkhand	Plantations have been initiated in 19 districts namely Bokaro, Chatra, Daltenganj, Devgarh, Dhanbad, Dumka, Garhwa, Godda, Giridih, Gumla, Hazaribag, Jamshedpur, Koderma, Pakur, Palamu, Ranchi, Sahibganj, Singbhum (east and west)
5	Gujarat	Plantations have initiated in 10 districts. Ahmednagar district more than 1,000 farmers are working with Govind Gramin Vikas Pratishthan for *Jatropha* planting an area of 2,500 acres. To date, more than 2 million *Jatropha* plants have been planted in the target area of the five villages of Vankute, Dhoki, Dhotre, Dhavalpuri, and Gajdipoor
6	Goa	Plantations have been initiated in Panaji, Padi, Ponda, and Sanguelim districts
7	Himachal Pradesh	Plantations have been initiated in Bilaspur, Nahan, Parvanu, Solan, and Unna districts

(continued)

Table 10.7 (continued)

No.	Name of state	Status of activities
8	Haryana	Plantations have been initiated in 11 districts namely, Ambala, Bhiwani, Faridabad, Gurgaon, Hisar, Jind, Jhajjar, Mohindergarh, Punchkula, Rewari, and Rohtak
9	Karnataka	A biofuel policy has been drafted by state government. Plantation has been initiated in 15 districts. Farmers in semiarid regions of Karnataka are also planting *Jatropha*. Since 2002, Labland Biodiesel, a Mysore-based private limited company, is active in biodiesel and *Jatropha* development
10	Kerala	Plantations have been initiated in Kottayam, Quilon, Trichur, and Thiruvananthapuram districts
11	Madhya Pradesh	Plantations have been initiated in 20 districts, namely Betul, Chhindwara, Guna, Hoshingabad, Jabalpur, Khandwa, Mand Saur, Mandla, Nimar, Ratlam, Raisena, Rewa, Shahdol, Shajapur, Shivpuri, Sagar, Satna, Shahdol, Tikamgarh, Ujjain, and Vidisha
12	Maharashtra	200 ha plantations were raised in Nasik and Aurangabad districts. In July 2006, Pune Municipal Corporation demonstrated biodiesel blended fuel in over 100 public buses. In September 2007, the Hindustan Petroleum Corporation Limited partnership with the Maharashtra State Farming Corporation Ltd. for a *Jatropha* based biodiesel venture
13	Orissa	Plantations have been initiated in 13 districts namely Bolangir, Cuttack, Dhenkanal, Ganiam, Gajapati, Jajapur, Koraput, Keonjhar, Kalahandi, Nowrangpur, Nawapra, Phulbani, and Puri
14	Punjab	Plantations have been initiated in 5 districts namely, Ferozpur, Gurdaspur, Hoshiarpur, Patiala, and Sangrur
15	Rajasthan	Plantations have been initiated in Udaipur, Kota, Sikar, Banswara, Chittor, and Churu districts
16	Tamil Nadu	The government has been promoting development of *Jatropha* through large scale entrepreneurs. To support contract farming of *Jatropha* in 20,000 ha, government allocated Rs. 400 million through primary Agriculture Cooperative Banks. The government has abolished purchase tax on *Jatropha*. Currently entrepreneurs established 1,000 acres area under *Jatropha* against the target of 20,000 ha
17	Uttarakhand	A biodiesel board has been established to coordinate *Jatropha* cultivation. Board also coordinate seed procurement, extraction, and transesterification. Along with MNRE government planned to electrify 500 villages with biodiesel

Recently, the Global Exchange for Social Investment (GEXSI 2008a, b) has conducted a detailed survey of status of Indian *Jatropha* plantations. They report that India *Jatropha* plantations fall into one of three types of ownership: private, public, and public–private partnerships with 31%, 31%, and 38% respectively. The total area under plantation estimated to be of 497,881 ha of which 84,000 ha is in Chhattisgarh, 33,000 ha is in Rajasthan, 20,277 ha is in Tamil Nadu, 16,715 ha is in Andhra

Table 10.8 Current status of commercial biodiesel production in India

No.	Organization/institution	Technology/raw material	Capacity
1	Southern Online Biotechnologies (P) Ltd.,[z] in Andhra Pradesh with Chemical Construction International Ltd, New Delhi[y]	The technology for the unit will be provided by Lurgi Life Science Engineering, Germany, along with their local partner, Chemical Constructions national Private Limited, New Delhi. Both oil expelling and transesterification units. Raw material supply—*Pongamia*, *Jatropha* and other raw materials like acid oils, distilled fatty acids, animal fatty acids and nonedible vegetable oils	Initial capacity of 30 tonnes of biodiesel/d, which is expandable to 100 tonnes/d. Current availability of seeds in the state is less than 4,000 tons
2	Maharashtra Energy Development Agency and Mint Biofuels, Pune[x]	Karanja oil-based biofuel	Initially had a capacity of 100 l/d, scaled up to 400 l/d
3	Gujarat Oelo Chem Limited (GOCL)[w]	From vegetable-based feedstock	Supplying to Indian Oil Corporation
4	Kochi Refineries Ltd. (KRL)[v]	From rubber seed oil	Capacity of 100 l/d
5	TeamSustain Ltd., Kochi; a division of US-based Dewcon Instruments Inc.[u]	Discussion in progress	
6	Shirke Biohealthcare Pvt. Ltd, Hinjewadi, Pune[t]	From *Jatropha* plant	To process 5,000 l
7	Jain Irrigation System Ltd[s]	Large-scale transesterification biodiesel plant with *Jatropha*	Capacity of 150,000 tons/d in Chattisgarh by 2008
8	Nova Bio Fuels Pvt. Ltd[r]	Transesterification biodiesel plant with *Jatropha* of Rs. 200 million	Capacity of 30 tons/d in Panipat in 2006
9	Naturol Bioenergy Limited, Andhra Pradesh[q]—a joint venture with Energea Gmbh (Austria) and Fe Clean Energy (United States), Kakinada	100% export-oriented unit to blend of palm oil, rapeseed, *Jatropha*, pongamia, and vegetable oil. Shipping biodiesel to the EU for blending the alternative fuel into gasoline and diesel	100,000 tonnes of biodiesel annually. 120,000 ha for *Jatropha* cultivation
10	Savoia Biodiesel plant, Ganapathipalayam, Tamil Nadu[p]	Transesterification biodiesel plant with *Jatropha*	
11	KTK German Bio Energies India[o]	Rubber seeds	Commercial production of biofuel in 2006

(continued)

Table 10.8 (continued)

No.	Organization/institution	Technology/raw material	Capacity
12	Mint Biofuels, Pune[a]	Pongamia seed based	400 l/d and 5,000 tonne of fuel/d
13	Sagar Jatropha Oil Extractions Private Limited, Vijayawada[m]	Jatropha oil extraction unit of Rs. 100 million	Jatropha oil is mixed with diesel to produce biodiesel
14	D1-Mohan Bio Oils Limited[l] (a joint venture of Mohan Breweries and Distilleries and U.K.based D1 Oils Plc)	One lakh hectares under Jatropha cultivation in Tamil Nadu	24 tonnes/d capacity
15	Classic Jatropha Oil (India) Ltd., Coimbatore[k]	Promoting cultivation of Jatropha in Tamil Nadu	
16	Bharat Renewable Energy and the government-owned Hindustan Petroleum[j]	40,000 ha of Jatropha cultivation	Million metric tons of biodiesel by 2015
17	Cleancities Biodiesel, Visakhapatnam[i]	Blend of palm oil, Jatropha oil and soya oil	Capacity of 250,000 tonnes
18	Emami Group, Kolkata[h]	Blending of waste cooking oil of Rs. 150 crores	
19	Alagarh Industries in Sivakasi (Tamil Nadu)[g]		5 tonnes/d capacity
20	D1-BP Fuel Crops, based in the UK, is a 50:50 Joint Venture between BP and D1 Oils[f]	Feedstock Jatropha and other nonedible oil seeds	
21	Newcastle-based D1 Oil Plc, along with Labland Biotech[e]	Developed 10,000 ha of Jatropha in India	
22	Aatmiya Biofuels Pvt Ltd, Gujarat[d]		Commercialized by March 2005 and now producing 1,000 l/d

[z] http://www.sol.net.in/bio_index.php cited 10 Nov 2008.
[y] http://cdm.unfccc.int/UserManagement/FileStorage/FS_686206579 cited 12 Dec 2008.
[x] http://www.mahaurja.com/ cited 10 Nov 2008.
[w] http://www.gujaratoleochem.com/product1.htm cited 12 Dec 2008.
[v] http://www.bharatpetroleum.com/refineries/refinerykochi_overview.asp?from=ref cited 12 Dec 2008.
[u] http://www.teamsustain.com/g_home.html cited 10 Nov 2008.
[t] http://www.shirkebiofuels.com/biodiesel.htm cited on 10 Feb 2009.
[s] http://www.jains.com/jatropha/Jatropha%20cultivation.htm cited 12 Dec 2008.
[r] http://novabiofuels.com/knowbiofuel.html cited 10 Nov 2008.
[q] http://www.naturol-bio.com/ cited 10 Feb 2009.
[p] http://www.savoiapower.com/biodiesel.html cited 10 Feb 2009.
[o] http://ecoworld.com/features/2006/04/06/indias-biodiesel-scene/ cited 12 Dec 2008.
[n] http://www.mintbiofuels.com/expanse.html cited 10 Feb 2009.
[m] http://www.tech2transfer.com/pdf/Article.pdf cited 10 Feb 2009.
[l] http://www.d1plc.com/globalIndia.php cited 12 Dec 2008.
[k] http://www.svlele.com/biodiesel_in_india.htm cited 10 Feb 2009.
[j] www.cleantech.com cited 12 Dec 2008.
[i] http://www.cleancities.in/ cited 09 Jan 2009.
[h] http://www.emamigroup.com/index.php cited 10 Feb 2009.
[g] http://www.unctad.org/en/docs/ditcted20066_en.pdf cited 09 Jan 2009.
[f] http://www.d1bpfuelcrops.com/ cited 10 Feb 2009.
[e] http://resourceguide.biospectrumasia.com/CompanyDetails.aspx?id=26 cited 10 Feb 2009.
[d] http://ecoworld.com/features/2006/04/06/indias-biodiesel-scene/ cited 09 Jan 2009.

Pradesh, 350 acres is in Uttaranchal, and 328 ha is in Haryana. Most of these crops are grown in Non irrigated land and 60% are planted in wastelands. It is also projected that India will have 1,179,760 ha of crop in 2010 and 1,861,833 ha in 2015. Recently concluded another study by German development institute (DIE 2008) confirmed that the biodiesel sector in India is different from elsewhere in the world. Biodiesel production is restricted to nonedible oil plants and not related to the price increase of edible oils. The focus is on Non intensive agricultural lands minimizing the competition between fuel and food. Biodiesel activity in rural areas can improve the food security as it provides additional income opportunities to the rural poor. The report also indicated that the biodiesel program address the five important development challenges such as energy supply, reduction of carbon dioxide emission, rural employment, rural energy securities, and protection of natural resources.

Constraints

The critical factors limiting the successful fuel alcohol program in India are restrictive government policies, availability of molasses, and price. All these factors are interlinked upon the historic control regimes and policies imposed on the sugar, molasses, and ethanol industry. There are several ministries involved in policymaking, regulation, promotion, and development for the biofuels sector. The ministry of food and consumer affairs controls sugar and molasses production. The ethanol for industrial use is controlled by ministry of industries and chemicals. The potable alcohol segment is controlled and managed by various state policies. The ministry of renewable energy has the overall policymaking role for promotion and development of biofuels. The ministry of petroleum and natural gas has the responsibility of marketing biofuels as well as the development, implementation of pricing, and procurement policy. The ministry of agriculture handles research and development for production of sugar, ethanol, and biofuel feedstock crops. The state governments control the licensing of new sugar factories and distilleries and their expansion. In addition, the state government controls the allotment of cane in their state and the distance of operation. The central government sets the policy regarding ethanol blending, but the state governments control the movement of molasses and often restrict molasses transport over state boundaries. The state governments also impose excise taxes on potable alcohol sales, a lucrative source of revenue. Foreign liquor imports are taxed at 150% or more, thus affording domestic potable alcohol the highest protection. The dynamics and complexity of managing the control regimes, regulations, and policies on fuel ethanol make it practically impossible for its sustainable production at present. The availability of molasses depends on cane area, cane yield, cane diverted to gur and khandasari, cane price, and sugar produced. Additionally, cane area and sugarcane production are subjected to the vicious cyclicality leading to shortages and surplus. Hence, molasses prices continue to fluctuate. Historical trend of cyclicality indicates that deficits/surplus gaps are widening compared to previous cycles. The surplus deficits of molasses expected to continue in coming years, unless government controls are removed. Recently, the ministry of food and consumer affairs amended the sugarcane control order allowing the direct use of sugarcane juice for sugar and ethanol production. Depressed crude oil prices and the increased market price of sugar led to lack of enthusiasm from the investors toward this choice.

The manufacturing cost of ethanol depends upon the price of molasses, taxes, and duties imposed by state and central government. The molasses price also depends on the cane price. The SMP of cane is regulated by central government over which state government announces the SAP. Hence, wide variation exists in cane and molasses price from state to state. The fuel ethanol market price depends upon the international crude oil price and various subsidies offered. Ethanol pricing in India is also complicated by differences in excise duty and sales tax across states. Central government is trying to rationalize ethanol sales tax across the country. The oil industry, however, is seeking parity between ethanol and the price of gasoline on an ex-refinery or import basis. India's petroleum ministry announced that it would appoint a Tariff Commission to fix an appropriate price for ethanol sourced from sugar mills. The sugar industry is seeking the parity in price with MTBE which alcohol substitutes as an oxygenating fuel. More significantly perhaps, there are still substantial differences in the profitability of potable alcohol in contrast to fuel alcohol in several states. Hence, alcohol-based biofuels production and use are neither encouraging nor remunerative as an automotive fuel.

In addition to the inconsistent government policies, availability of land, choice of crop, its yield, and the market price are the critical impediments encountered during the implementation program. Even though the government prepared an extensive plan in 2003, implementation of the program suffered widely because of the change in priorities and lack of effective coordination. The involvement of large number of agencies without responsibility and accountability made it very difficult to manage and coordinate. The multiple stake holders with conflicting objectives created incoherent views and confusion at all levels of implementation. In addition, various state governments initiated its own programs and policies either aligned with the central government or independently.

The project document identified 11.2 million hectares of land with specific categories for plantation. The quality and ownership of the land intended for the plantation continues to be under dispute. Many of the lands described in the plan are held by state government and managed by collaborative groups or own by selective community, such as *panchayats*. India's experience suggests collective ownership has been very difficult to manage for large scale commercial production. Even for private lands, the present land holding laws and tenancy act as stumbling blocks for large scale plantations. The expectation of *Jatropha*, a nonnative plant, providing high yields even on marginal and dry lands without inputs such as irrigation, fertilizers, and pesticide has not been materialized. This was mainly due to the lack of research data, amenability of *Jatropha* for large scale commercial plantation, and its inconsistent yield. Hence, the productivity and economic viability of this crop in India continued to be in question. Therefore, without the government subsidies, most of the farmers do not consider *jatropha* cultivation rewarding. As the focus and incentives was mainly on *Jatropha*, the native available potential oil seed bearing tress were neglected for research and commercial exploitation. The present price of biodiesel produced from *Jatropha* is not competitive with conventional diesel at current market price. Conventional diesel is heavily subsidized and present production cost of *Jatropha* oil is higher than the market price. In addition, committed subsidies, minimum support prices, and exemption from taxes are yet to be implemented.

The Way Forward

India's choice of feedstock for biofuel differs from rest of the world. Sugarcane–molasses-based biofuel and nonedible oil seed-based biodiesel make it an ideal fit for economic development, energy security, employment generation, poverty alleviation, and carbon dioxide mitigation without affecting the food supply. As the sugarcane industry is one of the largest rural industries, the bioethanol program is expected to improve rural agricultural income and generate additional employment for people associated directly or indirectly with the sugar industry. It also provides to opportunity to over come cyclicality of sugarcane, sugar, molasses, and alcohol production. Sustainable production enables market prices of sugar, molasses, and ethanol to stabilize and reduce drastic volatility experienced in past. Also, the wide fluctuations in the price of molasses, which is the main determining factor in the cost of fuel ethanol, can be brought under control. Even if sugar prices are depressed occasionally, factories can divert some of the sugarcane juice to ethanol production, thus bringing extra income, ensuring better and timely payment to the farmers. These will also encourage farmers to discontinue distress crop shifting to alternate crops every 2 to 3 yr. Steady market prices will increase attractiveness of biofuel ethanol business and drive much needed investments into the rural agriculture sector. Additional investments in farm mechanizations, drip irrigations and fertigations can bring substantial yield improvements leading to increased cane productions without an increase in dedicated land. Such an initiative provides additional benefits of carbon credits because of less energy use per unit area. However, the government needs to reform restrictive policies and controls to encourage sugar, molasses, ethanol, and fuel ethanol production, such as the deregulation of sugar and lifting ban on cross movements of molasses. Additional development of a consistent and coherent national policy, covering entire value chain from sugarcane to all its products, will ensure effective coordination and level playing field to all its stakeholders. There is also an urgent need to invest in long term research in second, third, and fourth generation biofuel technologies.

Biodiesel also requires similar policies dealing with its development integrating with its stake holders. There is a need to provide incentives for biodiesel programs until economic viability and profitability. To enable to compete with highly subsidized and volatile imported diesel, tax incentives and minimum support prices may be provided. To alleviate the existing problems, government needs to bring confidence building measure to its all stakeholders. Public and private partnerships needs to be encouraged in the development of plantations and production. Investments in research on native tree oil-bearing plants need to be carried out with long-term commitments for selection, large scale production, and commercial use.

A well-structured, centralized, coherent, and consistent biofuel policy at the national and state level is the immediate need of the hour. The sugarcane-based fuel ethanol and biodiesel from nonedible tree oils are continued to be ideal choices for India until new generation feed stock and technologies are developed. However, the new policy will have to strike a delicate balance of achieving the socioeconomic goals, food security, unemployment, energy security, and environment quality. India needs a leadership that is committed, coherent, and consistent with long-term vision to enable biofuel journey to reach its desired destination.

References

APGO (Andhra Pradesh Government Order). The response of farmers on yield of the plant. G.O. Rt. No. 148, 16.12.2006. http://goir.ap.gov.in/Reports.aspx; 2006.

Azam A. M.; Nahar N. M. Prospects and potential of fatty acid methyl esters of some non-traditional seed oils for use as bio-diesel in India. *Biomass Bioenergy* 294: 293–302; 2005. http://top25.sciencedirect.com/subject/energy/11/archive/7/

Bailey, R. Another inconvenient truth. Oxfam briefing paper. Oxfam International, June 2008. http://www.oxfam.org/pressroom/pressrelease/2008-06-25/another-inconvenient-truth-biofuels-are-not-answer2008.

Banse M.; Nowicki P.; vanMeijl H. Why are current food prices so high. In: Zuurbier P.; van de Vooren J. (eds) Sugarcane ethanol: Contributions to climate change mitigation and the environment. Wageningen Academic, Wageningen, pp 227–248; 2008. http://library.wur.nl/way/bestanden/clc/1880664.pdf

Bruce, E. H., Jamie, D. S., David, B. L. Canadian bio-diesel initiative: Aligning research needs and priorities with the emerging industry. Final report prepared for Natural Resources Canada http://www.greenfuels.org/bio-diesel/res/200408_BIOCAP_Bio-diesel04_Final.pdf2004.

CDP (Carbon Disclosure Project and World Wildlife Fund for Nature India). Carbon disclosure project report 2007: India http://www.cdproject.net/historic-reports.asp; 2007.

CEA (Central Electricity Authority) http://www.cea.nic/power-screports/executivesummary/2009-10/1-2.pdf; 2009.

CIL (Coal India Limited) http://coalindia.nic.in/pert5.pdf; 2009.

DEFRA (Department for Environment, food and Rural Affairs) A sustainability analysis of the Brazilian Ethanol, UNICAMP www.unica.com.br/download.asp?mmdCode=1A3D48D9-99E6-4B81-A863-5B9837E9FE39; 2008.

DIE (German Development Institute) Bio-diesel policies for rural development in India. Altenburg T, Dietz H, Hahl M, Nikolidakis N, Rosendahl C, Seelige K (eds) Bonn 27th May 2008; 2008.

Dufey, A. Biofuels production, trade and sustainable development: Emerging issues. Sustainable Markets Discussion Paper No. 2. London, International Institute for Environment and Development; 2006.

Dufey A.; Vermeulen S.; Vorley B. Biofuels: Strategic choices for commodity dependent developing countries. Common Fund for Commodities, Amsterdam 2007.

EIA (Energy Information Administration). International Energy outlook http://www.eia.doe.gov/ocaf/ieo/coal.html; 2008.

Ethanol India. http://www.ethanolindia.net/sugarind.html; 2009.

Faaji A.; Szwarc A.; Walter A. Demand for bio-ethanol for transport. In: Zuurbier P.; van de Vooren J. (eds) Sugarcane ethanol: Contributions to climate change mitigation and the environment. Wageningen Academic, Wageningen, pp 227–248; 2008.

FAO (Food and Agriculture Organization) The state of food and agriculture 2008: Biofuels: Prospects, risks and opportunities. FAO, Rome2008.

GEXSI (The Global Exchange for Social Investment) Global market study on Jatropha—final report. GEXSI, London; 2008a. http://www.*Jatropha*platform.org/downloads.htm.

GEXSI (The Global Exchange for Social Investment) Global market study on Jatropha—case studies. GEXSI, London; 2008b. http://www.*Jatropha*platform.org/case_studies.htm.

GOI (Government of India) India Vision 2020. Planning Commission, Government of India, New Delhi; 2000a. http://planningcommission.nic.in:80/plans/planrel/pl_vsn2020.pdf.

GOI (Government of India) Hydrocarbon Vision 2025 document. Planning Commission, Government of India, New Delhi; 2000b. http://planningcommission.nic.in:80/plans/annual-plan/ap2021pdf/ap2021ch8-1-3.pdf.

GOI (Government of India) Report of the commission on development of biofuel. Planning Commission, Government of India, New Delhi; 2003. http://planningcommission.nic.in/reports/genrep/cmtt_bio.pdf.

GOI (Government of India) Integrated energy policy: Report of the Expert Committee. Planning Commission, GOI (Government of India), New Delhi; 2006a. http://planningcommission.nic.in:80/reports/genrep/rep_intengy.pdf.

GOI (Government of India) Report on working group in automotive industry: 11th five year plan (2007–12). Ministry of Heavy Industry Public Enterprises, New Delhi; 2006b. http://planning-commission.nic.in:80/midterm/english-pdf/chapter-12.pdf.

GOI (Government of India) 11th five year (2007–12) plan document. Planning Commission, Government of India, New Delhi; 2007. http://planningcommission.nic.in/plans/plansrel/fiveyr/welcome.html.

Gonsalves J. An assessment of the biofuels industry in India. United Nations conference on trade and development. UN, Geneva 2006.

GTZ-TERI Liquid biofuels for transportation: India country study on potential and implications for sustainable agriculture and energy. The Energy and Resources Institute, India Habitat Centre, New Delhi2007.

IEA (International Energy Agency) Renewables in global energy supply: An IEA fact sheet. OECD/IEA, Paris; 2007a. Available at: http://iea.org/textbase/papers/2006/renewable_factsheet.pdf.

IEA (International Energy Agency) World energy outlook 2007: China and India insights. OECD/IEA, Paris; 2007b. http://www.iea.org/textbase/npsum/WEO2007SUM.pdf.

IEA (International Energy Agency) World Energy Outlook 2008. IEA, Paris 2008. http://www.worldenergyoutlook.org/docs/weo2008/WEO2008_es_english.pdf.

IEO (International Energy Outlook). http://www.eia.doc.gov/oiaf/ieo/pdf/Table 10.7.pdf; 2008.

IPCC (Intergovernmental Panel on Climate Change). Climate change 2007 synthesis report. http://www.ipcc.ch/ipccreports/ar4-syr.htm; 2007.

IPCC (Intergovernmental Panel on Climate Change) Climate change and water. Technical report, IPCC working group II; 2008. http://www.ipcc.ch/ipccreports/tp-climate-change-water.htm.

ISMA (Indian Sugar Mills Association) Hand book of Sugar statistics. Indian Sugar Mills Association, New Delhi2008.

Janulis P. Reduction of energy consumption in bio-diesel fuel life cycle. *Renew. Energy* 29: 861–871; 2004.

Koplow D. Biofuels—at what cost? Government support for ethanol and bio-diesel in the United States: 2007 update. Global subsidies initiative of the International Institute for Sustainable Development. International Institute for Sustainable Development, Geneva 2007.

KPMG The Indian Sugar Industry: Sector Roadmap 2017. KPMG International, India2007.

Kutas G.; Lindberg C.; Steenblik R. Biofuels—at what cost? Government support for ethanol and bio-diesel in the European Union: 2007 update. Global subsidies initiative of the International Institute for Sustainable Development. International Institute for Sustainable Development, Geneva 2007.

Martinot E. Renewables 2007—global status report. Worldwatch Institute for the Renewable Energy Policy Network for the 21st century (REN21). Worldwatch Institute, Washington, DC; 2008. Available at: http://www.ren21.net/globalstatusreport/default.asp.

Mitchell, D. A note on rising food prices. Policy research working paper, 4682. The World Bank, Development Prospects Group; 2008. http://www-wds.worldbank.org/external/default/WDSContentServer/WDSP/IB/2008/07/28/000020439_20080728103002/Rendered/PDF/WP4682.pdf.

MNRE (Ministry of New and Renewable Energy). Annual report 2007–08 chapter 3: Renewable energy for rural applications; 2008. http://mnes.nic.in/annualreport/2007_08_English/Chapter%203/chapter%203_1.htm

MNRE (Ministry of New and Renewable Energy). http://mnes.nic.in/; 2009.

MoC (Ministry of Coal). http://coal.nic.in/; http://coalindia.nic.in/ministry.htm; 2009.

MoCFA (Ministry of Consumer Affairs, Food and Public Distribution). Annual Report http://fcamin.nic.in/dfpd/EventListing.asp?Section=Annual%20Reportandid_pk=16andParentID=0; 2007.

MoCFA (Ministry of Consumer Affairs, Food and Public Distribution). Department of Food and Public Distribution http://fcamin.nic.in/dfpd_html/index.asp; 2009.

MoP (Ministry of Petroleum and Natural Gas). Basic statistics. http://petroleum.nic.in; 2004.

MoP (Ministry of Petroleum and Natural Gas). Basic statistics. http://petroleum.nic.in/petstat.pdf; 2009.

NCAER (National Council of Applied Economic Research. http://www.ncaer.org/research03.html; 2009.

NOVOD (National Oilseeds and Vegetable Oils Development Board) 3rd R and D Report on TBOs. Government of India, New Delhi; 2008. http://www.novodboard.com/3rd%20RandD-Report.pdf.

OCC (Office of Climate Change) Stern Review on economics of climate change. OCC, London; 2006. http://www.hm-treasurey.gov.uk/sternreview_report.htm. Accessed Jan 2009.

Ringwald A. India: Renewable energy trends. Centre for Social Markets, Kolkata; 2008. http://www.csmworld.org.

Rutz D.; Janseen R. Biofuel technology handbook. WIP Renewable Energies, Munich2007.

Searchinger T.; Heimlich R.; Houghton R. H.; Dong F.; Elobeid A.; Fabiosa J.; Togkoz S.; Hayes D.; Yu T. Use of US croplands for biofuels increases greenhouse gases through emissions from land use change. *Science* 319: 1238–140; 2008.

Steenblik R. Biofuels—at what cost? Government support for ethanol and bio-diesel in selected OECD countries: 2007 update. Global subsidies initiative of the International Institute for Sustainable Development. International Institute for Sustainable Development, Geneva2007.

Subramanian K. A.; Singal S. K.; Saxena M.; Singhal S. Utilization of liquid biofuels in automotive diesel engines: An Indian perspective. *Biomass Bioenergy* 29: 65–72; 2005.

UNICA Sugarcane in Brazil sustainable energy and climate. UNICA, São Paulo; 2008a. http://english.unica.com.br/multimedia.

UNICA Brazilian sugarcane ethanol: Get the facts rights and kill the myths. UNICA, São Paulo; 2008b. http://www.unica.com.br/download.asp.

World Bank Rising food prices: Policy options and World Bank response. World Bank, New York; 2008 http://siteresources.worldbank.org/NEWS/Resources/risingfoodprices_backgroundnote_apr08.pdf.

World Energy Outlook Fact Sheet: Global energy trends. OECD/IEA, Paris; 2008. http://www.worldenergyoutlook.org/docs/weo2008/fact_sheets_08.pdf.

Worldwatch Institute Biofuels for transportation: Global potential and implications for sustainable agriculture and energy in the 21st century. Worldwatch Institute, Washington, DC; 2007. http://www.worldwatch.org/node/4078.

Worldwatch Institute. Smart choices for biofuels. In: Earley J, McKeown A (eds). Worldwatch Institute, Washington, DC; 2009. http://www.worldwatch.org/smartchoicesforbiofuels?emc=elandm=203168andl=4andv=e96a58fc23.

Zuurbier P.; van de Vooren J. Introduction to sugarcane ethanol contributions to climate change mitigation and the environment. In: Zuurbier P.; van de Vooren J. (eds) Sugarcane ethanol: Contributions to climate change mitigation and the environment. Wageningen Academic, Wageningen, pp 227–248; 2008.

Chapter 11
Biofuels in China: Opportunities and Challenges

Feng Wang, Xue-Rong Xing, and Chun-Zhao Liu

Abstract With rapid economic development, energy consumption in China has tripled in the past 20 yr, exceeding 2.4 billion tons of standard coal in 2006. The search for new green energy as substitutes for the nonrenewable energy resources has become an urgent task. China has a variety of climates and is rich in potential biofuel plant species. Corn and cassava are used as the main raw materials for bioethanol production in China. At the end of 2005, bioethanol productivity had increased to 1.02 million tons produced by four companies, and bioethanol-blended petrol accounted for 20% of the total petrol consumption in China. According to the Mid- and Long-term Development Plan for Renewable Energy, the consumption of biodiesel in China will reach 0.2 million tons in 2010 and 2.0 million tons in 2020. This review is intended to provide an introduction to the distribution and development of biofuel crops and biofuel industry in China.

Keywords Biofuel • Bioethanol • Biodiesel • China • Energy crop • Fermentation

Introduction

In response to growing concerns about environmental pollution, energy security, and future oil supplies, the global community is seeking non-petroleum-based alternative fuels, along with more advanced energy technologies, such as fuel cells, to increase energy use efficiency (Semelsberger et al. 2006). According to the Food and

---

C.-Z. Liu (✉)
National Key Laboratory of Biochemical Engineering, Institute of Process Engineering, Chinese Academy of Sciences, Beijing, 100190, People's Republic of China
and
Graduate School of the Chinese Academy of Sciences, Beijing, 100049, People's Republic of China
e-mail: czliu@home.ipe.ac.cn

Agricultural Organization (FAO 2005) of the United Nations, fossil fuels such as coal, petroleum, and natural gas are currently the most important energy sources worldwide. China is the most populous country in the world and is ranked the third in land area. With rapid economic development, energy consumption in China has tripled in the past 20 yr, exceeding 2.4 billion tons standard coal in 2006 (NBSC 2007). The principal energy sources consumed in China are coal (69.4%) and petroleum (20.4%; Fig. 11.1). As reported by the National Bureau of Statistics of China (NBSC 2007), the percent of the total crude oil demand imported by China increased from 5.2% to 51.3% during the period between 1990 and 2006 (Table 11.1). Sustainable energy development in China has been limited because of energy deficit, energy consumption, and environmental consequences (Shao and Chu 2008).

Reducing China's dependence on fossil fuels and maintaining a safe and healthy environment is a high-priority (Nass et al. 2007) and increasingly urgent task (Shao and Chu 2008). Biofuels production from wood, charcoal, livestock manure, biogas, biohydrogen, bioalcohol, microbial biomass, agricultural waste and by-products, and energy crops are now being explored (FAO 2000; Xu et al. 2007). As a sustainable and renewable source of energy, biofuels will reduce the impact of rising petroleum prices, address environmental concerns, especially air pollution and greenhouse gases, and improve opportunities for farmers and rural communities (FAO 2005; Nass et al. 2007).

China has a variety of land types with different altitudes and rainfall and temperature conditions and is rich in plant species suitable for biofuel crops. Two of

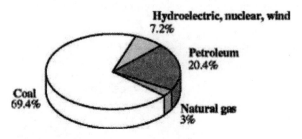

Fig. 11.1 China energy matrix in 2006. *Source*: National Bureau of Statistics of China (NBSC) 2007

Table 11.1 Production and import of crude oil in China

Year	Production (million ton)	Import (million ton)	Total consumption (million ton)	Import as percent of total consumption
1990	138.3	7.6	145.9	5.2
1995	150.1	36.7	186.8	19.7
2000	163.0	97.5	260.5	37.4
2004	175.9	172.9	348.8	49.6
2006	184.8	194.5	379.3	51.3

Source: National Bureau of Statistics of China (NBSC) 2007

these crops, corn and cassava, have been used as raw materials for bioethanol production. In order to stimulate the development of bioenergy in China, the National Development and Reform Commission of People's Republic of China (NDRCPRC) established the project Mid- and Long-term Development Plan for Renewable Energy (NDRCPRC 2007). In 2001, an attempt was made to produce bioethanol from corn and wheat and, in February 2004, the government of China made 10% ethanol blending with petrol mandatory in 27 cities of 9 provinces (Li 2008a).

Energy Crops for Biofuels Production in China

Crops for bioethanol production. Raw materials that contain appreciable amounts of sugar or materials that can be converted into sugar, such as starch or cellulose, can be fermented to produce bioethanol for use in gasoline engines (Tian et al. 2005; Malca and Freire 2006; Prasad et al. 2007; Balat et al. 2008). Bioethanol feedstocks can be classified as: (1) starch feedstocks (e.g., wheat, corn, and barley); (2) sugar feedstocks (e.g., sugar beet and sugarcane); and (3) cellulosic feedstocks (e.g., trees, forestry processing residues, and grasses) (Malca and Freire 2006). Because feedstocks typically account for greater than one third of the production costs, maximizing bioethanol yield is imperative (Dien et al. 2003). Therefore, the availability of raw materials is important for profitable bioethanol production. This section reviews several important crops, which are being used or are potentially useful for bioethanol production in China.

Corn, which is currently used to make about 90% of all US bioethanol, is expected to remain the predominant feedstock in the US, although its share likely will decline modestly by 2015 (Balat et al. 2008). Corn is also the main feedstock for bioethanol production in China now (Table 11.2), but the production cost of bioethanol in China is 1.5–2.0 times higher than that in the US. This difference might be due to the more advanced corn planting and ethanol fermentation technologies in the US (Yue et al. 2007). The single greatest and most volatile cost in the production of bioethanol from

Table 11.2 Status of bioethanol production in China

Production company	Output (million ton/yr)	Feedstock	Supply area/province
Jilin Fuel Alcohol Limited Company	0.30	Corn	Jilin, Liaoning
Heilongjiang Huarun Alcohol Limited Company	0.10	Corn	Heilongjiang
Henan Tianguan Enterprise Group Corporation	0.30	Wheat	Henan, Hubei, Hebei
Anhui Fengyuan Group Corporation	0.32	Corn	Anhui, Shandong, Jiangsu, Hebei

Source: Li 2008a

corn is the cost of the corn (Balat et al. 2008). In China, the increasing demand for corn caused variations in its price. In 2006 alone, the price of corn increased by 23% (Liu 2008). As a result, food industry and its related industries, such as food processing and livestock breeding, had a remarkable negative impact. Currently, corn is mainly grown in eastern regions, especially in the northeastern area (Fig. 11.2). The total corn production was 145.48 million tons in 2006 (Table 11.3) and ranked second in grain crops production in China (NBSC 2007). Jilin is the largest corn production province in China with a cultivation area of 2.805 million hectares and an output of 18.0 million tons/yr (NBSC 2007). It was estimated that the demand for corn in China would reach 204 million tons but that the output would be only 184 million tons by 2020 (Zhang and Hu 2000). The shortage of 20 million tons of corn demand would need to be met by import. Therefore, the main task for corn research is to increase corn production in order to meet the constantly rising demand in China. Different advanced biotechnologies, including the additions of transgenic traits, have been used for corn breeding; commercialization of transgenic maize would be an effective method to increase the supply of corn (Luo and Li 2006).

Fig. 11.2 Major areas of main biofuel crops in China. *Source*: National Bureau of Statistics of China (NBSC) 2007

Table 11.3 Cultivated area and production of main biofuel crops of China in 2006

Biofuel crops	Cultivated area (million ha)	Production (million ton)
Corn	26.971	145.48
Root crops	9.929	34.06
Sugarcane	1.495	99.78
Sugar beet	0.287	10.54
Rapeseed	6.888	12.65

Source: National Bureau of Statistics of China (NBSC) 2007

Root crops are another important agricultural produce in China and they are mainly cultivated in the southeast provinces (Fig. 11.2). The cultivated area and root crop production were stable during the last decade, reaching 9.929 million hectares and 34.06 million tons in 2006. Among the different types of root crops, the most promising for the bioethanol industry is cassava and sweet potato. Cassava starch is an inexpensive fermentable source compared to other starches. It is a tropical root crop planted in more than 80 countries (Sasson 1990). Fresh cassava has very high starch content, up to 30%. The content of sucrose is about 4%. Dried cassava has 80% fermentable substrate (Lin and Tanaka 2006). In the tropical and subtropical regions of China, cassava is the most ideal energy crop for bioethanol fermentation. In Guangxi province, the area planted in cassava reached 400,000 ha: 70% of the total planting area in the whole country (Li et al. 2008a).

The southern area of China produces large amounts of sugarcane, reaching 99.78 million tons/yr in 2006 (Table 11.3). The cultivated area increased more than threefold from 480,000 ha in 1980 to 1.495 million hectares in 2006, with a corresponding increase in sugarcane production from 22.8 million tons to 99.8 million tons (NBSC 2007). Sugarcane breeding technology has brought about improvement in varieties for sugarcane production. From the 1960s, three innovations were made in the plant varieties to improve sugarcane production and sugar yield from sugarcane (Zhang et al. 2008b). However, the sugarcane yield of 66.73 ton/ha was still lower than that of other countries, such as Brazil and the US (NBSC 2007; You et al. 2008). The biotechnology of sugarcane in China is still in its initial stage, and researchers have developed varieties with improved disease and insect resistance of sugarcane (Luo et al. 2003; Zhang et al. 2008b). Sugar beet is another biofuel possibility in China, and Xinjiang is the main production area accounting for more than half of the total sugar beet production in China (5.56 million ton). The area of sugar beet cultivation has decreased during recent years and is much smaller than that of sugarcane.

Sweet sorghum varieties were developed and used as an alternative sugar source in some areas where sugarcane could not be produced (Rooney 2004). Because of their high sugar content, sweet sorghums may also be amenable to the production of sugar for conversion to ethanol, using methodology similar to that used in sugarcane production (Rooney et al. 2007). Sweet sorghum has both drought tolerance and high water-use efficiency, so it can be produced in marginal environments where rainfall is limited and irrigation is either too expensive or would deplete water reserves.

Considering food security, a renewed interest in bioenergy from sweet sorghums is emerging in China. Research in China has resulted in four high-yielding sweet sorghum varieties adapted to drought and saline–alkaline regions with their sugar content of juice reaching about 20% (Qian 2007). The bottleneck for bioethanol production from sweet sorghum is storing it effectively, maintaining the sugar component of sweet sorghum for an adequate time (Yue et al. 2007). Recently, an elite variety of sweet sorghum, Lv Chunyuan, has been obtained by mutation breeding technology in China. This variety has good tolerance for barren or saline–alkaline soil, 2.3% higher in sugar content than sugarcane, and a shelf life of more than 8 mo (Qian 2007).

Lignocellulosic material, such as agricultural residues, wood, and energy crops, is an attractive feedstock for bioethanol production since it is the most abundant and renewable resource on the Earth (Lin and Tanaka 2006; Balat et al. 2008). Energy grass, such as *Miscanthus* spp. and *Panicum virgatum* L., is also a potential non-food-based feedstock in China since these crops have good characteristics of high biomass production, high potential for disease and pest resistance, drought and salinity tolerance, and good growth in marginal land (Xie et al. 2007; Li et al. 2008b). Ecological adaptability, growth and photosynthetic, and other physiological characteristics were investigated for *P. virgatum* L. in the Loess Plateau (Xu et al. 2001; Wang et al. 2006b; Lin et al. 2008), where the aboveground dry biomass was 13–16 tons/ha for lowland and 2,300–2,650 kg/ha for hilly areas between 2001 and 2002 (Xu et al. 2005). The fundamental research of *Miscanthus* spp. was mainly on karyotype, growth, and biological characteristics (Huang et al. 2007; Chen et al. 2008; Xi and Hong 2008).

Oil crops for biodiesel production. During this century, great progress has been made in biodiesel production in China. However, some barriers hinder biodiesel development in practice. Among these, the two important barriers are a stable supply of cheap feedstock and the high production cost of biodiesel (Sun and Liang 2008). Therefore, it is critical to seek the appropriate feedstocks for biodiesel production.

An energy crop, such as a vegetable oil crop, is an important raw material for biodiesel production. China is a major producer of rapeseed, reaching 12.65 million tons in 2006 (Table 11.3). Rapeseed can be cultivated as a winter or spring crop in China. The Yangtze River Basin is the main production area of rapeseed, which has an oil content up to 40% (Zhang and Li 2006).

In China, over 1,000 species of native woody oil plants containing more than 15% of oil, such as *Pistacia chinensis*, *Jatropha curcas* L., *Xanthoceras sorbifolia* Bunge, and *Garcinia multiflora* Khamp, are available to use as feedstock for biodiesel production (Sun et al. 2007). Although woody oil plants are grown in significant quantities in China, most are grown on a small scale (Li 2008b). Natural distributions of major woody oil plants in China have been reviewed by Shao and Chu (2008). Most of the main woody oil plants have some characteristics in common such as fast-growing tree species with a short growth cycle and long harvest period, easy to harvest, low-cost management, and highly adaptable to a wide range of environment.

More than four million hectares of oil-producing trees were planted in China during the past several years, yielding an annual seed output of more than 5.0 million tons (Li 2008b). In order to meet the increasing demand of raw materials for biodiesel production, a planting base of *Pistacia chinensis* with an area of more than 7,300 ha has been established in Hebei province. According to the National Energy Forest Construction Plan and the 11th 5-yr Biodiesel Raw Material Forest Base Program, 78.34×10^4 ha of energy forest demonstration base will be built in China, and including 40×10^4 ha of *J. curcas* L., 25×10^4 ha of *P. chinensis*, and 13.34×10^4 ha of *X. sorbifolia* Bunge. In 2007, construction was started on the first energy forest demonstration base of 4×10^4 ha, which is expected to supply 6×10^4 tons of raw materials for biodiesel production (Yin and Teah 2008).

Biofuels Industry in China

Bioethanol industry. After US and Brazil, China has the third largest bioethanol industry in the world. At the end of 2005, bioethanol production had increased to 1.02 million tons produced from four companies (Table 11.2), and bioethanol-blended petrol amounted to 20% of the total petrol consumption in China (Li 2008a; NDRCPRC *2008a*). A 5-yr plan for renewable energy development gave the projected demand for renewable energy required to meet 10% of the total energy consumption. According to estimates for the years 2006–2010, 2.0 million tons of bioethanol from nongrain feedstock must be produced directly, in addition to 1.02 million tons from grain resources (NDRCPRC 2008a). Bioethanol production in China is subsidized with the subsidy reaching about 204 million US$ per year. The subsidy per ton of bioethanol is 200.7 US$ in 2007, which is lower than in 2005 (275.3 US$/ton) and 2006 (238.0 US$/ton) because of the improvement of productivity and decline of production cost (Dong et al. 2007).

As shown in Table 11.2, corn is the major feedstock for bioethanol production at present. However, the first non-food-based bioethanol factory was established and put into production in Guangxi province recently. The factory produces 200,000 tons of bioethanol per year from cassava (NDRCPRC 2008b). Based on the current technologies, a projection of 1.0 billion ton bioethanol per year from cassava will produce 4,000 tons of flue gas, 0.25 million ton of boiler ash, 0.45 million ton of mash residues, and 10.0 million tons of waste water (Li et al. 2008a). Although some problems are yet to be resolved, bioethanol production from starch-containing raw materials, such as corn, wheat, and cassava, is available and commercialized in China.

The direct fermentation of sugarcane, sugar beets, and sweet sorghum syrup to produce ethanol requires the least costly pretreatment (Prasad et al. 2007). So they are more economical feedstocks than starches, lignocellulosic biomass, or urban and industrial wastes, which need costly pretreatment for conversion into fermentable substrates (Nikolov et al. 2000). As a result, sugarcane and sugar beet essentially are the preferred feedstocks for bioethanol (UNCTAD 2006).

All of the large sugar factories in China have the equipment to produce food alcohol, for which the fermentation feedstock is molasses produced from the sugar refining process (Liu et al. 2009). The bioethanol fermentation technology from sugar feedstocks has already matured in China (Li and Yu 2007), but fuel bioethanol production from sugarcane and sugar beet is not feasible. This is because the demand for sugar is so great in China due to its large population, and sugarcane and sugar beet remain the main resources for sugar production (Yi et al. 2007; Balat et al. 2008). Furthermore, sugar production from sugarcane and sugar beet is more profitable than that of bioethanol production for Chinese corporations at the moment (Yue et al. 2007).

Great effort has been made to develop bioethanol production from sweet sorghum in China (Wang et al. 2006a; Wang and Liu 2007), but all production systems are still in the pilot-scale phase (Yue et al. 2007). Two projects for bioethanol production from sweet sorghum have been put into operation in Shandong province and Heilongjiang province, respectively. One produces 400 tons/yr by submerged liquid fermentation of juice, and another produces 5,000 tons/yr by solid-state fermentation (Xu 2007).

Biodiesel production. The total diesel fuel consumption in China reached 140 million tons in 2005, of which 40 million tons were imported (Sun et al. 2007). Therefore, exploiting indigenous biodiesel to replace the amount imported is becoming more and more critical in China. The fundamental research on biodiesel production from vegetable oil was carried out in the 1980s. In 2003, biodiesel development was listed in the "National Science and Technology Industrialization Plan." In 2004, industrial exploitation of biodiesel production was started and supported by the Key Technological Research and Development Program of China, the Ministry of Science and Technology of People's Republic of China (Zhang 2007). According to the Mid- and Long-Term Development Plan for Renewable Energy, the consumption of biodiesel in China will reach 0.2 million ton in 2010 and 2.0 million tons in 2020 (NDRCPRC 2007).

In September 2001, the first biodiesel production factory was established in Hebei province with an annual output of 10,000 tons. This system uses primarily waste edible oil and seed oil of *P. chinensis* as feedstocks to produce biodiesel through a transesterification process. Establishment of this factory is a milestone in biodiesel promotion in China (Zhang 2007). In 2005, the annual production of biodiesel was about 85,000 tons in China, and this output increased sharply to about 0.2 million ton by 2006. In 2007, more biodiesel corporations have been established in China including a few foreign enterprises, and together they produced more than 1.0 million tons of biodiesel (Sun et al. 2007). As shown in Table 11.4, the main biodiesel

Table 11.4 Status of biodiesel industry in China

Corporation	Output (10^4 ton/yr)	Raw material	Production technology
Zhenghe Bioenergy Sources Limited Company	3	Waste oils, Seed oil of *Pistacia chinensis*	Chemical
China Biodiesel International Holding Company	4	Waste oil	Chemical
Gushan Environmental Energy Limited	17	Waste edible oils	Chemical
Hunan Tianyuan Bio-clean Energy Company	2	Waste edible oils	Chemical and enzymatic
Hunan Rivers Bioengineering Company	2	Waste oils	Enzymatic

Source: Sun and Liang 2008; Zhang et al. 2008a

production technology in China was the chemical process. In December 2006, the first enzymatic process for biodiesel production was implemented in Hunan with an output of 2×10^4 tons/yr (Liu et al. 2007).

Perspectives for Biofuel Development in China

According to the 11th 5-yr Plan of Renewable Energy Development in China, bioethanol production from non-food-based feedstocks will be the direction of fuel ethanol development in the future (Liu et al. 2006). Among the nongrain materials, sweet sorghum and cassava will be the most promising feedstocks for bioethanol production. Different feedstocks for bioethanol production in China and their comparative costs are shown in Table 11.5 (Tan and Wang 2006). The main raw material for biodiesel production in China is waste oil, including used vegetable oils (Sun et al. 2007). Because the components of waste oil are complex and inconsistent, this raw material can cause variation in the quality of biodiesel (Zhang et al. 2008a). In addition, these resources are limited and can not meet the increasing demand for the rapid development of the biodiesel industry. Rapeseed oil was suggested as a potential candidate material for biodiesel production (Xing 2007). However, China is short of edible vegetable oil and is the largest importer of edible oil in the world. Therefore, it is not appropriate to use edible vegetable oil for biodiesel production (Sun et al. 2007).

Biofuels, such as bioethanol and biodiesel, are recommended as a substitute for petroleum-based fuel because they are renewable, can be produced from domestic resource with an environmentally friendly emission profile, and are readily biodegradable. The annual output of biofuels in China has increased dramatically in the last few years. However, biofuel development in China is in its initial stage and there are still some significant challenges ahead in developing this area to a large-scale commercial reality. The main research challenges for bioethanol production include (1) the use of biotechnology to introduce new characteristics to bioethanol feedstock crops such as disease and pest resistance, drought and salinity tolerance,

Table 11.5 Cost of different feedstocks for bioethanol production in China

Feedstock	Unit price (US$ per ton feedstock)	Usage of feedstock (ton per ton bioethanol)	Feedstock cost (US$ per ton bioethanol)
Sugar beet	33.6–36.6	12.5	420–457.5
Sugarcane	26.3–29.2	12.0	315.6–350.4
Sweet sorghum	26.3	14.0	368.2
Japonica rice	175.5	2.5	438.7
Corn	131.6	3.0	394.8
Wheat	146.2–175.5	3.0	438.6–526.5
Cassava	43.9–52.6	8.0	351.2–420.8

Source: Tan and Wang 2006

and high yield; (2) the improvement and innovation of technologies for starch-based bioethanol such as fermentation of untreated feedstock, filtrate and recycling of clear broth, and waste treatment for bioethanol production from cassava; and (3) the development of new processes for bioethanol production from non-food-based feedstocks, such as lignocellulosic biomass. For biodiesel production, the main research challenges include (1) the development and promotion of cultivars and hybrids of conventional and potential oil-producing species; (2) the development of nonedible oil-producing plant species with high yield and good environmental adaptation; (3) the development of cultivation practices on nonedible woody oil-producing plants on arable and nonarable lands; (4) the evaluation and innovation in crop production systems including harvesting of nonedible oil species.

Conclusion

The main challenge for biofuel production in China is the establishment of biofuel production system which has a steady, low-cost, and non-food feedstock supply and highly efficient biofuel production technologies. Bioethanol production from non-food-based feedstocks and biodiesel production from nonedible woody oil-containing plants will be very promising developments. China is now in the development stage of establishing large-scale feedstock production bases and modern biofuel factories.

Acknowledgment This work is financially supported by the National Basic Research Program (973 Program) of China (no. 2007CB714301) and the Knowledge Innovation Program of the Chinese Academy of Sciences (no. KSCX1-YW-11-D1).

References

BalatM.;BalatH.;OzC.Progress in bioethanol processing. *Prog. Energy Combust. Sci*34: 551–573;2008.doi:10.1016/j.pecs.2007.11.001.

ChenS. F.;HeJ.;ZhouP. H.;YangS. G.The karyotypes of *Miscanthus sinensis* and M. floridulus. Acta Agric. Univ. Jiangxiensis30:123–126;2008.

DienB. S.;CottaM. A.;JeffriesT. W. Bacteria engineered for fuel ethanol production: current status. Appl. Microbiol. Biotechnol63:258–266;2003.doi:10.1007/s00253-003-1444-y.

DongD. D.;ZhaoD. Q.;LiaoC. P.;ChenX. S.;TangZ. X.Energy consumption of fuel ethanol production and review of energy-saving technologies.Chem. Ind. Eng. Prog26:1596–1601;2007.

Food and Agricultural Organization of the United Nations (FAO) FAO and bioenergy. FAO, Rome; 2000. http://www.fao.org/sd/Egdirect/EGre0055.htm.

Food and Agricultural Organization of the United Nations (FAO) Bioenergy. FAO, Rome; 2005. http://www.fao.org/docrep/meeting/009/j4313e.htm.

HuangP.;ZuoH. T.;HanL. B.Effect of water stress on the growth and biomass characteristics of amur silvergrass at the elongation stage. *Acta Agrestia. Sin* 15:153–157;2007.

LiZ. J.Development of biofuel ethanol industry: present situation, problems and policy suggestions.*Technol. Econ* 27:50–54;2008a.

LiZ. L.Production technology and development of biodiesel. *Chem. Eng. Des. Commun* 34:29–35;2008b.

LiZ. C.;HuangZ. M.;YangD. F.;ChenD.The harmful factors and countermeasure influencing development of cassava fuel-alcohol industry. *Renew. Energy Resour* 26:106–110;2008a.

LiG. Y.;LiJ. L.;WangY.;PanY. N.;DouG. Y.Research progress on the clean bio-energy production from high yield Panicum virgatum. *Pratacultural Sci* 25:15–21;2008b.

LiM. M.;YuS. J.Present status and the prospect of fuel ethanol production by sugarcane. *Liquor Mak. Sci. Technol* 156:111–113;2007.

LinC. S.;ChengX.;YangX. G.Ecological adaptability of introduced switchgrass in semi-arid loess hilly-gully areas, recommended as a bio-energy plant.*J. Southwest Agric. Univ*30:125–132;2008.

LinY.;TanakaS.Ethanol fermentation from biomass resources: current state and prospects.*Appl. Microbiol. Biotechnol*69:627–642;2006. doi:10.1007/s00253-005-0229-x.

LiuH. B.Developing of cellulosic ethanol production from nonfood-based biomass in China: challenges and strategies.*Chinese J. Bioprocess Eng*6:7–11;2008.

LiuH. J.;DuW.;LiuD. H.Progress of the biodiesel and 1,3-propanediol integrated production.*Process Chem*19:1185–1189;2007.

LiuW. F.;ZhangX. M.;ChenG. J.;LiuC. Z.Metabolic engineering for improving ethanol fermentation of xylose by yeasts and bacteria.*Chin. J. Proc. Eng*6:138–143;2006.

LiuC. Z.;WangF.;OuyangF.Ethanol fermentation in a magnetically fluidized bed reactor with immobilized *Saccharomyces cerevisiae* in magnetic particles.*Bioresour. Technol*100:878–882;2009.

LuoZ. F.;LiX. H.Research development and industrialization analysis of transgenic maize in China.*J. Maize Sci*14:4–6;2006.

LuoS. L.;ZhangsunD. T.;ChenR. K.Progress in sugarcane genetic transformation.*Biotechnol. Bull*2:9–13;2003.

MalcaJ.;FreireF.Renewability and life-cycle energy efficiency of bioethanol and bio-ethyl tertiary butyl ether (bioETBE): assessing the implications of allocation.*Energy*31:3362–3380;2006. doi:10.1016/j.energy.2006.03.013.

NassL. L.;PereiraP. A. A.;EllisD.Biofuels in Brazil: an overview.*Crop Sci*47:2228–2237;2007. doi:10.2135/cropsci2007.03.0166.

National Bureau of Statistics of China (NBSC) China statistical yearbook. NBSC, Beijing; 2007. http://www.stats.gov.cn/tjsj/ndsj/2007/indexch.htm.

National Development and Reform Commission of People's Republic of China (NDRCPRC) Mid- and long-term development plan for renewable energy. NDRCPRC, Beijing; 2007. http://www.ndrc.gov.cn/zcfb/zcfbtz/2007tongzhi/W020070904607346044110.pdf.

National Development and Reform Commission of People's Republic of China (NDRCPRC) The eleventh five-year plan of renewable energy development. NDRCPRC, Beijing; 2008a. http://www.ndrc.gov.cn/zcfb/zcfbtz/2008tongzhi/W020080318381136685896.pdf.

National Development and Reform Commission of People's Republic of China (NDRCPRC) Development of non-grain bioethanol industry in Guangxi province. NDRCPRC, Beijing; 2008b. http://www.ndrc.gov.cn/gyfz/zhdt/t20080509_210293.htm.

NikolovT.;BakolovaN.;PetrovaS.;BenadovaR.;SpasovS.;KolevD. An effective method for bioconversion of delignified waste-cellulose fibers from the paper industry with a cellulase complex.*Bioresour. Technol*71:1–4;2000.doi:10.1016/S0960-8524(99)00059-0.

PrasadS.;SinghA.;JainN.;JoshiH. C.Ethanol production from sweet sorghum syrup for utilization as automotive fuel in India.*Energy Fuels*21:2415–2420;2007.doi:10.1021/ef060328z.

QianB. Z.The status quo and prospects of fuel ethanol industry in China.*Sol. Energy*8:7–9;2007.

RooneyW. L. Sorghum improvement-integrating traditional and new technology to produce improved genotypes.Adv. Agron83:37–109;2004.doi:10.1016/S0065-2113(04)83002-5.

RooneyW. L.;BlumenthalJ.; BeanB.; MulletJ. E. Designing sorghum as a dedicated bioenergy feedstock. *Biofuels Bioprod. Biorefin*1:147–157;2007.doi:10.1002/bbb.15.

SassonA.Feeding tomorrow's world. UNESCO,Paris, pp500–510;1990.

SemelsbergerT. A.;BorupR. L.;GreeneH. L.Dimethyl ether (DME) as an alternative fuel.*J. Power Sour.*156:497–511;2006. doi:10.1016/j.jpowsour.2005.05.082.

ShaoH. B.;ChuL. Y.Resource evaluation of typical energy plants and possible functional zone planning in China.*Biomass Bioenerg*32:283–288;2008.

SunC.;LiangW.Status of biodiesel development and production in China.*SinoGlobal Energy*13:23–28;2008.

SunT.;DuW.;ChenX.;LiuD. H.Status and prospect of biodiesel industrialization in China.*Biotechnol. Bus*2:33–39;2007.

TanT. W.;WangF.Current development situation of bio-oil refining and its future prospect.*Mod. Chem. Ind*26:6–9;2006.

TianC. L.;GuoB.;LiuC. Z.Present situation and prospect of energy plants.*Chin. J. Biopro. Eng*3:14–19;2005.

United Nations Conference on Trade and Development (UNCTAD) Challenges and opportunities for developing countries in producing biofuels. UNCTAD publication, UNCTAD/DITC/COM/2006/15, Geneva, November 27, 2006.

WangF.;ChengX. Y.;WuT. X.;LiuW. F.;LiuC. Z.Ethanol fermentation of sweet sorghum stalk juice and its economic feasibility. Liquor-making.*Sci. Technol.*146:41–44;2006a.

WangF.;LiuC. Z.Solid state fermentation of sweet sorghum to ethanol.*Liquor Making Sci. Technol*158:36–40;2007.

WangH. M.;XuB. C.;LiF. M.;HeX. L.Preliminary study on growth response of *Panicum virgatum* L. to different sites in the Loess Plateau.*Res. Soil Water Conserv*13:91–93;2006b.

XiQ. G.;HongH.Description of an introduced plant *Miscanthus* x *giganteus*.*Pratacultural Sci*25:26–28;2008.

XieG. H.;GuoX. Q.;WangX.;DingR. E.;HuL.;ChengX.An overview and perspectives of energy crop resources.*Resour. Sci*29:74–80;2007.

XingY.Production status and development prospect of biodiesel.*Environ. Prot. Chem. Ind* 27:46–49;2007.

XuZ. Y.Current development situation of fuel ethanol and its future prospect.*Chem. Technol. Market* 30:18–23;2007.

XuG. W.;JiW. F.;WanY. H.;LiuC. Z.Energy production with light-industry biomass process residues rich in cellulose: a technical review.*Progr. Chem*19:1164–1176;2007.

XuB. C.;ShanL.;HuangZ. B.;LiD. Q.Study on the photosynthetic physiological characteristics of Panicum virgatum in loess hilly-gully region.*Acta. Bot. Boreal Occident. Sin* 21:625–630;2001.

XuB. C.;ShanL.;LiF. M.Aboveground biomass and water use efficiency of an introduced grass, *Panicum virgatum*, in the semiarid loess hilly-gully region.*Acta. Ecol. Sin*25:2206–2213;2005.

YiH. X.;WangZ.;ZhangL. W.Simple analysis on the exploitation value of beet.*China Beet Sugar*1:27–28;2007.

YinF. S.;TeahY. K.Status of bio-diesel industry in China.*Deterg. Cosmet*31:1–3;2008.

YouJ. F.;FangX. F.;ChenY. Z.;NongD. C.;FanB. N.Status and strategy research of the development of mechanization in sugarcane production in Guangxi.*Sugar Crop China*4:69–72;2008.

YueG. J.;WuG. Q.;HaoX. M.The status quo and prospects of fuel ethanol process technology in China.*Prog. Chem*19:1084–1090;2007.

ZhangD. H.;LiC. Q.Status and industrialization prospect of high-quality rapeseed.*Shanxi J. Agric. Sci*3:79–81;2006.

ZhangL. W.Present situation and problems of Chinese biodiesel industry.*China Oils Fats*32:12–15;2007.

ZhangS. H.;HuR. F.Innovating induce factors in hybrid maize breeding strategies.*J. Maize Sci*8:3–7;2000.

ZhangL. B.;LiC. Z.;OuR. M.;XiaoZ. H.;LiP. W.;LiD. X.Current situation and development prospect of biodiesel industry.*Hunan For. Sci. Technol*35:70–73;2008a.

ZhangY. B.;WuC. W.;LiuJ. Y.;ZhaoJ.Developmental status and suggestion of modern sugarcane industry.*Sugar Crop China*4:54–58;2008b.

Chapter 12
Genetic Modification of Lignin Biosynthesis for Improved Biofuel Production

Hiroshi Hisano, Rangaraj Nandakumar, and Zeng-Yu Wang

Abstract The energy in cellulosic biomass largely resides in plant cell walls. Cellulosic biomass is more difficult than starch to break down into sugars because of the presence of lignin and the complex structure of cell walls. Transgenic downregulation of major lignin genes led to reduced lignin content, increased dry matter degradability, and improved accessibility of cellulases for cellulose degradation. This review provides background information on lignin biosynthesis and focuses on genetic manipulation of lignin genes in important monocot species as well as the dicot potential biofuel crop alfalfa. Reduction of lignin in biofuel crops by genetic engineering is likely one of the most effective ways of reducing costs associated with pretreatment and hydrolysis of cellulosic feedstocks, although some potential fitness issues should also be addressed.

Keywords Biomass • Biofuel crops • Genetic engineering • Lignin modification

Introduction

Transgenic technology has greatly contributed to breakthroughs in plant improvement and is expected to play a crucial role in coming years in genetic modification of crops for biofuel production by modifying quantity or quality of biomass (Sánchez and Cardona 2007; Gressel 2008). Global industrialization, the increase in world population, and faster economic growth, especially in developing countries, call for continuous and steady increases in the demand for energy (Li et al. 2008; Yuan et al. 2008). High fluctuation in the global oil market, decrease in oil

Z.-Y. Wang (✉)
Forage Improvement Division, The Samuel Roberts Noble Foundation, 2510 Sam Noble Parkway, Ardmore, OK 73401, USA
and
BioEnergy Science Center (BESC), Oak Ridge, TN 37831, USA
e-mail: zywang@noble.org

reserves, global warming due to the emission of greenhouse effect gases, and other problems associated with the use of fossil fuels make the development of alternative sources of energy highly imperative to meet worldwide rising energy demands (Gray et al. 2006; Koonin 2006; Yuan et al. 2008).

In recent years, the exploitation of renewable and sustainable energy sources is taking center stage in science, research, media, and politics (Schubert 2006). Bioethanol, biodiesel, biomethanol, and other biofuels that can replace or can be mixed with fossil fuels are renewable energy sources (Gray et al. 2006; Ragauskas et al. 2006). To date, bioethanol production in the USA has been mainly based on the use of maize and other crops. However, high biomass producing non-grain crops like switchgrass or *Miscanthus* are being considered as a primary source of feedstocks to produce biofuels. Biofuels produced from these sources are called lignocellulosic biofuels, which represent an alternative fuel for future use (Yuan et al. 2008). These energy-rich cellulosic non-food or non-grain crops are mostly perennial grasses that can be grown in marginal lands with minimal nutrition inputs. Although cellulosic biofuel can overcome the limitations associated with starch-based ethanol production, the main obstacle is the high production cost incurred during the conversion process from the lignified plant cell walls which limits large-scale adoption of cellulosic ethanol production (Li et al. 2008). Other obstacles include the lack of infrastructure associated with harvest, transportation and storage of cellulosic biomass.

Lignocellulosic biomass is composed of cellulose, hemicellulose, and lignin; these are major components of the secondary cell walls of all vascular plants. Cellulose, consisting of glucose (6-carbon sugar) units linked by glycosidic bonds, is the most abundant substance on earth. Hemicellulose consists of 5-carbon sugars such as xylose or arabinose along with glucose. Hemicellulose forms complex cell wall network by cross-linking cellulose microfibrils with lignin (Rubin 2008). This complex network should be broken down for efficient biofuel production. The process of cellulosic biofuel production involves three major steps: (1) pretreatment of biomass feedstock; (2) hydrolysis and saccharification; and (3) fermentation of sugars into ethanol. After collection and processing of feedstocks, a pretreatment with acid or steam releases the polysaccharides. In the second step, the released complex polysaccharides are enzymatically converted into simple sugars by cellulase and hemicellulase enzymes. The final step converts the simple sugars into ethanol via microbial fermentation, as in the case of starch-based biofuel. However, the association of lignin with cellulose and hemicellulose has a negative impact in cellulosic ethanol production as it inhibits the release of polysaccharides during the pretreatment process and also absorbs the enzymes used for saccharification or reduces the accessibility of enzymes during the conversion process. The use of increased acidity or steam also reduces the efficiency of the saccharification and fermentation process at a later stage (Keating et al. 2006). The high cost incurred during processing is the major limiting factor in cellulosic biofuel production and makes the price of the cellulosic ethanol two- to threefold higher than starch-based ethanol (Sticklen 2006, 2008).

Although breeding plant biomass feedstock for reduced lignin content or increased biomass production will solve this problem (Bouton 2007), it will take a long time to achieve the goal. In this circumstance, modern biotechnological approaches offer great alternative opportunities to conventional plant breeding techniques to reduce the cost of cellulosic ethanol production (Gressel 2008). The genetic engineering approaches include up-regulation of cellulose and hemicellulose pathway enzymes or other enzymes involved in increasing plant biomass characteristics or production of recombinant cellulases or hemicellulases in plants (Ziegelhoffer et al. 1999; Ericksson et al. 2000; Biswas et al. 2006; Oraby et al. 2007; Ransom et al. 2007). These approaches will possibly compensate the reduced saccharification efficiency due to the presence of lignin or minimize the use of enzymes during saccharification (Sticklen 2008). A direct and effective approach is to down-regulate the enzymes involved in lignin biosynthesis to reduce lignin content or to modify its composition (Ralph et al. 2006; Chapple et al. 2007; Chen and Dixon 2007)

Lignin Biosynthesis

Lignin is a phenolic biopolymer of complex structure, synthesized by all plants. The deposition of lignin in the cell wall is considered critical for plant growth and development (Dixon et al. 2001; Rogers and Campbell 2004). The biosynthesis of lignin begins with the synthesis of cinnamic acid from the amino acid phenylalanine by phenylalanine ammonia lyase (PAL) in the cytosol. Lignin is made up of three main *p*-hydroxycinammyl alcohol precursors or monolignols, namely *p*-coumaryl, coniferyl, and sinapyl alcohols, which later undergo dehydrogenative polymerizations by peroxidase (PER) and laccase (LAC) to form *p*-hydroxyphenyl (H), guaiacyl (G) and syringyl (S) lignin, respectively (Weng et al. 2008). The relative proportion of each lignin unit varies with species, plant parts, and maturity. The biosynthetic pathway to lignin has been under constant revision during the past decade, mainly as a result of genetic and transgenic studies. These studies question the *in vivo* specificities of the monolignol pathway enzymes as initially extrapolated from *in vitro* studies (Chen et al. 2006). The current view of the general pathway of lignin biosynthesis in higher plants is shown in Fig.12.1 (Chen et al. 2006; Li et al. 2008). Many studies have proposed that the following enzymes are required for monolignol biosynthesis through phenylpropanoid pathway: phenylalanine ammonia lyase; cinnamate 4-hydroxylase (C4H); 4-coumarate-CoA ligase (4CL); cinnamoyl CoA reductase (CCR); hydroxycinnamoyl CoA: shikimate hydroxycinnamoyl transferase (HCT); coumarate 3-hydroxylase (C3H); caffeoyl CoA 3-*O*-methyltransferase (CCoAOMT); ferulate 5-hydroxylase (F5H); caffeic acid 3-*O*-methyltransferase (COMT); and cinnamyl alcohol dehydrogenase (CAD). In addition, transcription factors like MYB, LIM, and NAC genes are thought to be coordinately regulating the expression of these genes for lignin biosynthesis (Rogers and Campbell 2004). The functions of many lignin genes have been well-studied in several plant species, especially in dicot plants using either mutants or transgenic plants. With the availability of information of genes involved in lignin biosynthesis and by taking advantage of developments in

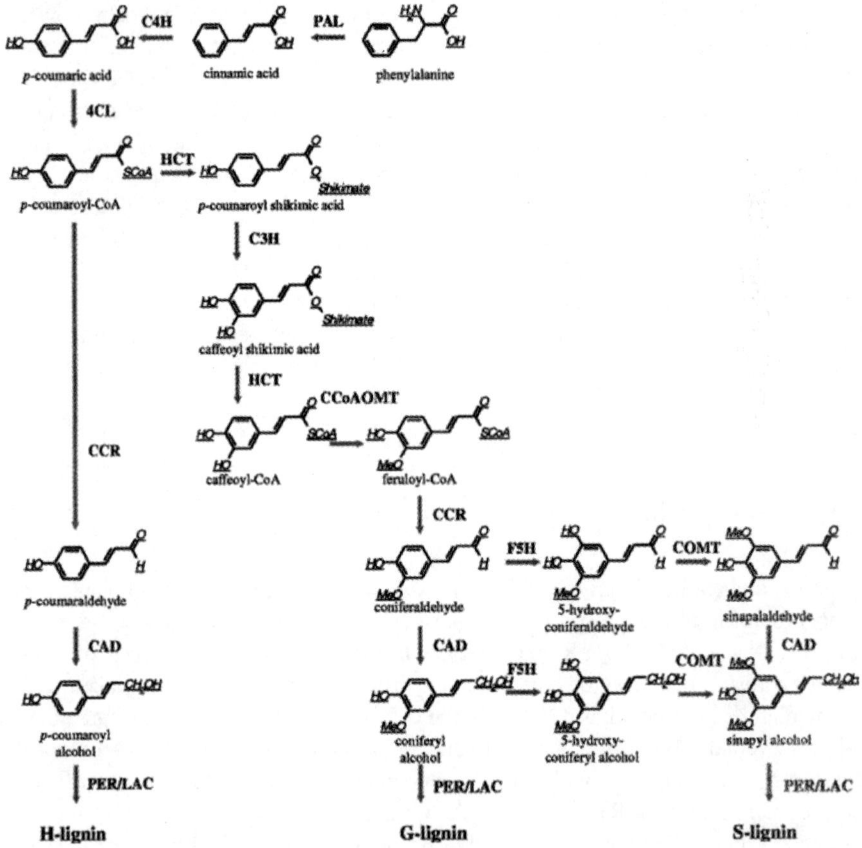

Fig. 12.1 One of the current views of lignin biosynthetic pathway. The enzymes involved in the pathway are: phenylalanine ammonia lyase (*PAL*), cinnamate 4-hydroxylase (*C4H*), 4-coumarate-CoA ligase (*4CL*), cinnamoyl-CoA reductase (*CCR*), hydroxycinnamoyl-CoA: shikimate hydroxycinnamoyl transferase (*HCT*), coumarate 3-hydroxylase (*C3H*), caffeoyl CoA 3-*O*-methyltransferase (*CCoAOMT*), ferulate 5-hydroxylase (*F5H*), caffeic acid 3-*O*-methyltransferase (*COMT*), cinnamyl alcohol dehydrogenase (*CAD*), peroxidase (*PER*), and laccase (*LAC*)

plant transformation technology, it is now possible to modify or reduce lignin content in biofuel crops by overexpression, down-regulation, or suppression of genes involved in either lignin synthesis, regulation, or polymerization (Li et al. 2003; Ralph et al. 2006; Chen and Dixon 2007). The extent of lignin reduction or modification depends on the kind of gene which is down-regulated. For example, the down-regulation of the upstream genes like C3H, HCT, or 4CL leads to reduction in lignin content, while the down-regulation of F5H and COMT resulted in changes of S/G ratio (Weng et al. 2008). Although many studies and findings were first reported in non-feedstock model plants such as tobacco and *Arabidopsis* (Zhou et al. 2009), it is assumed that similar approaches can be applied to cellulosic feedstock crops as the lignin pathway is conserved among plant species.

Plant Transformation and Gene Regulation Methods for Lignin Modification

Because plant genetic engineering plays a major role in lignin modification, availability or establishment of a well-defined, highly efficient transformation system for feedstock crops is an important prerequisite for the successful manipulation of lignin pathway genes to modify the quality or quantity of biomass (Gressel 2008). Since most of the cellulosic feedstock crops are perennial grasses, which are considered recalcitrant for transformation procedures, the choice of transformation method will also have a great impact. To date, genetic transformation of plants has been performed by two main methods: *Agrobacterium*-mediated transformation and particle bombardment. The *Agrobacterium* method was originally used for dicotyledonous (dicot) plants such as tobacco, alfalfa, and poplar because these plants are natural hosts for *Agrobacterium*. After extensive studies, the *Agrobacterium* method has been extended to various monocotyledonous (monocot) plants including some feedstock species such as maize and switchgrass (Ishida et al. 1996; Somleva et al. 2002). The early reports on grass transformation were mainly based on particle bombardment (Conger et al. 1993; Denchev et al. 1997; Wang and Ge 2006). Considering the advantages of *Agrobacterium*-mediated transformation (lower copy number, fewer rearrangements of the transgene), it is the method of choice for transforming biofuel crops (Somleva et al. 2002).

As far as the modification of lignin genes is concerned, constant or specific "knockdown" and "overexpression" techniques are being used (Capell and Christou 2004). Initially, reduced lignin plants were identified from natural or chemically induced (e.g., ethylmethane sulfonate treatment) mutants. With the development of genetic transformation techniques, antisense, RNA interference (RNAi), and virus-induced gene silencing (VIGS) have been used to knock down or silence the target gene(s) (Lu et al. 2003; Chen et al. 2006; Chen and Dixon 2007). Antisense method is to introduce and express RNAs equivalent to an antisense strand of the mRNA of target genes. RNAi usually involves stable transformation with a gene construct that, when expressed, produces a small double-stranded RNA homologous to a portion of the target gene sequence. This is usually generated via an inverted repeat of the short target sequence interrupted by a plant intron sequence (Wesley et al. 2001). This approach has been effectively used for modifying a number of plant traits through targeted down-regulation of a specific gene or genes (Miki et al. 2005; Dixon et al. 2007). Overexpression of target genes sometimes causes cosuppression, in which the endogenous gene is silenced. Since RNAi or antisense may not totally abolish expression of the gene, the technique is sometimes referred to as a "knockdown" to distinguish it from "knockout" procedures in which expression of a gene is entirely eliminated. RNAi technology has emerged as an attractive tool to study the gene functions in plants through genetic engineering. VIGS takes advantage of an endogenous defense mechanism against viral infection and is used for high throughput tests of gene functions in plants (Lu et al. 2003). However, VIGS is a transient system which does not lead to stable integration of the transgenes.

Lignin Modification in Monocots

Because of the negative correlation between lignin and forage digestibility, lignin modification has been one of the breeding goals in grasses to improve feed quality. In recent years, several monocot species have been recognized as major candidates of biomass materials for cellulosic ethanol production; lignin modification in these species has received much attention (Stewart 2007).

Monocot mutants with reduced lignin. Maize is not only used for food; it has also been widely used as silage for animal production. Moreover, maize stover and cobs are considered important sources of biomass for cellulosic ethanol production. The brown-midrib (*bm*) mutants of maize, which differ in quality and quantity of lignin from normal genotypes, were known as pioneering models to study lignifications and digestibility (Kuc and Nelson 1964, Rook et al. 1977, Barriere et al. 2004, Guillaumie et al. 2007). It is a simple recessive trait that phenotypically produces a reddish-brown pigmentation associated with lignified tissues (Cherney et al. 1991). Lignin contents in *bm* genotypes are consistently lower than their normal counterparts. *In vitro* digestibility of *bm* genotypes has been consistently higher than normal (Cherney et al. 1991). Activities of several enzymes involved in lignification differed between *bm* and normal genotypes (Grand et al. 1985; Cherney et al. 1991; Pillonel et al. 1991). However, differences in enzyme activities were not consistent across species and genotypes, indicating that different modifications of the lignification pathway may result in a similar *bm* phenotype.

Maize *bm3* mutant is severely deficient in OMT activity, with only 10% of the activity found in normal plants (Grand et al. 1985). Two independent maize *bm3* mutations were analyzed concerning their effects on expression of *COMT* gene (Vignols et al. 1995). By sequencing the *COMT* clones obtained from the *bm3-1* and *bm3-2* maize, the *bm3-1* allele was found to arise from an insertional event producing a *COMT* mRNA altered in both size and amount, and the *bm3-2* was resulted from a deletion of part of the *COMT* gene (Vignols et al. 1995). These results demonstrated that mutations at the *COMT* gene lead to the *bm3* phenotype.

In maize *bm1* genotypes, CAD activity was significantly reduced by 60–70% in stem tissue (Halpin et al. 1998). A *CAD* cDNA was isolated and used as a probe to map the location of the *CAD* gene. The *CAD* gene is located very closely to the known location of *bm1* and co-segregates with the *bm1* locus in two independent recombinant inbred populations. The results strongly suggest that maize *bm1* directly affects expression of the *CAD* gene (Halpin et al. 1998).

Functions of *bm2* and *bm4* mutants are still unknown, although both mutants showed reduced lignin content of 15–25%, especially G lignin unit (Marita et al. 2003, Barriere et al. 2004). Guillaumie et al. (2007) showed that *bm2* and *bm4* mutations could affect regulatory genes involved in the regulation, polymerization, or transportation of coniferaldehyde in maize tissues. In addition to maize, brown-midrib mutants have been induced in other monocots such as sorghum and pearl millet that also showed a red-brown color of midribs with modified lignin composition and improved digestibility (Cherney et al. 1991, Barriere and Argiller 1993).

The *bm* mutants are good candidates for studying the relationship between lignin reduction and biofuel production.

Transgenic maize with modified lignin. There were two reports on generating transgenic maize through the particle bombardment method for lignin modification. Piquemal *et al.* (2002) produced COMT down-regulated transgenic maize by the antisense approach and showed decreased lignin contents in the transgenic plants. In this case, *COMT* antisense sequence was driven by maize alcohol dehydrogenase 1 (*Adh1*) promoter which showed good expression in vascular tissues and lignifying sclerenchyma. One transgenic line carrying antisense *COMT* had only 15–30% residual COMT activity and showed similar phenotype as the maize *bm3* mutant. Further analyses revealed that several transcription factor genes, cell signaling genes, transport and detoxification genes, genes involved in cell wall carbohydrate metabolism, and genes encoding cell wall proteins were differentially expressed and mostly over-expressed in COMT-deficient plants (Guillaumie et al. 2008). A separate study obtained similar results by introducing a sorghum *O*-methyltransferase (*OMT*) antisense construct into maize (He et al. 2003). The transgenic plants showed red-brown midrib phenotype with reduced OMT activity by 60% and reduced lignin contents by an average of 17%. Digestibility was significantly improved in transgenic plants by 2% in leaves and 7% in stems (He et al. 2003). The studies demonstrated the feasibility of using transformation technology to modify lignin biosynthetic pathway and to alter the lignin profile in maize. Further research is needed to evaluate bioethanol production from maize stover with reduced lignin.

Transgenic tall fescue with modified lignin. Tall fescue, a predominant cool-season grass in the USA, has been used as a model species to study lignin deposition at defined developmental stages (Chen et al. 2002). Lignification of cell walls of tall fescue increased drastically from elongation stage to reproductive stage. The relative S lignin content and S/G ratio increased when plants matured, while relative G and H lignin content decreased during the same period. It seemed that G lignin was deposited at the early stage of plant growth, and S lignin was preferentially deposited at the later developmental stage (Chen et al. 2002). It has been noted that the aromatic composition of lignin of monocot plants is characterized by the presence of H unit (Iiyama and Lam 2001). However, H unit only comprises a small portion of total lignin when compared with S and G lignin in most monocot species (Baucher et al. 1998).

Transgenic tall fescue plants with down-regulated CAD and COMT were obtained by particle bombardment of embryogenic cell cultures (Chen et al. 2003, 2004). Transgenic *CAD* plants showed reduced lignin content and altered S/G ratio. There were no significant changes in levels of celluloses, hemicelluloses, neutral sugar composition, *p*-coumaric acid, or ferulic acid in the transgenics. The *CAD*-transgenic plants showed a significant increase in dry matter digestibility of 7.2–9.5% (Chen et al. 2004). Similarly, transgenic tall fescue down-regulated with *COMT* gene showed substantially reduced levels of *COMT* transcripts, significantly reduced COMT activity, reduced lignin content, and increased dry matter digestibility (Chen et al. 2004). These results indicated that down-regulation of lignin genes could lead to development of grass germplasm with improved forage quality and improved characteristics for bioethanol production.

Transgenic Alfalfa with Modified Lignin

Alfalfa is a perennial forage crop that has also been proposed as a feedstock for biofuel purposes. Lignin genes have been systematically down-regulated in alfalfa (Guo et al. 2001a; Reddy et al. 2005; Shadle et al. 2007). The first lignin gene down-regulated in alfalfa was *CAD* (Baucher et al. 1999). Reduction of the CAD enzyme was associated with a red coloration of the stem. Although lignin quantity remained unchanged, lignin composition was altered, and the rate of disappearance of dry matter *in situ* was increased (Baucher et al. 1999). Later, Guo et al. (2001a) produced transgenic alfalfa plants down-regulated with COMT and CCoAOMT. The cDNA sequences of these genes in sense and antisense orientations were controlled by the vascular-specific bean PAL promoter. Strong down-regulation of COMT resulted in decreased lignin content, a reduction in total guaiacyl (G) lignin units, a near total loss of syringyl (S) units in monomeric and dimeric lignin degradation products, and appearance of low levels of 5-hydroxy guaiacyl units and a novel dimer. In contrast, strong down-regulation of CCoAOMT led to reduced lignin levels, a reduction in G units without reduction in S units, and increases in β-5 linked dimers of G units (Guo et al. 2001a). Analysis of rumen digestibility of alfalfa forage revealed improved digestibility of forage from COMT down-regulated plants but a greater improvement in digestibility following down-regulation of CCoAOMT (Guo et al. 2001b).

Several cytochrome P450 enzymes, C3H, C4H, and F5H, were subsequently down-regulated in transgenic alfalfa (Reddy et al. 2005). Down-regulation of C4H, C3H, or F5H produced plants with greatly reduced lignin without significant impact on composition, lignin-rich in *p*-hydroxyphenyl (H) units, or lignin rich in G-units with reduced S content, respectively. There was a strong negative relationship between lignin content and forage digestibility, but no relationship between lignin composition and digestibility was detected in the transgenic lines (Reddy et al. 2005). Down-regulation of a recently discovered enzyme, HCT, resulted in strongly reduced lignin content and striking changes in lignin monomer composition, with predominant deposition of 4-hydroxyphenyl units in the lignin (Shadle et al. 2007). Vascular structure was impaired in the strongly down-regulated lines, and forage digestibility was increased by up to 20% (Shadle et al. 2007).

Lignin Modification and Cellulosic Ethanol Production

The relationships between lignin content/composition and chemical/enzymatic saccharification were first convincingly documented by Chen and Dixon (2007). They analyzed transgenic alfalfa down-regulated with different antisense gene constructs. Lignin content of mature stems decreased in the order: F5H and control (most lignin) > COMT and CCoAOMT > C4H, C3H, and HCT (lowest lignin level). Down-regulation of genes early in the pathway (C4H, C3H, and HCT) was most effective at reducing lignin content, in some cases leading to plants that contain less than half the lignin present in wild type. Plants with the least lignin had the highest total carbohydrate levels in untreated biomass, reflecting compensation for the reduction in lignin level on a mass balance basis. The amount of carbohydrate released by acid pretreatment increased in proportion to the reduction in lignin levels. A strong negative correlation between lignin content and sugar released by enzymatic hydrolysis was observed. Some lines showed two- to threefold greater yield of monosaccharides (the substrates for ethanol production) compared with wild-type materials (Chen and Dixon 2007).

The results from transgenic alfalfa demonstrated that genetic reduction of lignin content effectively overcame cell wall recalcitrance to bioconversion. For ethanol production, the current paradigm is that biomass must first be subjected to a costly pretreatment to make cell walls accessible to enzymes. However, several transgenic HCT and C3H alfalfa lines produced greater amounts of sugar from untreated biomass than that obtainable from pretreated biomass of control plants. Thus, it may be possible to reduce or eliminate the pretreatment step by using biomass from low-lignin transgenic plants, thereby greatly reducing the cost of biofuel production. Moreover, the simplified process without harsh chemical pretreatment allows for taking advantage of other traits, such as *in planta* expression of enzymes to increase enzymatic processing efficiency (Chen and Dixon 2007).

Conclusions

The energy in plant biomass largely resides in plant cell walls. The major obstacle for ethanol production from cellulosic feedstocks is the high cost of obtaining sugars from cell walls (Boudet et al. 2003). Because of the presence of lignin and the complex structure of cell walls, cellulosic biomass is more difficult than starch to break down into sugars. Transgenic down-regulation of major lignin genes led to reduced lignin content, increased dry matter degradability, and improved accessibility of cellulases for cellulose degradation. Wall polysaccharides of lignin-down-regulated plants were more easily hydrolyzed, and the proportion of released sugars in the transgenic material was much greater than that of the control. Thus, improvements of downstream procedures can be achieved by genetically redesigning the properties of the feedstocks. Furthermore, since lignin degradation products are known inhibitors of ethanol fermentation, reduction of lignin may help to reduce the degradation products that inhibit the fermentation process.

Although there have been many reports on down-regulation of lignin biosynthesis in dicot species (e.g. *Arabidopsis*, tobacco, alfalfa, poplar), only limited information is available in monocots (corn and tall fescue). To date, no public information on lignin modification is available in the major biofuel crops, switchgrass and *Miscanthus*. The lack of reports on successful modification of lignin in these monocot species is mainly due to the difficulties in obtaining transgenics and identifying transgenic plants having changes in lignin. The biosynthetic pathways to lignin monomers are conserved across species, and knowledge gained from dicots should be applicable to monocots. There is an urgent need to systematically characterize lignin genes in monocot species and to develop strategies to improve ethanol production for the major biofuel crops.

Reduction of lignin in the biofuel crops by genetic methods is likely one of the most effective/economic ways of reducing costs associated with pretreatment and hydrolysis of lignocellulosic feedstocks. However, some potential negative issues should also be addressed. For example, some lignin-down-regulated alfalfa lines had reduced biomass production. Although increases in fermentable sugar production could compensate for the decreases in biomass, high productivity of biofuel crops is a basic requirement for the industry. Future studies are needed to break up the negative relationship between lignin reduction and biomass productivity. The problem seems solvable by combining lignin modification with other approaches, such as manipulation of cellulose and hemicellulose biosynthesis and deposition. It is also important to evaluate the impact of cell wall manipulation on plant structure and tolerance to biotic and abiotic stresses. The development of new cultivars with optimized biomass yield and quality will greatly benefit the biofuel industry.

Acknowledgments This work was supported by the BioEnergy Science Center and the Samuel Roberts Noble Foundation. The BioEnergy Science Center is supported by the Office of Biological and Environmental Research in the DOE Office of Science.

References

Barriere Y. O.; Argiller O. Brown-midrib genes of maize: A review. *Agronomie* 13: 865–876; 1993. doi:10.1051/agro:19931001.

Barriere Y.; Ralph J.; Mechin V.; Guillaumie S.; Grabber J. H.; Argillier O.; Chabbert B.; Lapierre C. Genetic and molecular basis of grass cell wall biosynthesis and degradability: II. Lessons from brown-midrib mutants. *C R Biol* 327: 847–860; 2004. doi:10.1016/j.crvi.2004.05.010.

Baucher M.; Bernard-Vailhe M. A.; Chabbert B.; Besle J. M.; Opsomer C.; Van Montagu M.; Botterman J. Downregulation of cinnamyl alcohol dehydrogenase in transgenic alfalfa (*Medicago sativa* L.) and the effect on lignin composition and digestibility. *Plant Mol Biol* 39: 437–447; 1999. doi:10.1023/A:1006182925584.

Baucher M.; Monties B.; Van Montagu M.; Boerjan W. Biosynthesis and genetic engineering of lignin. *Crit Rev Plant Sci* 17: 125–197; 1998. doi:10.1016/S0735-2689(98)00360-8.

Biswas G.; Ransom C.; Sticklen M. Expression of biologically active *Acidothermus cellulolyticus* endoglucanase in transgenic maize. *Plant Sci* 171: 617–623; 2006. doi:10.1016/j.plantsci.2006.06.004.

Boudet A. M.; Kajita S.; Grima-Pettenati J.; Goffner D. Lignins and lignocellulosics: a better control of synthesis for new and improved uses. *Trends Plant Sci* 8: 576–581; 2003. doi:10.1016/j.tplants.2003.10.001.

Bouton J. H. Molecular breeding of switchgrass as a bioenergy crop. *Curr Opin Gen Develop* 17: 553–558; 2007. doi:10.1016/j.gde.2007.08.012.

Chapple C.; Ladisch M.; Melian R. Loosening lignin's grip on biofuel production. *Nat Biotechnol* 25: 746–747; 2007. doi:10.1038/nbt0707-746.

Capell T.; Christou P. Progress in plant metabolic engineering. *Curr Opin Biotechnol* 15: 148–154; 2004. doi:10.1016/j.copbio.2004.01.009.

Chen F.; Dixon R. A. Lignin modification improves fermentable sugar yields for biofuel production. *Nat Biotechnol* 25: 759–761; 2007. doi:10.1038/nbt1316.

Chen F.; Reddy M. S. S.; Temple S.; Jackson L.; Shadle G.; Dixon R. A. Multi-site genetic modulation of monolignol biosynthesis suggests new routes for formation of syringyl lignin and wall-bound ferulic acid in alfalfa (*Medicago sativa* L.). *Plant J* 48: 113–124; 2006. doi:10.1111/j.1365-313X.2006.02857.x.

Chen L.; Auh C.; Chen F.; Cheng X. F.; Aljoe H.; Dixon R. A.; Wang Z. Y. Lignin deposition and associated changes in anatomy, enzyme activity, gene expression and ruminal degradability in stems of tall fescue at different developmental stages. *J Agric Food Chem* 50: 5558–5565; 2002. doi:10.1021/jf020516x.

Chen L.; Auh C.; Dowling P.; Bell J.; Lehmann D.; Wang Z. Y. Transgenic down-regulation of caffeic acid *O*-methyltransferase (COMT) led to improved digestibility in tall fescue (*Festuca arundinacea*). *Funct Plant Biol* 31: 235–245; 2004. doi:10.1071/FP03254.

Chen L.; Auh C. K.; Dowling P.; Bell J.; Chen F.; Hopkins A.; Dixon R. A.; Wang Z. Y. Improved forage digestibility of tall fescue (*Festuca arundinacea*) by transgenic down-regulation of cinamyl alcohol dehydrogenase. *Plant Biotechnol J* 1: 437–449; 2003. doi:10.1046/j.1467-7652.2003.00040.x.

Cherney J. H.; Cherney D. J. R.; Akin D. E.; Axtell J. D. Potential of brown-midrib, low-lignin mutants for improving forage quality. *Adv Agron* 46: 157–198; 1991. doi:10.1016/S0065-2113(08)60580-5.

Conger, B. V.; Songstad, D. D.; McDaniel, J. K.; Bond, J. Genetic transformation of *Dactylis glomerata* by microprojectile bombardment. *Proc XVII Intl Grasslands Congr.* 1034–1036; 1993.

Denchev P. D.; Songstad D. D.; McDaniel J. K.; Conger B. V. Transgenic orchardgrass (*Dactylis glomerata*) plants by direct embryogenesis from microprojectile bombarded leaf cells. *Plant Cell Rep* 16: 813–819; 1997. doi:10.1007/s002990050326.

Dixon R. A.; Bouton J. H.; Narasimhamoorthy B.; Saha M.; Wang Z. Y.; May G. D. Beyond structural genomics for plant science. *Adv. Agron* 95: 77–161; 2007. doi:10.1016/S0065-2113(07)95002-6.

Dixon R. A.; Chen F.; Gua D.; Parvathi K. The biosynthesis of monolignols: a "metabolic grid" or independent pathways to guaiacyl and syringyl units? *Phytochemistry* 57: 1069–1084; 2001. doi:10.1016/S0031-9422(01)00092-9.

Ericksson M. E.; Israelsson M.; Olsson O.; Moritiz T. Increased gibberellin biosynthesis in transgenic trees promotes growth, biomass production and xylem fiber length. *Nat Biotechnol* 18: 784–788; 2000. doi:10.1038/77355.

Grand C.; Parmentier P.; Boudet A.; Boudet A. M. Comparison of lignins and of enzymes involved in lignification in normal and brown midrib (bm3) mutant corn seedlings. *Physiologie Végétale* 23: 905–911; 1985.

Gray K. A.; Zhao L.; Emptage M. *Bioethanol Curr Opin Chem Biol* 10: 141–146; 2006. doi:10.1016/j.cbpa.2006.02.035.

Gressel J. Transgenics are imperative for biofuel crops. *Plant Sci* 174: 246–263; 2008. doi:10.1016/j.plantsci.2007.11.009.

Guillaumie S.; Goffner D.; Barbier O.; Martinant J. P.; Pichon M.; Barrière Y. Expression of cell wall related genes in basal and ear internodes of silking *brown-midrib-3*, caffeic acid *O*-methyltransferase (COMT) down-regulated and normal maize plants. *BMC Plant Biol* 8: 71; 2008. doi:10.1186/1471-2229-8-71.

Guillaumie S.; Pichon M.; Martinant J. P.; Bosio M.; Goffner D.; Barriere Y. Differential expression of phenylpropanoid and related genes in brown-midrib bm1, bm2, bm3, and bm4 young near-isogenic maize plants. *Planta* 226: 235–250; 2007. doi:10.1007/s00425-006-0468-9.

Guo D.; Chen F.; Inoue K.; Blount J. W.; Dixon R. A. Downregulation of caffeic acid 3-*O*-methyltransferase and caffeoyl CoA 3-*O*-methyltransferase in transgenic alfalfa: impacts on lignin structure and implications for the biosynthesis of G and S lignin. *Plant Cell* 13: 73–88; 2001a.

Guo D.; Chen F.; Wheeler J.; Winder J.; Selman S.; Peterson M.; Dixon R. A. Improvement of in-rumen digestibility of alfalfa forage by genetic manipulation of lignin *O*-methyltransferases. *Transgenic Res* 10: 457–464; 2001b. doi:10.1023/A:1012278106147.

Halpin C.; Holt K.; Chojecki J.; Oliver D.; Chabbert B.; Monties B.; Edwards K.; Barakate A.; Foxon G. A. Brown-midrib maize (bm1)—a mutation affecting the cinnamyl alcohol dehydrogenase gene. *Plant J* 14: 545–553; 1998. doi:10.1046/j.1365-313X.1998.00153.x.

He X.; Hall M. B.; Gallo-Meagher M.; Smith R. L. Improvement of forage quality by downregulation of maize *O*-methyltransferase. *Crop Sci* 43: 2240–2251; 2003.

Iiyama K.; Lam T. B. T. Structural characteristics of cell walls of forage grasses: their nutritional evaluation for ruminants. *Asian-Austr J Anim Sci* 14: 862–879; 2001.

Ishida Y.; Saito H.; Ohta S.; Hiei Y.; Komari T.; Kumashiro T. High efficiency transformation of maize (Zea mays L.) mediated by *Agrobacterium tumefaciens*. *Nat Biotechnol* 14: 745–750; 1996. doi:10.1038/nbt0696-745.

Keating J. D.; Panganiban C.; Mansfield S. D. Tolerance and adaptation of ethanologenic yeasts to lignocellulosic inhibitory compounds. *Biotechnol Bioeng* 93: 1196–1206; 2006. doi:10.1002/bit.20838.

Koonin S. E. Getting serious about biofuels. *Science* 311: 435; 2006. doi:10.1126/science.1124886.

Kuc J.; Nelson O. E. The abnormal lignins produced by the brown midrib mutants of maize. I. The brown midrib mutant. *Arch Biochem Biophys* 105: 103; 1964. doi:10.1016/0003-9861(64)90240-1.

Li X.; Weng J. K.; Chapple C. Improvement of biomass through lignin modification. *Plant J* 54: 569–581; 2008. doi:10.1111/j.1365-313X.2008.03457.x.

Li Y.; Kajita S.; Kawai S.; Katayama Y.; Morohoshi N. Down-regulation of an anionic peroxidase in transgenic aspen and its effect on lignin characteristics. *J Plant Res* 116: 175–182; 2003. doi:10.1007/s10265-003-0087-5.

Lu R.; Martin-Hernandez A. M.; Peart J. R.; Malcuit I.; Baulcombe D. C. Virus-induced gene silencing in plants. *Methods* 30: 296–303; 2003. doi:10.1016/S1046-2023(03)00037-9.

Marita J. M.; Vermerris W.; Ralph J.; Hatfield R. D. Variations in the cell wall composition of maize brown midrib mutants. *J Agric Food Chem* 51: 1313–1321; 2003. doi:10.1021/jf0260592.

Miki D.; Itoh R.; Shimamoto K. RNA silencing of single and multiple members in a gene family of rice. *Plant Physiol* 138: 1903–1913; 2005. doi:10.1104/pp.105.063933.

Oraby H.; Venkatesh B.; Dale B.; Ahamd R.; Ransome C.; Oehmke J.; Sticklen M. B. Enhanced conversion of plant biomass into glucose using transgenic rice-produced endoglucanase for cellulosic ethanol. *Trans Res* 16: 739–749; 2007. doi:10.1007/s11248-006-9064-9.

Pillonel C.; Mulder M. M.; Boon J. J.; Forster B.; Binder A. Involvement of cinnamyl-alcohol dehydrogenase in the control of lignin formation in Sorghum bicolor L. *Moench Planta* 185: 538–544; 1991.

Piquemal J.; Chamayou S.; Nadaud I.; Beckert M.; Barrière Y.; Mila I.; Lapierre C.; Rigau J.; Puigdomenech P.; Jauneau A.; Digonnet C.; Boudet A. M.; Goffner D.; Pichon M. Downregulation of caffeic acid *O*-methyltransferase in maize revisited using a transgenic approach. *Plant Physiol* 130: 1675–1685; 2002. doi:10.1104/pp.012237.

Ragauskas A. J.; Williams C. K.; Davison B. H.; Britovsek G.; Cairney J.; Eckert C. A.; Frederick W. J.; Hallett J. P.; Leak D. J.; Liotta C. L.; Mielenz J. R.; Murphy R.; Templer R.; Tschaplinski T. The path forward for biofuels and biomaterials. *Science* 311: 484–489; 2006. doi:10.1126/science.1114736.

Ralph J.; Akiyama T.; Kim H.; Lu F.; Schatz P. F.; Marita J. M.; Ralph S. A.; Reddy M. S. S.; Chen F.; Dixon R. A. Effects of coumarate 3-hydroxylase down-regulation on lignin structure. *J Biol Chem* 281: 8843–8853; 2006. doi:10.1074/jbc.M511598200.

Ransom C.; Venkatesh B.; Dale B.; Biswas G.; Sticklen M. B. Heterologous *Acidothermus cellulolyticus* 1,4-β-endoglucanase E1 produced within the corn biomass converts corn stover into glucose. *Applied Biochem Biotech* 140: 137–219; 2007. doi:10.1007/s12010-007-9053-3.

Reddy M. S.; Chen F.; Shadle G.; Jackson L.; Aljoe H.; Dixon R. A. Targeted down-regulation of cytochrome P450 enzymes for forage quality improvement in alfalfa (*Medicago sativa* L.). *Proc. Natl Acad. Sci. USA* 102: 16573–16578; 2005. doi:10.1073/pnas.0505749102.

Rogers L. A.; Campbell M. M. The genetic control of lignin deposition during plant growth and development.. *New Phytologist* 164: 17–30; 2004. doi:10.1111/j.1469-8137.2004.01143.x.

Rook J. A.; Muller L. D.; Shank D. B. Intake and digestibility of brown-midrib corn silage by lactating dairy cows. *J Dairy Sci* 60: 1894–1904; 1977.

Rubin E. M. Genomics of cellulosic biofuels. *Nature* 454: 841–844; 2008. doi:10.1038/nature07190.

Sánchez O. J.; Cardona C. A. Trends in biotechnological production of fuel ethanol from different feedstocks. *Bioresour Technol* 99: 5270–5295; 2007. doi:10.1016/j.biortech.2007.11.013.

Schubert C. Can biofuels finally take center stage? *Nature Biotechnol* 24: 777–784; 2006. doi:10.1038/nbt0706-777.

Shadle G.; Chen F.; Reddy M. S. S.; Jackson L.; Nakashima J.; Dixon R. A. Down-regulation of hydroxycinnamoyl CoA:shikimate hydroxycinnamoyl transferase in transgenic alfalfa affects lignification, development and forage quality. *Phytochemistry* 68: 1521–1529; 2007. doi:10.1016/j.phytochem.2007.03.022.

Somleva M. N.; Tomaszewski Z.; Conger B. V. *Agrobacterium*-mediated genetic transformation of switchgrass.. *Crop Sci* 42: 2080–2087; 2002.

Stewart C. N. J. Biofuels and biocontainment. *Nature Biotechnol* 25: 283–284; 2007. doi:10.1038/nbt0307-283.

Sticklen M. B. Plant genetic engineering to improve biomass characterization for biofuels. *Curr Opin Biotechnol* 17: 315–319; 2006. doi:10.1016/j.copbio.2006.05.003.

Sticklen M. B. Plant genetic engineering for biofuel production: towards affordable cellulosic ethanol. *Nat Rev Genet* 9: 433–443; 2008. doi:10.1038/nrg2336.

Vignols F.; Rigau J.; Torres M. A.; Capellades M.; Puigdomenech P. The brown midrib3 (bm3) mutation in maize occurs in the gene encoding caffeic acid *O*-methyltransferase. *Plant Cell* 7: 407–416; 1995.

Wang Z. Y.; Ge Y. Recent advances in genetic transformation of forage and turf grasses. *In Vitro Cell Develop. Biol Plant* 42: 1–18; 2006. doi:10.1079/IVP2005726.

Weng J. K.; Li X.; Bonawitz N. D.; Chapple C. Emerging strategies of lignin engineering and degradation for cellulosic biofuel production. *Curr Opin Biotechnol* 19: 66–172; 2008. doi:10.1016/j.copbio.2008.02.014.

Wesley S. V.; Helliwell C. A.; Smith N. A.; Wang M. B.; Rouse D. T.; Liu Q.; Gooding P. S.; Singh S. P.; Abbott D.; Stoutjesdijk P. A.; Robinson S. P.; Gleave A. P.; Green A. G.; Waterhouse P. M. Construct design for efficient, effective and high-throughput gene silencing in plants. *Plant J* 27: 581–590; 2001. doi:10.1046/j.1365-313X.2001.01105.x.

Yuan J. S.; Tiller K. H.; Al-Ahmad H.; Stewart N. R.; Stewart N. C. Plants to power: bioenergy to fuel the future. *Trends Plant Sci* 13: 421–429; 2008. doi:10.1016/j.tplants.2008.06.001.

Ziegelhoffer T.; Will J.; Austin-Phillips S. Expression of bacterial cellulase gene in transgenic alfalfa (*Medicago sativa* L), potato (*Solanum tuberosum* L) and tobacco (*Nicotiana tobacum*). *Mol Breed* 5: 309–318; 1999. doi:10.1023/A:1009646830403.

Zhou J.; Lee C.; Zhong R.; Ye Z. H. MYB58 and MYB63 are transcriptional activators of the lignin biosynthetic pathway during secondary cell wall formation in Arabidopsis. *Plant Cell* 21: 248–266; 2009. doi:10.1105/tpc.108.063321.

Chapter 13
Commercial Cellulosic Ethanol: The Role of Plant-Expressed Enzymes

Manuel B. Sainz

Abstract The use and production of biofuels has risen dramatically in recent years Bioethanol comprises 85% of total global biofuels production, with benefits including reduction of greenhouse gas emissions and promotion of energy independence and rural economic development. Ethanol is primarily made from corn grain in the USA and sugarcane juice in Brazil. However, ethanol production using current technologies will ultimately be limited by land availability, government policy, and alternative uses for these agricultural products. Biomass feedstocks are an enormous and renewable source of fermentable sugars that could potentially provide a significant proportion of transport fuels globally. A major technical challenge in making cellulosic ethanol economically viable is the need to lower the costs of the enzymes needed to convert biomass to fermentable sugars. The expression of cellulases and hemicellulases in crop plants and their integration with existing ethanol production systems are key technologies under development that will significantly improve the process economics of cellulosic ethanol production.

Keywords Biofuels • Cellulosic ethanol • Cellulases • Pretreatment • Biomass • Enzymes

The Rise of Biofuels

The global production and use of biofuels has increased dramatically in recent years, from 18.2 billion liters in 2000 to 60.6 billion liters in 2007, with about 85% of this being bioethanol (Coyle 2007). Bioethanol production from first-generation

M.B. Sainz (✉)
Syngenta Centre for Sugarcane Biofuels Development, Centre for Tropical Crops and Biocommodities, Queensland University of Technology, Brisbane Queensland,
2 George Street, 4001, Australia
e-mail: manuel.sainz@syngenta.com

technologies is projected to increase to 113.6 billion liters by 2022 (Goldemberg and Guardabassi 2009). Several primary drivers underlie the increase in biofuels. One is the increasing uncertainty of petroleum supplies in the face of rising demand from emerging economies and the decline in known reserves. These reserves are primarily located in regions with governments that are unstable or unfriendly to Western democracies, making the long-term petroleum supply subject to political developments. A second factor is that uncertainty in petroleum supplies has led to government programs promoting biofuels and accomplishing two main policy goals: energy independence and support for rural economies. Significantly, the USA, Brazil, the European Union, and China together account for about 90% of global biofuels production, a direct result of government support in these countries (Coyle 2007). Thirdly, concerns over global warming and greenhouse gas emissions associated with fossil fuel usage have contributed to increasing interest in biofuels that reduce carbon emissions or are at least carbon-neutral. The use of bioethanol is estimated to reduce greenhouse gas emissions by approximately 30% to 85% compared to gasoline, depending on whether corn or sugarcane feedstock is used (Fulton et al. 2004). Finally, biofuels are unique among available alternative energy sources in their general compatibility with our existing liquid transport fuel infrastructure despite the potential for corrosion of existing pipelines by high concentrations of ethanol.

The USA and Brazil are currently the primary producers of fuel ethanol, producing 49.6% and 38.3% of 2007 global production, respectively (http://www.ethanolrfa.org/industry/statistics/#E, accessed 7 Jan 2009). US bioethanol production is almost entirely from maize (corn) starch, which is converted to fermentable glucose by the addition of amylase and glucoamylase enzymes. In 2007, 24.6 billion liters of ethanol was produced in the USA, yet this comprised only 3.2% of gasoline consumption on an energy-equivalent basis (Tyner 2008). Increases in agricultural commodity prices in 2007 and early 2008 were blamed by some on the increasing use of grain for biofuels. However, economic factors such as increasing demand in emerging economies and supply restrictions caused by weather, low carryover stocks, and years of low investment in agricultural R&D have been identified as the major contributors to increases in grain prices (Gressel 2008; Anonymous 2008). Further evidence for biofuel production's minor role in driving commodity prices is that despite continued expansion in corn ethanol production, commodity prices decreased in late 2008. Nonetheless, it is estimated that restrictions on available acreage and price pressures will limit the contribution of grain-based ethanol to the US liquid transport fuel mix to less than 8% of gasoline consumption on an energy-equivalent basis (Tyner 2008).

Sugarcane juice is the preferred feedstock in Brazil, accounting for about 80% of production, with the remainder being sugarcane molasses (Sanchez and Cardona 2008). Sugarcane bioethanol production is expected to increase in Brazil with the construction of new mills and associated plantings, from 19 billion liters in 2007 to 36 billion liters in 2013, and ethanol is expected to continue providing approximately 50% of Brazil's transport fuel needs (Goldemberg and Guardabassi 2009). It is predicted that sugarcane ethanol production in Brazil will ultimately reach

79.5 billion liters in 2022, and increases of 50–100% in global sugarcane tonnage are expected in the coming decade (Kline et al. 2008). However, sugarcane ethanol production using current technologies will eventually be limited by the same agro-economic factors that restrict grain-based ethanol production: the lack of suitable land and competing demand by alternative uses, in this case sugar production.

New technologies are required if biofuels are to significantly contribute to planetary energy needs and the reduction of greenhouse gas emissions. Despite meeting about one third of the increase in global oil demand in recent years (Lavelle and Garber 2008), in 2006, biofuels represented merely 0.8% of total global energy usage (Martinot et al. 2007), or only about 2% of our transport fuel (Koonin 2006). Fermentative production of ethanol and other alcohols from lignocellulosic materials represent the most attractive option for continued expansion of biofuel production. Some of the benefits include a high efficiency of carbohydrate recovery compared to other technologies, the possibilities for technological improvement afforded by biotechnology, and lower capital costs (Wyman et al. 2005a; Carroll and Somerville 2009).

This review addresses the role of plant-expressed enzymes and other developing technologies in enabling the commercial viability of cellulosic ethanol. The emphasis is on approaches and technologies with significant promise in reducing production costs to make cellulosic ethanol competitive with first-generation ethanol production and in identifying areas for further research to address the interrelated technical challenges of converting biomass to fermentable sugars.

Cellulosic Ethanol: Economic Aspects

The major economic barrier to viable commercial cellulosic ethanol production are high production costs, estimated to be between US$102 and 123 per barrel (Tyner 2008), or more than US$50 per gallon (US$0.66 per liter; Coyle 2007). Feedstock, enzymes, and processing costs, together with capital expenses associated with new plants, all combine to make cellulosic ethanol production using current technologies expensive in comparison to first-generation bioethanol. In addition, since there are no commercial-scale cellulosic ethanol production plants currently in operation, techno-economic process models rely on estimates, laboratory experiments, or, at best, pilot-scale plants (Galbe et al. 2007), increasing the investment risk profile.

Thus, it is not surprising that the first commercial cellulosic ethanol plants currently under construction take advantage of pre-collected feedstocks, proximity to existing infrastructure, and government funding. One such plant now under construction is termed Project Liberty (http://www.slideshare.net/rhapsodyingreen/project-liberty/, accessed 4 Jan 2009). It is being built by Poet, a major US corn ethanol producer, at a cost of US$200 million at the site of an existing corn ethanol plant in Emmetsburg, Iowa. The plant intends to use corn cobs and fiber for cellulosic ethanol production, with process economics improved by fractionation technologies to produce higher value co-products in addition to 94.6 million liters of cellulosic ethanol annually. The project benefits from a US$80 million grant from the US Department of Energy under the Energy Independence and Security Act of 2007, which calls for the production of 379 million liters per yr of cellulosic ethanol by 2010 (Greer 2008). More recently, Verenium announced that they intend to build the first commercial cellulosic ethanol plant in Florida (see http://phx.corporate-ir.net/phoenix. zhtml?c=81345&p=RssLanding&cat=news&id=1244987, accessed 27 Jan 2009).

The US government alone provided over US$1 billion in funding for cellulosic ethanol projects in 2007 (Waltz 2008). In addition to government funds in the USA and elsewhere, large oil and automobile companies are supporting cellulosic ethanol research by commercial and academic technology providers. Royal Dutch Shell has a 50% stake in Iogen (http://www.iogen.ca/), a company that has been running a large pilot-scale cellulosic ethanol plant in Ottawa, Canada and is planning a commercial cellulosic ethanol facility. British Petroleum has a joint venture with Verenium (http://www.verenium.com/), gaining access to their cellulosic ethanol technology, and is also providing US$500 million for biofuel research at the Energy Biosciences Institute (http://www.energybiosciencesinstitute.org/), an academic consortium including the University of California at Berkeley, the Lawrence Berkeley National Lab, and the University of Illinois. General Motors is also interested and has invested in two cellulosic ethanol technology companies, Mascoma (http://www.mascoma.com/) and Coskata (http://www.coskata.com/).

Other countries are also investing in cellulosic ethanol. China is the third largest ethanol producer globally, with 2.2 billion liters of production in 2007 (http://www.ethanolrfa.org/industry/statistics/#E, accessed 7 Jan 2008). China Alcohol Resources Corporation, the second largest ethanol producer in China, has a demonstration cellulosic ethanol plant under continuous operation, producing 6.4 million liters annually based on SunOpta (http://www.sunopta.com/bioprocess/index.aspx) technology. The Chinese government has committed to spending US$500 million on cellulosic ethanol research (Waltz 2008).

Cellulosic Ethanol: Challenges

The reason for increasing investment is that despite significant challenges, cellulosic ethanol has the potential to sustainably supply a significant proportion of our transport fuel needs. Importantly, cellulosic ethanol can significantly reduce greenhouse gas emissions compared to fossil fuels (Wang et al. 2007). Worldwide, biofuel production potential from agricultural crop residues alone are estimated at 30% of global gasoline consumption (Kim and Dale 2004; Koonin 2006). Research activity in cellulosic ethanol technology has accelerated as a result of increasing interest and funding. Below, the progress and challenges in key research areas are summarized (Table 13.1), with a focus on pretreatment/enzymatic hydrolysis as the most promising approach in the near term (Wyman et al. 2005a).

Feedstocks. Feedstock costs represent a major portion of first-generation ethanol production, ranging from 37% for sugarcane in Brazil to 40–50% for corn grain

13 Commercial Cellulosic Ethanol: The Role of Plant-Expressed Enzymes

Table 13.1 Major research areas, progress and challenges in cellulosic ethanol development

Area	Description	Progress	Challenges
Feedstocks	Use and modification of biomass sources: agricultural, forestry or municipal wastes, or dedicated energy crops	Initial analyses of feedstock yields and collection costs; compositional analyses; research into cell wall biosynthesis and chemistry	Reducing collection/feedstock costs; determination of desired feedstock characteristics; genetic modification of feedstocks to maximize value
Pretreatment	Mechanical and chemical treatments to facilitate conversion of lignocellulosic biomass to fermentable sugars	Evaluation of effectiveness of different pre-treatment processes on variety of feedstocks; characterization of inhibitors of downstream processes	Reducing capital expenses and input costs; reducing energy inputs; recycling/usage of waste streams; process integration
Enzymatic hydrolysis	Enzymatic conversion of cellulose and hemicellulose polymers to fermentable sugars	Reduction in cost of cellulase enzymes; understanding of *T. reesei* and *A. niger* cellulases	High enzyme costs; poor activity/long incubation times; optimized enzyme mixtures for specific feedstocks/processes
Fermentation	Conversion of fermentable sugars to ethanol or other fuels and bio-products	Characterization of C5/C6 sugar fermenting organisms; analysis of tolerance to inhibitors in fermentation	Organisms with rapid growth, improved tolerance to inhibitors and fermentation of multiple sugars under industrial conditions
Process engineering	Engineering designs to enable economic biomass processing at commercial scale	Process models constructed, tested and revised	Optimized process integration; incorporating best (sometimes proprietary) data into models

ethanol in the USA (Coyle 2007). The costs of feedstock for cellulosic ethanol production run from US$30 to US$90 per metric ton (Galbe et al. 2007), or about a third of production costs (Wyman 2007). A large proportion of feedstock costs (up to US$25 per ton) are attributable to the harvesting and transportation of bulky biomass feedstock (Rath 2007). Process economics thus favor pre-collected feedstock such as sugarcane bagasse over agricultural residues (corn stover and grain straws) and dedicated energy crops where delivered costs to a centralized processing facility need to be considered.

The major components of plant biomass (lignocellulose) are cellulose, hemicellulose, and lignin; in dicots, pectins are also important. Cellulose is a linear homopolymer of $\beta1,4$-linked cellobiose (glucose dimer) subunits. The primary chains are organized into hydrogen-bonded layers, then into compact 4- to 5-nm elementary fibrils of about 100 chains, and finally into 7- to 30-nm micrcofibrils consisting of cellulose elementary fibrils within a hemicellulose matrix and coated with lignin (Zhang and Lynd 2004). Cellulose exists primarily in crystalline form in the plant cell wall, interspersed with more disorganized (amorphous) regions, and is insoluble in water and commonly used solvents.

Plant hemicelluloses are heterogeneous branched polymers of pentose (C5) and hexose (C6) sugars whose composition varies according to species. For example, glucuronoarabinoxylans (GAXs) are the primary hemicelluloses in grasses and are composed of C5 sugars, including a $\beta1,4$-linked xylose backbone, together with arabinose and glucuronic acid. Another major hemicellulose component in grasses are the mixed linkage or β-glucans, polymers of glucose with both $\beta1,3$ and $\beta1,4$ linkages (Vogel 2008). While GAXs fill in the space between cellulose microfibrils and provide structural rigidity, the β-glucans tightly coat the microfibril (Carpita et al. 2003).

Lignin is a complex phenylpropanoid heteropolymer, a network of coumaryl, coniferyl, and sinapyl alcohols that acts as a glue to link and strengthen the polysaccharide components (Jorgensen et al. 2007). Lignin from grasses differs from that in dicots in containing significant quantities of ρ-hydroxyphenyl acid monomers in the structure and of ferulic acid attached to GAX that may serve as nucleation sites for lignin polymerization (Vogel 2008). Unlike cellulose and hemicellulose, the complex structure and diversity of chemical bonds in lignin make enzymatic deconstruction difficult (Weng et al. 2008).

Sugarcane bagasse and corn stover are attractive feedstocks due to their availability in proximity to existing bioethanol production plants. Sugarcane bagasse in particular is already present in substantial quantities at existing sugar and ethanol mills and has a lower ash content (5.0%) compared to other agricultural residues such as corn stover (11.6%; US Department of Energy 2009). The average composition of sugarcane bagasse is approximately 39% cellulose, 23% hemicellulose (with 89% of this being xylan), and 24% lignin. By comparison, corn stover has a lower average lignin content (19%) than bagasse. Corn stover is 35% cellulose and has about the same amount of hemicellulose as sugarcane bagasse. Corn stover and grain straw feedstocks for cellulosic ethanol production will incur collection and transport costs and thus have a higher cost basis than sugarcane bagasse.

The ideal feedstock composition for production of cellulosic biofuels and other bio-based products is currently uncertain. Optimal feedstock composition depends on the economics of the processing technologies used, the value of potential co-products, waste disposal costs, and the costs associated with engineering solutions to problematic constituents such as ash. Feedstock composition can be modified by genetic modification and classical breeding approaches. For example, considerable progress has been made in understanding the lignin biosynthesis pathway in diverse plant species (Li et al. 2008), and there is evidence that changes in lignin content and/or composition can facilitate hydrolysis of cellulose and hemicellulose to fermentable sugars. Although decreases in lignin can be accompanied by increases in cellulose content (Li et al. 2003; Chen and Dixon 2007), concerns remain about possible effects on structural strength (standability), resistance to diseases and insect pests, and biomass yield (Chapple et al. 2007; Weng et al. 2008). In addition, it may be that in future biorefineries, lignin-derived products such as chemical feedstocks would have greater value than biofuels derived from fermentable sugars. Nonetheless, lignin modification of biomass feedstocks to facilitate production of biofuels represents a promising avenue for ongoing cellulosic ethanol research (Dunn-Coleman et al. 2001; Houghton et al. 2006; Buanafina et al. 2008; Carroll and Somerville 2009).

A better understanding of cell wall synthesis and the complex interactions between its different components will likely provide information on how to genetically modify additional biomass components to reduce recalcitrance to hydrolysis (Himmel et al. 2007). For example, although a number of biosynthetic enzymes are known, it is not currently possible to genetically engineer major changes in plant cell walls, such as the amount or composition of hemicellulose, as can be done for lignin (McCann and Carpita 2008). In addition, depending on the feedstock, other biomass components may be important. For example, despite being the most abundant agricultural biomass feedstock globally (Kim and Dale 2004), rice straw has a high silica content that acts to inhibit enzymatic hydrolysis and may cause other problems in biomass processing (Gressel 2008).

Pretreatment. The goals of commercial pretreatment include improving the efficiency of subsequent enzymatic hydrolysis and fermentation, maximizing the recovery of fermentable sugars, and minimizing costs associated with energy and chemical inputs, inhibitor removal, and waste stream disposal (Galbe and Zacchi 2007; Jorgensen et al. 2007). Pretreatment is a critical step, consuming about 18% of production costs and impacting the efficiency of downstream processes as well (Yang and Wyman 2008). Capital costs are typically high for most pretreatment technologies due to the need for expensive corrosion-resistant materials or for specialized recovery and/or waste disposal systems (Eggeman and Elander 2005).

The resistance of biomass substrates to enzymatic hydrolysis is termed recalcitrance and is due to a number of factors. Lignin sterically hinders enzyme access to substrates by coating the cellulose microfibrils and in addition binds irreversibly to proteins, reducing enzymatic activity (Yang and Wyman 2006). Hemicellulose also hinders cellulose hydrolysis, and lignocellulosic modifications such as acetylation and the presence of other compounds such as ash can negatively affect enzymatic action (Himmel et al. 2007). Reduction of biomass particle size and an increase in the porosity of the material can help reduce recalcitrance to enzymatic hydrolysis by facilitating access to substrates by hydrolytic enzymes. Biomass recalcitrance can be reduced by chemical as well as physical pretreatment methods.

The mechanisms by which pretreatments alter biomass structure and composition to improve downstream enzymatic hydrolysis differ depending on the methodology employed. Among the possible mechanisms are improved substrate access by removal or modification of hemicellulose and lignin, by increasing porosity, and by changing the degree of polymerization or the crystallinity of the cellulose. Mechanical treatments such as hammer-milling to a fine particle size or extrusion of the biomass (Litzen et al. 2006) facilitate downstream chemical pretreatments and enzymatic hydrolysis, but are likely too energy- and cost-intensive to be commercially viable. Chemical pretreatments are often combined with high temperatures (typically 100–200°C) and pressures and sometimes with rapid explosive decompression. Diverse pretreatment methodologies have been researched using a variety of feedstocks, and each has advantages and disadvantages (Table 13.2), although all seem to obtain high sugar yields from corn stover feedstock (Wyman et al. 2005b).

Acids and alkalis are the two major pretreatment chemicals used. Acid-based pretreatments act primarily by hydrolyzing hemicellulose, which is efficiently converted first to oligomers then to monomeric pentose sugars with increasing temperature and pressure. Dilute acid hydrolysis (typically using sulfuric acid) is among the most extensively studied methods and is thought to be the closest to commercialization (Jorgensen et al. 2007). Hot water or steam pretreatments similarly rely on the generation of organic acids during the process to affect additional hydrolysis and thus can be thought of as mildly acidic auto-hydrolyses. There is a trade-off in reaction conditions, since acid-based pretreatments of increasing temperature and pressure lead to the loss of sugars and the generation of inhibitors of downstream processes, especially fermentation. These inhibitors include acetic acid and other organic acids; aldehyde lignin derivatives; and furfural and 5-hydroxymethylfurfural, furan degradation products of pentose and hexose sugars, respectively (Almeida et al. 2007). Further degradation of furans can generate formic acid, a cellulase inhibitor (Panagiotou and Olsson 2007). On the other hand, lower temperature acid-based pretreatments minimize the formation of inhibitors but can lead to poor enzymatic hydrolysis due to residual hemicellulose. Lignin is often partially melted and then redistributed under acid pretreatment conditions, improving access by cellulases, but complete removal improves subsequent enzymatic hydrolysis (Ohgren et al. 2007).

Alkali pretreatments with commercial potential include lime and ammonia fiber explosion (AFEX) technologies. In contrast to acid-based methodologies, alkali pretreatments remove lignin rather than the hemicellulose, and thus, inhibitor formation is minimized. AFEX technology has several promising features, including high solids loading, no separate liquid output stream, low temperature/energy input, and the potential for efficient ammonia recovery (Wyman et al. 2005a). However, hemicellulases as well as cellulases need to be included in the enzymatic hydrolysis

Table 13.2 Features of selected pretreatment technologies

Pretreatment	Description	Advantages	Issues
Dilute acid hydrolysis	Dilute (0.5–3%) H_2SO_4 at 130–200°C/3–15 atm pressure	Low operating costs; extensively researched; highly efficient hemicellulose hydrolysis; broadly applicable to different feedstocks	Low solids loading (~5%); pH neutralization required (cost, waste disposal); formation of inhibitors of downstream processes (washing required); loss of sugars; lignin binding by cellulases slows hydrolysis
Steam explosion	160–240°C/6–34 atm pressure	High (30%) solids loading possible; efficient hemicellulose hydrolysis; broadly applicable to different feedstocks	Low xylose recovery; generation of inhibitors of downstream processes, washing required; lignin binding by cellulases slows enzymatic hydrolysis
Ammonia fiber explosion (AFEX)	Anhydrous ammonia –NH_3/biomass 1:1 at 70–90°C/15–20 atm pressure, followed by rapid decompression	Very high (60%) solids loading possible; no liquid stream; no loss of fermentable sugars; no inhibitors formed; NH_3 is recoverable; residual ammonia is N source in fermentation	Safety hazards of dealing with ammonia; need for hemicellulases to complete conversion to C5 sugars; mixed C5/C6 sugar hydrolysate; suitable only for agricultural feedstocks (not wood)
Lime	0.05–0.15 g $Ca(OH)_2$ per gram biomass at either 70–130°C/1–6 atm pressure (1–2 h), or under ambient conditions (wk)	Removes acetyl groups and ~1/3 of lignin, improving enzymatic hydrolysis; recoverable by CO_2 addition + lime kiln	Need for hemicellulases to complete conversion to C5 sugars; mixed C5/C6 sugar hydrolysate; suitable only for agricultural feedstocks

step to obtain high fermentable sugar yields, and the presence of both C5 and C6 sugars in the resultant hydrolysate requires a microorganism capable of fermenting both types of sugars efficiently. Additional work needs to be done to validate alkali-based methodologies at pilot plant scale.

Pretreatment options need to be considered in the context of the overall process, including the nature of the output streams into enzymatic hydrolysis; the enzymes used in hydrolysis; the type of fermentation, including the microorganism(s) used; and waste stream handling (Wyman et al. 2005b). Two-stage pretreatments using different process conditions or methodologies may ultimately prove to be most effective in maximizing efficient recovery of fermentable sugars and other valuable by-products (Kim and Lee 2006).

Enzymes. Cellulases and hemicellulases belong to the large glycosyl hydrolase family of enzymes. Cellulases are one to two orders of magnitude less efficient than other polysaccharidases in this family, such as amylases (Zhang and Lynd 2004), primarily due to the difficulty of hydrolyzing solid crystalline cellulose as opposed to a soluble substrate. In addition to binding irreversibly to lignin, cellulases can bind unproductively to cellulose. While this effect can be mitigated by the addition of surfactants such as Tween20™, surfactants can act as inhibitors of downstream fermentation (Sun and Cheng 2002).

Complete cellulosic hydrolysis requires at least three key enzymes: an endoglucanase, an exoglucanase, and a β-glucosidase. Endo-1,4-β-D-glucanases (EG; EC3.2.1.4) carry out the first step by hydrolyzing internal β-1,4 glucosidic bonds in the cellulose polymer. This action frees up ends that are attacked by exo-1,4-β-D-glucanases or cellobiohydrolases (CBH; EC3.2.1.91) that processively move along the cellulose chain, cleaving cellobiose units. Cellobiohydrolases come in two forms, CBH1 and CBH2, which work from the reducing and non-reducing ends of the cellulose polymer, respectively. Finally, 1,4-β-D-glucosidases (BG; EC3.2.1.21) hydrolyze cellobiose to glucose to relieve product inhibition of the cellobiohydrolases and generate fermentable sugar.

Reduction in the costs of enzymes used in lignocellulosic hydrolysis is a key issue in commercial cellulosic ethanol production. The high cost of enzymatic hydrolysis is due to the poor activity of cellulases. Although production costs are similar, 40 to 100 times more enzyme is needed to digest cellulose as compared to starch, on a mass basis (Merino and Cherry 2007). In addition to requiring about 15–25 kg of enzymes per ton of biomass (Houghton et al. 2006; Taylor et al. 2008), long incubation times add to the capital cost of vessels for enzymatic hydrolysis. Despite the significant cost reductions that have been reported by researchers from the National Renewable Energy Laboratory (NREL), Novozymes and Genencor (http://www.nrel.gov/awards/2004hrvtd.html; Anonymous 2005), the cost of cellulases for biomass hydrolysis remains high. Enzymes needed for maize grain ethanol production cost US$2.64–5.28 per cubic meter (=1,000 l) of ethanol produced (Houghton et al. 2006), whereas cellulase enzymes for a commercial process are projected to cost about US$79.25 per cubic meter of ethanol (Lynd et al. 2008), or at least 20–40 times more, depending on whether enzymes are produced on-site or purchased from commercial suppliers (Somerville 2007;

Sanchez and Cardona 2008). Enzymes thus comprise an estimated 20–40% of cellulosic ethanol production costs.

Hemicellulases reflect the diversity of hemicellulose itself, but include endo-1,4-β-D-xylanases (EC3.2.1.8), which hydrolyze the internal bonds in the xylan chain, and 1,4-β-D-xylosidases (EC3.2.1.37), which attack the non-reducing end of the polymer, releasing xylose. GAX side chains are attacked by several enzymes, including α-L-arabinofuranosidases (EC3.2.1.55), α-glucuronidases (EC3.2.1.139), acetyl xylan esterases (EC3.1.1.72), and feruloyl and ρ-coumaric acid esterases (Jorgensen et al. 2007). Access to the GAX xylan backbone is facilitated by removal of ester linkages and the arabinose and glucuronic acid side groups.

A large number of fungi and bacteria produce cellulases and hemicellulases (Sun and Cheng 2002). The best-studied microbial cellulase producers are *Trichoderma* spp., particularly *Trichoderma reesei*. The composition of *T. reesei*-secreted cellulases varies according to substrate, but is typically about 60% CBH1, 20% CBH2, 12% EG2 (Zhang and Lynd 2004) and roughly 0.5% BG1 and BG2 (Merino and Cherry 2007). In addition to two cellobiohydrolases, five endoglucanases, and two β-glucosidases (with additional putative cellulases), *T. reesei* has four endoxylanases (Jorgensen et al. 2007). Despite difficulties in production, most commercial cellulases are produced by *T. reesei* fermentation (Taherzadeh and Karimi 2007).

Cellulases act synergistically with hemicellulases and other enzymes in breaking down plant cell wall material. The addition of hemicellulases and pectinases to commercial cellulase mixture can increase the yield of fermentable sugars, in part by boosting yields of cellulose hydrolysis (Berlin et al. 2007). The optimal combination of enzymes to affect hydrolysis depends on the nature of the substrate and the interactions between the individual enzymes. As an example, the activity of standard *T. reesei* cellulase mixtures can be significantly increased by supplementation with β-glucosidases or with heterologous glycosyl hydrolases of unknown function (Merino and Cherry 2007; Taherzadeh and Karimi 2007). Merino and Cherry (2007) describe experiments to determine the optimal ratio of arabinofuridases to xylosidase to hydrolyze an arabinoxylan substrate. Intriguingly, a number of anaerobic cellulolytic bacteria feature large extracellular enzyme complexes of cellulases, hemicellulases, and pectinases called cellulosomes. Cellulosomes are up to 5–7 mDa in size and act together to digest plant cell walls (Murashima et al. 2003; Doi 2008). The composition of cellulosomes is modular and changes according to the nature of the substrate encountered.

Hydrolysis and fermentation. The rate and extent of enzymatic hydrolysis is affected by the pretreatment method, substrate concentration and accessibility, enzyme activity (loading), and reaction conditions such as pH, temperature, and mixing (Merino and Cherry 2007; Taherzadeh and Karimi 2007). Different strategies for enzymatic hydrolysis and ethanolic fermentation have been developed to address specific process engineering issues (Table 13.3). These strategies require different levels of additional fermentation technology development. Separate hydrolysis and fermentation (SHF) effectively addresses the current significantly different temperature optima for enzymatic hydrolysis (45–50°C) and fermentation (30–35°C)

Table 13.3 Selected hydrolysis and fermentation strategies

Name	Description	Features
SHF: Separate hydrolysis and fermentation	Enzymatic hydrolysis and fermentation done sequentially in different vessels	Hydrolysis and fermentation at respective optimal conditions; enzyme product inhibition; separate treatment of C5 and C6 sugar streams
SSF: Simultaneous saccharification and fermentation	Enzymatic hydrolysis and fermentation done simultaneously in same vessel	Compromise in conditions for optimal hydrolysis and fermentation; improved rates and yields; separate treatment of C5 and C6 sugar streams
HHF: Hybrid hydrolysis and fermentation	Enzymatic hydrolysis and fermentation done roughly sequentially in same vessel	Hydrolysis continues after shift to fermentation conditions; process optimization difficult; separate treatment of C5 and C6 sugar streams
NSSF: Non-isothermal simultaneous saccharification and fermentation	Enzymatic hydrolysis and fermentation done roughly simultaneously in different vessels	Hydrolysis and fermentation at respective optimal conditions; process optimization difficult; separate treatment of C5 and C6 sugar streams
SSCF: Simultaneous saccharification and co-fermentation	Like SSF, only both C5 and C6 sugars are fermented in same vessel	Fewer vessels, lower capital costs; requires engineered microorganism optimized for efficient C5/C6 fermentation
CBP: Consolidated bioprocessing	Enzymatic hydrolysis and fermentation carried out in single vessel by single or combination of microorganisms	Fewer vessels, lower capital costs; requires engineered microorganism optimized for enzyme production and C5/C6 fermentation

by carrying out these reactions sequentially. Although SHF offers the advantage of cell culture recycling, increasing enzymatic product inhibition during the course of hydrolysis impacts productivity. Simultaneous saccharification and fermentation (SSF) addresses this issue by having the fermenting microorganisms consume the products of hydrolysis (cellobiose and glucose), lowering their concentration and thus their inhibitory effects on cellulase activity. However, this requires a compromise on temperature optima, or the use of less efficient thermotolerant fermentation organisms. Ethanol generated in SSF can help reduce the risk of culture contamination, and the use of a single vessel helps to reduce capital costs. SSF has been extensively studied at the laboratory and pilot plant scale.

Two variants of SSF, non-isothermal simultaneous saccharification and fermentation (NSSF) and simultaneous saccharification and co-fermentation (SSCF), are less well studied but promising (Taherzadeh and Karimi 2007). In NSSF, hydrolysis and fermentation occur in two vessels at their respective temperature optima, but simultaneously, the effluent from the hydrolysis reaction is shunted to the fermenter during the process. Despite higher capital costs, this technology has the potential to significantly reduce enzyme loadings and/or retention times. SSCF refers to enzymatic hydrolysis and co-fermentation of C5 and C6 sugars in a single vessel. The challenge here is identifying microorganisms that can carry out this fermentation efficiently, such as the engineered *Zymononas mobilis* strain used by the US NREL in their techno-economic process analysis (Aden et al. 2002, Aden 2008). One potential problem is that most commonly used fermentation organisms strongly prefer glucose as a carbon source, leading to inefficient or underutilization of xylose and other C5 sugars (Wyman et al. 2005b). Fermentation of xylose to ethanol, separately or by co-fermentation with C6 sugars, is crucial to cellulosic ethanol process economics (Merino and Cherry 2007).

Consolidated bioprocessing (CBP), also known as direct microbial conversion, is a technology in which one or more microorganisms carry out enzymatic hydrolysis and ethanolic fermentation simultaneously in a single vessel (Taherzadeh and Karimi 2007). Despite the potential for significant cost reductions, the technology faces challenging technical hurdles. CBP requires strains that can efficiently (in terms of yield and rates) convert glucose and C5 sugars to ethanol, effectively express multiple cellulases, exhibit robustness under industrial conditions, and for which the requisite genetic and metabolic pathway background knowledge exists (Chang 2007). No such strain has yet been reported even at lab scale, but progress is being made on CBP strain engineering (Lynd et al. 2005; Jorgensen et al. 2007).

Microorganisms are a key component of the technology used in different fermentation regimes. The potential of a number of different microorganisms has been studied, including yeasts such as the naturally pentose fermenting yeast *Pichia stipitis* (Agbogbo and Coward-Kelly 2008), mesophilic bacteria such as *Klebsiella oxytoca* (Ingram et al. 1999), and cellulolytic thermophilic anaerobes such as *Clostridium thermocellum* (Lu et al. 2006). *Saccharomyces cerevisiae* remains the gold standard in industrial ethanol production, attaining a production rate of 170 g ethanol per liter per hour on glucose under optimal laboratory conditions (Cheryan and Mehaia 1984). The ability of *S. cerevisiae* to tolerate low pH and rapidly produce ethanol helps to prevent contamination, and it has good tolerance to ethanol and other inhibitors, making it a strong candidate for further development and commercialization in cellulosic ethanol production (Almeida et al. 2007). An alternative fermentation microorganism is the mesophilic Gram-negative bacterium *Zymomonas mobilis*. Although not as hardy as industrial *S. cerevisiae* strains, the ethanol yield of *Z. mobilis* per unit of fermented glucose is 5–10% higher than that of *S. cerevisiae* due to its unique glucose metabolism (Lin and Tanaka 2006). Both species have been engineered to metabolize pentose sugars (Zhang et al. 1995; Ho et al. 1998). *Escherichia coli* has also been studied for use in lignocellulosic fermentation and has been engineered to produce and tolerate ethanol levels as high as 7.5 g/L from xylose or glucose under laboratory conditions (Yomano et al. 1998). However, ethanol concentrations produced from lignocellulosic hydrolysate fermentation are typically higher (10–35 g/L; Taherzadeh and Karimi 2007).

Expression of Enzymes in Crops

Efficient enzymatic hydrolysis of biomass substrates remains an economic and technical challenge in the development of cellulosic ethanol (Himmel et al. 2007; Wyman 2007). Reducing the costs of enzymes used in the process is crucial to favorable cellulosic ethanol process economics and commercialization (Stephanopoulos 2007). Given that loadings have been extensively optimized, improvements in enzyme performance or reduction in enzyme production costs will be required. Improved cellulases with higher specific activity, reduced allosteric inhibition, and improved tolerance to high temperatures and specific pH optima can be achieved using protein engineering methodologies. An example of a promising approach is the screening of cell-surface-tethered mutant enzyme libraries on solid lignocellulosic substrate for enhanced activity (Zhang et al. 2006). Production costs can be reduced by expressing cellulases and hemicellulases in crop plants (Sticklen 2008; Taylor et al. 2008), getting around the capital and operating costs associated with fermentation. Combining the two approaches with multiple improved enzymes expressed in crop plants would dramatically improve cellulosic ethanol process economics.

Plants provide a significantly lower cost alternative to fermentation for the production of industrial enzymes and can have additional processing benefits. Syngenta (www.syngenta.com) will soon be marketing the first commercial crop-produced enzyme product designed for corn grain ethanol production. Corn Event 3272 expresses an α-amylase gene with an improved temperature and pH profile. Amylase is used in corn grain ethanol production to convert starch to glucose for fermentation. The amylase in Event 3272 replaces commercial amylase enzymes produced by fermentation, with additional processing benefits that reduce production costs (Syngenta 2009). The product also features the environmental benefits of estimated 10% reductions in both processing water and greenhouse gas emissions. While these benefits are likely to vary depending on plant configuration, adoption of transgenic corn amylase technology has the potential to significantly improve the efficiency and environmental footprint of the US corn ethanol industry.

Significant advances in the expression of cellulases and hemicellulases in crop plants have occurred over the past 15 yr (Table 13.4). Some of the earliest work was the expression of mixed-linkage glucanase and endoglucanase in cultured barley cells, with the objective of reducing viscosity in beer brewing (Phillipson 1993; Aspegren et al. 1995). Expression of a chimeric, codon-optimized mixed linkage glucanase in barley demonstrated the ability to express the enzyme in grain for brewing and animal feed applications (Jensen et al. 1996).

In the mid-1990s, work done at Syngenta provided the first example of the expression of active cellulases in plants. Two EG and a CBH sourced from *Thermonospora fusca* (since renamed *Thermobifida fusca*) were expressed from both constitutive and inducible promoters and were targeted to either the cytoplasm, the vacuole (using an appropriate targeting sequence), or to the chloroplast using direct transformation of the organelle (Lebel et al. 2008). Nuclear transformants

Table 13.4 Expression of cellulases and hemicellulases in plants

Enzyme	Enzyme source	Host	Targeting	Reference	Comments
1,3-1,4-β-Glucanase	*Bacillus* spp.; chimeric	Barley aleurone protoplasts[z]	Apoplast; cytoplasm	Phillipson 1993	Thermostable mixed-linkage glucanase
EG1 endoglucanase	*Trichoderma reesei*	Barley cultured cells[y]	Apoplast	Aspegren et al. 1995	Secreted into culture medium; could not transform tobacco
XynZ xylanase	*Clostridium thermocellum*	*N. tabacum* cv Samsum NN	Apoplast	Herbers et al. 1995	Thermostable; normal plant growth
XYLD-A xylanase	*Ruminococcus flavefaciens*	*N. tabacum* cv Samsum NN	Apoplast	Herbers et al. 1996	*xynD*-encoded protein domain; xylanase
XYLD-C 1,3-1,4-β-glucanase	*Ruminococcus flavefaciens*	*N.tabacum* cv Samsum NN	Apoplast	Herbers et al. 1996	*xynD*-encoded protein domain; mixed-linkage glucanase
1,3-1,4-β-Glucanase	*Baccillus* spp.; chimeric	*H. vulgare* cv Golden Promise	Seed aleurone	Jensen et al. 1996	Thermostable codon-optimized mixed-linkage glucanase
EG1 endoglucanase	*Ruminococcus albus* F-40	*N. tabacum* BY2 cultured cells	Cytoplasm	Kawazu et al. 1996	Also has xylosidase activity; no effect on cell growth
E2 cellobiohydrolase	*Thermonospora fusca*	Tobacco, maize and wheat[x]	Various[w]	Lebel et al. 2008[v]	Provisional patent applications filed in 1996 and 1997
E1 endoglucanase	*Thermonospora fusca*	Tobacco, maize and wheat[x]	Various[w]	Lebel et al. 2008[v]	Provisional patent applications filed in 1996 and 1997
E5 endoglucanase	*Thermonospora fusca*	Tobacco, maize and wheat[3]	Various[w]	Lebelx et al. 2008[v]	Provisional patent applications filed in 1996 and 1997
XynC xylanase	*Neocallimastix patriciarum*	*B. napus*	Seed oil body	Liu et al. 1997	Oleosin fusion protein; enzyme works immobilized

(continued)

Table 13.4 (continued)

Enzyme	Enzyme source	Host	Targeting	Reference	Comments
XynB xylanase	Clostridium stercorarium	N. tabacum BY2 suspension cells		Sun et al. 1997	Intracellular and culture supernatant enzyme activity
CBH1 cellobiohydrolase	Trichoderma reesei	Tobacco[u]		Dai et al. 1999	Normal growth
EG1 endoglucanase	Ruminococcus albus F-40	N. tabacum cv Xanthi nc	Cytoplasm	Kawazu et al. 1999	Apoplast targeting deletion needed to get transformants
EG1 endoglucanase	Trichoderma reesei	Barley[t]	Seed	Nuutila et al. 1999	No enzymatic assay data; catalytic domain only
E2 endoglucanase	Thermonospora fusca	Alfalfa, tobacco and potato[s]		Ziegelhoffer et al. 1999	Thermostable; activity in dried leaf material
E3 cellobiohydrolase	Thermonospora fusca	Alfalfa, tobacco and potato[s]		Ziegelhoffer et al. 1999	Thermostable; activity in dried leaf material
E1 endoglucanase	Acidothermus cellulolyticus	Tobacco[u]	Chloroplast	Dai et al. 2000a	Normal growth/photosynthesis; truncated; dry leaf activity
E1 endoglucanase	Acidothermus cellulolyticus	Potato and tobacco[r]	Various[q]	Dai et al. 2000b	Cellulose binding domain (CBD) deletion retained activity
1,3-1,4-β-Glucanase	Bacillus spp.; chimeric	H. vulgare cv Golden Promise	Endosperm Vacuole	Horvath et al. 2000	Codon-optimized; targeted to protein bodies
XynA xylanase	Neocallimastix patriciarum	H. vulgare cv Golden Promise	Seed endosperm	Patel et al. 2000	Modified xylanase; stable activity in storage
E1 endoglucanase	Acidothermus cellulolyticus	Arabidopsis and BY2 cells[p]	Apoplast	Zeigler et al. 2000	Thermostable; high expression and activity in Arabidopsis
E1 endoglucanase	Acidothermus cellulolyticus	Potato and tobacco[o]	Various[n]	Hooker et al. 2001	Activity in dried leaf material
E1 and CD endoglucanases	Acidothermus cellulolyticus	N. tabacum cv W38	Various[m]	Ziegelhoffer et al. 2001	Higher activity from catalytic domain (CD) version

Enzyme	Source organism	Plant host	Localization	Reference	Notes
CelE endoglucanase	Clostridium thermocellum	N. tabacum cv Samsum NN	Cytoplasm; apoplast	Abdeev et al. 2003	Pleiotropic phenotypic alterations observed
E1 and CD endoglucanases	Acidothermus cellulolyticus	N. tabacum cv W38	Chloroplast	Jin et al. 2003	Spacer length impacts chloroplast import
XynA xylanase	Clostridium thermocellum	O. sativa cv Notohikari		Kimura et al. 2003	Activity in rice leaves and dessicated grain
Cel-Hyb1 endoglucanase	Chimeric[l]	H. vulgare cv Golden Promise	Seed endosperm	Xue et al. 2003	Codon-optimized; heritable; mixed linkage glucanase
CD endoglucanase	Acidothermus cellulolyticus	N. tabacum cv W38	Apoplast	Teymouri et al. 2004	AFEX pretreatment causes substantial loss of activity
E1 endoglucanase	Acidothermus cellulolyticus	Tobacco[a]	Various[k]	Dai et al. 2005	Apoplast best, vacuole worst targeting for E1 activity
CD endoglucanase	Acidothermus cellulolyticus	Z. mays Hi II	Apoplast	Biswas et al. 2006	
XYL2 xylanase	Trichoderma reesei	A. thaliana Columbia	Various[j]	Hyunjong et al. 2006	Higher expression and activity with dual targeting
E1 endoglucanase	Acidothermus cellulolyticus	Z. mays Hi II	Various[i]	Hood et al. 2007	Codon-optimized; high seed ER, vacuole activity; heritable
CBH1 cellobio-hydrolase	Trichoderma koningii	Z. mays Hi II	Various[i]	Hood et al. 2007	Codon-optimized; high seed ER, cell wall activity; heritable
CD endoglucanase	Acidothermus cellulolyticus	O. sativa cv Taipei 309	Apoplast	Oraby et al. 2007	Activity on pretreated corn stover and rice straw
CD endoglucanase	Acidothermus cellulolyticus	Zea mays Hi II	Apoplast	Ransom et al. 2007	Activity on pretreated corn stover
E1 endoglucanase	Acidothermus cellulolyticus	Duckweed[h]		Sun et al. 2007	Low expression of proteolytically derived CD

(continued)

Table 13.4 (continued)

Enzyme	Enzyme source	Host	Targeting	Reference	Comments
Cel6A endoglucanase	*Thermobifida fusca*	Tobacco[g]	Chloroplast	Yu et al. 2007a, b	Chloroplast transformation; homoplastomic events
Cel6B cellobio-hydrolase	*Thermobifida fusca*	Tobacco[g]	Chloroplast	Yu et al. 2007a, b	Chloroplast transformation; homoplastomic events
CD endoglucanase	*Acidothermus cellulolyticus*	*Z. mays* Hi II	Various[f]	Mei et al. 2009	

[z] *Hordeum vulgare* cv Himalaya
[y] *Hordeum vulgare* cv Pokko suspension cells
[x] *Nicotiana tabacum* cv Xanthi nc and nahG transgenic derivative
[w] Constitutive: cytoplasm, vacuole; inducible: cytoplasm, vacuole; chloroplast transformed/nuclear inducible
[v] First publication date 23 May 2002 as US2002/0062502 A1
[u] *Nicotiana.tabacum* cv Petit Havana SR1
[t] *Hordeum vulgare* cv Kymppi and cv Golden Promise
[s] *Mendicago sativa* RSY27; *Nicotiana tabacum* W38; *Solanum tuberosum* PI 203900
[r] *Nicotiana tabacum* cv Petit Havana SR1; *Solanum tuberosum* cv Desiree and cv FL1607
[q] Constitutive: apoplast; green tissue: chloroplast and vacuole
[p] *Arabidopsis thaliana* Columbia; *Nicotiana tabacum* Bright Yellow 2 (BY2) suspension cells
[o] *Nicotiana tabacum* cv Petit Havana SR1; *Solanum tuberosum* cv Desiree
[n] Constitutive: cytoplasm and apoplast; green tissue: chloroplast and vacuole
[m] Constitutive: apoplast, chloroplast and cytoplasm
[l] *Neocallimastix patriciarum* CelA (activity) + *Piromyces* sp. Cel6G (thermotolerance) by gene shuffling
[k] Constitutive: apoplast; chloroplast, vacuole and endoplasmic reticulum; green tissue: chloroplast
[j] Green tissue: cytoplasm, chloroplast, peroxisome, chloroplast + peroxisome
[i] Seed embryo-preferred: apoplast, endoplasmic reticulum and vacuole
[h] *Lemna minor* 8627
[g] *Nicotiana tabacum* cv Samsun and cv 22X-1 (cv K327 non-nicotine derivative)
[f] Green tissue: cytoplasm, endoplasmic reticulum and mitochondria

of tobacco, corn, and wheat were generated that exhibited chemically induced cellulase expression. Interestingly, tobacco transgenics with a T7 promoter driving EG expression in the chloroplast were crossed with nuclear transformants that had the chemically inducible tobacco PR-1a promoter driving the T7 RNA polymerase gene fused to a chloroplast targeting sequence. The doubly transgenic lines exhibited chemical induction of chloroplast-expressed EG. The Lebel et al. (2008) patent thus demonstrates cellulase expression in plants and highlights the use of targeting sequences and inducible promoters among the tools available to do so.

The stability of plant-expressed enzymes is important to their use in industrial processes. Seeds represent one attractive storage option for enzymes. Horvath et al. (2000) demonstrated expression of codon-optimized hemicellulases fused to a signal peptide mediating secretion. Additional research in barley (Patel et al. 2000; Xue et al. 2003) and in rice (Kimura et al. 2003) provided evidence that hemicellulases were stable in stored grain. Hydrolytic enzyme activity can also be stable upon drying and when frozen as a crude extract. Stable *Acidothermus cellulolyticus* E1 endoglucanase activity was demonstrated in dried leaf material from tobacco (Ziegelhoffer et al. 1999; Dai et al. 2000a; Teymouri et al. 2004) and alfalfa (Ziegelhoffer et al. 1999) and in frozen crude extracts (Sticklen 2006). Stability, extraction, and storage are important considerations for the effective use of plant-expressed enzymes in industrial processing. Directed research and economic analyses in this area could help determine the optimal crops and tissues to target for enzyme expression and whether this varies depending on the biomass feedstock in question.

A thermostable, apoplast-targeted xylanase was the first hemicellulase to be expressed in whole tobacco plants, with normal growth being observed (Herbers et al. 1995). The strategy of expressing thermostable cellulases and hemicellulases continues to be used today to get around the problem that enzyme activity at physiological temperatures could pose for the plant. The use of thermostable enzymes in cellulosic ethanol production is a promising approach for other reasons, including improved enzyme activity and stability, lower viscosity, ease in extraction, and enhanced flexibility in process configurations (Viikari et al. 2007; Taylor et al. 2008). However, it is currently unclear how thermostable cellulases and hemicellulases would be integrated into an overall biomass processing scheme, suggesting that additional research is needed.

Cellulases and hemicellulases can be expressed in plants as truncated fully active proteins. The *xynD* gene of a ruminant microorganism was successfully expressed as separate, apoplast-targeted xylanase and mixed-linkage glucanase domains in tobacco (Herbers et al. 1996). Subsequently, *A. cellulolyticus* endoglucanase E1 was found in truncated but active form in tobacco and in duckweed, consistent with proteolytic removal of the cellulose binding domain (CBD), leaving just the catalytic domain (CD; Dai et al. 2000a; Sun et al. 2007). Expression of *A. cellulolyticus* endoglucanase CD had equal or higher activity than the full-length protein when expressed in potato or tobacco (Dai et al. 2000b; Ziegelhoffer et al. 2001), and apoplast-targeted CD has also been successfully expressed in maize (Biswas et al. 2006; Ransom et al. 2007). Importantly, although the enzyme is not able to survive a mild AFEX pretreatment (Teymouri et al. 2004), rice- and maize-expressed

A. cellulolyticus CD has activity on pretreated corn stover and rice straw (Oraby et al. 2007; Ransom et al. 2007).

Most experiments on the expression of cellulases in plants have focused on endoglucanases, with a predominance of experiments with thermostable enzymes such as *A. cellulolyticus* E1 and the CD derivative of this enzyme. Kawazu et al. (1996) demonstrated expression of an EG from a ruminant microorganism in tobacco suspension cells, followed by expression in whole tobacco plants (Kawazu et al. 1999). *T. reesei* cellobiohydrolase (CBH1) expression in plants has been reported in tobacco (Dai et al. 1999) and in maize seed (Hood et al. 2007); in addition, expression of thermostable cellobiohydrolases have been reported in tobacco (Ziegelhoffer et al. 1999; Yu et al. 2007a), alfalfa, and potato (Ziegelhoffer et al. 1999). Given the need for relatively large amounts of CBH activity required to hydrolyze pretreated biomass cellulose, high-level expression of these enzymes in plants is needed.

The importance of subcellular targeting in the expression of cell wall hydrolyzing enzymes in plants has been confirmed in numerous reports. Early reports targeted cell wall hydrolases to the cytoplasm, vacuole, and/or the apoplast. More recently, experiments with transgenic plants bearing different constructs targeting the same protein to different subcellular compartments have been used to compare expression levels. Cellulases and hemicellulases have been successfully targeted to chloroplasts (Dai et al. 2000a, b, 2005; Hooker et al. 2001; Ziegelhoffer et al. 2001; Jin et al. 2003; Hyunjong et al. 2006), vacuoles (Dai et al. 2000b, 2005; Hooker et al. 2001; Lebel et al. 2008), peroxisomes (Hyunjong et al. 2006), mitochondria (Mei et al. 2009), and the endoplasmic reticulum (ER; Dai et al. 2005; Mei et al. 2009). In addition, direct chloroplast transformation can be used to obtain very high level expression of cellulases in homoplastomic tobacco (Yu et al. 2007a, b; Lebel et al. 2008).

In seeds, codon-optimized *A. cellulolyticus* E1 EG and *Trichoderma* spp. CBH1 were each targeted to either the apoplast, the vacuole, or the ER using an embryo-preferred promoter, and the expression and activity of the expressed enzymes was assayed (Hood et al. 2007). While the activity of both enzymes was high for the ER-targeted versions, they exhibited differences; EG had activity when targeted to the vacuole but not when targeted to the apoplast (cell wall), while the reverse was true for CBH1. These results suggest that expression optimization through targeting will likely depend on the enzyme class and individual characteristics of the protein in question. In other experiments, the highest xylanase activity was observed in plants with an enzyme with dual targeting to both the chloroplast and the peroxisome, relative to those targeted to either compartment individually, suggesting multiple targeting as a strategy to maximize expression (Hyunjong et al. 2006). Additional research on targeting multiple cellulases and hemicellulases to subcellular compartments would help resolve if there are limitations to this strategy, such as whether localization in certain compartments can interfere with enzyme accumulation in others.

High level enzyme expression in an agricultural setting is the primary challenge to delivering on the promise of inexpensive, plant-produced cellulases, and hemicellulases for cellulosic ethanol production. *A. cellulolyticus* EG has been expressed as high as 2% total soluble protein (TSP) in corn stover (Biswas et al. 2006; Mei et al. 2009), but it is estimated that 10% TSP is needed for complete hydrolysis (Sticklen 2008).

In contrast, both EG and CBH1 constituted about 16–18% TSP in maize seed in the best expressing lines (Hood et al. 2007), but these represented only around 0.05% of dry grain weight. A yield of 150 bu/A of 15.5% moisture corn is equivalent to 7.97 t/ha dry weight (Graham et al. 2007). Taking the expression levels reported by Hood et al. (2007), the enzyme yield would be 7.97 t×0.05%≈4 kg enzyme per hectare. Assuming a 1:1 grain/stover harvest index and that 30% of the corn stover (2.4 t/ha) could be sustainably harvested (Graham et al. 2007), and a conservative hydrolytic enzyme requirement of 15 kg per ton biomass (Houghton et al. 2006), a minimum estimated enzyme yield of ~36 kg/ha would enable the grain to provide the enzymes required to process the stover from the same hectarage.

The actual target yields for different plant-expressed enzymes will depend on the amount and proportions of different enzymes needed for efficient biomass hydrolysis, which in turn depends on the nature of the pretreated feedstock and the individual activities of the chosen enzymes, taken together with any losses incurred during extraction and storage prior to use. Much progress has been made in demonstrating the utility of subcellular targeting for high level expression of cell wall hydrolyzing enzymes in plants, but more remains to be done in plant gene expression tools, including the use of strong, tissue-specific, or inducible promoters; transcriptional, translational, and intronic enhancers; and codon optimization (Streatfield 2007). In addition, developing technologies such as monocot chloroplast transformation (Bock 2007), mini-chromosomes (Carlson et al. 2007; Yu et al. 2007b; see also Chromatin Inc. website, http://www.chromatininc.com/), and INPACT (Dale et al. 2001, 2004; see also Farmacule Bioindustries website, http://www.farmacule.com/) could potentially provide additional options for boosting plant-expressed enzyme production to target levels.

Conclusions

Cellulosic ethanol can make a substantial contribution to our future energy needs and is projected to be even more environmentally friendly than first-generation biofuels, with the potential to reduce greenhouse gas emissions by an estimated 85% compared to gasoline (Fulton et al. 2004). Enzymatic hydrolysis represents the most attractive approach for biomass to biofuels conversion in the near term, with the promise of continuing improvements through biotechnology (Wyman et al. 2005a). Analysis of cellulosic ethanol process economics identifies the conversion of lignocellulosic biomass into fermentable sugars as the key technical challenge in reducing the costs of cellulosic ethanol production (Lynd et al. 2008). Over the long term, the problem of biomass recalcitrance will be addressed by improving biomass feedstock yield and processing characteristics and developing improved organisms for fermentation. The new commercial-scale cellulosic ethanol plants coming online will almost certainly give rise to improved engineering designs, driving down investments that are currently about three times higher per unit of ethanol produced than an equivalent maize grain ethanol plant (Galbe et al. 2007). However, to make cellulosic ethanol economically viable, our immediate research priorities need to be to lower the cost and improve the effectiveness of cellulases and to develop pretreatment technologies compatible with an optimized, integrated process, especially downstream enzymatic hydrolysis and fermentation.

A major theme in the research and development of cellulosic ethanol is that the individual steps in turning biomass to biofuels are inextricably linked, and research improvements need to be undertaken in the context of an integrated process. Because so much of the technology is under development, many interdependent variables need to be considered before focusing on a set of research goals. Towards this end, the Syngenta Centre for Sugarcane Biofuels Develoment (SCSBD) was formed to develop technologies enabling economically viable production of cellulosic ethanol from sugarcane bagasse (Sainz and Dale 2009). The SCSBD is a public-private partnership between Syngenta and the Queensland University of Technology (QUT) in Brisbane, Australia, with support from the Queensland state government. The approach is to express improved cellulases at high levels and in a controlled fashion in transgenic sugarcane and to use these enzymes in optimized processes designed to be integrated into existing sugar mills. The SCSBD project team consists of QUT and Syngenta scientists and engineers working together on solutions along a process stream that includes bagasse pretreatment and plant-expressed enzyme production to enzymatic hydrolysis and fermentation. The project plans to make use of a sugarcane bagasse cellulosic ethanol pilot plant being built in Mackay, Queensland with support from the Australian federal government.

The SCSBD example is relevant because in the future, success in improving the cost basis of cellulosic ethanol production will likely require increasingly integrated technical solutions drawing from diverse disciplines, including agronomy, plant breeding, and microbiology, in addition to biotechnology, enzymology, and engineering. Process modeling will be important in guiding research by identifying the most promising areas for improvement in reducing costs to make cellulosic ethanol production economically viable.

Acknowledgments Thanks to my colleagues Nancy Nye and Sherrica Morris for help in accessing references and to Karen Bruce, Terry Stone, John Steffens, Tom Bregger, Larry Gasper, and Scott Betts for helpful comments on the manuscript. My apologies to those colleagues whose publications I have not been able to include due to space constraints. The Syngenta Centre for Sugarcane Biofuels Development is supported by Syngenta, the Queensland University of Technology, Farmacule Bioindustries and by a grant from the National and International Research Alliances Program of the Queensland State government.

References

Abdeev R. M.; Goldenkova I. V.; Musiychuk K. A.; Piruzian E. S. Expression of a thermostable bacterial cellulase in transgenic tobacco plants. *Russ. J. Genet.* 393: 376–382; 2003. doi: 10.1023/A:1023227802029.

Aden, A. Biochemical production of ethanol from corn stover: 2007 state of technology model. Technical report NREL/TP-510-43205, National Renewable Energy Laboratory, Golden, CO USA. http://www.nrel.gov/docs/fy08osti/43205.pdf (accessed 8 Jan 2009); 2008.

Aden, A.; Ruth, M.; Ibsen, K.; Jechura, J.; Neeves, K.; Sheehan, J.; Wallace, B.; Montague, L.; Slayton, A.; Lukas, J. Lignocellulosic biomass to ethanol process design and economics utilizing co-current dilute acid prehydrolysis and enzymatic hydrolysis for corn stover. Technical report NREL/TP-510-32438 (http://www.nrel.gov/docs/fy02osti/32438.pdf (accessed 8 Jan 2009); 2002.

Agbogbo F. K.; Coward-Kelly G. Cellulosic ethanol production using the naturally occurring xylose-fermenting yeast, *Pichia stipitis*. *Biotechnol. Lett.* 30: 1515–1524; 2008. doi:10.1007/s10529-008-9728-z.

Almeida J. R. M.; Modig T.; Petersson A.; Hahn-Hagerdal B.; Liden G.; Gorwa-Grauslund M. F. Increased tolerance and conversion of inhibitors in lignocellulosic hydrolysates by *Saccharomyces cerevisiae*. *J. Chem. Technol. Biotechnol.* 82: 340–349; 2007. doi:10.1002/jctb.1676.

Anonymous. Enzyme contract concludes successfully. Ethanol Producer June 2005. http://www.ethanolproducer.com/article.jsp?article_id = 201 (accessed 26 Jan 2009); 2005.

Anonymous. Food prices: fact versus fiction. Biofuels International 2(5). http://biofuels-news.com/content_item_details.php?item_id = 126 (accessed 4 Jan 2009); 2008.

Aspegren K.; Mannonen L.; Ritala A.; Puupponen-Pimia R.; Kurten U.; Salmenkallio-Marttila M.; Kauppinen V.; Teeri T. H. Secretion of a heat-stable fungal β-glucanase from transgenic, suspension-cultured barley cells. *Mol. Breed* 1: 91–99; 1995. doi:10.1007/BF01682092.

Berlin A.; Maximenko V.; Gilkes N.; Saddler J. Optimization of enzyme complexes for lignocellulosic hydrolysis. *Biotechnol. Bioeng.* 972: 287–296; 2007. doi:10.1002/bit.21238.

Biswas G. C. G.; Ransom C.; Sticklen M. Expression of biologically active *Acidothermus cellulolyticus* endoglucanase in transgenic maize plants. *Plant Sci.* 7: 617–623; 2006. doi:10.1016/j.plantsci.2006.06.004.

Bock R. Plastid biotechnology: prospects for herbicide and insect resistance, metabolic engineering and molecular farming. *Curr. Opin. Biotechnol.* 18: 100–106; 2007. doi:10.1016/j.copbio.2006.12.001.

Buanafina M. M. O.; Langdon T.; Hauck B.; Dalton S.; Morris P. Expression of a fungal ferulic acid esterase increases cell wall digestibility of tall fescue (*Festuca arundinacea*). *Plant Biotechnol. J.* 6: 1–17; 2008. doi:10.1111/j.1467-7652.2007.00317.x.

Carlson S. R.; Rudgers G. W.; Zieler H.; Mach J. M.; Luo S.; Grunden E.; Krol C.; Copenhaver G. P.; Preuss D. Meiotic transmission of an *in vitro*-assembled autonomous maize minichromosome. *PLoS Genet.* 310: 1965–1974; 2007. doi: 10.1371/journal.pgen.0030179.

Carpita N. C.; Defernez M.; Findlay K.; Wells B.; Shoue D. A.; Catchpole G.; Wilson R. H.; McCann M. C. Cell wall architecture of the elongating maize coleoptile. *Plant Physiol.* 1272: 551–565; 2003. doi:10.1104/pp.010146.

Carroll A.; Somerville C. Cellulosic biofuels. *Annu. Rev. Plant Biol.* 60: 165–182; 2009.

Chang M. C. Y. Harnessing energy from plant biomass. *Curr. Opin. Chem. Biol.* 11: 677–684; 2007. doi:10.1016/j.cbpa.2007.08.039.

Chapple C.; Ladisch M.; Meilan R. Loosening lignin's grip on biofuel production. *Nat. Biotechnol.* 257: 746–748; 2007. doi:10.1038/nbt0707-746.

Chen F.; Dixon R. A. Lignin modification improves fermentable sugar yields for biofuel production. *Nat. Biotechnol.* 257: 759–761; 2007. doi:10.1038/nbt1316.

Cheryan M.; Mehaia M. A. Ethanol production in a membrane recycle bioreactor. Conversion of glucose using *Saccharomyces cerevisiae*. *Process Biochem.* 19: 204–208; 1984.

Coyle, W. The future of biofuels: a global perspective. Amber Waves 5(5). Available online at http://www.ers.usda.gov/AmberWaves/November07/Features/Biofuels.htm (accessed 7 Jan 2009); 2007.

Dai Z.; Hooker B. S.; Anderson D. B.; Thomas S. R. Expression of *Acidothermus cellulolyticus* endoglucanase E1 in transgenic tobacco, biochemical characteristics and physiological effects. *Transgenic Res.* 9: 43–54; 2000a. doi: 10.1023/A:1008922404834.

Dai Z.; Hooker B. S.; Anderson D. B.; Thomas S. R. Improved plant-based production of E1 endoglucanase using potato: expression optimization and tissue targeting. *Mol. Breed.* 6: 277–285; 2000b. doi:10.1023/A:1009653011948.

Dai Z.; Hooker B. S.; Quesenberry R. D.; Gao J. Expression of *Trichoderma reesei* exo-cellobiohydrolase I in transgenic tobacco leaves and calli. *Appl. Biochem. Biotechnol.* 77–79: 689–699; 1999. doi:10.1385/ABAB:79:1-3:689.

Dai Z.; Hooker B. S.; Quesenberry R. D.; Thomas S. R. Optimization of *Acidothermus cellulolyticus* endoglucanase (E1) production in transgenic tobacco plants by transcriptional, post-transcription and post-translation modification. *Transgenic Res.* 14: 627–643; 2005. doi:10.1007/s11248-005-5695-5.

Cellulases of mesophilic microorganisms—cellulosome and noncellulosome producers. *Ann. N.Y. Acad. Sci.* 1125: 267–279; 2008. doi: 10.1196/annals.1419.002.

Dale, J. L.; Dugdale, B.; Hafner, G. J.; Hermann, S. R.; Becker, D. K. A construct capable of release in closed circular form from a larger nucleotide sequence permitting site specific expression and/or developmentally regulated expression of selected genetic sequences. International Patent Application WO 01/72996 A1; 2001.

Dale, J. L.; Dugdale, B.; Hafner, G. J.; Hermann, S. R.; Becker, D. K.; Harding, R. M.; Chowpongpang, S. Construct capable of release in closed circular form from a larger nucleotide sequence permitting site specific expression and/or developmentally regulated expression of selected genetic sequences. US Patent Application US2004/0121430 A1; 2004.

Dunn-Coleman, N.; Landgon, T.; Morris, P. Manipulation of the phenolic acid content and digestibility of plant cell walls by targeted expression of genes encoding cell wall degrading enzymes. US patent application US2003/0024009 A1; 2001.

Eggeman T.; Elander R. T. Process and economic analysis of pretreatment technologies. *Bioresour. Technol.* 96: 2019–2025; 2005. doi:10.1016/j.biortech.2005.01.017.

Fulton, L.; Howes, T.; Hardy, J. Biofuels for transport—an international perspective. International Energy Agency, Paris, France. Available online at http://www.iea.org/textbase/nppdf/free/2004/biofuels2004.pdf (accessed 23 Jan 2009); 2004.

Galbe M.; Sassner P.; Wingren A.; Zacchi G. Process engineering economics of bioethanol production. *Adv. Biochem. Eng. Biotechnol.* 108: 303–327; 2007. doi: 10.1007/10_2007_063.

Galbe M.; Zacchi G. Pretreatment of lignocellulosic materials for efficient bioethanol production. *Adv. Biochem. Eng. Biotechnol.* 108: 41–65; 2007. doi:10.1007/10_2007_070.

Goldemberg J.; Guardabasi P. Are biofuels a feasible option? *Energy Policy* 37: 10–14; 2009. doi:10.1016/j.enpol.2008.08.031.

Graham R. L.; Nelson R.; Sheehan J.; Perlack R. D.; Wright L. L. Current and potential U.S. corn stover supplies. *Agron. J.* 99: 1–11; 2007. doi:10.1016/S0065-2113(04)92001-9.

Greer, D. Commercializing Cellulosic Ethanol. Biocycle 49(11):47. Available online at http://www.jgpress.com/archives/_free/001764.html (accessed 4 Jan 2009); 2008.

Gressel J. Transgenics are imperative for biofuel crops. *Plant Sci.* 174: 246–263; 2008. doi:10.1016/j.plantsci.2007.11.009.

Herbers K.; Flint H. J.; Sonnewald U. Apoplastic expression of the xylanase and β (1-3,1-4) glucanase domains of the *xyn* D gene from *Ruminococcus flavefaciens* leads to functional polypeptides in transgenic tobacco plants. *Mol. Breed.* 2: 81–87; 1996. doi: 10.1007/BF00171354.

Herbers K.; Wilke I.; Sonnewald U. A thermostable xylanase from *Clostidium thermocellum* expressed at high levels in the apoplast of transgenic tobacco has no detrimental effects and is easily purified. *Nat. Biotechnol.* 13: 63–66; 1995. doi:10.1038/nbt0195-63.

Himmel M. E.; Ding S.-Y.; Johnson D. K.; Adney W. S.; Nimlos M. R.; Brady J. W.; Foust T. D. Biomass recalcitrance: engineering plants and enzymes for biofuels production. *Science* 315: 804–807; 2007. doi:10.1126/science.1137016.

Ho N. W. Y.; Chen Z.; Brainard A. P. Genetically engineered *Saccharomyces* yeast capable of effective cofermentation of glucose and xylose. *Appl. Environ. Microbiol.* 645: 1852–1859; 1998.

Hood E. E.; Love R.; Lane J.; Bray J.; Clough R.; Pappu K.; Drees C.; Hood K. R.; Yoon S.; Ahmad A.; Howard J. A. Subcellular targeting is a key condition for high level accumulation of cellulase protein in transgenic maize seed. *Plant Biotechnol. J.* 5: 709–719; 2007. doi:10.1111/j.1467-7652.2007.00275.x.

Hooker B. S.; Dai Z.; Anderson D. B.; Quesenberry R. D.; Ruth M. F.; Thomas S. R. Production of microbial cellulases in transgenic crop plants. In: HimmelM. E.; BakerJ.O.; SaddlerJ. N. (eds) Glycosyl hydrolases for biomass conversion. American Chemical Society, Washington, DC, pp 55–90; 2001.

Horvath H.; Huang J.; Wong O.; Kohl E.; Okita T.; Kannagara C. G.; von Wettstein D. The production of recombinant proteins in transgenic barley grains. *Proc. Natl. Acad. Sci. U. S. A.* 974: 1914–1919; 2000. doi:10.1073/pnas.030527497.

Houghton, J.; Weatherwax, S.; Ferrell, J. Breaking the biological barriers to cellulosic ethanol: a joint research agenda. US Department of Energy Office of Science and Office of Energy Efficiency and Renewable Energy 206 pp; 2006.

Hyunjong B.; Lee D.-S.; Hwang I. Dual targeting of a xylanase to chloroplasts and peroxisomes as a means to increase protein accumulation in plant cells. *J. Exp. Bot.* 571: 161–169; 2006. doi:10.1093/jxb/erj019.

Ingram L. O.; Aldrich H. C.; Borges A. C. C.; Causey T. B.; Martinez A.; Morales F.; Saleh A.; Underwood S. A.; Yomano L. P.; York S. W.; Zaldivar J.; Zhou S. Enteric bacterial catalysts for fuel ethanol production. *Biotechnol. Prog.* 15: 855–866; 1999. doi:10.1021/bp9901062.

Jensen L. G.; Olsen O.; Kops O.; Wolf N.; Thomsen K. K.; von Wettstein D. Transgenic barley expressing a protein-engineered, thermostable (1-3,1-4)-β-glucanase during germination. *Proc. Natl. Acad. Sci. U. S. A.* 93: 3487–3491; 1996. doi:10.1073/pnas.93.8.3487.

Jin R.; Richter S.; Zhong R.; Lamppa G. K. Expression and import of an active cellulase from a thermophilic bacterium into the chloroplast both *in vitro* and *in vivo*. *Plant Mol. Biol.* 51: 493–507; 2003. doi:10.1023/A:1022354124741.

Jorgensen H.; Kristensen J. B.; Felby C. Enzymatic conversion of lignocellulose into fermentable sugars: challenges and opportunities. *Biofuels Bioprod. Biorefin.* 1: 119–134; 2007. doi:10.1002/bbb.4.

Kawazu T.; Ohta T.; Ito K.; Shibata M.; Kimura T.; Saaka K.; Ohmiya K. Expression of a *Ruminococcus albus* cellulase gene in tobacco suspension cells. *J. Ferment. Bioeng.* 823: 205–209; 1996. doi:10.1016/0922-338X(96)88809-X.

Kawazu T.; Sun J.-L.; Shibata M.; Kimura T.; Saaka K.; Ohmiya K. Expression of a bacterial endoglucanase gene in tobacco increases digestibility of its cell wall fibers. *J. Biosci. Bioeng.* 884: 421–425; 1999. doi:10.1016/S1389-1723(99)80220-5.

Kim S.; Dale B. E. Global potential biofuel production from wasted crops and crop residues. *Biomass Bioenerg.* 26: 361–375; 2004. doi:10.1016/j.biombioe.2003.08.002.

Kim T. H.; Lee Y. Y. Fractionation of corn stover by hot-water and aqueous ammonia treatment. *Bioresour. Technol.* 97: 224–232; 2006. doi:10.1016/j.biortech.2005.02.040.

Kimura T.; Mizutani T.; Tanaka T.; Koyama T.; Sakka K.; Ohmiya K. Molecular breeding of transgenic rice expressing a xylanase domain of the *xynA* gene from *Clostridium thermocellum*. *Appl. Microbiol. Biotechnol.* 62: 374–379; 2003. doi:10.1007/s00253-003-1301-z.

Kline, K. L.; Oladosu, G. A.; Wolfe, A. K.; Perlack, R. D.; Dale, V. H.; McMahon, M. Biofuel feedstock assessment for selected countries. Report ORNL/TM-2007/224 prepared by Oak Ridge National Laboratory for the US Department of Energy (DOE). Available online at http://www.osti.gov/bridge/purl.cover.jsp;jsessionid=FD7D1D5D71C9ACFE53C44AE856766653?purl=/924080-y8ATDg/ (accessed 4 Jan 2009); 2008.

Koonin S. E. Getting serious about biofuels. *Science* 311: 435; 2006. doi:10.1126/science.1124886.

Lavelle, M.; Garber, K. Fixing the food crisis. US News and World Report, May 19 2008, pp 36–42; 2008.

Lebel E. G.; Heifetz P. B.; Ward E. R.; Uknes S. J. Transgenic plant expressing a cellulase. US Patent 7361806 B2; 2008.

Li L.; Zhou Y.; Cheng X.; Sun J.; Marita J. M.; Ralph J.; Chiang V. L. Combinatorial modification of multiple lignin traits in trees through multigene cotransformation. *Proc. Natl. Acad. Sci. U. S. A.* 1008: 4939–4944; 2003. doi:10.1073/pnas.0831166100.

Li X.; Weng J.-K.; Chapple C. Improvement of biomass through lignin modification. *Plant J.* 54: 569–581; 2008. doi:10.1111/j.1365-313X.2008.03457.x.

Lin Y.; Tanaka S. Ethanol fermentation from biomass resources: current state and prospects. *Appl. Microbiol. Biotechnol.* 69: 627–642; 2006. doi:10.1007/s00253-005-0229-x.

Litzen, D.; Dixon, D.; Gilcrease, P.; Winter, R. Pretreatment of biomass for ethanol production. US Patent Application US2006/0141584 A1; 2006.

Liu J.-H.; Selinger L. B.; Cheng K.-J.; Beauchemin K. A.; Moloney M. M. Plant seed oil-bodies as an immobilization matrix for recombinant xylanase from the rumen fungus *Neocallimastix patriciarum*. *Mol. Breed.* 3: 463–470; 1997. doi:10.1023/A:1009604119618.

Lu Y.; Zhang Y.-H. P.; Lynd L. R. Enzyme-microbe synergy during cellulose hydrolysis by *Clostridium thermocellum*. *Proc. Natl. Acad. Sci. U. S. A.* 10344: 16165–16169; 2006. doi:10.1073/pnas.0605381103.

Lynd L. R.; Laser M. S.; Bransby D.; Dale B. E.; Davison B.; Hamilton R.; Himmel M.; Keller M.; McMillan J. D.; Sheehan J.; Wyman C. E. How biotech can transform biofuels. *Nat. Biotechnol.* 262: 169–172; 2008. doi: 10.1038/nbt0208-169.

Lynd L. R.; van Zyl W. H.; McBride J. E.; Laser M. Consolidated bioprocessing of cellulosic biomass: an update. *Curr. Opin. Biotechnol.* 16: 577–583; 2005. doi:10.1016/j.copbio.2005.08.009.

Martinot, E. et al. Renewables 2007: World Status Report. Renewable energy policy network for the 21st century (REN21) and Worldwatch Institute, Washington, DC, USA. Available online at http://www.worldwatch.org/files/pdf/renewables2007.pdf (accessed 15 Jan 2009); 2007.

McCann M. C.; Carpita N. C. Designing the deconstruction of plant cell walls. *Curr. Opin. Plant Biol.* 11: 314–320; 2008. doi:10.1016/j.pbi.2008.04.001.

Mei C.; Park S. H.; Sabzikar R.; Qi C.; Ransom C.; Sticklen M. Green tissue-specific production of a microbial endo-cellulase in maize (*Zea mays* L.) endoplasmic-reticulum and mitochondria converts cellulose to fermentable sugars. *J. Chem. Technol. Biotechnol* 84: 689–695; 2009.

Merino S. T.; Cherry J. Progress and challenges in enzyme development for biomass utilization. *Adv. Biochem. Eng. Biotechnol.* 108: 95–120; 2007. doi:10.1007/10_2007_066.

Murashima K.; Kosugi A.; Doi R. H. Synergistic effects of cellulosmal xylanase and cellulases from *Clostridium cellulovorans* on plant cell wall degradation. *J. Bacteriol.* 1855: 1518–1524; 2003. doi:10.1128/JB.185.5.1518-1524.2003.

Nuutila A. M.; Ritala A.; Skadsen R. W.; Mannonen L.; Kauppinen V. Expression of fungal thermotolerant endo-1,4-β-endoglucanase in transgenic barley seeds during germination. *Plant Mol. Biol.* 41: 777–783; 1999. doi: 10.1023/A:1006318206471.

Ohgren K.; Bura R.; Saddler J.; Zacchi G. Effect of hemicellulose and lignin removal on enzymatic hydrolysis of steam pretreated corn stover. *Bioresour. Technol.* 98: 2503–2520; 2007. doi:10.1016/j.biortech.2006.09.003.

Oraby H.; Venkatesh B.; Dale B.; Ahmad R.; Ransom C.; Oehmke J.; Sticklen M. Enhanced conversion of plant biomass into glucose using transgenic rice-produced endoglucanase for cellulosic ethanol. *Transgen. Res.* 16: 739–749; 2007. doi:10.1007/s11248-006-9064-9.

Panagiotou G.; Olsson L. Effect of compounds released during pretreatment of wheat straw on microbial growth and enzymatic hydrolysis rates. *Biotechnol. Bioeng.* 962: 250–258; 2007. doi:10.1002/bit.21100.

Patel M.; Johnson J. S.; Brettell R. L. S.; Jacobsen J.; Xue G. -P. Transgenic barley expressing a fungal xylanase gene in the endosperm of developing grains. *Mol. Breed.* 6: 113–123; 2000. doi:10.1023/A:1009640427515.

Phillipson B. A. Expression of a hybrid (1-3,1-4)-β-glucanase in barley protoplasts. *Plant Sci.* 91: 195–206; 1993. doi:10.1016/0168-9452(93)90142-M.

Ransom C.; Balan V.; Biswas G.; Dale B.; Crockett E.; Sticklen M. Heterologous *Acidothermus cellulolyticus* 1,4-β-endoglucanase E1 produced within corn biomass converts corn stover into glucose. *Appl. Biochem. Biotechnol.* 136–140: 207–219; 2007. doi:10.1007/s12010-007-9053-3.

Rath, A. Focus on raw materials could reap big rewards. Bioenergy Business July/August 2007, pp 16–18; 2007.

Sainz, M. B.; Dale, J. Towards cellulosic ethanol from sugarcane bagasse. Proceedings of the Australian Society of Sugar Cane Technologists 31: 18–23; 2009.

Sanchez O. J.; Cardona C. A. Trends in biotechnological production of fuel ethanol from different feedstocks. *Bioresour. Technol.* 99: 5270–5295; 2008. doi:10.1016/j.biortech.2007.11.013.

Somerville C. Biofuels. *Curr. Biol.* 174: R115–R119; 2007. doi:10.1016/j.cub.2007.01.010.

Stephanopoulos G. Challenges in engineering microbes for biofuels production. *Science* 315: 801–804; 2007. doi:10.1126/science.1139612.

Sticklen M. Plant genetic engineering to improve biomass characteristics for biofuels. *Curr. Opin. Biotechnol.* 17: 315–319; 2006. doi:10.1016/j.copbio.2006.05.003.

Sticklen M. Plant genetic engineering for biofuel production: towards affordable cellulosic ethanol. *Nat. Rev. Genet.* 9: 433–443; 2008. doi:10.1038/nrg2336.

Streatfield S. J. Approaches to achieve high-level heterologous protein production in plants. *Plant Biotechnol. J.* 5: 2–15; 2007. doi:10.1111/j.1467-7652.2006.00216.x.

Sun J.; Kawazu T.; Kimura T.; Karita S.; Sakka K.; Ohmiya K. High expression of the xylanase B gene from *Clostridium stercorarium* in tobacco cells. *J. Ferment. Bioeng.* 843: 219–223; 1997. doi: 10.1016/S0922-338X(97)82057-0.

Sun Y.;Cheng J. Hydrolysis of lignocellulosic materials for ethanol production: a review. *Bioresour. Technol.* 83: 1–11; 2002. doi:10.1016/S0960-8524(01)00212-7.

Sun Y.; Cheng J. J.; Himmel M. E.; Skory C. D.; Adney W. S.; Thomas S. R.; Tisserat B.; Nishimura Y.; Yamamoto Y. T. Expression and characterization of *Acidothermus cellulolyticus* E1 endoglucanase in transgenic duckweed *Lemna minor* 8627. *Bioresour. Technol.* 98: 2866–2872; 2007. doi:10.1016/j.biortech.2006.09.055.

Syngenta. The potential of corn amylase for helping to meet US energy needs. http://www.syngenta.com/en/corporate_responsibility/pdf/Fact%20Sheet%20The%20Potental%20of%20CA%20for%20Helping%20Meet%20US%20Energy%20Needs.pdf (accessed 7 February 2009); 2009.

Taherzadeh M. J.; Karimi K. Enzyme-based hydrolysis processes for ethanol from lignocellulosic materials: a review. *BioResources* 24: 707–738; 2007.

Taylor L. E.; Dai Z.; Decker S. R.; Brunecky R.; Adney W. S.; Ding S.-Y.; Himmel M. E. Heterologous expression of glycosyl hydrolases in planta: a new departure for biofuels. *Trends Biotechnol.* 268: 413–424; 2008. doi:10.1016/j.tibtech.2008.05.002.

Teymouri F.; Alizadeh H.; Laureano-Perez L.; Dale B.; Sticklen M. Effects of ammonia fiber explosion treatment on activity of endoglucanase from *Acidothermus cellulolyticus* in transgenic plant. *Appl. Biochem. Biotechnol.* 113–116: 1183–1191; 2004. doi:10.1385/ABAB:116:1-3:1183.

Tyner W. The US ethanol and biofuels boom: its origins, current status and future prospects. *Bioscience* 587: 646–653; 2008. doi:10.1641/B580718.

US Department of Energy. Biomass program: biomass feedstock composition and property database. Available online at http://www1.eere.energy.gov/biomass/feedstock_databases.html (accessed 5 February 2009); 2009.

Viikari L.; Alapuranen M.; Puranen T.; Vehmaanpera J.; Siika-aho M. Thermostable enzymes in lignocellulose hydrolysis. *Adv. Biochem. Eng. Biotechnol.* 108: 121–145; 2007. doi:10.1007/10_2007_065.

Vogel J. Unique aspects of the grass cell wall. *Curr. Opin. Plant Biol.* 11: 301–307; 2008. doi:10.1016/j.pbi.2008.03.002.

Waltz E. Cellulosic ethanol booms despite unproven business models. *Nat. Biotechnol.* 261: 8–9; 2008. doi:10.1038/nbt0108-8.

Wang M.; Wu M.; Huo H. Life-cycle energy and greenhouse gas emission impacts of different corn ethanol plant types. *Environ. Res. Lett.* 2: 024001; 2007(13 pp).

Weng J. K.; Li X.; Bonawitz N. D.; Chapple C. Emerging strategies of lignin engineering and degradation for cellulosic biofuel production. *Curr. Opin. Biotechnol.* 19: 166–172; 2008. doi:10.1016/j.copbio.2008.02.014.

Wyman C. E. What is (and is not) vital in advancing cellulosic ethanol. *Trends Biotechnol.* 254: 153–157; 2007. doi:10.1016/j.tibtech.2007.02.009.

Wyman C. E.; Dale B. E.; Elander R. T.; Holtzapple M.; Ladisch M. R.; Lee Y. Y. Coordinated development of leading biomass pretreatment technologies. *Bioresour. Technol.* 96: 1959–1966; 2005a. doi:10.1016/j.biortech.2005.01.010.

Wyman C. E.; Dale B. E.; Elander R. T.; Holtzapple M.; Ladisch M. R.; Lee Y. Y. Comparative sugar recovery data from laboratory scale application of leading pretreatment technologies to corn stover. *Bioresour. Technol.* 96: 2026–2032; 2005b. doi: 10.1016/j.biortech.2005.01.018.

Xue G. P.; Patel M.; Johnson J. S.; Smyth D. J.; Vickers C. E. Selectable marker-free transgenic barley producing a high level of cellulase (1,4-β-glucanase) in developing grains. *Plant Cell. Rep.* 21: 1088–1094; 2003. doi:10.1007/s00299-003-0627-4.

Yang B.; Wyman C. E. BSA Treatment to enhance enzymatic hydrolysis of cellulose in lignin containing substrates. *Biotechnol. Bioeng.* 944: 611–617; 2006. doi: 10.1002/bit.20750.

Yang B.; Wyman C. E. Pretreatment: the key to unlocking low-cost cellulosic ethanol. *Biofuels Bioproc. Biorefin.* 2: 26–40; 2008. doi:10.1002/bbb.49.

Yomano L. P.; York S. W.; Ingram L. O. Isolation and characterization of ethanol-tolerant mutants of *Escherichia coli* KO11 for fuel ethanol production. *J. Ind. Microbiol. Biotechnol.* 20: 132–138; 1998. doi:10.1038/sj.jim.2900496.

Yu L.-X.; Gray B. N.; Rutzke C. J.; Walker L. P.; Wilson D. B.; Hanson M. R. Expression of thermostable microbial cellulases in the chloroplasts of nicotine-free tobacco. *J. Biotechnol.* 131: 362–369; 2007a. doi:10.1016/j.jbiotec.2007.07.942.

Yu W.; Han F.; Gao Z.; Vega J. M.; Birchler J. A. Construction and behavior of engineered minichromosomes in maize. *Proc. Natl. Acad. Sci. U. S. A.* 10421: 8924–8929; 2007b. doi:10.1073/pnas.0700932104.

Zeigler M. T.; Thomas S. R.; Danna K. J. Accumulation of a thermostable endo-1,4-β-D-glucanase in the apoplast of *Arabidopsis thaliana* leaves. *Mol. Breed.* 6: 37–46; 2000. doi:10.1023/A:1009667524690.

Zhang M.; Eddy C.; Deanda K.; Finkelstein M.; Picataggio S. Metabolic engineering of a pentose metabolism pathway in ethanologenic *Zymomonas mobilis*. *Science* 267: 240–243; 1995. doi:10.1126/science.267.5195.240.

Zhang Y.-H. P.; Himmel H. E.; Mielenz J. R. Outlook for cellulase improvement: screening and selection strategies. *Biotechnol. Adv.* 24: 452–481; 2006. doi:10.1016/j.biotechadv.2005.10.002.

Zhang Y.-H. P.; Lynd L. R. Toward an aggregated understanding of enzymatic hydrolysis of cellulose: non-complexed cellulase systems. *Biotechnol. Bioeng.* 887: 797–824; 2004. doi:10.1002/bit.20282.

Ziegelhoffer T.; Raasch J. A.; Austin-Phillips S. Dramatic effects of truncation and sub-cellular targeting on the accumulation of recombinant microbial cellulase in tobacco. *Mol. Breed.* 8: 147–158; 2001. doi:10.1023/A:1013338312948.

Ziegelhoffer T.; Will J.; Austin-Phillips S. Expression of bacterial cellulase genes in transgenic alfalfa (*Mendicago sativa* L.), potato (*Solanum tuberosum* L.) and tobacco (*Nicotiana tabacum* L.). *Mol. Breed.* 5: 309–318; 1999. doi:10.1023/A:1009646830403.

Chapter 14
Integrated Biorefineries with Engineered Microbes and High-value Co-products for Profitable Biofuels Production

William Gibbons and Stephen Hughes

Abstract Corn-based fuel ethanol production processes provide several advantages which could be synergistically applied to overcome limitations of biofuel processes based on lignocellulose. These include resources such as equipment, manpower, nutrients, water, and heat. The fact that several demonstration-scale biomass ethanol processes are using corn as a platform supports this viewpoint. This report summarizes the advantages of first-generation corn-based biofuel processes and then describes the technologies, advantages, and limitations of second-generation lignocellulose-based biofuel systems. This is followed by a discussion of the potential benefit of fully integrating first- and second-generation processes. We conclude with an overview of the technology improvements that are needed to enhance the profitability of biofuel production through development of an integrated biorefinery. A key requirement is creation of industrially robust, multifunctional ethanologens that are engineered for maximum ethanol production from mixed sugars. In addition to ethanol, combined biorefineries could also be the source of valuable co-products, such as chemicals and plastics. However, this will require expression systems that produce high-value co-products. Advantages of this approach are that (1) such strains could be used for bioconversion in any part of the combined biorefinery and (2) using one recombinant organism with many additions should simplify the process of obtaining necessary FDA approval for feed products produced by or containing recombinant organisms.

Keywords First-generation biofuels • Second-generation biofuels • Biorefinery • Recombinant ethanologens

W. Gibbons (✉)
Biology/Microbiology Department, South Dakota State University, DM 201, Brookings, SD 57007, USA
e-mail: william.gibbons@sdstate.edu

Introduction

The corn-based fuel ethanol industry in the USA has shown remarkable growth since its beginnings in the late 1970s. Between 1980 and 2000, annual production soared from 175 million gallons to 1.63 billion gallons (Renewable Fuels Association 2009). Production of this first-generation biofuel in 2008 exceeded nine billion gallons and now represents about 2–3% of the transportation fuel usage in the USA (Hill et al. 2006; Westcott 2007). The USA produces about 50% of the global fuel ethanol supply, while Brazil accounts for almost 40% of the supply using sugarcane as a feedstock (FAPRI 2008, FAPRI 2009). Production of fuel ethanol and the feed co-product distillers' dried grains with solubles used approximately 14% of the US corn crop in 2006 (Westcott 2007). It is estimated that if all the remaining corn were converted to ethanol, the total would replace only 12% of gasoline usage (Hill et al. 2006).

The Energy Independence and Security Act (EISA 2007), signed into law in the USA in December 2007, mandates minimum levels of domestic use of classes of biofuels, with a target of 36 billion gallons of renewable fuel use by 2022. In addition, the Renewable Fuel Standards (RFS; Energy Policy Act 2005) require increasing amounts of conventional (first generation), advanced (second generation), and biodiesel fuel production to achieve the target of 36 billion gallons. By 2017, no more than 15 billion gallons of corn-based (conventional) ethanol can count toward the RFS amount of 24 billion gallons, and by 2022, approximately 21 of the target of 36 billion gallons must be derived from second-generation biofuels (FAPRI 2008).

To achieve the ethanol production goals established by the EISA and RFS, it will be necessary to add second-generation biofuels to the US ethanol industry portfolio. Second-generation biofuels include ethanol produced from lignocellulosic materials, such as agricultural and forestry residues, solid waste, and dedicated perennial woody and herbaceous energy crops. While these feedstocks are abundant, their composition and characteristics present significant technical and economic challenges to the scientists developing ethanol production processes (Sims et al. 2008). In some cases, these challenges could be mitigated by developing integrated biorefinery operations that process both first- and second-generation feedstocks. Several companies have recognized the benefits of integrated biorefineries and are using corn-based ethanol facilities as test platforms to develop their second generation technologies to process biomass (http://www.projectliberty.com).

This chapter summarizes first-generation biofuel production, in particular corn-based processes, along with the advantages that this corn-based industry provides. Then, a description of the technologies, advantages, and limitations of second-generation biofuels is presented, followed by a discussion of the potential synergies of fully integrating first- and second-generation processes. In conclusion, the technology improvements that are necessary to enhance the profitability of biofuel production are described.

First-Generation Biofuel Production from Corn

The current corn-based ethanol industry arose out of the oil embargoes of the 1970s. Since that time, several improvements have been made in the design, engineering, and operation of fuel ethanol production facilities. This has resulted in increased efficiencies and reduced costs (Wheals et al. 1999). For example, yields have increased from 2.5–2.6 gallons ethanol/bushel corn to 2.7–2.8 gallons/bushel (Bothast and Schlicher 2005), while the energy consumed to produce a gallon of ethanol and dry co-products has fallen from 67,768 to 45,802 Btu (Shapouri et al. 1995; 2004). Net water use per gallon of ethanol produced has fallen from 5.8 to 4.2 gallons (Phillips et al. 2007). The most recently designed facilities produce 2.8 gallons of ethanol per bushel of corn, consuming 20,000 Btu for ethanol production, 12,000 Btu for co-product drying, and 3.5 gallons water/gallon ethanol. The objective of the Department of Energy Biomass Program (Perlack et al. 2005) is to incorporate new technologies that use residues and intermediates from existing processes to increase yields of biomass-derived liquid fuels, primarily ethanol. This near-term strategy will lay the technical foundation and build expertise for the use of lignocellulosic feedstocks.

Corn ethanol production in the USA is carried out by one of three methods: dry grind, dry mill, or wet mill. These methods largely differ in how the corn grain is initially processed. In the dry grind process, the kernels are ground into a relatively fine powder, which is then mixed with water and alpha-amylase enzymes for liquefaction. This process converts the starch into shorter chain glucose polymers called dextrins during a 15- to 60-min holding time at temperatures of 90°C to 100°C. The dry milling process subjects the kernels to a series of roller mills followed by screening stages to separate the germ, fiber, and endosperm. This reduces the load of non-fermentables in the conversion process and recovers food-grade corn oil (Bothast and Schlicher 2005). The endosperm is then similarly mixed with water and liquefied as above. The wet-milling process involves soaking and swelling the kernels in a dilute acid solution prior to a series of milling, density separation, and washing steps to separate the fiber, germ, protein, and starch. After separation, the starch slurry is then liquefied. The wet milling process is more complex and capital-intensive than dry grind/mill, but produces a range of higher value products including gluten feed, gluten meal, high fructose corn syrup, corn steep liquor, corn oil, and germ.

For all processes, the step following liquefaction is called saccharification. This process uses glucoamylase enzymes to convert the dextrins to fermentable sugars. Some facilities carry out a separate saccharification step at 50°C to 60°C for 1–2 h. However, most facilities conduct a simultaneous saccharification and fermentation (SSF). Here, the yeast inoculum (*Saccharomyces cerevisiae*) and glucoamylase enzyme are added at the same time, and the resulting mash is held at 28°C to 32°C for 48–72 h. The primary advantages of SSF are reduced contamination risks and reduced capital expenses for separate saccharification tanks. Yeast is the most desirable organism for ethanol production from corn because of its high ethanol productivity and tolerance and its robustness for industrial processing conditions. Large-scale industrial fermentation processes typically result in ethanol concentrations of 12% to 15% on a volume basis. After fermentation, the "beer" is distilled. The distillation residue, called whole stillage, is then centrifuged to recover yeast and any corn solids, which are used as livestock feed. A portion of the liquid from centrifugation (thin stillage) is recycled to replace 10% to 40% of the water needed at the start of the process. Thin stillage also recycles nutrients from yeast lysed during distillation and cooking back to the fermentation step.

To improve economic performance, the corn-based ethanol industry is pursing several strategies. Many dry grind operations are installing distillers' grains fractionation technologies to recover oil from distillers' grains (Parkin et al. 2007). This oil could be used for food or biodiesel production, while the low-oil distillers' grains would have higher value as cattle feed. Technologies to recover fiber from distillers' grains will provide an additional source of fermentable carbohydrates, while the reduced fiber content will provide a higher protein feed more suitable for poultry or swine rations. Other dry grind facilities are retrofitting with dry mill equipment to provide higher value fractionated co-products. In some cases, the non-starch fractions are still used as livestock feed, but in others, they are used for human foods or food-grade zein polymers. Still other facilities are adding fermentation systems to produce additional fermentation products from thin stillage or are exploring technologies to capture value from carbon dioxide generated during fermentation (Kheshgi and Prince 2005). Improved enzymes and yeast strains are also being developed.

As second-generation technologies are developed and made cost-effective, corn-based ethanol production will still remain a significant contributor to the US ethanol supply due to increased corn yields. Corn yield increases have been attributed over the years to improved hybrids and also to biotech traits (Troyer 2006; Edgerton 2009). Insect-resistant maize has a documented 5% yield advantage (vs control) in the USA and Canada and as high as 15% and 24% in South Africa and the Philippines, respectively (Barfoot and Brookes 2008). Parallel with this is the clear pesticide reduction with insect resistant traits and increase in "no-till" farming associated with herbicide-tolerant traits leading to reduced diesel fuel necessary for farming, increased carbon sequestration, and increased income (higher yield and lower input costs) for the producer (Barfoot and Brookes 2008). Corn yields will continue to rise due to breeding and biotechnology advances such as water use efficiency (Nelson et al. 2007; Castiglioni et al. 2008) and impending insect control (Baum et al. 2007). Through these advances, along with improved management practices such as no-till farming, it is quite likely that corn-based ethanol will continue to similarly expand (Edgerton 2009).

Technological advancements are also occurring within existing ethanol facilities. In fact, these existing facilities, because of their equipment, manpower, nutrient, water, and heat resources, provide a valuable platform upon which to develop additional processes. For example, Poet is using an existing dry mill corn facility in Emmettsburg, Iowa as the base upon which to add their corn cob-to-ethanol process (http://www.projectliberty.com). Other cellulose demonstration facilities are following this example. One could envision that corn ethanol facilities could also add upstream systems to handle sugar-based crops and downstream systems to use or sequester carbon dioxide from fermentation. The advantages of the current corn-based ethanol industry include:

1. Infrastructure for corn production and logistics are well-established.
2. Projected increases in corn yields with reduced inputs (such as fertilizer and tillage) should keep feedstock price reasonable.
3. Process produces fuel and also high-value protein feed and corn oil (for food or biodiesel);
4. Conversion process improvements have reduced energy and water use; costs of inputs have been minimized.
5. Corn mash is a nutrient-rich medium with few inhibitory components.
6. Saccharification and fermentation are typically carried out simultaneously.
7. Low enzyme cost permits adding excess enzyme to counter the less than optimal temperature of SSF.
8. Due to the density of corn starch, it is possible to achieve high ethanol concentrations in fermented beer.
9. Glucose is the sole sugar.
10. Traditional submerged fermentation is possible because of the relatively low viscosity.
11. Yeast are very robust to industrial conditions and can tolerate both high sugar and ethanol concentrations.

Technologies and Limitations of Second-Generation Biofuels

Second-generation biofuel feedstocks are commonly accepted to include only lignocellulosic resources such as trees, forestry and crop residues, grasses, and municipal solid waste. Sugar-containing crops, with the exception of sugarcane in Brazil, have been largely overlooked. However, these feedstocks offer an interesting opportunity, since they contain both simple sugars and lignocellulose. In general, both sugar and lignocellulosic feedstocks present several challenges with regard to biofuel production:

1. Nutrient content: These feedstocks typically contain little protein or other nutrients (besides carbohydrates) that fermentation microbes will require; hence, nutrient supplementation may be required.
2. Harvesting/logistics: Some systems have already been developed for sugar crops (beets, cane) and some lignocellulosic feedstocks (wood, grasses/residues). However, improvements are needed to reduce harvest and transportation costs.
3. Storage: Sugar crops are especially difficult to store without spoilage, but this can also be a problem with lignocellulosic feedstocks.
4. Pretreatment: For sugar crops, pretreatment is not needed, unless the lignocellulose component is also targeted. Lignocellulose pretreatments are often sufficiently intensive to generate inhibitory compounds.

5. Saccharification: For sugar crops, saccharification is not needed. However, to access sugars in lignocellulose, current deconstructing enzymes are more expensive and less effective than corresponding enzymes used with starch. It will also be necessary to conduct simultaneous saccharification and fermentation with lignocellulose due to the feedback inhibition of the enzymes.
6. Sugar type: Sucrose in sugar crops is readily fermentable; however, lignocellulosic feedstocks contain a mixture of five- and six-carbon sugars, not all of which can be fermented simultaneously by wild-type microbes.
7. Current fermentation microbes: Currently used ethanologens lack tolerance to inhibitors generated during lignocellulose pretreatment and cannot ferment mixed sugars. These microbes also cannot operate at the 50–60°C optimal temperature of saccharification enzymes.
8. Engineered fermentation microbes: Engineered microbes that can ferment mixed sugars may require regulatory approval, especially if fermentation residue is used in animal feeds. Such microbes will also need sufficient robustness to industrial conditions, high sugar and ethanol concentrations, and other variables.
9. Sugar concentration/fermentation: For both sugar and biomass crops, the lower bulk density limits the sugar concentration that can be achieved in a liquid medium if the solids are left in the broth. Thus, traditional submerged fermentation is less practical due to viscosity issues. Either non-fermentable solids must be removed and sugars concentrated or alternative solid state conversion systems must be developed.

Sugar-based feedstocks have been used for fuel ethanol production for as long as corn, with sugarcane in Brazil being the leading contributor (FAPRI 2008). Year-round harvesting, use of stalks and leaves (bagasse) to generate process steam and electricity, and use of liquid effluent (vinasse) as a fertilizer and irrigation supply to the cane fields are among its advantages (Kojima and Johnson 2005). According to Goldemberg (2007), sugarcane yields in Brazil are ~560 gallons/acre, with a total production of 5.2 billion gallons in 2007 (FAPRI 2008). Other crops with excellent yield potential include sugar or fodder beets (400–600 gallons/acre; Shapouri et al. 2006) and sweet sorghum (average of 20 gallons/ton of stalks; Gnansounou et al. 2005). Besides yield potential, the other main advantage of sugar crops is that they are directly fermentable, without the need for pretreatment or saccharification. The chief limitations of sugar crops include agronomic inputs, limited geographical adaptation, seasonal availability, and low bulk density and sugar concentration.

Compared to most forms of lignocellulosic biomass, annually seeded sugar (and starch) crops both require greater inputs for fertilizer, fuel, water, machinery, labor, and other agronomic inputs. In the USA, sugarcane production would be limited to the southern tier of states and, even then, would only be seasonally available. Sugar beets, fodder beets, and sweet sorghum can be grown throughout a broader area where suitable rainfall occurs; however, these crops would also have a short harvest window.

Seasonal availability is one of the biggest drawbacks of sugar crops. While sugar beets are stored in ventilated and covered piles from October through April in North Dakota and Minnesota, sugar loss is a significant problem during the spring. Conventional ensilage storage of sugarcane or sweet sorghum is not a workable solution due to the very high conversion of sugars to organic acids, which would subsequently inhibit yeast fermentation. The other option is immediate sugar recovery (via pressing or extraction), followed by lime precipitation and filtration to remove non-fermentable solids. The solution is then evaporated to concentrate the sugar for storage. Unfortunately, this large capital investment for sugar recovery and concentration would be idle for 6–9 months of the year. An alternative would be to use the sugar crop on a seasonal basis, using low-cost storage that would maintain acceptable feedstock quality. The facility would then need to operate the remainder of the year with other more storable starch or cellulosic feedstocks.

A final disadvantage of sugar-based crops is their low bulk density and sugar concentration. These crops contain a significant proportion of water and fiber which make both storage and transportation problematic. Moreover, it is not practical to simply grind these feedstocks and blend them in water to conduct submerged fermentation. So much water would be necessary to achieve a flowable solution that final ethanol titers would be unacceptably low (Gibbons et al. 1984, 1986). The alternative of extracting sugars or pressing out juice would remove non-fermentable solids and permit use of traditional fermentors. However, as pointed out previously, the capital and energy expense of these alternatives could be cost-prohibitive. An alternative could be solid-state fermentation in which the moist solid substrate is directly inoculated with yeast in a reactor designed to handle solids.

Lignocellulosic biomass includes a diverse range of potential feedstocks. The USDOE/USDA billion-ton study (Perlack et al. 2005) estimated that the annual availability in the USA is 368 million tons of forest biomass and 998 million tons of agricultural lignocellulose: a total of 1,366 million tons. Forest biomass is widely available in the eastern US, along with the Rocky Mountain States and Northwest coast. Woody biomass is dense, highly storable, and can be harvested year-round. Crop residues are available in the Midwest; however, only a portion of these residues can be removed to prevent soil erosion and fertility problems (Wilts et al. 2004). Other potential drawbacks of crop residues include a short harvest window (when farmers are more concerned with grain harvesting), low bulk density, and storage stability. Perennial grasses offer the advantages of high yield per acre and low production costs but, like crop residues, have low bulk density and storage issues. The lignocellulosic fraction of municipal solid waste may be a very desirable biofuel feedstock because it often has a negative value, has already been subjected to some degree of delignification, and is available year-round at centralized locations.

Conversion of sugar or lignocellulosic feedstocks into biofuels also presents several challenges. Issues regarding feedstock supply and uniformity are especially critical for both types of feedstocks, as is the ability to process low bulk density biomass. The recalcitrance of lignocellulose to hydrolysis adds an additional level of complexity (Vermerris et al. 2007). Effective pretreatments are needed to liberate the cellulose and hemicellulose fibers from the protective lignin sheath, exposing them for enzymatic saccharification into monosaccharides (Saha 2003). A variety of pretreatment technologies are in development (Varga et al. 2003). Acidic methods generally depolymerize the hemicellulose and make the cellulose accessible to enzymatic treatment. Alkaline methods generally depolymerize the lignin, making the hemicellulose and cellulose accessible to enzymatic treatment. For saccharification, efficient and inexpensive cellulase and hemicellulase enzymes are needed to generate fermentable sugars for microbes that can simultaneously ferment the resulting five- and six-carbon sugars. While dramatic improvements have been made in cellulose-degrading organisms that have lowered the cost of enzymes considerably in laboratory trials (Dien et al. 2003; Wilkins et al. 2008), significant work is still required to engineer a microorganism for commercial use that will rapidly produce and tolerate high ethanol concentrations at industrial temperature conditions and will use both five- and six-carbon sugars. Ethanol recovery procedures must also be improved since lower concentrations of ethanol are expected to be obtained from these low bulk density feedstocks.

Synergies of First- and Second-Generation Processes

To achieve national mandates for biofuel production, a broad range of biomass sources will be required as feedstocks. A study conducted by De la Torre Ungarte et al. (2007) evaluated the ability of the agricultural sector to meet the goal of supplying 10% of the US fuel demand with biofuels by 2020. Their model predicted that by 2011, approximately 30.5 million dry tons of corn stover, 32.5 million dry tons of switchgrass, and 2.32 billion bushels of corn would produce 11.24 billion gallons of ethanol. Their analysis ends at 2014, at which time nearly half the projected ethanol demand is met through the use of biomass feedstocks, producing 16.73 billion gallons of ethanol.

Several groups have noted that a key to sustainable, cost-effective bioethanol is process integration (Cardona and Sanchez 2007). As demonstrated by the petroleum industry, integration offers the advantages of reduced energy costs, smaller and fewer process units, and combined biological and downstream steps (Cardona and Sanchez 2007). This integration must also include low-cost biomass from all sources, new processing methods to obtain high yields of liquid fuels, and production of other higher value products. These biorefineries will use engineered microbes, integrate fermentation processes to use waste streams, improve production efficiency, recycle water and energy, and reduce emissions.

A conceptual diagram of an advanced biorefinery using corn and lignocellulose as the basic feedstocks is presented in Fig.14.1. Sugar-based crops could also be incorporated for use when seasonably available. In addition to ethanol, this biorefinery is capable of producing high-value co-products such as specialized animal feed, organic chemicals, bioplastic monomers, and biodiesel. Lignin provides the bulk of the energy for the combined heat and power systems to generate process steam and electricity (Hobbs 2008). Residual corn nutrients, in the form of steep water and/or thin stillage, balance the low nutrient value of lignocellulose. Fiber fractions from corn or sugar crops can be integrated into the lignocellulose process for saccharification, while sugar concentrations can be optimized by blending sugar streams from the three feedstocks.

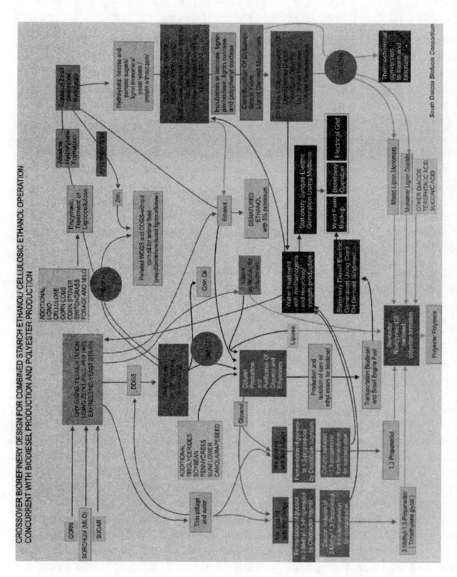

Fig. 14.1 Diagram of an advanced crossover biorefinery combining corn-based dry grind ethanol production (*medium blue*) with lignocellulose-based ethanol production (*light blue*) capable of producing high-value co-products (specialized animal feed, bioplastic monomers, and biodiesel). Inputs are shown in *yellow*; outputs are in *light green*. Column processes are in *gray*. Red switch circles indicate points at which the process path can be directed to preferentially select one of the legs. *Black* designates water recycling and production of heat and electricity

Technology Improvements to Enhance Biofuel Production Economics

To achieve the goal of integrated biorefineries that can use a broad range of feedstocks profitably, the following technology improvements are necessary:

1. More efficient pretreatment processes that open the structure of biomass (and/or generate fractionated component streams) with reduced production of inhibitory byproducts;
2. Consolidated bioprocessing microbe that produces its own lignocellulose deconstructing enzymes (some or all) and then can ferment all sugars to ethanol simultaneously;
3. Ability to conduct simultaneous saccharification/fermentation at elevated temperatures (>50°C) to maximize conversion rate and minimize contamination;
4. Ability of microbe to tolerate inhibitory byproducts produced during lignocellulose pretreatment;
5. Ability of microbe to tolerate industrial conditions (robustness to change) and tolerate high ethanol concentrations;
6. Ability to handle low bulk density of the feedstock which limits ethanol concentration;
    a. Submerged SSF using a membrane-based system to recover ethanol by pervaporation integrated with fermentation; or
    b. Solid-state SSF using novel reactor design with ethanol recovery via steam stripping.

Considerable research has been conducted to improve the characteristics of bioprocessing microbes. To date, most of this effort has focused on traits that are thought to be more amenable to manipulation, such as engineering microbes to express enzymes capable of deconstructing lignocelluloses (Lynd et al. 2005; Kumar et al. 2008) or expanding their ability to use all the sugars in biomass hydrolysates (Van Maris et al. 2006). On the other hand, characteristics such as activity at elevated temperatures, tolerance to inhibitors, and robustness are more difficult to impart. Currently, a comparison of ethanologens is underway to determine the organism best suited for further fuel ethanol development (Hahn-Hägerdal et al. 2007). At this point, yeast have several distinct advantages over bacterial ethanologens. For example, *S. cerevisiae* is a robust organism with high tolerance to inhibitors such as furfural and methyl furfural, as well as to the ethanol produced. *S. cerevisiae*'s tolerance to 18–20% ethanol, using glucose as a substrate, is significantly higher than the yeast *Pichia stipitis* (now named *Scheffveromyces stipitis*), which yields a maximum of 6% ethanol on glucose with limited oxygenation. Yeast are also far better at tolerating low pH than many bacteria. A comparison of microbes that are currently used or are under consideration for use as ethanologens is presented in Table 14.1 (Kurtzman 2009).

Strategies for producing key improvements in the microbes used for the production of fuel ethanol include:

Table 14.1 Microbes currently used or under consideration for use as ethanologens

Ethanologen trait table	*Saccharomyces cerevisiae*	*Scheffersomyces stipitis* (formerly *Pichia stipitis*)	*Candida shahatae* or *Pachysolen tannophilus*	*Kluyveromyces marxianus*	*Escherichia coli* (FBR2)	*Zymomonas mobilis* (Zm4)
Sugars metabolized	Glucose, sucrose; maltose, galactose, fructose, trehalose, isomaltose, raffinose, maltotriose, ribose, glucuronic acid and have been engineered to use lactose, xylose, arabinose	Glucose, sucrose; maltose, galactose, fructose, trehalose, isomaltose, raffinose, maltotriose, ribose, glucuronic acid, lactose, xylose, arabinose, cellobiose, rhamnose, fucose, sorbose and maltotetrose	Glucose, sucrose, maltose, galactose, fructose, raffinose, xylose, arabinose	Glucose, sucrose, maltose, galactose, fructose, trehalose, isomaltose, raffinose, maltotriose, xylose, arabinose, lactose	Glucose, sucrose, maltose, galactose, fructose, glucuronic acid, galacturonic acid xylose, arabinose, mannose	Glucose, sucrose, maltose, galactose, lactose, fructose, xylose, arabinose, mellibiose, raffinose, mannose
Sugars fermented	Glucose, sucrose; maltose, galactose, fructose, trehalose, isomaltose, raffinose, maltotriose	Glucose, sucrose; maltose, galactose, fructose, trehalose, isomaltose, raffinose, maltotriose, xylose, arabinose	Glucose, sucrose, maltose, galactose, fructose, trehalose, isomaltose, raffinose, maltotriose, xylose, arabinose	Glucose, sucrose, maltose, fructose, xylose, arabinose	Glucose, sucrose, maltose, galactose, fructose, xylose, arabinose, mannose	Glucose, sucrose, maltose, galactose, lactose; fructose, xylose, arabinose, mellibiose, raffinose, mannose

(continued)

Table 14.1 (continued)

Ethanologen trait table	Saccharomyces cerevisiae	Scheffersomyces stipitis (formerly Pichia stipitis)	Candida shahatae or Pachysolen tannophilus	Kluyveromyces marxianus	Escherichia coli (FBR2)	Zymomonas mobilis (Zm4)
Temperature for growth	<44°C	26–35°C	10–40°C	<40°C	<49°C	27–37.5°C
pH range	3.0–8.0	4.0–7.5	3.0–75	4.8–6.3	4.8–6.3	5.5–6.8
Ethanol production/tolerance	15–21%/<22–23%	4.4–6.0%/<10%	3.5–3.8%/<4.6–4.8% (<2 h)	6.0–11.1%/<22.5%	4.38%/<5%	8.1–10.5%/<15%
Crabtree type and/or metabolic flux	Crabtree positive uses facilitated diffusion of glucose	Crabtree negative	Crabtree negative (glucose taken up by proton symport)	Crabtree negative	Sugars to pyruvate using an engineered Entner-Doudoroff pathway	Sugars to pyruvate using the Entner-Doudoroff pathway
Genome sequence accession number flexgene library additions	Max-Planck-Institute for Biochemistry (MIPS), Martinsried/Munich Germany Sequence Maintained At MIPS Comprehensive Yeast Genome Database (CYGD)	DOE JGI data sequence completed	Neither strain fully sequenced	Not fully sequenced	The Genome Center at the University at Wisconsin maintains sequenced	DOE JGI data sequence not completed
FDA-CVM status	Several recombinant yeast have GRAS status with FDA	One recombinant Pichia pastoris strain has GRAS status with FDA should be applicable to P. stipitis	Not GRAS	Not GRAS	Only K12 type has GRAS status	Not GRAS

1. *Engineering one microbe with the ability to ferment mixed sugars.*

    - Recombinant yeast strains that use glucose and xylose have been engineered, but they do not ferment xylose to ethanol in high yield. Although ethanol production from xylose is limited, the xylose could be used for cell growth, allowing all the glucose to be used solely for fermentation to ethanol.
    - Yeast strains that were systematically transformed with yeast libraries were screened to find key genes that regulate xylose and arabinose use.
    - Stable yeast strains were engineered using these gene sets to allow fermentation of cellulosic hydrolysates that contain arabinose and xylose in addition to glucose fructose, mannose, and galactose, which are the principal constituents of hard and soft woods, switchgrass, corn stover, and corn cobs.
    - A stable industrial cell line will be established to produce a highly durable background strain (GMAX; Fig.14.2) to use in cellulosic ethanol production.
    - Genes encoding lignocellulose-deconstructing enzymes could also be inserted into the yeast strain engineered for xylose use. The resulting yeast strain will be selected for tolerance to processing temperatures, to presence of inhibitors, and to high ethanol concentration.
    - This same strain also allows production of ethanol from sugarcane, sugar beets, and sweet sorghum. This flexibility will be needed to allow use of dedicated energy crops in biorefineries where ethanol is produced from both lignocellulose and sucrose, as for sugarcane with bagasse and sucrose or for sugar beets with high concentrations of sucrose and cellulosic waste.
    - Starch operations could use this same yeast for ethanol production and also to process corn stover or in the wet mill process to use pericarps for cellulosic ethanol production.
    - The GMAX yeast strain can provide an expression setting to produce low cost enzymes for the production processes in the combined biorefinery.

2. *Engineering the microbe to produce multiple products or using additional microbes.*

    - Yeast strains can also be engineered to express lipase to catalyze ethanol transesterification of triglycerides to form corn oil ethyl esters for biodiesel. Many low-cost enzymes such as lipases for transesterification or hydrolases for saccharification have been shown to be functional in yeast (Den Haan et al. 2007).
    - Fermentation of glycerol and thin stillage obtained from the process (Fig.14.1) to 2-methyl-1,3-propanediol and 1,3 propanediol will require the additional microbes *Citrobacter freundii* and *Clostridium butyricum*, respectively. The propanediol moieties can be used for condensation polymers with terephthalic acid or with lignol moieties from lignin to produce bioderived plastics.
    - Expression of saccharification enzymes in the GMAX yeast would drastically reduce the requirement for addition of enzymes in cellulosic ethanol production. The economic analysis presented in Table14.2 shows that the advanced biorefinery has negligible enzyme costs.
    - Existing refineries can use these lignocellulosic add-ons to generate high-value co-products and use low-revenue feed items from the process for higher

value plastics and food-grade polymers. For example, zein can be expressed as homozein types with desired polymer properties. The product can be added as a high-value nutritional feed supplement and pelletized for shipment. The nutrient value of animal feed can also be improved by using the yeast to express proteins having high proportions of essential amino acids.

3. *Adding desired metabolic capabilities, using routes such as mutagenesis or addition of multiple FLEXGene collections followed by selection to develop tolerance to ethanol and inhibitors.*

- Various routes of making xylitol are possible with yeast expressing xylose isomerase and xylitol dehydrogenase. The metabolic pathway involving these enzymes directs some xylose into production of this sweetener to make the high-nutrient cattle feed more palatable.
- Further enhancements of the GMAX strains will require multigene additions or changes. One method of accomplishing these multigene changes would require the mutagenesis of the strain using treatment with chemicals or radiation. The strength of the treatment would determine how many multigene changes are produced. Resulting strains would be screened for survival on medium containing inhibitors or high amounts of ethanol and incubated anaerobically at high temperatures.
- A second approach would be to take the stable GMAX yeast strain and combine the four-plasmid expression with multiple fungal or bacterial expression libraries simultaneously for multigene interactions. Both these routes would require ultrahigh-throughput screening, as the factorial combinations of genes involved in this approach is enormous.
- Further metabolic engineering is necessary after addition of genes for any of the enzymes as these will inherently impact cell growth, and metabolic adjustments will be required to maintain industrial doubling times and superior ethanol production.

4. *Determining the most logical sequence to add the desired traits.*

- Selected optimized genes or gene sets will be added to the improved GMAX strain and the activity of the resulting strain evaluated after each addition.
- Additional selected genes will be added to the GMAX strain if evaluation demonstrates that the fermentative pathways in *S. cerevisiae* are not disrupted, so use of glucose remains intact and that the genes for pentose use are still providing the capacity for the strain to metabolize xylose and arabinose.

Conclusion

A comparison of the actual costs and profits of dedicated ethanol production facilities versus the potential costs and profits of a combined biorefinery is presented in Table 2 (McAloon 2009; USDA Economic Research Service 2009; USDA National Agricultural Statistical Service 2009).

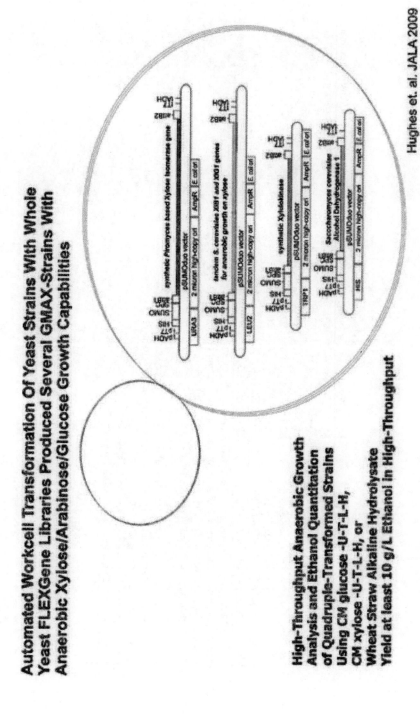

Fig. 14.2 Schematic of recombinant GMAX yeast strain

Table 14.2 Comparison of costs and profits for dedicated versus combined refineries

Ethanol refinery cost comparison	Wet mill	Dry mill/grind	Sugarcane	Lignocellulose	Crossover refinery[s]
Avg. cost of plant[z]	$233.84 million	$115.5 million	$62.5 million	>$375.00 million	>$200.00 million
Lifespan of plant[y]	>60 yr	30–60 yr	40–60 yr	Continuous	Continuous
Price feedstock[x] (mt)	$188.46	$188.46	$42.00	$95.00	$<50.00
Production costs[w]	$0.88/gallon	$0.71/gallon	$0.54/gallon	Experimental	Experimental
Cost enzymes[v]	$0.06/gallon	$0.06/gallon	<$0.01/gallon	$0.30/gallon	Potentially $0
Total ethanol profit[u]	$2.56 billion	$16.77 billion	$46.65 billion	$0.051 billion	>$70 billion
Total co-product profit[t]	$5.05 billion	$9.80 billion	$6.64 billion	Experimental	>$100 billion

[z] Based on 2008 adjusted prices for construction materials
[y] Time projections made at time of construction and expected amount of ethanol produced in planning stages
[x] Conservative prices for 2008 average price levels for first quarter
[w] Cost for plant operation 2008 feedstock prices for sucrose operations in Brazil, or for wet and dry mill starch operations in Midwest United States
[v] AT 2008 Novozyme and Danisco commercial price levels
[u] Value represents world production levels for the year or theoretical production for year 2008 in US dollars based on Chicago board of trade ethanol average price in third quarter 2008
[t] For products of representative plants used in study (ref)
[s] For concept plant form using shared utilities and operational staff (Gibbons et al. 1984, 1986).

The first step toward achieving a cost-effective, sustainable fuel ethanol biorevcommercially feasible. A combined biorefinery will be the source of valuable co-products, such as chemicals and plastics, to reduce dependence on petroleum. It will require creation of industrially robust, multifunctional recombinant ethanologenic organisms. Energy and personnel sharing strategies will be used in the combined biorefinery to reduce costs. In the long term, production of ethanol from biomass is likely to become commercially viable. Moving into lignocellulosic feedstocks with optimized recombinant yeast strains will provide large quantities of fuel ethanol via flexible worldwide usage at the local level of all available biomass to establish a self-sustaining biofuel industry.

An important part of achieving profitability in a combined biorefinery is the ethanologenic organism. Once an ideal ethanologen is engineered for maximum ethanol production from the mixture of sugars, it will be necessary to adapt the strain for industrial robustness. It is also essential for profitability that an expression system is incorporated into this strain which will produce high-value co-products. A key advantage of this approach is that this strain could be used for bioconversion in any part of the crossover biorefinery. Using one recombinant organism with many additions should simplify the process of obtaining necessary FDA approval for feed products produced by or containing recombinant organisms. This new generation of platform biorefineries will produce transportation fuels and provide electricity and plastics, supplanting many of the current uses of fossil fuel. Thus, reducing the cost of ethanol production in a mixed or crossover biorefinery will secure sustainable co-production of renewable energy for transportation and electrical generation, which in turn will also decrease our dependence on fossil fuel.

References

Barfoot P.; Brookes G. Global impact of biotech crops: socio-economic and environmental effects, 1996–2006. *AgBioForum* 111: 21–38; 2008.

Baum J. A.; Bogaert T.; Clinton W.; Heck G. R.; Feldmann P.; Ilagan O.; Johnson S.; Plaetinck G.; Munyikwa T.; Pleau M.; Vaughn T.; Roberts J. Control of coleopteran insect pests through RNA interference. *Nat. Biotechnol.* 25: 1322–1326; 2007. doi:10.1038/nbt1359.

Bothast R. J.; Schlicher M. A. Biotechnological processes for conversion of corn into ethanol. *Appl. Microbiol. Biotechnol.* 67: 19–25; 2005. doi:10.1007/s00253-004-1819-8.

Cardona C. A.; Sanchez O. J. Fuel ethanol production: process design trends and integration opportunities. *Bioresour. Technol.* 98: 2415–2457; 2007. doi:10.1016/j.biortech.2007.01.002.

Castiglioni P.; Warner D.; Bensen R. J.; Anstrom D. C.; Harrison J.; Stoecker M.; Abad M.; Kumar G.; Salvador S.; D'Ordine R.; Navarro S.; Back S.; Fernandes M.; Targolli J.; Dasgupta S.; Bonin C.; Luethy M. H.; Heard J. E. Bacterial RNA chaperones confer abiotic stress tolerance in plants and improved grain yield in maize under water-limited conditions. *Plant. Physiol.* 147: 446–455; 2008. doi:10.1104/pp.108.118828.

De La Torre Ugarte D. G.; English B. C.; Hellwinckel C. M.; Menard R. J.; Walsh M. E. Economic implications to the agricultural sector of increasing the production of biomass feedstocks to meet biopower, biofuels, and bioproduct demands. Institute of Agriculture, Department of Agricultural Economics, University of Tennessee, Knoxville, TN; 2007.

Den Haan R.; Rose S. H.; Lynd L. R.; Van Zyl W. H. Hydrolysis and fermentation of amorphous cellulose by recombinant *Saccharomyces cerevisiae*. *Metab. Eng.* 9: 87–94; 2007. doi:10.1016/j.ymben.2006.08.005.

Dien B. S.; Cotta M. A.; Jeffries T. W. Bacteria engineered for fuel ethanol production: current status. *Appl. Microbiol. Biotechnol.* 63: 258–266; 2003. doi:10.1007/s00253-003-1444-y.

Edgerton M. D. Increasing crop productivity to meet global needs for feed, food, and fuel. *Plant Physiol.* 149: 7–13; 2009. doi:10.1104/pp.108.130195.

Energy Independence and Security Act of 2007. H.R. 6, 110th US Congress, Washington, DC; 2007.

Energy Policy Act of 2005. Public Law 109–58, 109th US Congress, Washington, DC; 2005.

Food and Agricultural Policy Research Institute. FAPRI Agricultural Outlook 2007. Department of Economics, Iowa State University. World Biofuels; 2008.

Food and Agricultural Policy Research Institute. FAPRI Agricultural Outlook 2008. Department of Economics, Iowa State University. World Biofuels; 2009.

Gibbons W. R.; Westby C. A.; Dobbs T. L. A continuous, farm-scale solid-phase fermentation process for production of fuel ethanol and protein feed production from fodder beets. *Biotechnol. Bioeng.* 26: 1098–1107; 1984. doi:10.1002/bit.260260913.

Gibbons W. R.; Westby C. A.; Dobbs T. Intermediate-scale, semicontinuous solid-phase fermentation process for production of fuel ethanol from sweet sorghum. *Appl. Env. Microbiol.* 51: 115–122; 1986.

Gnansounou E.; Dauriat A.; Wyman C. E. Refining sweet sorghum to ethanol and sugar: economic trade-offs in the context of North China. *Bioresour. Technol.* 96: 985–1002; 2005. doi:10.1016/j.biortech.2004.09.015.

Goldemberg J. Ethanol for a sustainable energy future. *Science* 315: 808–810; 2007. doi:10.1126/science.1137013.

Hahn-Hägerdal B.; Karhumoa K.; Fonseca C.; Spencer-Martins I.; Gorwa-Grauslund M. F. Towards industrial pentose-fermenting yeast strains. *Appl. Microbiol. Biotechnol.* 74: 937–953; 2007. doi:10.1007/s00253-006-0827-2.

Hill J.; Nelson E.; Tilman D.; Polasky S.; Tiffany D. Environmental, economic, and energetic costs and benefits of biodiesel and ethanol biofuels. *Proc. Natl. Acad. Sci. U. S. A.* 103: 11206–11210; 2006. doi:10.1073/pnas.0604600103.

Hobbs, A. Biomass and biofuels—successful implementation of CHP. The role of CHP in Florida's Energy Future, Jacksonville, FL; 2008.

Kheshgi H. S.; Prince R. C. Sequestration of fermentation CO_2 from ethanol production. *Energy* 30: 1865–1871; 2005. doi:10.1016/j.energy.2004.11.004.

Kojima, M.; Johnson, T. Potential for biofuels for transport in developing countries. The International Bank for Reconstruction and Development, The World Bank, Energy Sector Management Assistance Programme Report; 2005.

Kumar R.; Singh S.; Singh O. V. Bioconversion of lignocellulosic biomass: biochemical and molecular perspectives. *J. Ind. Microbiol. Biotechnol.* 35: 377–391; 2008. doi:10.1007/s10295-008-0327-8.

Kurtzman, C. P. Management and genetic characterization of agricultural and biotechnological microbial resources. Accession Numbers 408614 and 413230. Microbial Genomics and Bioprocessing Research Unit, USDA, ARS, NCAUR; 2009.

Lynd L. R.; Van Zyl W. H.; McBride J. E.; Laser M. Consolidated bioprocessing of cellulosic biomass: an update. *Curr. Opin. Biotechnol.* 16: 577–583; 2005. doi:10.1016/j.copbio.2005.08.009.

McAloon, A. Supergroup chemical engineering and cost evaluations of biofuels processes. USDA, ARS, Eastern Regional Research Center; 2009.

Nelson D. E.; Repetti P. P.; Adams T. R.; Creelman R. A.; Wu J.; Warner D. C.; Anstrom D. C.; Bensen R. J.; Castiglioni P. P.; Donnarummo M. G.; Hinchey B. S.; Kumimoto R. W.; Maszle D. R.; Canales R. D.; Krolikowski K. A.; Dotson S. B.; Gutterson N.; Ratcliffe O. J.; Heard J. E. Plant nuclear factor Y (NF-Y) B subunits confer drought tolerance and lead to improved corn yields on water-limited acres. *Proc. Natl. Acad. Sci.* 104: 16450–16455; 2007. doi:10.1073/pnas.0707193104.

Parkin, G.; Weyer, P.; Just, C. L. Riding the bioeconomy wave: smooth sailing or rough water for the environment and public health? Proceedings of the 2007 Iowa Water Conference—Water and Bioenergy. Iowa State Center, Ames, I.A.; 2007.

Perlack, R. D.; Wright, L. L.; Turhollow, A. F.; Graham, R. L.; Stokes, B. J.; Erbach, D. C. Biomass as feedstock for a bioenergy and bioproducts industry: the technical feasibility of a billion-ton annual supply. A joint study sponsored by US Department of Energy and US Department of Agriculture; 2005.

Phillips, S.; Aden, A.; Jechura, J.; Dayton, D.; Eggeman, T. Thermochemical ethanol via indirect gasification and mixed alcohol synthesis of lignocellulosic biomass. Technical Report NREL/TP-510-41168. National Renewable Energy Laboratory, Golden, C.O.; 2007.

Renewable Fuels Association. Ethanol industry statistics. http://www.ethanolrfa.org/industry/statistics. Cited 25 Jan 2009; 2009.

Saha B. C. Hemicellulose bioconversion. *J. Ind. Microbiol. Biotechnol.* 30: 279–291; 2003. doi:10.1007/s10295-003-0049-x.

Shapouri, H.; Duffield, J.; McAloon, A.; Wang, M. The 2001 net energy balance of corn–ethanol. USDA/Economic Research Service, Office of Energy, USDA/USDOE Report. Washington, DC; 2004.

Shapouri, H.; Duffield, J. A.; Graboski, M. S. Estimating the net energy balance of corn–ethanol. USDA/Economic Research Service, Office of Energy, Agriculture Economic Report No. 721. Washington, DC; 1995.

Shapouri, H.; Salassi, M.; Fairbanks, J. N. The economic feasibility of ethanol production from sugar in the United States. Washington, Office of Energy Policy and New Uses, Office of the Chief Economist, United States Department of Agriculture/Baton Rouge, Louisiana State University; 2006.

Sims R.; Taylor M.; Saddler J.; Mabee W. From 1st to 2nd generation biofuel technologies—an overview of current industry and RD&D activities. International Energy Agency, Paris, France; 2008.

Troyer F. A. Adaptedness and heterosis in corn and mule hybrids. *Crop Sci.* 46: 528–543; 2006. doi:10.2135/cropsci2005.0065.

USDA Economic Research Service. Economics of various cropping systems economics study of biofuels production. http://www.ers.usda.gov;/ Cited 25; Jan 2009.

USDA National Agricultural Statistical Service. Volume and cost of agricultural products each year. http://www.nass.usda.gov; Cited 25; Jan 2009.

Van Maris A. J. A.; Abbott D. A.; Bellissimi E.; van den Brink J.; Kuyper M.; Luttik M. A. H.; Wisselink H. W.; Scheffers W. A.; van Dijken J. P.; Pronk J. T. Alcoholic fermentation of carbon sources in biomass hydrolysates by *Saccharomyces cerevisiae*: current status. *Antonie van Leeuwenhoek* 90: 391–418; 2006. doi:10.1007/s10482-006-9085-7.

Varga E.; Schmidt A. S.; Reczey K.; Thomsen A. B. Pretreatment of corn stover using wet oxidation to enhance enzymatic digestibility. *Appl. Biochem. Biotechnol.* 104: 37–50; 2003. doi:10.1385/ABAB:104:1:37.

Vermerris W.; Saballos A.; Ejeta G.; Mosier N. S.; Ladisch N. R.; Carpita N. C. Molecular breeding to enhance ethanol production from corn and sorghum stover. *Crop Sci.* 47: S142–S153; 2007. doi:doi:10.2135/cropsci2007.04.0013IPBS.

Westcott, P. C. Ethanol expansion in the United States: how will the agricultural sector adjust? FDS-07D-01. United States Department of Agriculture, Economic Research Service; 2007.

Wheals A. E.; Basso L. C.; Alves D. M. G.; Amorim H. V. Fuel ethanol after 25 yr. *TIBTECH* 17: 482–487; 1999. doi:10.1016/S0167-7799(99)01384-0.

Wilkins M. R.; Mueller M.; Eichling S.; Banat I. M. Fermentation of xylose by the thermotolerant yeast strains *Kluveromyces marxianus* IMB2, IMB4, and IMB5 under anaerobic conditions. *Process Biochem.* 43: 346–350; 2008. doi:10.1016/j.procbio.2007.12.011.

Wilts A. R.; Reicosky D. C.; Allmaras R. R.; Clapp C. E. Long-term corn residue effects: harvest alternatives, soil carbon turnover, and root-derived carbon. *Soil. Sci. Soc. Am. J.* 68: 1342–1351; 2004.

Chapter 15
Biodiesel Production, Properties, and Feedstocks

Bryan R. Moser

Abstract Biodiesel, defined as the mono-alkyl esters of vegetable oils or animal fats, is an environmentally attractive alternative to conventional petroleum diesel fuel (petrodiesel). Produced by transesterification with a monohydric alcohol, usually methanol, biodiesel has many important technical advantages over petrodiesel, such as inherent lubricity, low toxicity, derivation from a renewable and domestic feedstock, superior flash point and biodegradability, negligible sulfur content, and lower exhaust emissions. Important disadvantages of biodiesel include high feedstock cost, inferior storage and oxidative stability, lower volumetric energy content, inferior low-temperature operability, and in some cases, higher NO_x exhaust emissions. This chapter covers the process by which biodiesel is prepared, the types of catalysts that may be used for the production of biodiesel, the influence of free fatty acids on biodiesel production, the use of different monohydric alcohols in the preparation of biodiesel, the influence of biodiesel composition on fuel properties, the influence of blending biodiesel with other fuels on fuel properties, alternative uses for biodiesel, and value-added uses of glycerol, a co-product of biodiesel production. A particular emphasis is placed on alternative feedstocks for biodiesel production. Lastly, future challenges and outlook for biodiesel are discussed.

Keywords Alternative feedstocks • Biodiesel • Fatty acid FAME • Fuel properties • Methanolysis • Transesterification

Disclaimer: Product names are necessary to report factually on available data; however, the USDA neither guarantees nor warrants the standard of the product, and the use of the name by USDA implies no approval of the product to the exclusion of others that may also be suitable.

B.R. Moser (✉)
United States Department of Agriculture, Agricultural Research Service, National Center for Agricultural Utilization Research, 1815 N University St, Peoria, IL 61604, USA
e-mail: Bryan.Moser@ars.usda.gov

In Vitro Cell.Dev.Biol.—Plant (2009) 45: 229–266, DOI 10.1007/s11627-009-9204-z,
Published online: 25 March 2009. Editors: Prakash Lakshmanan and David Songstad;
The Society for In Vitro Biology 2009.

Introduction

Biodiesel is defined by ASTM International as a fuel composed of monoalkyl esters of long-chain fatty acids derived from renewable vegetable oils or animal fats meeting the requirements of ASTM D6751 (ASTM 2008a). Vegetable oils and animal fats are principally composed of triacylglycerols (TAG) consisting of long-chain fatty acids chemically bound to a glycerol (1,2,3-propanetriol) backbone. The chemical process by which biodiesel is prepared is known as the transesterification reaction and involves a TAG reacting with a short-chain monohydric alcohol normally in the presence of a catalyst at elevated temperature to form fatty acid alkyl esters (FAAE) and glycerol (Fig.15.1). The conversion of TAG to biodiesel is a stepwise process whereby the alcohol initially reacts with TAG as the alkoxide anion to produce FAAE and diacylglycerols (DAG, reaction [1], Fig.15.1), which react further with alcohol (alkoxide) to liberate another molecule of FAAE and generate monoacylglyerols (MAG, reaction [2], Fig.15.1). Lastly, MAG undergo alcoholysis to yield glycerol and FAAE (reaction [3], Fig.15.1), with the combined FAAE collectively known as biodiesel. Three moles of biodiesel and one mole of glycerol are produced for every mole of TAG that undergoes complete conversion. The transesterification reaction is reversible, although the reverse reaction (production of MAG from FAAE and glycerol, for instance) is negligible largely because glycerol is not miscible with FAAE, especially fatty acid FAME (FAME) when using methanol as the alcohol component. The

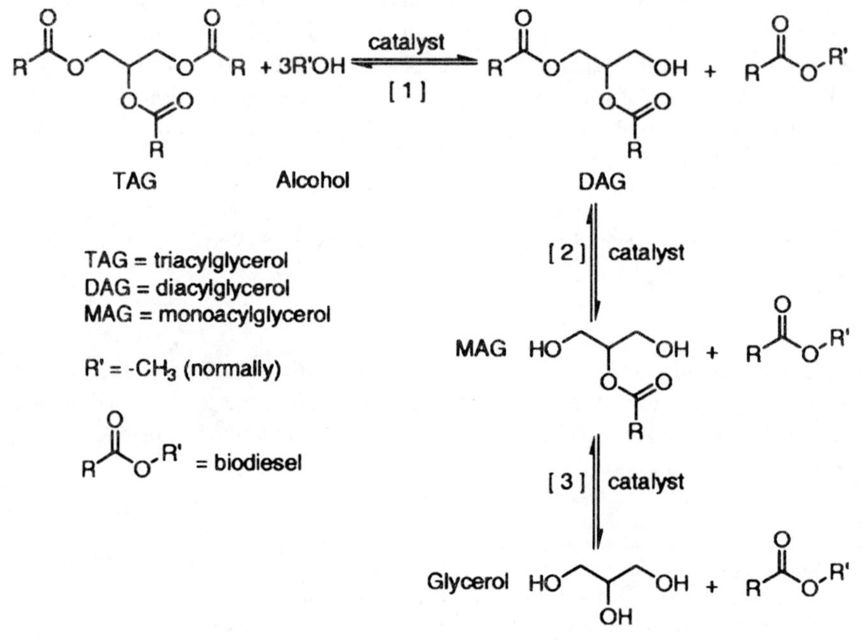

Fig. 15.1 Transesterification of triacylglycerols to yield fatty acid alkyl esters (biodiesel)

reaction system is biphasic at the beginning and the end of biodiesel production, as methanol and vegetable oil and glycerol and FAME are not miscible. Methanol is most commonly used in the commercial production of biodiesel, since it is generally less expensive than other alcohols, but ethanol prevails in regions such as Brazil where it is less expensive than methanol. Other alcohols aside from methanol and ethanol are also of interest for biodiesel production because FAAE produced from higher alcohols have different fuel properties in comparison to methyl or ethyl esters (Knothe 2005). A more detailed discussion of the influence of ester head group on the fuel properties of biodiesel will be presented in a subsequent section of this chapter.

Inexpensive homogenous alkaline base catalysts such as sodium or potassium hydroxide or methoxide are typically used in the commercial preparation of biodiesel from refined or treated oils. The classic alcoholysis conditions described by Freedman et al. (1984) include TAG reacting with an excess of six molar equivalents of methanol (with respect to TAG) and 0.5 weight percent (wt.%) alkali catalyst (with respect to TAG) at 60°C for 1 h to produce fatty acid FAME (FAME, biodiesel) and glycerol. The chemical composition of biodiesel is dependent upon the feedstock from which it is produced, as vegetable oils and animal fats of differing origin have dissimilar fatty acid compositions (Table 15.1). The fatty ester composition of biodiesel is essentially identical to that of the parent oil or fat from which it was produced.

Table 15.1 Typical fatty acid composition (wt.%)[a] of a number of common feedstock oils[b] and fats that may be used for biodiesel production

Fatty acid[c]	CO	PO	SBO	SFO	COO	CSO	CCO	CF	BT
C6:0							1		
C8:0							7		
C10:0							7		
C12:0							47		1
C14:0		1				1	18	1	4
C16:0	4	45	11	6	11	23	9	25	26
C18:0	2	4	4	5	2	2	3	6	20
C20:0									
C22:0				1					
C16:1						1		8	4
C18:1	61	39	23	29	28	17	6	41	28
C18:2	22	11	54	58	58	56	2	18	3
C18:3	10		8	1	1			1	
C20:1	1								
Other									14

[a]From Gunstone and Harwood (2007); trace amounts (<1%) of other constituents may also be present
[b]*CO* canola (low erucic acid rapeseed oil) oil, *PO* palm oil, *SBO* soybean oil, *SFO* sunflower oil, *COO* corn oil (maize), *CSO* cottonseed oil, *CCO* coconut oil, *CF* chicken fat, *BT* beef tallow
[c]*C6:0* methyl caproate, *C8:0* methyl caprylate, *C10:0* methyl caprate, *C12:0* methyl laurate, *C14:0* methyl myristate, *C16:0* meythyl palmitate, *C18:0* methyl stearate, *C20:0* methyl arachidate, *C22:0* methyl behenate, *C16:1* methyl palmitoleate, *C18:1* methyl oleate, *C18:2* methyl linoleate, *C18:3* methyl linolenate, *C20:1* methyl eicosenoate

A recent report (International Grains Council 2008) indicated that rapeseed oil was the predominant feedstock for worldwide biodiesel production in 2007 (48%, 4.6 million metric tons, MMT). The remaining oils primarily included soybean (22%, 2.1 MMT) and palm (11%, 1.0 MMT), with the rest (19%, 1.8 MMT) distributed among other unspecified vegetable oils and animal fats. The leading vegetable oils produced worldwide during the 2008 fiscal year (October 1, 2008 to September 30, 2009) were palm (43.20 MMT), soybean (37.81 MMT), rapeseed (19.38 MMT), and sunflower (11.68 MMT) oils (USDA 2008). Not surprisingly, vegetable oil production and biodiesel feedstock usage are intimately related. Feedstocks for biodiesel production vary with location according to climate and

Table 15.2 ASTM D6751 biodiesel fuel standard

Property	Test method	Limits	Units
Flash point (closed cup)	ASTM D93	93 min[a]	°C
Alcohol control One of the following must be met:			
1. Methanol content	EN 14110	0.2 max[a]	% volume
2. Flash point	ASTM D93	130.0 min	°C
Water and sediment	ASTM D2709	0.050 max	% volume
Kinematic viscosity, 40°C	ASTM D445	1.9–6.0	mm^2/s
Sulfated ash	ASTM D874	0.020 max	% mass
Sulfur[b]	ASTM D5453	0.0015 max (S15) 0.05 max (S500)	% mass (ppm)
Copper strip corrosion	ASTM D130	No. 3 max	
Cetane number	ASTM D613	47 min	
Cloud point	ASTM D2500	report	°C
Cold soak filterability	ASTM D7501	360 max[c]	s
Carbon residue	ASTM D4530	0.050 max	% mass
Acid value	ASTM D664	0.50 max	mg KOH/g
Free glycerin	ASTM D6584	0.020	% mass
Total glycerin	ASTM D6584	0.240	% mass
Oxidation stability	EN 14112	3.0 min	h
Phosphorous content	ASTM D4951	0.001 max	% mass
Sodium and potassium, combined	EN 14538	5 max	ppm
Calcium and magnesium, combined	EN 14538	5 max	ppm
Distillation temperature, Atmospheric equivalent temperature, 90% recovered	ASTM D1160	360 max	°C

[a]For all Tables: min refers to minimum and max refers to maximum
[b]The limits are for Grade S15 and Grade S500 biodiesel, with S15 and S500 referring to maximum allowable sulfur content (ppm)
[c]B100 intended for blending into petrodiesel that is expected to give satisfactory performance at fuel temperatures at or below −12°C shall comply with a maximum cold soak filterability limit of 200 s

15 Biodiesel Production, Properties, and Feedstocks

availability. Generally, the most abundant commodity oils or fats in a particular region are the most common feedstocks. Thus, rapeseed and sunflower oils are principally used in Europe for biodiesel production, palm oil predominates in tropical countries, and soybean oil and animal fats are most commonly used in the USA (Demirbas 2006). However, even combining these feedstocks does not suffice to fully replace the volume of conventional petroleum diesel fuel (petrodiesel). Therefore, exploration of additional feedstocks for biodiesel production has been continuously gaining significance, as discussed later.

Biodiesel standards are in place in a number of countries in an effort to ensure that only high-quality biodiesel reaches the marketplace. The two most important fuel standards, ASTM D6751 (ASTM 2008a) in the United States and EN 14214 (European Committee for Standardization, CEN) (CEN 2003a) in the European Union, are summarized in Tables 15.2 and 3, respectively. In addition, a

Table 15.3 European Committee for Standardization EN 14214 biodiesel fuel standard

Property	Test method(s)	Limits	Units
Ester content	EN 14103	96.5 min	% (mol/mol)
Density, 15°C	EN ISO 3675, EN ISO 12185	860–900	kg/m^3
Kinematic viscosity, 40°C	EN ISO 3104, ISO 3105	3.5–5.0	mm^2/s
Flash point	EN ISO 3679	120 min	°C
Sulfur content	EN ISO 20846, EN ISO 20884	10.0 max	mg/kg
Carbon residue (10% distillation residue)	EN ISO 10370	0.30 max	% (mol/mol)
Cetane number	EN ISO 5165	51 min	
Sulfated ash	ISO 3987	0.02 max	% (mol/mol)
Water content	EN ISO 12937	500 max	mg/kg
Total contamination	EN 12662	24 max	mg/kg
Copper strip corrosion (3 h, 50°C)	EN ISO 2160	1	Degree of corrosion
Oxidation stability, 110°C	EN 14112	6.0 min	h
Acid value	EN 14104	0.50 max	mg KOH/g
Iodine value	EN 14111	120 max	g I_2/100 g
Linolenic acid content	EN 14103	12.0 max	% (mol/mol)
Polyunsaturated (≥4 double bonds) FAME	EN 14103	1 max	% (mol/mol)
Methanol content	EN 14110	0.20 max	% (mol/mol)
MAG content	EN 14105	0.80 max	% (mol/mol)
DAG content	EN 14105	0.20 max	% (mol/mol)
TAG content	EN 14105	0.20 max	% (mol/mol)
Free glycerol	EN 14105 EN 14106	0.020 max	% (mol/mol)
Total glycerol	EN 14105	0.25 max	% (mol/mol)
Group I metals (Na, K)	EN 14108 EN 14109	5.0 max	mg/kg
Group II metals (Ca, Mg)	EN 14538	5.0 max	mg/kg
Phosphorous content	EN 14107	10.0 max	mg/kg

Table 15.4 ASTM D7467 biodiesel-petrodiesel blend (B6–B20) fuel standard

Property	Test method	Limits	Units
Biodiesel content	ASTM D7371	6–20	% volume
Flash point (closed cup)	ASTM D93	52 min	°C
One of the following must be met			
1. Cetane index	ASTM D976	40 min	
2. Aromaticity	ASTM D1319	35 max	% volume
Water and sediment	ASTM D2709	0.050 max	% volume
Kinematic viscosity, 40°C	ASTM D445	1.9–4.1	mm^2/s
Sulfur content[a]	ASTM D5453	15 max (S15)	ppm
	ASTM D2622	500 max (S500)	
Copper strip corrosion	ASTM D130	No. 3 max	
Cetane number	ASTM D613	40 min	
Ramsbottom Carbon residue	ASTM D524	0.35 max	% mass
Acid value	ASTM D664	0.30 max	mg KOH/g
Oxidation stability	EN 14112	6.0 min	h
Ash content	ASTM D482	0.01 max	% mass
Lubricity, HFRR, 60°C	ASTM D6079	520 max	μm
Cloud point or LTFT/CFPP	ASTM D2500, D4539, D6371	only guidance provided	°C
Distillation temperature, 90% recovered	ASTM D86	343 max	°C

[a]The limits are for Grade S15 and Grade S500 biodiesel, with S15 and S500 referring to maximum allowable sulfur content (ppm)

petrodiesel–biodiesel blend standard, ASTM D7467 (ASTM 2008b), was recently introduced that covers blends of biodiesel in petrodiesel from 6 to 20 vol.%, and is summarized in Table 15.4. The petrodiesel standards ASTM D975 (ASTM 2008c) and EN 590 (CEN 2004) allow for inclusion of up to 5 (B5) and 7 (B7) vol.% biodiesel, respectively, as a blend component in petrodiesel. The heating oil standard ASTM D396 (ASTM 2008d) specifies that up to B5 is permissible as well. In Europe, biodiesel (B100; 100% biodiesel) may be directly used as heating oil, which is covered by a separate standard, EN 14213 (CEN 2003b). In the cases of ASTM D7467, D975, and D396, the biodiesel component must satisfy the requirements of ASTM D6751 before inclusion in the respective fuels. Correspondingly, in the European Union, biodiesel must be satisfactory according to EN 14214 before inclusion in petrodiesel, as mandated by EN 590.

Advantages and Disadvantages of Biodiesel

Biodiesel has attracted considerable interest as an alternative fuel or extender for petrodiesel for combustion in compression–ignition (diesel) engines. Biodiesel is miscible with petrodiesel in any proportion and possesses several technical advantages over ultra-low sulfur diesel fuel (ULSD, <15 ppm S), such as inherent lubricity, low toxicity, derivation from a renewable and domestic feedstock, superior flash point and biodegradability, negligible sulfur content, and lower overall exhaust emissions. Important disadvantages of biodiesel include high feedstock cost, inferior storage and oxidative stability, lower volumetric energy content, inferior low-temperature operability versus petrodiesel, and in some cases, higher NO_x exhaust emissions (DeOliveira et al. 2006; Knothe 2008). Many of these deficiencies can be mitigated through cold flow improver (Chiu et al. 2004; Soriano et al. 2005, 2006; Sern et al. 2007; Hancsok et al. 2008; Moser et al. 2008; Moser and Erhan 2008;) and antioxidant (Mittelbach and Schober 2003; Loh et al. 2006; Tang et al. 2008) additives, blending with petrodiesel (Benjumea et al. 2008; Moser et al. 2008), and/or reducing storage time (Bondioli et al. 2003). Additional methods to enhance the low-temperature performance of biodiesel include crystallization fractionation (Dunn et al. 1997; Kerschbaum et al. 2008;) and transesterification with long- or branched-chain alcohols (Lee et al. 1995; Foglia et al. 1997; Wu et al. 1998). Strategies to improve the exhaust emissions of biodiesel, petrodiesel, and blends of biodiesel in petrodiesel include various engine or after-treatment technologies such as selective catalytic reduction (SCR), exhaust gas recirculation (EGR), diesel oxidation catalysts, and NO_x or particulate traps (McGeehan 2004; Knothe et al. 2006). However, feedstock acquisition currently accounts for over 80% of biodiesel production expenses, which is a serious threat to the economic viability of the biodiesel industry (Paulson and Ginder 2007; Retka-Schill 2008). One potential solution to this problem is employment of alternative feedstocks of varying type, quality, and cost. These feedstocks may include soapstocks, acid oils, tall oils, used cooking oils, and waste restaurant greases, various animal fats, non-food vegetable oils, and oils obtained from trees and microorganisms such as algae. However, many of these alternative feedstocks may contain high levels of free fatty acids (FFA), water, or insoluble matter, which affect biodiesel production.

Influence of Free Fatty Acids on Biodiesel Production

Feedstock quality in large part dictates what type of catalyst or process is needed to produce FAAE that satisfies relevant biodiesel fuel standards such as ASTM D6751 or EN 14214. If the feedstock contains a significant percentage of FFA (>3 wt.%), typical homogenous alkaline base catalysts such as sodium or potassium hydroxide or methoxide will not be effective as a result of an unwanted side reaction (reaction [1], Fig.15.2) in which the catalyst reacts with FFA to form soap (sodium salt of fatty acid) and water (or methanol in the case of sodium methoxide), thus irreversibly quenching the catalyst and resulting in an undesirable mixture of FFA, unreacted TAG, soap, DAG, MAG, biodiesel, glycerol, water, and/or methanol (Lotero et al. 2005). In fact, base-catalyzed transesterification will not occur or will be significantly retarded if the FFA content of the feedstock is 3 wt.% or greater (Canakci and Van Gerpen 1999, 2001). For instance, nearly quantitative yields of biodiesel are achieved with homogenous alkaline base catalysts in cases where the FFA content of the feedstock is 0.5 wt.% or less (Naik et al. 2008). However, the

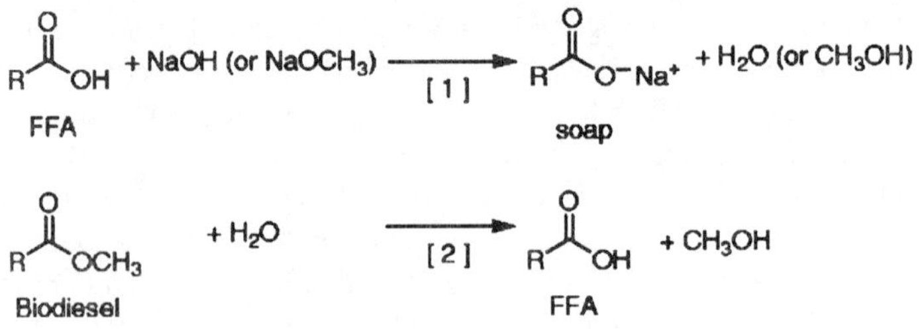

Fig. 15.2 Formation of soap from reaction of free fatty acids (FFA) with catalyst (reaction [1]) and hydrolysis of biodiesel (reaction [2]) to yield FFA and methanol

yield of biodiesel plummets to 6% with an increase in FFA content to 5.3 wt.% (Naik et al. 2008). A further complicating factor of high FFA content is the production of water upon reaction with homogenous alkaline base catalysts (reaction [1], Fig.15.2). Water is particularly problematic because, in the presence of any remaining catalyst, it can participate in hydrolysis of biodiesel to produce additional FFA and methanol (reaction [2], Fig.15.2).

A common approach in cases where the FFA content of a feedstock is in excess of 1.0 wt.% (Freedman et al. 1984; Mbaraka et al. 2003; Zhang et al. 2003; Wang et al. 2005) is a two-step process in which mineral acid pretreatment of the feedstock to lower its FFA content is followed by transesterification with homogenous alkaline base catalysts to produce biodiesel. In a typical mineral acid pretreatment procedure, FFA are esterified to the corresponding FAME in the presence of heat, excess methanol, and sulfuric acid catalyst (Ramadhas et al. 2005; Nebel and Mittelbach 2006; Veljkovic et al. 2006; Issariyakul et al. 2007; Kumartiwari et al. 2007; Sahoo et al. 2007; Meng et al. 2008; Naik et al. 2008; Rashid et al. 2008a). The two-step procedure readily accommodates high FFA-containing low-cost feedstocks for the preparation of biodiesel (Canakci and Van Gerpen 1999, 2001, 2003a). For example, the FFA content of crude jatropha oil (*Jatropha curcas*) was lowered to less than 1 from 14 wt.% upon treatment at 60°C for 88 min with 1.43 vol.% H_2SO_4 catalyst and 28 vol.% methanol. Subsequent alkaline-catalyzed (0.5 wt.% KOH) transesterification at 60°C for 30 min with 20 vol.% methanol yielded the corresponding FAME in essentially quantitative yield (Kumartiwari et al. 2007). In another example, the acid value (AV) of tung oil (from the "tung" tree, *Vernicia fordii*) was reduced from 9.55 to 0.72 mg KOH/g using Amberlyst-15 (20.8 wt.%), a heterogeneous acid catalyst, and methanol (7.5:1 molar ratio of methanol to tung oil) at 80°C for 2 h. Tung oil FAME were then prepared with a final purity of 90.2 wt.% employing 0.9 wt.% KOH and methanol (6:1 molar ratio of methanol to tung oil) at 80°C for 20 min (Park et al. 2008a).

Despite the added capital costs associated with production, the integrated two-step process is being increasingly applied to prepare biodiesel from low-cost feedstocks with high FFA content with good results (Lotero et al. 2005). Table15.5 lists a number of recent examples of biodiesel prepared from feedstocks with high FFA content. Other potential strategies for the production of biodiesel from feedstocks with high FFA content include feedstock purification such as refining, bleaching,

15 Biodiesel Production, Properties, and Feedstocks

Table 15.5 Examples of biodiesel production from feedstocks high in free fatty acids (FFA)

Feedstock	FFA (wt %)	Pretreatment method	Catalyst for transesterification	R^a	Yield (wt %)	Ref
Pongamia pinnata	Up to 20	H_2SO_4	KOH	Me	97	Naik et al. 2008
Moringa oleifera	2.9/0.953[b]	H_2SO_4	NaOCH$_3$	Me	n.r.[c]	Rashid et al. 2008a
Jatropha curcas	14/<1	H_2SO_4	KOH	Me	99+	Kumartiwari et al. 2007
Madhuca indica	20	None	*Pseudomonas cepacia*	Et	96+[d]	Kumari et al. 2007
Nicotiana tabacum	35/<2	H_2SO_4	KOH	Me	91	Veljkovic et al. 2006
Calophyllum inophyllum	22/<2	H_2SO_4	KOH	Me	85	Sahoo et al. 2007
Zanthoxylum bungeanum	45.5/1.16[b]	None	H_2SO_4	Me	98	Zhang and Jiang 2008
Hevea brasiliensis	17/<2	H_2SO_4	NaOH	Me	n.r.	Ramadhas et al. 2005
Heterotrophic microalgal	8.97[b]	None	H_2SO_4	Me	n.r.	Miao and Wu 2006
Acid oil	59.3	None	H_2SO_4	Me	95	Haas et al. 2003
Fat from meat and bone meal	11	H_2SO_4	KOH	Me	45.7	Nebel and Mittelbach 2006
Brown grease	40/<1	Diarylammonium catalysts	NaOCH$_3$	Me	98+[d]	Ngo et al. 2008
Waste cooking oil	7.25/<1[b]	H_2SO_4	NaOH	Me	90[d]	Meng et al. 2008
Waste fryer grease	5.6	H_2SO_4	KOH	Me/Et	90+	Issariyakul et al. 2007
Tung oil	9.55/0.72[b]	Amberlyst-15	KOH	Me	90.2	Park et al. 2008a, b
Tall oil	100%	None	HCl	Me	n.r.	Demirbas 2008
Sorghum bug oil	10.5	None	H_2SO_4	Me/Et	77.4/97.6	Mariod et al. 2006

[a] R refers to ester head group. Me methyl, Et ethyl.
[b] Acid value (mg KOH/g) was given instead of FFA. In cases where two values are given, the first value is prior to pretreatment and the second is after.
[c] Not reported
[d] Conversion to esters (wt %) is provided instead of yield

and deodorization to remove FFA and other undesirable materials, if present (Zappi et al. 2003). However, feedstock refining further increases production costs as a result of the additional equipment, time, and manpower that are required. Lastly, the employment of catalysts that are not destroyed by FFA in the production of biodiesel is another alternative to the methods listed above.

Catalysts for Biodiesel Production

Biodiesel is produced commercially using homogenous alkaline base catalysts such as sodium (or potassium) hydroxide or methoxide because the transesterification reaction is generally faster, less expensive, and more complete with these materials than with acid catalysts (Boocock et al. 1996a). The biodiesel industry currently uses sodium methoxide, since methoxide cannot form water upon reaction with alcohol such as with hydroxides (*see* Fig.15.2; Zhou and Boocock 2006a). Other alkoxides, such as calcium ethoxide, have also effectively catalyzed biodiesel production, albeit with higher methanol and catalyst requirements (Liu et al. 2008). The homogenous alkaline base-catalyzed transesterification reaction is about 4,000 times faster than the corresponding mineral acid-catalyzed process (Reid 1911; Srivastava and Prasad 2000). Furthermore, alkaline base-catalyzed reactions are performed at generally lower temperatures, pressures, and reaction times and are less corrosive to industrial equipment than mineral acid-catalyzed methods. Therefore, fewer capital and operating costs are incurred by biodiesel production facilities in the case of the alkaline base-catalyzed transesterification method (Freedman et al. 1986; Demirbas 2008). However, the homogenous acid-catalyzed reaction holds an important advantage over the base-catalyzed method in that the performance of acid catalysts is not adversely influenced by the presence of FFA. In fact, acids can simultaneously catalyze both esterification and transesterification (Haas et al. 2003; Lotero et al. 2005; Miao and Wu 2006; Demirbas 2008; Zhang and Jiang 2008). For instance, FAME were prepared from acid oil, which consisted of 59.3 wt.% FFA, by mineral acid-catalyzed transesterification at 65°C for 26 h with H_2SO_4 (1.5:1 molar ratio of catalyst to oil) and methanol (15:1 molar ratio of methanol to oil) in 95 wt.% purity. The remaining products consisted of FFA (3.2 wt.%), TAG (1.3 wt.%), and DAG (0.2 wt.%) (Haas et al. 2003).

A wide range of catalysts may be used for biodiesel production, such as homogenous and heterogeneous acids and bases, sugars, lipases, ion exchange resins, zeolites, and other heterogeneous materials. A recent exotic example is that of KF/Eu_2O_3, which was used to prepare rapeseed oil FAME with 92.5% conversion efficiency (Sun et al. 2008). In general, acids are more appropriate for feedstocks high in FFA content. Homogenously catalyzed reactions generally require less alcohol, shorter reaction times, and more complicated purification procedures than heterogeneously catalyzed transesterification reactions. Heterogeneous lipases are generally not tolerant of methanol, so production of ethyl or higher esters is more common with enzymatic methods. For a recent comprehensive review on catalysts used for biodiesel preparation, please *see* Narasimharao et al. (2007).

Noncatalytic transesterification of biodiesel may be accomplished in supercritical fluids such as methanol, but a very high pressure (45–65 bar), temperature (350°C), and amount of alcohol (42:1 molar ratio) are required (Saka and Kusdiana 2001; Demirbas 2003, 2005, 2006; Kusdiana and Saka 2004). Advantages of supercritical transesterification versus various catalytic methods are that only very short reaction times (4 min, for instance) are needed, and product purification is simplified because there is no need to remove a catalyst. Disadvantages of this approach include limitation to a batch-wise process, elevated energy and alcohol requirements during production, and increased capital expenses and maintenance associated with pressurized reaction vessels (Saka and Kusdiana 2001; Demirbas 2003, 2005, 2006; Kusdiana and Saka 2004).

Alcohols Used in the Production of Biodiesel

As previously mentioned, methanol is the most common alcohol used in the production of biodiesel. Other alcohols may also be used in the preparation of biodiesel, such as ethanol, propanol, iso-propanol, and butanol (Freedman et al. 1984, 1986; Schwab et al. 1987; Ali and Hanna 1994; Lee et al. 1995; Peterson et al. 1996; Foglia et al. 1997; Wu et al. 1998; Nimcevic et al. 2000; Canakci and Van Gerpen 2001; Lang et al. 2001; Encinar et al. 2002, 2007; Zhou et al. 2003; Wang et al. 2005; Mariod et al. 2006; Meneghetti et al. 2006; Yao and Hammond 2006; Dantas et al. 2007; Issariyakul et al. 2007; Kulkarni et al. 2007; Alamu et al. 2008; Domingos et al. 2008; Georgogianni et al. 2008; Lima et al. 2008; Rodrigues et al. 2008; Stavarache et al. 2008). Ethanol is of particular interest primarily because it is less expensive than methanol in some regions of the world, and biodiesel prepared from bio-ethanol is completely bio-based. Butanol may also be obtained from biological materials (Qureshi et al. 2008a, b), thus yielding completely bio-based biodiesel as well. Methanol, propanol, and iso-propanol are normally produced from hydrocarbon materials such as methane obtained from natural gas in the case of methanol.

Methanolysis. The classic reaction conditions for the methanolysis of vegetable oils or animal fats are 6:1 molar ratio of methanol to oil, 0.5 wt.% alkali catalyst (with respect to TAG), 600+ rpm, 60°C reaction temperature, and 1 h reaction time to produce FAME and glycerol (Freedman et al. 1984). A number of recent studies described optimal reaction conditions for biodiesel production from various feedstocks using response surface methodology (RSM). Parameters that are normally optimized to produce the highest yield of biodiesel include catalyst type and amount, reaction time and temperature, amount of alcohol, and/or agitation intensity. Please refer to Table15. 6 for a summary of recent examples of biodiesel process optimization employing RSM. In addition to the studies listed in Table15.6 are the following: Park et al. (2008a), Rashid and Anwar (2008a), Yuan et al. (2008), Wang et al. (2008), Cetinkaya and Karaosmanoglu (2004), Antolin et al. 2002. A representative example of reaction conditions optimized by RSM is the work of

Table 15.6 Recent examples of optimization of reaction conditions[a] for production of biodiesel from various feedstocks using response surface methodology

Feedstock oil or fat	Catalyst (wt %)	Temp (°C)	MeOH	rpm	Time (min)	Yield (wt.%)	Ref
Pork lard	1.26 KOH	65	7.5:1	n.r.[b]	20	97.8[c]	Jeong et al. 2009
Rapeseed	1.0 KOH	65	6:1	600	120	95–96	Rashid and Anwar 2008a
Sunflower	1.0 NaOH	60	6:1	600	120	97.1	Rashid et al. 2008b
Safflower	1.0 NaOCH$_3$	60	6:1	600	120	98	Rashid and Anwar 2008b
Jojoba	1.35 KOH	25	6:1	600	60	83.5	Bouaid et al. 2007
Rice bran	0.75 NaOH	55	9:1	n.r.	60	90.2	Sinha et al. 2008
Waste cooking oil	1.0 NaOH	50	9:1	n.r.	90	89.9[c]	Meng et al. 2008
Jatropha curcas	0.55 KOH	60	5:1	n.r.	24	99	Kumartiwari et al. 2007
Madhuca indica	0.70 KOH	60	6:1	n.r.	30	98	Ghadge and Raheman 2006
Pongamia pinnata	1.0 KOH	65	6:1	360	180	97–98	Meher et al. 2006b
Brassica carinata	1.2 KOH	25	6:1	600	60	97	Vicente et al. 2005
Used frying oil	1.1 NaOH	60	7:1	600	20	88.8	Leung and Guo 2006
canola	1.0 NaOH	40	6:1	600	60	93.5	Leung and Guo 2006
Cottonseed	1.07 KOH	25	20:1[d]	600	30	98	Joshi et al. 2008a, b
Raphanus sativus	0.6 NaOH	38	11.7:1[d]	n.r.	60	99.1	Domingos et al. 2008

[a]*temp* temperature of the reaction, *MeOH* mole ratio of methanol to oil, *rpm* (rotations per min) agitation intensity, *time* how long the reaction was conducted
[b]Not reported
[c]Conversion to esters (wt %) is provided instead of yield
[d]Ethanol was used to produce the corresponding ethyl esters

Kumartiwari et al. (2007) in which *Jatropha curcas* oil FAME were produced (after acid pretreatment) using 0.55 wt.% KOH, 60°C reaction temperature, 5:1 molar ratio of methanol to oil, and 24 min reaction time to provide biodiesel in 99% yield. The reaction parameters do not vary by a significant amount from the classic reaction conditions elucidated by Freedman et al. 1984.

The transesterification reaction employing methanol commences as two immiscible phases as a result of the very low solubility of TAG in methanol (Boocock et al. 1998; Stavarache et al. 2003, 2008; Zhou et al. 2003; Mao et al. 2004; Mahajan et al. 2006, 2007; Zhou and Boocock 2006a, b; Doell et al. 2008). Illustrative of this point is the fact that only 7.5 g of soybean oil is soluble in 1 L of methanol at 30°C (Boocock et al. 1996b). The polar homogenous alkali catalyst is exclusively dissolved in the polar methanol phase at the beginning of the reaction and does not come into contact with the TAG phase unless sufficient agitation is introduced. Stirring of sufficient magnitude causes TAG to transport into the methanol phase where it is rapidly converted into FAME and glycerol, as depicted in Fig.15.1. The rate at which FAME are produced during the transesterification reaction is thus controlled by mass-transfer limitations, which results in a lag time before conversion to FAME begins (Boocock et al. 1998; Zhou and Boocock 2006b; Doell et al. 008). Once the DAG and MAG intermediates are formed in sufficient quantity during the transesterification reaction, they serve as surfactants that improve mass transfer of TAG into the methanol phase. The reaction eventually transforms into another biphasic system that consists of ester-rich (FAME) and glycerol-rich phases. The alkali catalyst is preferentially soluble in the more polar glycerol-rich phase, which may result in a retardation of the rate of reaction (Mao et al. 2004). The glycerol-rich phase facilitates purification by settling to the bottom of the reaction vessel when agitation is ceased.

As a result of the biphasic nature of the reaction mixture, there is a lag time at the beginning of the methanolysis reaction before FAME begins to form, after which the reaction speeds up, but then quickly decelerates (Freedman et al. 1984; Darnoko and Cheryan 2000). Addition of co-solvents such as tetrahydrofuran (THF) significantly accelerates the production of FAME as a result of the formation of a monophasic as opposed to a biphasic reaction mixture (Boocock et al. 1998; Mahajan et al. 2006; Doell et al. 2008). However, the molar ratio of methanol to oil must be increased to at least 25:1, which results in additional solvent recovery during purification. Other possibilities for accelerating the methanolysis reaction are microwave (Breccia et al. 1999) or ultrasonic (Stavarache et al. 2003, 2008) irradiation.

Ethanolysis. The classic conditions for ethanolysis of vegetable oils or animal fats are 6:1 molar ratio of ethanol to oil, 0.5 wt.% catalyst (with respect to TAG), 600+ rpm, 75°C reaction temperature, and 1 h reaction time to produce fatty acid ethyl esters (FAEE) and glycerol (Freedman et al. 1984). Ethyl esters have been prepared from a number of feedstocks for use or evaluation as potential biodiesel fuels (Schwab et al. 1987; Peterson et al. 1996; Foglia et al. 1997; Wu et al. 1998; Nimcevic et al. 2000; Lang et al. 2001; Lee et al. 2002; Encinar et al. 2002, 2007; Zhou et al. 2003; Mariod et al. 2006; Meneghetti et al. 2006; Dantas et al. 2007; Issariyakul et al. 2007; Kucek et al. 2007; Kulkarni et al. 2007; Kumari et al. 2007; Moreira et al. 2007; Alamu et al. 2008; Domingos et al. 2008; Georgogianni et al. 2008; Hamad et al. 2008; Joshi et al. 2008, 2009; Lima et al. 2008; Rodrigues et al. 2008; Rosa et al. 2008; Stavarache et al. 2008). In addition, mixtures of methyl and ethyl esters have been reported whereby the transesterification reaction was conducted with both methanol and ethanol (Issariyakul et al. 2007; Kulkarni et al. 2007; Joshi et al. 2009). As in the case of methanolysis, the ethanolysis reaction has been optimized using RSM (Kucek et al. 2007; Domingos et al. 2008; Joshi et al. 2008). Please refer to Table15.6 for two recent examples from the literature. A representative example is that of the ethanolysis of crude *Raphanus sativus* oil (Domingos et al. 2008) in which 0.60 wt.% NaOH, 11.7:1 molar ratio of ethanol to oil, 38°C reaction temperature, and a 1-h reaction time afforded the corresponding FAEE in 99.1% yield. The reaction temperature and amount of ethanol in this case varied considerably from the conditions initially reported by Freedman et al. (1984).

Ethanolysis proceeds at a slower rate than methanolysis because of the higher reactivity of the methoxide anion in comparison to ethoxide. As the length of the carbon chain of the alkoxide anion increases, a corresponding decrease in nucleophilicity occurs, resulting in a reduction in the reactivity of ethoxide in comparison to methoxide (Sridharan and Mathai 1974). An example of this phenomenon is the transesterification (at 25 and 75°C) of canola oil with a 1:1 mixture of ethanol to methanol (to provide an overall molar ratio of alcohol to oil of 6:1) that results in 50% more methyl than ethyl esters (Kulkarni et al. 2007). Another example is the transesterification of canola oil at 25 and 75°C with a 1:1 mixture of ethanol to methanol that results in methyl to ethyl ester ratios of 2.7:1 and 1.3:1, respectively (Joshi et al. 2009). These results indicate that FAME are preferentially formed at both ambient and elevated reaction temperatures, but the preference is diminished at elevated temperatures. Even though the formation of FAEE is comparatively slow, the overall rate of formation of esters is faster than with methanol alone due to the better solubility of TAG in a mixture of methanol and ethanol, which results in a reduction of mass transfer limitations (Kulkarni et al. 2007). For example, ultrasonically assisted transesterification of *Melia azedarach* (syringa) oil is complete after 40 and 20 min and with methanol and ethanol respectively (Stavarache et al. 2008).

The base-catalyzed formation of FAEE is more complicated than the production of FAME. Specifically, the formation of stable emulsions during ethanolysis is problematic during subsequent purification (Korus et al. 1993; Zhou et al. 2003; Zhou and Boocock 2006a). In the case of methanolysis, these emulsions quickly and easily separate to form a lower glycerol-rich and an upper FAME-rich phase after agitation of the reaction has ceased. In ethanolysis, these emulsions are much more stable and complicate separation and purification of biodiesel (Zhou et al. 2003; Zhou and Boocock 2006a). Ethanol is less polar than methanol, so it is slightly more miscible with TAG at ambient temperature than methanol, but mechanical agitation during the transesterification reaction is once again required to facilitate sufficient mass transfer between phases (Kulkarni et al. 2007).

Butanolysis. The classic conditions for butanolysis of vegetable oils or animal fats are 6:1 molar ratio of butanol to oil, 0.5 wt.% catalyst (with respect to TAG), 600+ rpm, 114°C reaction temperature, and 1 h reaction time to produce fatty acid butyl esters and glycerol (Freedman et al. 1984). Butyl esters have been prepared from a variety of feedstocks for use or evaluation as potential biodiesel fuels (Freedman et al. 1986; Schwab et al. 1987; Ali and Hanna 1994; Foglia et al. 1997; Nimcevic et al. 2000; Lang et al. 2001; Zhou and Boocock 2006a, b; Rodrigues et al. 2008). To date, the butanolysis reaction has not yet been optimized by RSM.

Butanol is completely miscible with vegetable oils and animal fats due to its lower polarity versus methanol and ethanol (Boocock et al. 1996a). Consequently, transesterification reactions employing butanol are monophasic throughout (Zhou and Boocock 2006a, b). The monophasic nature of butanolysis influences the rate and extent of the reaction. There are no mass transfer limitations in the case of butanolysis, since all reactants and catalysts are contained in a single phase. As a result, the initial rate of butanolysis is considerably faster than that of methanolysis. For example, the yield of esters after 1 min is 88 wt.% in the case of butanolysis (114°C reaction temperature) but only 78 wt.% for methanolysis (60°C; Schwab et al. 1987). Another study found that butanolysis (30°C) was 50% complete after only 15 s of reaction, and 60% and 63.5% complete after 90 and 150 s, respectively. However, methanolysis (40°C) was only 55% complete after 10 min (Freedman et al. 1986). In a more recent example, 15.4 wt.% of TAG remained after 3 min of butanolysis as opposed to 84.4 wt.% in the case of methanolysis (Zhou and Boocock 2006a). At up to 40% conversion to alkyl esters, methanolysis is 12–16 times slower than butanolysis if lag time in the case of methanolysis is ignored and even slower if it is not ignored (Boocock et al. 1996a; Freedman et al. 1986). The difference in reactivity would be even more striking had the reactions in the above example been performed at similar temperatures (methanolysis was conducted at 40°C as opposed to 30°C for butanolysis).

Because the reactions depicted in Fig.15.1 for the conversion of TAG into alkyl esters are reversible, the monophasic nature of butanolysis affects the extent of reaction. In the case of methanolysis, glycerol separation from FAME severely curtails the unwanted reverse reactions. In the case of butanolysis, the reverse reactions are more likely to occur because all materials are in contact throughout the reaction. The monophasic nature of butanolysis also complicates purification of the resultant butyl esters, as gravity separation of glycerol at the conclusion of the reaction is not possible. The weaker nucleophilicity of butoxide versus methoxide is another factor that affects the extent of reaction. Although butanolysis proceeds at a faster initial rate than methanolysis, the final conversion to products after 1 h reaction (114°C and 60°C reaction temperatures, respectively) is 96 wt.% versus 98 wt.% for methanolysis (Schwab et al. 1987). In addition, after 1 h (at 23°C), 14.4 wt.% of bound glycerol (TAG + DAG + MAG) remained, whereas only 11.7 and 7.2 wt.% remained in the cases of methanolysis and ethanolysis, respectively (Zhou and Boocock 2006b). In summary, the butanolysis reaction is monophasic throughout, which results in a faster initial rate of reaction but may yield lower overall conversion to butyl esters in comparison to methyl or ethyl esters.

Influence of Biodiesel Composition on Fuel Properties

The fatty ester composition, along with the presence of contaminants and minor components, dictates the fuel properties of biodiesel fuel. Because each feedstock has a unique chemical composition, biodiesel produced from different feedstocks will in turn have different fuel properties. Important properties of biodiesel that are directly influenced by fatty ester composition and the presence of contaminants and minor components include low-temperature operability, oxidative and storage stability, kinematic viscosity, exhaust emissions, cetane number, and energy content. In the context of biodiesel, minor components (Fig.15.3) are defined as naturally occurring species

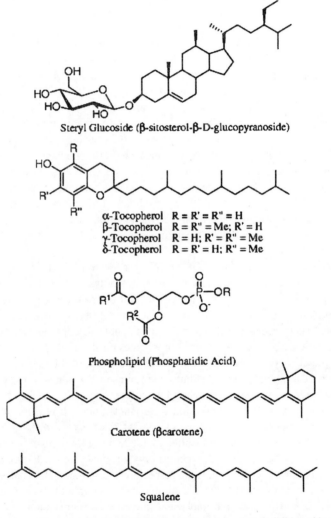

Fig. 15.3 Representative examples steryl glucosides, tocopherols, phospholipids and hydrocarbons that may be found in biodiesel

found in vegetable oils and animal fats and may include tocopherols, phospholipids, steryl glucosides (also called sterol glucosides, steryl glycosides, sterol glycosides, or phytosterols), chlorophyll, fat soluble vitamins, and hydrocarbons (such as alkanes, squalene, carotenes, and polycyclic aromatic hydrocarbons; Gunstone 2004). Contaminants are defined as incomplete or unwanted reaction products, such as FFA, soaps, TAG, DAG, MAG, alcohol, catalyst, glycerol, metals, and water.

Low-temperature operability. Low-temperature operability of biodiesel is normally determined by three common parameters: cloud point (CP; ASTM D2500 or D5773), pour point (PP; ASTM D97 or D5949), and cold filter plugging point (CFPP; ASTM D6371). For a list of low-temperature specifications contained within ASTM D6751 and EN 14214, please refer to Tables 15.2 and 15.3. A fourth parameter, the low-temperature flow test (LTFT; ASTM D4539), is less user friendly and less commonly used. The cloud point is defined as the temperature at which crystal growth is large enough (diameter ≥ 0.5 μm) to be visible to the naked eye. At temperatures below the CP, larger crystals fuse together and form agglomerations that eventually become extensive enough to prevent pouring of the fluid. The lowest temperature at which the fluid will pour is defined as the PP. The CFPP is defined as the lowest temperature at which a given volume of biodiesel completely flows under vacuum through a wire mesh filter screen within 60 s. The CFPP is generally considered to be a more reliable indicator of low-temperature operability than CP or PP, since the fuel will contain solids of sufficient size to render the engine inoperable due to fuel filter plugging once the CFPP is reached (Dunn and Bagby 1995; Dunn et al. 1996; Park et al. 2008b). It should be noted that it is inappropriate to measure CP, PP, and CFPP of chemically pure compounds (pure methyl oleate, for instance). Instead, determination of melting point (mp) as a means to measure low-temperature operability is appropriate for chemically pure compounds. The CP, PP, CFPP, and LTFT attempt to quantify low-temperature behavior of complex mixtures (such as biodiesel) that are composed of constituents that generally exhibit a relatively wide range of mp (or freezing points, fp). For instance, in the case of CP, it is very likely that initial crystal growth that occurs upon cooling of biodiesel to sub-ambient conditions is primarily a result of higher melting constituents undergoing crystallization.

The low-temperature behavior of chemical compounds is dictated by molecular structure. Structural features such as chain length, degree of unsaturation, orientation of double bonds, and type of ester head group strongly influence mp of individual chemical constituents of biodiesel. As can be seen from Table 15.7, as the chain length of saturated FAME is increased from 12 (C12:0, methyl laurate, mp 5°C) to 18 (C18:0, methyl stearate, mp 39°C) carbons, a corresponding increase in mp is observed. With respect to unsaturation, compounds of similar chain length but increasing levels of unsaturation display lower mp, as evidenced by the mp of C18:0 (mp 39°C), C18:1 (methyl oleate, mp −20°C), C18:2 (methyl linoleate, mp −35°C), and C18:3 (methyl linolenate, mp −52°C). The orientation of the double bond(s) is another important factor influencing mp. Nearly all naturally occurring unsaturated fatty acids contain *cis*-oriented double bonds. However, *trans*-isomers may be chemically introduced through catalytic partial hydrogenation.

Table 15.7 Melting point (mp), standard heat of combustion ($\Delta_c H^o$), kinematic viscosity (v, 40°C), oil stability index (OSI, 110°C), cetane number (CN), and lubricity (Lub, 60°C) of fatty acid alkyl esters (FAAE) commonly found in biodiesel

FAAE[a]	mp[b] (°C)	$\Delta_c H^o$ (MJ/mol; MJ/kg)[c]	v[d] (mm²/s)	OSI[e] (h)	CN[f]	Lub[g] (μm)
C12:0 ME	5	8.14/37.97	2.43	>40	67	416
C12:0 EE	−2		2.63	>40		
C14:0 ME	19	10.67/39.45	3.30	>40		353
C14:0 EE	12		3.52	>40		
C16:0 ME	31	10.67/39.45	4.38	>40	86	357
C16:0 EE	19		4.57		93	
C16:1 ME	−34	10.55/39.30	3.67	2.1	51	246
C16:1 EE	−37					
C18:0 ME	39	11.96/40.07	5.85	>40	101	322
C18:0 EE	32		5.92	>40	97	
C18:0 BE	28		7.59		92	
C18:1 ME	−20	11.89/40.09	4.51	2.5	59	290
C18:1 EE	−20		4.78	3.5	68	
C18:1 BE	−26		5.69		62	303
C18:2 ME	−35	11.69/39.70	3.65	1.0	38	236
C18:2 EE			4.25	1.1	40	
C18:3 ME	−52	11.51/39.34	3.14	0.2	23	183
C18:3 EE			3.42	0.2	27	

[a]Refer to Table15.1 for explanation of notation. *ME* methyl ester, *EE* ethyl ester, *BE* butyl ester
[b]Values are from Anonymous (2007) or Knothe (2008)
[c]Values from Moseret et al. (2009)
[d]Values from Knothe and Steidley (2005a)
[e]Values from Knothe (2008) or Moser (2008b)
[f]Values from Knothe et al. (2003) or Knothe (2008)
[g]Values from Knothe and Steidley (2005b)

Constituents that contain *trans-* double bonds exhibit higher mp than the corresponding *cis-* isomers. For example, the mp of methyl elaidate (methyl 9*E*-octadecenoate) is 13°C (Anonymous 2007), which is considerably higher than the *cis* isomer (methyl oleate, mp −20°C). With respect to ester head group, the mp of FAAE is generally lower with larger alkyl head groups up to about eight carbons. For instance, methyl, ethyl, and butyl stearates have mp of 39°C, 32°C, and 28°C, respectively. Refer to Table15.8 for a summary of the effects of various structural features on the low-temperature operability of FAAE.

Applying the above trends to biodiesel, one can see a pronounced effect in the case of ester head group: the CP of soybean oil methyl, ethyl, isopropyl, and 2-butyl esters is −2, −2, −9 and −12°C, respectively (Lee et al. 1995). The CP of beef tallow methyl, ethyl, propyl, and butyl esters is 17°C, 15°C, 12°C, and 9°C, respectively (Foglia et al. 1997). The examples of soybean oil (14% saturated FAEE) and tallow (25% saturated FAEE) esters are illustrative of the influence of double bond content on low-temperature operability: soybean oil alkyl esters contain a lower amount of saturated FAAE than do tallow esters, which results in superior low-temperature properties. Another example of the influence of saturated FAME on low-temperature

Table 15.8 Effect of fatty acid alkyl ester structural features on melting point (mp), oxidative stability (OSI), kinematic viscosity (v), standard heat of combustion ($\Delta_c H^\circ$), cetane number (CN), and lubricity (Lub)

Structural feature	mp	OSI	v	$\Delta_c H^\circ$	CN	Lub
Chain length	↑[a]	↑	↑	↑	↑	↓
Number of double bonds	↓	↓	↓	↓	↓	↓
cis double bond(s)[b]	↓	↓	↓	–[c]	?[d]	?[d]
Larger ester head groups	↓	–[c]	↑	↑	–[c]	–[c]

[a] ↑ higher numeric value, ↓ lower numeric value. For mp, v, and lubricity, ↓ indicates that the property is improved. For OSI, $\Delta_c H^\circ$, and CN, ↓ indicates that the property is negatively impacted
[b] In comparison to the corresponding *trans* isomer
[c] Negligible or no impact
[d] Effect has not been reported

operability is that of soybean and palm oil FAME (SME and PME), which have PP values of 0°C and 18°C, respectively, and CFPP of −4°C and 12°C, respectively (Moser 2008a). The influence of double bond orientation is evidenced by comparison of the CP and PP values of SME (0°C and −2°C) with *trans*-containing partially hydrogenated SME (3°C and 0°C; Moser et al. 2007).

Oxidative stability. Oxidative stability of biodiesel is determined through measurement of the oil stability index (OSI) by the Rancimat method (EN 14112). The Rancimat method indirectly measures oxidation by monitoring the gradual change in conductivity of a solution of water caused by volatile oxidative degradation products that have been transported via a stream of air (10 L/h) from the vessel (at 110°C) containing the biodiesel sample. The OSI is mathematically determined as the inflection point of a computer-generated plot of conductivity (μS/cm) of distilled water versus time (h). The units for OSI are normally expressed in hours. Biodiesel fuels with longer OSI times are more stable to oxidation than samples with shorter values. For a list of oxidative stability specifications contained within ASTM D6751 and EN 14214, please refer to Tables15.2 and 15.3. The American Oil Chemists' Society (AOCS) official method Cd 12b-92 is nearly identical to EN 14112 and provides essentially interchangeable data.

Oxidation initiates in the case of lipids at methylene carbons allylic to sites of unsaturation (Frankel 2005). Autoxidation of lipids, including biodiesel, produces free radicals through hydrogen abstraction in the presence of various initiators such as light, heat, peroxides, hydroperoxides, and transition metals (Fig.15.4). These free radicals further react exothermically with molecular oxygen to produce peroxides, which in turn react with unoxidized lipids to produce additional free radicals (Frankel 2005). Generally, the rate-limiting step in the autoxidation of lipids is initial hydrogen abstraction. Accelerated methods to determine oxidative stability of lipids, such as EN 14112 or AOCS Cd 12b-92, use elevated temperatures to greatly accelerate initial hydrogen abstraction. The EN 14112 method increases the amount of oxygen available to react with lipid-free radicals by pumping air at a constant rate through the sample. Products that ultimately form through oxidation of lipids may include aldehydes, shorter-chain fatty acids, other oxygenated species

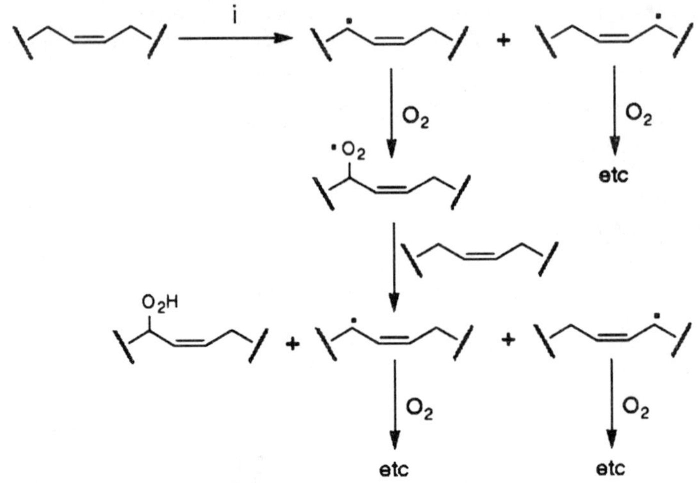

Fig. 15.4 Simplified pathway of initial lipid autoxidation (*i* initiator)

(such as ketones), and polymers (Frankel 2005). The rate of autoxidation is dependant on the number and location of methylene-interrupted double bonds contained within FAAE. Materials that contain more methylene carbons allylic to sites of unsaturation (such as polyunsaturated esters) are particularly vulnerable to autoxidation, as evidenced by the relative rates of oxidation of the unsaturates: 1, 41, and 98 for ethyl esters of oleic, linoleic, and linolenic acids (Holman and Elmer 1947). Not only will easily oxidized biodiesel fail oxidative stability requirements contained within ASTM D6751 and EN 14214 (*see* Tables 15.2 and 15.3), but oxidative degradation negatively impacts AV and kinematic viscosity (Plessis et al. 1985; Mittelbach and Gangl 2001; Bondioli et al. 2003), both of which are specified in ASTM D6751 and EN 14214.

Structural features such as degree of unsaturation, double bond orientation, chain length, and type of ester head group influence the OSI of FAAE. Esters that are otherwise similar but contain a greater number of methylene-interrupted double bonds undergo oxidative degradation at progressively faster rates (Holman and Elmer 1947; Knothe and Dunn 2003; Frankel 2005; Knothe 2008; Moser 2009a), which is confirmed through examination of the OSI values (Table 15.7) of FAME of stearic (>40 h), oleic (2.5 h), linoleic (1.0 h), and linolenic (0.2 h) acids. Applying these results to biodiesel, superior oxidative stability is expected from biodiesel fuels prepared from feedstocks relatively high in saturated fatty acid content or relatively low in polyunsaturated fatty acid content. For example, PME (50 wt.% saturated FAME, 11% polyunsaturated FAME, Table 15.1) are considerably more stable to oxidation than SME (15% saturates, 62% polyunsaturates) according to EN 14112 (Moser 2008a; Park et al. 2008b).

Double bond orientation (*cis* versus *trans* isomers) also impacts oxidative stability, as seen by the contrasting OSI values of FAME of oleic (2.5 h) and elaidic (7.7 h) acids (Moser 2009a). This is not surprising, as it is known that *trans* isomers are generally thermodynamically more stable than the corresponding *cis* isomers. Alkyl substituents are separated by a greater distance in *trans* than they are in *cis* isomers, thus keeping the influence of steric hinderance to a minimum. Applying these observations to biodiesel, fuels that contain at least some *trans* constituents exhibit enhanced stability to oxidation according to EN 14112 than biodiesel fuels that contain a similar number of entirely *cis* double bonds. According to a recent study (Moser et al. 2007), partially hydrogenated SME (7.7% *trans* FAME, 16.4% saturated FAME, 44.7% polyunsaturated FAME) yielded an OSI value of 6.2 h (110°C, AOCS Cd 12b-92) versus 2.3 h for SME (0% *trans*, 16.9% saturated, 59.8% polyunsaturated).

Due to the nature of the AOCS and EN 14112 methods, a bias is introduced whereby compounds of higher molecular weight (MW; i.e., longer chain length) will appear to be more stable to oxidation than compounds of similar double bond content but lower MW. The parameters of the method are the cause of this bias: a specified mass (3.0 g in the case of EN 14112) of material is required as opposed to a specified molar amount. Compounds of longer chain length will have fewer molecules in the experimental sample, which will reduce the number of double bonds available for oxidation versus molecules of similar double bond content but with lower MW. Illustrative of this point are the FAME of palmitoleic (MW 268.44), oleic (MW 296.49), and gondoic (C20:1; MW 324.54) acids, which have similar double bond content but increasing OSI values of 2.1, 2.5, and 2.9 h, respectively (Table 15.7 and Moser 2009a). Concomitant with these results is the variation in ester head group among otherwise similar compounds. Ethyl oleate (MW 310.52) appears to have greater oxidative stability than methyl oleate, as evidenced by OSI values of 3.5 and 2.5 h, respectively. The effect of ester head group on OSI is less pronounced for polyunsaturated compounds such as methyl and ethyl linoleates and methyl and ethyl linolenates. The increased double bond content of these materials results in an accelerated rate of oxidation that essentially masks the aforementioned MW effect on OSI (Moser 2009a). Applying these results to biodiesel, one may expect improved oxidative stability according to EN 14112 with the use of FAEE (or larger ester head groups) as opposed to FAME with feedstocks not particularly high in polyunsaturated FAME content. Analogously, biodiesel produced from feedstocks with generally longer chain lengths but similar double bond content may also be more stable to oxidation according to EN 14112. To date, these suppositions remain unexplored in the scientific literature. Refer to Table 15.8 for a summary of the influence of various structural features on oxidative stability of FAAE.

Oxidative stability and low-temperature operability are normally inversely related: structural factors that improve oxidative stability adversely influence low-temperature operability and vice versa. Ester head group is the only exception to this relationship, as larger ester head groups tend to improve both low-temperature performance and oxidative stability.

Kinematic viscosity. Kinematic viscosity is the primary reason why biodiesel is used as an alternative fuel instead of neat vegetable oils or animal fats. The high kinematic viscosities of vegetable oils and animal fats ultimately lead to operational problems such as engine deposits when used directly as fuels (Knothe and

Steidley 2005a). The kinematic viscosity of biodiesel is approximately an order of magnitude less than typical vegetable oils or animal fats and is slightly higher than petrodiesel. For instance, the kinematic viscosities (40°C) of soybean oil, SME, and ULSD are 31.49, 4.13, and 2.32 mm^2/s, respectively (Moser et al. 2008, 2009). Kinematic viscosity is determined following ASTM D445 at 40°C. For a list of kinematic viscosity specifications contained within ASTM D6751 and EN 14214, please refer to Tables 15.2 and 15.3. Several structural features influence the kinematic viscosities of FAAE, such as chain length, degree of unsaturation, double bond orientation, and type of ester head group. As seen from Table 15.7, factors such as longer chain length and larger ester head group result in increases in kinematic viscosity. For example, FAME of lauric, myristic, palmitic, and stearic acids have kinematic viscosities of 2.43, 3.30, 4.38, and 5.85 mm^2/s, respectively. Furthermore, methyl, ethyl, and butyl esters of stearic acid exhibit kinematic viscosities of 5.85, 5.92, and 7.59 mm^2/s, respectively. Increasing the degree of unsaturation results in a decrease in kinematic viscosity, as evidenced by the methyl esters of stearic (5.85 mm^2/s), oleic (4.51 mm^2/s), linoleic (3.65 mm^2/s), and linolenic (3.14 mm^2/s) acids. Double bond orientation also impacts kinematic viscosity, as seen by comparison of methyl elaidate (5.86 mm^2/s) and methyl linoelaidate (5.33 mm^2/s) to the corresponding *cis* isomers (methyl oleate and linoleate; Knothe and Steidley 2005a). Refer to Table 15.8 for a summary of the effects of various structural features on kinematic viscosity of FAAE.

The influence of ester head group is indicated by comparison of the kinematic viscosities ethyl and isopropyl esters of tallowate (5.2 and 6.4 mm^2/s, respectively) (Wu et al. 1998). The methyl and ethyl esters of canola oil have kinematic viscosities of 3.9 and 4.4 mm^2/s, respectively (Kulkarni et al. 2007). The influence of saturated FAME on viscosity is illustrated by comparison of SME (4.12 mm^2/s) to PME (4.58 mm^2/s; Moser 2008a). The impact of double bond orientation is demonstrated by comparison of SME to partially hydrogenated SME (5.0 mm^2/s; Moser et al. 2007).

Exhaust emissions. Exhaust emissions regulated by title 40, section 86 of the U.S. Code of Federal Regulations (CFR) include oxides of nitrogen (NO_x), PM (particulate matter), THC (total hydrocarbons), and carbon monoxide (CO). Procedures outlined in title 40, section 86, subpart N of the CFR describe how regulated exhaust emissions are measured. Neither ASTM D6751 nor EN 14214 contains specifications relating to exhaust emissions. Combustion of biodiesel (B100) in diesel engines results in an average increase in NO_x exhaust emissions of 12% and decreases in PM, THC, and CO emissions of 48%, 77%, and 48%, respectively, relative to petrodiesel (Graboski and McCormick 1998; Choi and Reitz 1999; EPA 2002 ; Song et al. 2002; Hess et al. 2005, 2007). For B20 blends of SME in petrodiesel, NO_x emissions are increased by 0–4% versus neat petrodiesel, but PM, THC, and CO emissions are reduced by 10%, 20%, and 11%, respectively (Hess et al. 2007; EPA 2002). Another study demonstrated that combustion of B5 and B20 blends of ULSD in a modern diesel engine equipped with EGR showed no significant difference in NO_x emissions from that of neat ULSD (Williams et al. 2006).

The increase in NO_x emissions with combustion of biodiesel and in some cases biodiesel–petrodiesel blends is of concern in environmentally sensitive areas such as national parks and urban centers. Reduction of smog-forming NO_x exhaust emissions to levels equal to or lower than those observed for petrodiesel is essential for universal acceptance of biodiesel.

The Zeldovich (thermal) and Fenimore (prompt) mechanisms are two pathways by which NO_x may be produced during combustion of biodiesel or petrodiesel (Miller and Bowman 1989; McCormick et al. 2003; Hess et al. 2005; Fernando et al. 2006). Thermal NO_x is formed at high temperatures in the combustion chamber of diesel engines when atmospheric molecular oxygen (O_2) combines with atmospheric molecular nitrogen (N_2). To reduce the rate of NO_x formation by this mechanism, the flame temperature inside of the combustion chamber must be reduced (Fernando et al. 2006). Prompt NO_x is produced by a complex pathway whereby hydrocarbon radicals react with N_2 to form species that subsequently react with O_2 to form NO_x. However, prompt NO_x has only been detected in the laboratory, and none has been observed during combustion of petrodiesel in diesel engines (Newhall and Shahed 1971). Consequently, the Zeldovich mechanism is responsible for NO_x exhaust emissions during combustion of petrodiesel (Newhall and Shahed 1971; Heywood 1998) and biodiesel (Fernando et al. 2006; Ban-Weiss et al. 2007) in diesel engines. A third mechanism largely irrelevant to biodiesel is the fuel NO_x pathway whereby nitrogen that is chemically bound to the fuel combines with excess O_2 during combustion (Fernando et al. 2006). Biodiesel does not normally contain nitrogen, so NO_x formation by the fuel NO_x mechanism is negligible.

Several engine or after-treatment technologies, such as EGR, SCR, diesel oxidation catalysts, and NO_x or particulate traps, may reduce NO_x exhaust emissions of biodiesel and blends with petrodiesel (McGeehan 2004). A factor influencing NO_x emissions is the chemical nature of FAAE that constitute biodiesel. Specifically, decreasing the chain length and/or increasing the number of double bonds (i.e., higher iodine value (IV)) of FAAE results in an increase in NO_x emissions (McCormick et al. 2001; Szybist et al. 2005; Knothe et al. 2006). The chemical composition of biodiesel varies according to the feedstock from which it is prepared. As a result, biodiesel obtained from feedstocks of significantly different compositions will exhibit different NO_x exhaust emissions behavior, as evidenced by a comparison of beef tallow FAME (IV 54; +0.0% increase in NO_x over petrodiesel), lard FAME (IV 63; +3.0%), chicken fat FAME (IV 77; +2.4%), and SME (IV 129; +6.2%; Wyatt et al. 2005). Enrichment of SME with up to 76% methyl oleate resulted in the elimination of the increase in NO_x exhaust emissions normally observed with B20 blends (Szybist et al. 2005). Additional studies have confirmed the influence of differing FAME compositions among various biodiesel fuels on NO_x exhaust emissions of B100 and blends with petrodiesel (Haas et al. 2001; Canakci and Van Gerpen 2003b; Kocak et al. 2007).

Cetane number. Cetane number (CN) is determined in accordance with ASTM D613 and is one of the primary indicators of diesel fuel quality. It is related to the ignition delay time a fuel experiences once injected into the combustion chamber. Generally, shorter ignition delay times result in higher CN and vice versa. Hexadecane, also known as cetane (trivial name), which gives the cetane scale its name, is the high-quality reference standard with a short ignition delay time and an arbitrarily assigned CN of 100. The compound 2,2,4,4,6,8,8-heptamethylnonane is the low-quality reference standard with a long ignition delay time and an arbitrarily assigned CN of 15 (Knothe et al. 1997). The end points of the CN scale indicate that branching negatively impacts CN. Also influencing CN is chain length, which yields higher CN for longer chain compounds, as seen in Table 15.7 in the case of the FAME of lauric (CN 67), palmitic (CN 86), and stearic (CN 101) acids. Increasing levels of unsaturation negatively impact CN, as evidenced by the FAME of stearic (101), oleic (59), linoleic (38), and linolenic (23) acids. The effect of ester head group on CN is not well-defined. Generally, it would be expected that CN would increase with increasing ester head group size, but this is not necessarily the case. For instance, the CN of methyl, ethyl, propyl, and butyl esters of oleic acid are 59, 68, 59, and 62 (Knothe et al. 2003). The concomitant influences of chain length and unsaturation on CN of biodiesel is evident by comparison of PME (62, DeOliveira et al. 2006) and SME (54, Knothe et al. 2006). Refer to Table 15.8 for a summary of the effects of various structural features on CN of FAAE. For a list of CN specifications contained within ASTM D6751 and EN 14214, please refer to Tables 15.2 and 15.3.

It is important to point out that the cetane index (ASTM D976 or D4737) is not applicable to biodiesel. The cetane index is used in the case of middle distillate fuels (i.e., ULSD) as an estimation of CN and is calculated from density, API gravity, and boiling range. However, biodiesel has dramatically different distillation qualities (e.g., much higher boiling range) than diesel fuels, thus rendering the equation used to calculate the cetane index inapplicable to biodiesel. Nevertheless, the scientific literature contains numerous erroneous examples of calculated cetane index values for biodiesel, which should be viewed with skepticism.

Heat of combustion. Heat of combustion of liquid hydrocarbons is determined according to ASTM D240, but it is not specified in ASTM D6741 nor EN 14214. Heat of combustion is the thermal energy that is liberated upon combustion, so it is commonly referred to as energy content. The heats of combustion of diesel fuel, B20 SME, and SME are 46.7, 43.8, and 38.1 MJ/kg, respectively (DeOliveira et al. 2006). As can be seen, as the biodiesel component in the fuel is increased from 0% to 100%, a concomitant decrease in energy content is observed.

Factors that influence the energy content of biodiesel include the oxygen content and carbon to hydrogen ratio. Generally, as the oxygen content of FAAE is increased, a corresponding reduction in energy content is observed. As seen in Table 15.7, methyl stearate possesses greater energy content (11.96 MJ/mol or 40.07 MJ/kg) than methyl laurate (8.14 MJ/mol or 37.97 MJ/kg). Therefore, the energy content of FAAE is directly proportional to chain length (Knothe 2008; Moser 2009b), since longer-chain FAAE contain more carbons but a similar number of oxygen atoms.

Another important factor is the carbon to hydrogen ratio. Generally, FAAE of similar chain length but lower carbon to hydrogen ratios (i.e., more hydrogen) exhibit greater energy content. For instance, FAME with 18 carbons in the fatty acid backbone include FAME of stearic (largest hydrogen content), oleic, linoleic, and linolenic (smallest hydrogen content) acids. As seen in Table 15.7, methyl stearate possesses the highest energy content and methyl linolenate the lowest. Therefore, lower energy content is obtained from progressively greater levels of unsaturation with otherwise similar chain length. Biodiesel fuels with larger ester head groups (such as ethyl, propyl, or butyl) are expected to have higher energy content as a result of their greater carbon to oxygen ratios. Refer to Table 15.8 for a summary of the effects of various structural features on energy content of FAAE. The heats of combustion of PME (37.86 MJ/kg) and SME (38.09 MJ/kg) are somewhat similar, but PME has a greater amount of saturated (increases energy content) and shorter-chain (reduces energy content) FAME than does SME (DeOliveira et al. 2006).

Lubricity. Lubricity is determined at 60°C in accordance to ASTM D6079 using a high-frequency reciprocating rig instrument. During the course of the experiment, a ball and disk are submerged in a liquid sample and rubbed at 60°C against each other for 75 min at a rate of 50 Hz to generate a wear scar. At the conclusion of the experiment, the maximum length of the wear scar is determined, and this value represents the lubricity of the sample. Shorter wear scar values indicate has superior lubricity versus samples with wear scars. Lubricity is not prescribed in ASTM D6751 or EN 14214. However, the petrodiesel standards, ASTM D975 and EN 590, contain maximum allowable wear scar limits of 520 and 460 µm, respectively. Biodiesel possesses inherently good lubricity, especially when compared to petrodiesel (Drown et al. 2001; Hughes et al. 2002; Goodrum and Geller 2005; Hu et al. 2005; Knothe and Steidley 2005b; Bhatnagar et al. 2006; Moser et al. 2008). For instance, the lubricity of ULSD (without lubricity-enhancing additives) is 551 µm, whereas SME displays a substantially lower value of 162 µm (Moser et al. 2008b). In another example, the lubricities of additive-free ULSD and SME were 651 and 129 µm, respectively (Knothe and Steidley 2005b). As seen from the above two examples, additive-free ULSD fails to satisfy either of the petrodiesel fuel standards (ASTM D975 and EN 590) with respect to lubricity. The poor lubricity of petrodiesel requires lubricity-enhancing additives or blending with another fuel of sufficient lubricity to achieve satisfactory lubricity. The reason for the poor lubricity of ULSD is not exclusively the removal of sulfur-containing compounds during hydrodesulfurization, but rather that polar compounds with other heteroatoms such as oxygen and nitrogen are also removed (Barbour et al. 2000; Dimitrakis 2003; Knothe and Steidley 2005b). As such, biodiesel serves as an excellent lubricity-enhancing additive for ULSD. For example, B2 and B20 blends of SME in ULSD improve lubricity from 551 to 212 and 171 µm, respectively (Moser et al. 2008).

Various structural features such as the presence of heteroatoms, chain length, and unsaturation influence lubricity of biodiesel. Biodiesel fuels possess at least two oxygen atoms that in large part explains their enhanced lubricities over typical hydrocarbon-containing petrodiesel fuels (Knothe and Steidley 2005b). Another structural feature that is known to influence lubricity is chain length. As seen in Table 15.7, compounds of increasing chain length generally display increasingly superior lubricity values (shorter wear scars), although the correlation is not perfect. For instance, the lubricities of the FAME of lauric, myristic, palmitic, and stearic acids are 416, 353, 357, and 246 µm, respectively. Increasing levels of unsaturation also impart superior lubricity to biodiesel, as evidenced by the FAME of stearic (322 µm), oleic (290 µm), linoleic (236 µm), and linolenic (183 µm) acids. Refer to Table 15.8 for a summary of the effects of various structural features on the lubricities of FAAE. Among biodiesel fuels prepared from a variety of feedstocks, some variability exists between fuels, but they are all essentially in the range of 120–200 µm, which is superior to neat ULSD (without additives; Drown et al. 2001; Bhatnagar et al. 2006; Moser et al. 2008).

Contaminants. Contaminants in biodiesel may include methanol, water, catalyst, glycerol, FFA, soaps, metals, MAG, DAG, and TAG. Methanol contamination in biodiesel is indirectly measured through flash point determination following ASTM D93. Both ASTM D6751 and EN 14214 contain flash point specifications, which can be found in Tables 15.2 and 15.3. If biodiesel is contaminated with methanol, it will fail to meet the minimum flash point specified in relevant fuel standards. Methanol contamination normally results from insufficient purification of biodiesel following the transesterification reaction.

Water is a major source of fuel contamination. While fuel leaving a production facility may be virtually free of water, once it enters the existing distribution and storage network, it will come into contact with water as a result of environmental humidity (Knothe et al. 2005). Water in biodiesel causes three serious problems: corrosion of engine fuel system components, promotion of microbial growth, and hydrolysis of FAME. Consequently, both ASTM D6751 and EN 14214 contain strict limits on water content in biodiesel (*see* Tables 15.2 and 3). Water may be present in biodiesel as either dissolved or free water. Dissolved water, which is measured by the Karl Fisher titration method in EN 14214 according to EN ISO 12937 (Table 15.3), is water that is soluble in biodiesel. Free water, which is measured by a centrifugation method (ASTM D2709) in ASTM D6751 (Table 15.2), arises after biodiesel becomes saturated with water, resulting in a separate water phase.

Residual homogeneous alkali catalyst may be present in biodiesel through insufficient purification following the transesterification reaction. The catalyst can be detected through combined sodium and potassium determination (EN 14538). Calcium and magnesium, determined according to EN 14538, may also be introduced into biodiesel during purification through washing with hard water or through the use of drying agents (such as magnesium sulfate) that contain these metals. Both ASTM D6751 and EN 14214 contain limits for combined sodium and potassium and combined calcium and magnesium contents. The primary problem associated with metal contamination is elevated ash production during combustion (Knothe et al. 2005). *See* Tables 15.2 and 3 for the limits for combined sodium and potassium and combined calcium and magnesium specifications found in ASTM D6751 and EN 14214.

Glycerol may be present in insufficiently purified biodiesel, which is determined using ASTM D6584 or EN 14105. Both ASTM D6751 and EN 14214 contain limits for the maximum allowable levels of glycerol permitted in biodiesel (*see* Tables 15.2 and 15.3). Glycerol is suspected of contributing to engine deposit formation during combustion (Knothe et al. 2005).

Bound glycerol (MAG + DAG + TAG) is limited in ASTM D6751 by the total glycerol (total glycerol = free + bound glycerol) specification (Table 15.2) and in EN 14214 (Table 15.3) by individual MAG, DAG, and TAG limits, along with a total glycerol requirement. In the case of ASTM D6751, total glycerol is quantified by GC according to ASTM D6584. Bound glycerol in biodiesel results from incomplete conversion of TAG into FAAE during the transesterification reaction and may cause carbon deposits on fuel injector tips and piston rings of diesel engines during combustion (Knothe et al. 2005). The presence of bound glycerol in biodiesel may also influence low-temperature operability, kinematic viscosity, and lubricity. As in the case of FFA, bound glycerol has a beneficial effect on lubricity, which is apparent from the wear scar data (ASTM D6079) of MAG (139 µm), DAG (186 µm), TAG (143 µm), and methyl oleate (290 µm; Knothe and Steidley 2005b). However, the presence of saturated bound glycerol (saturated fatty acids bound to glycerol) has a deleterious effect on low-temperature operability of biodiesel due to its very low solubility in FAME. As a result, high temperatures are required to keep them from crystallizing. For example, 0.5 wt.% 1-monostearin and dipalmitin in SME result in increases in CP from −6°C to 22°C and 11°C, respectively (Yu et al. 1998). However, a similar amount of monoolein does not adversely affect CP of biodiesel. In summary, saturated MAG are most problematic with respect to low-temperature operability of biodiesel, followed by saturated DAG and TAG. Lastly, bound glycerol adversely affects kinematic viscosity of biodiesel, as evidenced by the contrasting kinematic viscosities of triolein (32.94 mm^2/s; Knothe and Steidley 2005a) and methyl oleate (4.51 mm^2/s).

Free fatty acids may be present in biodiesel that was prepared from a feedstock with high FFA content or may be formed during hydrolysis (*see* Fig. 15.2) of biodiesel in the presence of water and catalyst. The presence of FFA is determined by AV according to ASTM D664. Strict upper limits on AV are specified in ASTM D6751 and EN 14214 (*see* Tables 15.2 and 15.3). The presence of FFA in biodiesel may impact other important fuel properties such as low-temperature performance, oxidative stability, kinematic viscosity, and lubricity. In addition to the issues with soap formation, FFA are known to act as pro-oxidants (Miyashita and Takagi 1986; Frankel 2005), so the presence of FFA in biodiesel may also negatively impact oxidative stability. FFA also have significantly higher kinematic viscosities and mp than the corresponding FAME, as evidenced by comparison of oleic acid (19.91 mm^2/s; 12°C) and methyl oleate (4.51 mm^2/s; −20°C) (Knothe and Steidley 2005a; Anonymous 2007). In the case of lubricity, FFA are beneficial, as seen by the wear scar data obtained by (ASTM D6079, 60°C) for oleic acid (0 µm; Knothe and Steidley 2005b) and methyl oleate (290 µm).

Minor components. Minor components in biodiesel may include tocopherols, phospholipids, steryl glucosides, chlorophyll, fat soluble vitamins, and hydrocarbons.

The quantities of these components depend on the feedstock from which the biodiesel is prepared, how the biodiesel is purified, and the degree of pre-processing (refining, bleaching, deodorization, degumming, etc) that is performed on the feedstock prior to transesterification. Many minor components, such as tocopherols (Mittelbach and Schober 2003; Dunn 2005a, *b*; Liang et al. 2006; Frohlich and Schober 2007; Bostyn et al. 2008; Moser 2008b; Tang et al. 2008), may serve as beneficial antioxidants (Frankel 2005). Chlorophylls, on the other hand, act as sensitizers for photo-oxidation (Gunstone 2004).

The presence of steryl glucosides (*see* Fig.15.3) in biodiesel is problematic because they result in significant low-temperature operability issues at temperatures above CP (Bondioli et al. 2008; Hoed et al. 2008; Moreau et al. 2008). Steryl glucoside precipitates, which form after production of biodiesel, cause engine failure due to fuel filter plugging (Lee et al. 2007; Pfalzgraf et al. 2007). Many steryl glucosides have mp in excess of 240°C and are insoluble in biodiesel (Lee et al. 2007). If present in sufficient concentration, over time, they precipitate to form solids above the CP of the fuel. This problem is especially prevalent in soybean and palm oil-based biodiesel fuels, as the parent oils contain relatively high natural levels of steryl glucosides (Hoed et al. 2008). Steryl glucoside levels are not limited in either ASTM D6751 or EN 14214, but a recent update to ASTM D6751 includes a new test method (ASTM D7501) that is designed to detect the presence of precipitates that form above the CP of the fuel, termed the "cold soak filtration test". Either GC (Bondioli et al. 2008; Hoed et al. 2008) or high-performance liquid chromatography (Moreau et al. 2008) may be used to detect and quantify the presence of steryl glucosides. Archer Daniels Midland Company recently filed a patent application for a method to remove steryl glucosides whereby the biodiesel sample is cooled, followed by filtration through diatomaceous earth (Lee et al. 2006). Another patent application also describes a process to remove steryl glucosides and other precipitates by cold treatment and filtration (Danzer et al. 2007).

Alternative Feedstocks for Biodiesel Production

A rapid increase in biodiesel production capacity and governmental mandates for alternative fuel usage around the world in the last several years has necessitated the development of alternative biodiesel feedstocks, as it does not appear possible to meet the increased production capacity and mandated demand with traditional sources of biodiesel (soybeans, rapeseed/canola, palm, and various greases and used cooking oils, for instance). The U.S. National Biodiesel Board estimates that biodiesel production capacity for 2009 in the U.S. was 2.69 billion gallons per year in comparison to a mere 23 million gallons per year in 2003. Production capacity in the 27 countries that comprise the European Union in 2009 was estimated by the European Biodiesel Board to be 12.94 billion gallons per yr versus 1.4 billion gallons per yr in 2006. The continued drive for energy sustainability and independence among energy-consuming countries, governmental mandates for alternative fuel usage and increased global production capacity contribute to the need for alternative sources of biodiesel fuel. For example, the U.S. Energy Independence and Security Act of 2007 includes a renewable fuels standard (RFS2) that mandates that 12.95 billion gallons of renewable fuels are to be blended into energy supplies by the end of 2010, of which 11.35 billion gallons is to be ethanol for use in gasoline. The other 1.6 billion gallons of renewable fuels must be "advanced biofuels," of which 650 million gallons is to be biomass-based diesel fuel (which includes biodiesel). By 2012, one billion gallons of biomass-based diesel is required under the RFS2 and the total renewable fuel requirement by 2022 will be 36 billion gallons. Lastly, the high prices of commodity vegetable oils and animal fats have made exploration of economical alternative non-food feedstocks an important research topic.

Desirable characteristics of alternative oilseed feedstocks for biodiesel production include adaptability to local growing conditions (rainfall, soil type, latitude, etc.), regional availability, high oil content, favorable fatty acid composition, compatibility with existing farm infrastructure, low agricultural inputs (water, fertilizer, pesticides), definable growth season, uniform seed maturation rates, potential markets for agricultural by-products, and the ability to grow in agriculturally undesirable lands and/or in the off-season from conventional commodity crops. Biodiesel fuels prepared from feedstocks that meet at least a majority of the above criteria will hold the most promise as alternatives to petrodiesel. In general, there are four major biodiesel feedstock categories: algae, oilseeds, animal fats, and various low-value materials such as used cooking oils, greases, and soapstocks. For recent comprehensive reviews on biodiesel produced from algae, please *see* Meng et al. 2009, Li et al. 2008, and Chisti 2007. Table 15.9 contains reaction conditions for the preparation of biodiesel from alternative feedstocks that are not already listed in Table 15.5. Tables 15.10 and 11 contain the fatty acid profiles of the alternative feedstock oils. Cases where feedstock(s) are missing from Tables 15.9 (or 5) and/or 10 and 11 are the result of a lack of information reported in the scientific literature.

Oilseeds. Traditional oilseed feedstocks for biodiesel production predominately include soybean, rapeseed/canola, palm, corn, sunflower, cottonseed, peanut, and coconut oils. Current agronomic efforts are focused on increasing feedstock supply by increasing yields. Both molecular breeding and biotechnology have the potential to increase the rate of yield growth for many crops, including corn and soybeans. For additional information, *see* Eathington et al. (2007). Alternative feedstocks normally arise out of necessity from regions of the world where the above materials are not locally available or as part of a concerted effort to reduce dependence on imported petroleum. For instance, non-edible *Jatropha curcas* (Jatropha) oil has recently attracted considerable interest as a feedstock for biodiesel production in India and other climatically similar regions of the world (Mohibbeazam et al. 2005; Kumartiwari et al. 2007; Kalbande et al. 2008). The Jatropha tree is a perennial poisonous oilseed shrub (up to 5 m) belonging to the Euphorbiaceae family whose seeds contain up to 30 wt.% oil that can be found in tropical and subtropical regions such as Central America, Africa, the Indian subcontinent, and other countries in Asia (Mohibbeazam et al. 2005). Because Jatropha oil contains a relatively high percentage of saturated fatty acids (34 wt.%, *see* Table 15.10), the corresponding

Table 15.9 Production of biodiesel from various alternative feedstocks

Feedstock	Oil (wt.%)	FFA (wt.%)	Catalyst (wt.%)	Temp (°C)	MeOH	Time (min)	Yield (wt.%)	Ref[a]
Melia azedarach	10	2.8	1.0 NaOH	36	9:1[b]	40	63.8	1
Balanites aegyptiaca	47	N/A	1.7 KOH	rt	6:1	60	90	2
Terminalia catappa	49	Trace	0.2 mol % NaOCH$_3$	65	6:1	60	93	3
Asclepias syiaca	20–25	N/A	1.0 NaOCH$_3$	60	6:1	60	99+	4
Cynara cardunculus	25	N/A	1.0 NaOH	75	12:1[c]	120	94	5
Camelina sativa	31	1.0	1.5 KOH	rt	6:1	60	98	6
Carthamus tinctorius	35–45	0.2	1.0 NaOCH$_3$	60	6:1	90	98	7
Sesamum indicum	44–58	N/A	0.5 NaOH	60	6:1	120	74	8
Sclerocarya birrea	N/A	2.1	18.4 H$_2$SO$_4$	60	22:1	180	77	9
Cucurbita pepo	45	0.27	NaOH	65	6:1	60	97.5	10
Melon bug	45	3.0	18.4 H$_2$SO$_4$	60	22:1	180	79	9
Soybean soapstock	50	>90	500+ H$_2$SO$_4$	35	30:1	120	99+	11
Municipal sludge	Up to 30	Up to 65	5.0 H$_2$SO$_4$	75	12:1	N/A	14.5	12

[a] *1* Stavarache et al. 2008, *2* Chapagain et al. 2009, *3* dos Santos et al. 2008, *4* Holser and Harry-O'Kuru 2006, *5* Encinar et al. 2002, *6* Frohlich and Rice 2005, *7* Rashid and Anwar 2008b, *8* Saydut et al. 2008, *9* Schinas et al. 2009, *10* Mariod et al. 2006, *11* Haas 2005, *12* Mondala et al. 2009
[b] Mole ratio of methanol to oil
[c] Ethanol instead of methanol

Table 15.10 Fatty acid profile (wt.%) of biodiesel fuels prepared from various alternative oilseed feedstocks

Feedstock	14:0	16:0	16:1	18:0	18:1	18:2	18:3	20:0	20:1	22:0	22:1	other	Ref[a]
Jatropha curcas	1.4	11.3		17.0	12.8	47.3		4.7				5.5	1
Pongamia pinnata		10.6		6.8	49.4	19.0		4.1	2.4	5.3		2.4[b]	2
Madhuca indica	1.0	17.8		14.0	46.3	17.9		3.0					3
Melia azedarach		10.1		3.5	21.8	64.1	0.4	0.2	0.3				4
Moringa oleifera		6.5		6.0	72.2	1.0		4.0	2.0	7.1			5
Nicotiana tabacum	0.1	11.0		3.3	14.5	69.5	0.7					0.9	6
Balanites aegyptiaca		13.7		11.0	43.7	31.5							7
Terminalia catappa		35.0		5.0	32.0	28.0							8
Hevea brasiliensis		10.2		8.7	24.6	39.6	16.3						9
Asclepias syiaca		5.9	6.8	2.3	34.8	48.7	1.2	0.2					10
Zanthoxylum bungeanum		10.6	5.2	1.4	32.1	25.6	24.1					1.0	11
Rice bran		18.8		2.4	43.1	33.2	0.6	0.7					12
Raphanus sativus		5.7		2.2	34.5	17.8	12.5	1.0	10.0		16.4		13
Brassica carinata [c]		5.3			10.0	24.6	16.5				43.6		14
Calophyllum inophyllum		5.4		0.2	43.2	36.0	15.2						15
Cynara cardunculus		12.0		13.0	34.1	38.3	0.3						16
Camelina sativa		14.0		3.0	25.0	56.0							17
Carthamus tinctorius		5.4		2.6	14.3	14.3	38.4	0.3	16.8	1.4	2.9	2.0	17
Sesamum indicum		6.9		2.0	14.2	76.0							18
Vernicia fordii		11.0		7.0	43.0	35.0						4.0	19
Sclerocarya birrea		3.0		2.2	8.6	11.5			0.8	8.4		58.3[d]	20
Cucurbita pepo		14.2	0.2	8.8	67.3	5.9		0.4	0.1			0.1	21
	0.1	12.5	0.2	5.4	37.1	43.7	0.2	0.4	0.1			0.2	22

[a] *1* Kumartiwari et al. 2007, *2* Bringi 1987, *3* Singh and Singh 1991, *4* Stavarache et al. 2008, *5* Rashid et al. 2008a, *6* Usta 2005, *7* Chapagain et al. 2009, *8* dos Santos et al. 2008, *9* Ramadhas et al. 2005, *10* Holser and Harry-O'Kuru 2006, *11* Zhang and Jiang 2008, *12* Sinha et al. 2008, *13* Domingos et al. 2008, *14* Dorado et al. 2004, *15* Sahoo et al. 2007, *16* Encinar et al. 2002, *17* Frohlich and Rice 2005, *18* Rashid and Anwar 2008b, *19* Elleuch et al. 2007, *20* Park et al. 2008a, *21* Mariod et al. 2006, *22* Schinas et al. 2009
[b] 2.4 wt.% 24:0
[c] Both high erucic acid (top) and low erucic acid (bottom) varieties are shown
[d] 63.8 wt.% eleostearic acid (18:3); 0.5 wt.% heneicosanoic acid (21:0); 1.3 wt.% unknown

Table 15.11 Fatty acid profile (wt.%) of biodiesel fuels prepared from animal fats and other low-value alternative feedstocks

Feedstock	14:0	16:0	16:1	18:0	18:1	18:2	18:3	20:0	20:1	22:0	24:0	other	Ref[a]
Salmon	6.8	14.9	6.1	3.2	15.6	2.1	11.5					39.8	1
Melon bug		30.9	10.7	3.5	46.6	3.9						2.4	2
Sorghum bug		12.2	1.0	7.3	40.9	34.5						0.1	2
Pork lard		26.4		12.1	44.7	12.7	1.0					3.1	3
Beef tallow	3.1	23.8	4.7	12.7	47.2	2.6	0.8					5.1[b]	4
Chicken fat	0.7	20.9	5.4	5.6	40.9	20.5						6.0[c]	4
Waste frying oil	1.0	30.7	0.6	5.7	40.5	19.1	0.2	0.6	0.4	0.3	0.4	0.3	5
Waste cooking oil	16.3	10.6		3.3	8.2	2.0						59.7[d]	6
Waste cooking oil		16.0		5.2	34.3	40.8							7
Used frying oil	0.9	20.4	4.6	4.8	52.9	13.5	0.8	0.1	0.8			0.1	8
Used cooking oil	0.2	11.9	0.2	3.8	31.3	50.8		0.3		0.5	0.2		9
Waste frying oil		8.4	0.2	3.7	34.6	50.5	0.6	0.4	0.4	0.8	0.3		10
Soybean soapstock		17.2		4.4	15.7	55.6	7.1						11
Yellow grease	2.4	23.2	3.8	13.0	44.3	7.0	0.7						12
Brown grease	1.7	22.8	3.1	12.5	42.4	12.1	0.8						12

[a] *1* Chiou et al. 2008, *2* Mariod et al. 2006, *3* Jeong et al. 2009, *4* Wyatt et al. 2005, *5* Predojevic 2008, *6* Phan and Phan 2008, *7* Meng et al. 2008, *8* Leung and Guo 2006, *9* Cetinkaya and Karaosmanoglu 2004, *10* Dias et al. 2008, *11* Haas 2005, *12* Canakci and Van Gerpen 2001
[b] 1.3 wt.% 14:1; 0.5 wt.% 15:0; 1.1 wt.% 17:0; 2.2 wt.% unknown
[c] 0.1 wt.% 14:1; 3.8 wt.% unknown
[d] 8.8 wt.% 8:0; 6.2 wt.% 10:0; 44.7 wt.% 12:0

FAME exhibit relatively poor low-temperature operability, as evidenced by a PP value of 2°C (Kumartiwari et al. 2007). The exhaust emission characteristics of FAME from Jatropha oil are summarized by Banapurmath et al. (2008). See Table 15.5 for an example of parameters used in the production of biodiesel from Jatropha oil.

Another non-edible feedstock of Indian origin is *Pongamia pinnata* (Karanja or Honge), which is a medium-sized (18 m) deciduous tree that grows fast in humid and subtropical environments and matures after 4 to 7 yr to provide fruit that contains one to two kidney-shaped kernels (Mohibbeazam et al. 2005). The oil content of Karanja kernels ranges between 30 and 40 wt.% (Karmee and Chadha 2005; Mohibbeazam et al. 2005). The primary fatty acid found in Karanja oil is oleic acid (45–70 wt.%; Table 15.10), followed by linoleic, palmitic, and stearic acids (Bringi 1987; Karmee and Chadha 2005; Naik et al. 2008). The low-temperature operability of the corresponding FAME is superior to that of Jatropha oil FAME as a result of the relatively high percentage of oleic acid in Karanja oil, as evidenced by CP and PP values of −2°C and −6°C, respectively (Srivastava and Verma 2008). The exhaust emissions characteristics of Karanja oil FAME are summarized by Raheman and Phadatare (2004) and Banapurmath et al. (2008). See Table 15.5 for an example of parameters used in the production of biodiesel from Karanja oil. *P. pinnata* may also be referred to as *P. glabra* in the scientific literature (Raheman and Phadatare 2004; Sarma et al. 2005) where it is sometimes referred to as koroch.

Madhuca indica, commonly known as Mahua, is a tropical tree found largely in the central and northern plains and forests of India. It belongs to the Sapotaceae family and grows quickly to approximately 20 m in height, possesses evergreen or semi-evergreen foliage, and is adapted to arid environments (Ghadge and Raheman 2006; Kumari et al. 2007). Non-edible fruit is obtained from the tree in 4 to 7 yr and contains one to two kidney-shaped kernels (Mohibbeazam et al. 2005). The oil content of dried Mahua seeds is around 50 wt.% (Bhatt et al. 2004). Mahua oil is characterized by an FFA content of around 20 wt.% (Ghadge and Raheman 2005, 2006) and a relatively high percentage of saturated fatty acids (Table 15.10) such as palmitic (17.8 wt.%) and stearic (14.0 wt.%) acids (Singh and Singh 1991). The remaining fatty acids are primarily distributed among unsaturated components such as oleic (46.3 wt.%) and linoleic (17.9 wt.%) acids (Singh and Singh 1991). The relatively high percentage of saturated fatty acids (35.8 wt.%, *see* Table 15.10) found in Mahua oil results in relatively poor low-temperature properties of the corresponding FAME, as evidenced by a PP value of 6°C (Ghadge and Raheman 2005, 2006). See Table 15.5 for an example of reaction parameters used in the preparation of biodiesel from Mahua oil.

Melia azedarach, commonly referred to as syringa or Persian lilac, is a deciduous tree that grows between 7 and 12 m in height in the mahogany family of Meliaceae that is native to India, southern China, and Australia (Stavarache et al. 2008). The oil content of dried syringa berries, which are poisonous, is around 10 wt.% (Stavarache et al. 2008). Syringa oil is characterized by a high percentage of unsaturated fatty acids (Table 15.10) such as oleic (21.8 wt.%) and linoleic

(64.1 wt.%) acids. Other constituents that are present in greater than 1 wt.% are saturated species such as palmitic (10.1 wt.%) and stearic (3.5 wt.%) acids (Stavarache et al. 2008). Physical properties of syringa oil FAME include an IV of 127, a kinematic viscosity (40°C) of 4.37 mm^2/s, and a specific gravity of 0.894 g/mL (Stavarache et al. 2008). *See* Table 15.9 for an example of reaction parameters used in the preparation of biodiesel from syringa oil.

Moringa oleifera, commonly known as Moringa, is an oilseed tree that grows in height from 5 to 10 m and is the most widely known and distributed of the Moringaceae family (Rashid et al. 2008a). The Moringa tree, indigenous to sub-Himalayan regions of northwest India, Africa, Arabia, southeast Asia, the Pacific and Caribbean islands, and South America, is now also distributed in the Philippines, Cambodia, and Central and North America. It thrives best in a tropical insular climate and is plentiful near the sandy beds of rivers and streams. The fast-growing, drought-tolerant Moringa tree can tolerate poor soil, a wide rainfall range (25–300+ cm per year), and soil pH from 5.0 to 9.0 (Palada and Changl 2003; Rashid et al. 2008a). When fully mature, dried seeds contain between 33% and 41% wt.% oil, which is high in oleic acid (>70%, *see* Table 15.10) and is commercially known as "ben oil" or "behen oil" due to the presence of 7.1 wt.% behenic acid (Rashid et al. 2008a). As a result of the relatively high behenic acid content of moringa oil, the FAME display relatively poor low-temperature properties, as evidenced by CP and PP values of 18°C and 17°C, respectively (Rashid et al. 2008a). In addition, moringa oil FAME exhibit a high CN of 67, one of the highest for a biodiesel fuel. *See* Table 15.5 for an example of reaction parameters used for the preparation of biodiesel from moringa oil.

Nicotiana tabacum, commonly referred to as tobacco, is a common oilseed plant with pink flowers and green capsules containing numerous small seeds that is grown in a large number of countries around the world (Usta 2005). The leaf of the plant is a commercial product and is used in the production of cigarettes and other tobacco-containing products. The oil content of the seeds, which are by-products of tobacco leaf production, ranges from 36 to 41 wt.% (Usta 2005). Tobacco seed oil normally contains more than 17 wt.% FFA (Veljkovic et al. 2006) and is high in linoleic acid (69.5 wt.%; Table 15.10), with oleic (14.5 wt.%) and palmitic (11.0 wt.%) acids also present in significant quantities. As a result of the relatively high linoleic acid content of tobacco seed oil, the corresponding FAME display a relatively low kinematic viscosity (3.5 mm^2/s; Usta 2005) in comparison to most other biodiesel fuels. In addition, the FAME have a CN of 51 and an IV of 112 (Usta 2005). *See* Table 15.5 for an example of reaction parameters used in the preparation of biodiesel from tobacco seed oil.

Balanites aegyptiaca, commonly known as desert date, is a small tree that grows up to 10 m that belongs to the Zygophyllaceae family. Common uses of the desert date include timber and food, although the fruit is shunned as a food source due to its undesirable taste unless other food is scarce (Chapagain et al. 2009). The tree is highly adapted to growing in arid regions in Africa and Asia and produces as many as 10,000 fruits per annum in optimal growing conditions (Chapagain et al. 2009). Each fruit weighs between 5 and 8 g and contains a kernel that constitutes 8 to 12 wt.% of the overall fruit mass (Chapagain et al. 2009). The kernel seeds, which are high in unsaturated fatty acids such as oleic (43.7 wt.%; Table 15.10) and linoleic (31.5 wt.%) acids, contain around 50 wt.% oil (Chapagain et al. 2009). Other common constituents found in desert date oil include saturated species such as palmitic (13.7 wt.%) and stearic (11.0 wt.%) acids. The physical properties of the resultant FAME include a CN of 53.5, IV of around 100, CP of 3–7°C, and CFPP of 1–3°C (Chapagain et al. 2009). See Table 15.9 for an example of reaction parameters used in the preparation of biodiesel from desert date oil.

Terminalia catappa, commonly known in Brazil as castanhola, is a large tropical tree belonging to the Combretaceae family growing up to 35 m in height (dos Santos et al. 2008). The tree primarily grows in freely drained, well-aerated, sandy soils and is tolerant of strong winds, salt spray, and moderate levels of salinity in the root zone. The fruit is produced from about 3 yr of age and contains a very hard kernel with an edible almond (Thomson and Evans 2006; dos Santos et al. 2008). The seeds, which are high in palmitic (35.0 wt.%; Table 15.10), oleic (32.0 wt.%), and linoleic (28.0 wt.%) acids, contain around 49 wt.% oil (dos Santos et al. 2008). The physical properties of the resultant FAME include an IV of 83 and a kinematic viscosity (40°C) of 4.3 mm^2/s (dos Santos et al. 2008). See Table 15.9 for an example of reaction parameters used in the preparation of biodiesel from *T. catappa* oil.

Hevea brasiliensis, commonly referred to as the rubber tree, belongs to the family Euphorbiaceae and is of major economic importance because it is the primary source of natural rubber and a sap-like extract (known as latex) can be collected and used in various applications (Ramadhas et al. 2005). Growing up to 34 m in height, the tree requires heavy rainfall and produces seeds weighing from 2 to 4 g that do not currently have any major industrial applications. The oil content of the seeds, which may contain up to 17 wt.% FFA, ranges from 40 to 50 wt.% and is high in unsaturated constituents (Table 15.10) such as linoleic (39.6 wt.%), oleic (24.6 wt.%), and linolenic (16.3 wt.%) acids (Ramadhas et al. 2005). Other fatty acids found in rubber seed oil include saturated species such as palmitic (10.2 wt.%) and stearic (8.7 wt.%) acids. The physical properties of the resultant rubber seed oil FAME include CP and PP values of 4°C and −8°C, respectively, and a kinematic viscosity (40°C) of 5.81 mm^2/s (Ramadhas et al. 2005). See Table 15.5 for an example of reaction parameters used in the preparation of biodiesel from rubber seed oil.

Vernicia fordii, commonly known as tung, is a small- to medium-sized (up to 20 m in height) deeciduous tree belonging to the Euphorbecaeae family that is native to China, Burma, and Vietnam. The fruit is a hard, woody pear-shaped drupe that is 4 to 6 cm long and 3 to 5 cm in diameter that contains four to five large seeds. Tung oil principally contains the unusual conjugated fatty acid eleostearic acid (9,11,13-octadecatrienoic acid; 63.8 wt.%; Table 15.10), with linoleic (11.5 wt.%), oleic (8.6 wt.%) and behenic (8.4 wt.%) acids also present in significant quantities (Park et al. 2008a). The physical properties of tung oil FAME include a CFPP of −11°C, OSI (110°C) value of 0.5 h, and kinematic viscosity (40°C) of 9.8 mm^2/s (Park et al. 2008a). Refer to Table 15.5 for an example of reaction parameters used in the preparation of biodiesel from tung oil.

Asclepias syriaca, commonly referred to as common milkweed, belongs to the family Asclepiadaceae and is native to the northeast and north–central United States where it grows on roadsides and in undisturbed habitat (Holser and Harry-O'Kuru 2006). Generally considered a nuisance by farmers, it is a herbaceous perennial oilseed plant that grows up to 2 m in height and produces a marketable fiber along with a white latex. The small flowers (1–2 cm in diameter) are grouped in several spherical umbels and contain numerous seeds that are attached to long, white flossy hairs. The dried seed contains 20 to 25 wt.% oil, which is composed mostly of unsaturated fatty acids (Table 15.10) such as linoleic (48.7 wt.%), oleic (34.8 wt.%), and palmitoleic (6.8 wt.%) acids (Holser and Harry-O'Kuru 2006). The majority of the remaining fatty acids are distributed among palmitic (5.9 wt.%) and stearic (2.3 wt.%) acids. The physical properties of the resultant milkweed seed oil FAME include CP and PP values of −1°C and −6°C, respectively, an OSI (100°C according to AOCS Cd 12b-92) value of 1.5 h, and a kinematic viscosity (40°C) of 4.6 mm^2/s (Holser and Harry-O'Kuru 2006). *See* Table 15.9 for an example of reaction parameters used in the preparation of biodiesel from common milkweed seed oil.

Zanthoxylum bungeanum is a small (3–7 m), fast growing, deciduous tree found throughout China that is a member of the Rutaceae family (Yang et al. 2008). The fruit of *Z. bungeanum* consists of a dark seed and a red shell, which is used as a culinary spice in China and is particularly popular in Sichuan cuisine (Sichuan pepper). The seed is a by-product of spice production and contains between 24 and 31 wt.% inedible oil that is high in FFA content (25 wt.%) (Yang et al. 2008; Zhang and Jiang 2008). As seen from Table 15.10, *Z. bungeanum* seed oil contains a large percentage of unsaturated fatty acids such as oleic (32.1 wt.%), linoleic (25.6 wt.%), and linolenic (24.1 wt.%) acids, with palmitic acid (10.6 wt.%) also present in significant quantity (Zhang and Jiang 2008). The physical properties of the resultant FAME include CP and CFPP values of 2°C and a kinematic viscosity (40°C) of 4.0 mm^2/s (Yang et al. 2008). Refer to Table 15.5 for an example of reaction parameters used in the preparation of biodiesel from *Z. bungeanum* seed oil.

Rice bran is a by-product of rice milling in the production of refined white rice from brown rice and is common in countries such as China and India. Rice bran contains about 15 to 23 wt.% inedible oil that is high in oleic (43.1 wt.%; Table 15.10), linoleic (33.2 wt.%), and palmitic (18.8 wt.%) acids, and has a relatively low FFA content (2.8 wt.%; Sinha et al. 2008). The physical properties of rice bran oil FAME include CP and PP values of 9°C and −2°C, respectively, a CN of 63.8, and a kinematic viscosity (40°C) of 5.29 mm^2/s (Sinha et al. 2008). Refer to Table 15.6 for an example of reaction parameters used in the preparation of biodiesel from rice bran oil.

Raphanus sativus, commonly known as radish, is a perennial oilseed plant of the Brassicacease family that is widely grown and consumed throughout the world. The radish has considerable low-temperature tolerance, grows rapidly, and requires few agricultural inputs (Domingos et al. 2008). The seeds contain between 40 to 54 wt.% of inedible oil that primarily consists of oleic (34.5 wt.%; Table 15.10), linoleic

(17.8 wt.%), erucic (16.4 wt.%), linolenic (12.5 wt.%), and arachidic (10.0 wt.%) acids (Domingos et al. 2008). The physical properties of ethyl esters prepared from radish seed oil include an OSI (110°C) value of 4.65 h, kinematic viscosity (40°C) of 4.65 mm^2/s, IV of 105, and a CFPP value of −2°C (Domingos et al. 2008). Refer to Table 15.6 for an example of reaction parameters used in the preparation of biodiesel from radish seed oil.

Brassica carinata, commonly known as Ethiopian or Abyssinian mustard, is an oilseed plant of the Brassicacease family that is well-adapted to marginal lands and can tolerate arid regions such as Ethiopia (Dorado et al. 2004). *B. carinata* is of interest as a potential biodiesel crop because it displays better agronomic performance in areas with unfavorable growing conditions than *B. napus* (rapeseed; Cardone et al. 2002). Natural *B. carinata* contains up to 43.6 wt.% erucic acid. However, varieties of *B. carinata* with no erucic acid have been developed (Dorado et al. 2004). Refer to Table 15.10 for the fatty acid composition of *B. carinata* varieties that do and do not contain erucic acid. Physical properties of the corresponding FAME (with no erucic acid) include CP, PP, and CFPP values of −9°C, −6°C, and −9°C, respectively, an IV of 138, and a kinematic viscosity (40°C) of 4.83 mm^2/s (Dorado et al. 2004). The exhaust emissions of FAME from *B. carinata* oil are summarized by Cardone et al. 2002. Refer to Table 15.6 for an example of reaction parameters used in the preparation of biodiesel from *B. carinata* seed oil.

Camelina sativa, commonly known as false flax or gold-of-pleasure, is a spring annual broadleaf oilseed plant of the Brassicaceae family that grows well in temperate climates. Camelina, unlike soybean, is cold weather tolerant and is well-adapted to the northern regions of North America, Europe, and Asia (Frohlich and Rice 2005). The oil content of seeds from camelina can range from 28 to 40 wt.% and is characterized by low FFA content (1.0 wt.%) and high levels of unsaturated fatty acids (Table 15.10; Frohlich and Rice 2005). Linolenic (38.4 wt.%), gondoic (16.8 wt.%), linoleic (14.3 wt.%), and oleic (14.3 wt.%) acids comprise the majority of the fatty acids found in camelina oil (Frohlich and Rice 2005). The IV of camelina oil FAME is a rather high 155 as a result of the high unsaturated fatty acid content of the parent oil (Frohlich and Rice 2005). Other physical properties of camelina oil FAME include CP, CFPP, and PP values of 3°C, −3°C, and −4°C, respectively (Frohlich and Rice 2005). Refer to Table 15.9 for an example of reaction parameters used in the preparation of biodiesel from *C. sativa* seed oil.

Calophyllum inophyllum, commonly known as Polanga, is an inedible oilseed ornamental evergreen tree belonging to the Clusiaceae family that is found in tropical regions of India, Malaysia, Indonesia, and the Philippines. Typically growing up to 25 m in height, the Polanga tree produces a slightly toxic fruit that contains a single, large seed (Sahoo et al. 2007). The oil obtained from polanga seeds is high in FFA content (up to 22 wt.%) and unsaturated species (Table 15.10) such as linoleic (38.3 wt.%) and oleic (34.1 wt.%) acids (Sahoo et al. 2007). The remaining fatty acids include stearic (13.0 wt.%) and palmitic (12.0 wt.%) acids with a trace amount of linoleic acid (0.3 wt.%) also present (Sahoo et al. 2007).

Physical properties of polanga oil FAME include CP and PP values of 13°C and 4°C, respectively, a flash point of 140°C, and a kinematic viscosity (40°C) of 4.92 mm^2/s (Sahoo et al. 2007). The exhaust emissions of FAME from polanga oil are summarized by Sahoo et al. (2007). Refer to Table 15.5 for an example of reaction parameters used in the preparation of biodiesel from polanga seed oil.

Cynara cardunculus, commonly known as cardoon, is a Mediterranean herbaceous perennial thistle with a characteristic rosette of large spiny leaves and branched flowering stems. It belongs to the Asteraceae family and is cultivated for its branched leaf petioles, which are regarded as a great delicacy in many parts of the Mediterranean (Raccuia and Melilli 2007). *C. cardunculus* seed oil normally contains around 25 wt.% FFA (Encinar et al. 2002) and is high in linoleic acid (56.0 wt.%; Table 15.10), with oleic (25.0 wt.%), palmitic (14.0 wt.%), and stearic (3.0 wt.%) acids also present in significant quantities (Encinar et al. 2002). As a result of the relatively high linoleic acid content of *C. cardunculus* seed oil, the corresponding FAME display a relatively low kinematic viscosity (40°C) of 3.56 mm^2/s, CP, and PP values of −1°C and -3°C, respectively, and a flash point of 175°C (Encinar et al. 2002). Ethyl esters from *C. cardunculus* seed oil display slightly improved CP, PP, and flash point values of −3°C, −6°C, and 188°C, respectively (Encinar et al. 2002). Refer to Table 15.9 for an example of reaction parameters used in the preparation of FAME from *C. cardunculus* seed oil.

Carthamus tinctorius, commonly known as safflower, is a highly branched, herbaceous, thistle-like annual oilseed plant belonging to the Asteraceae family that has long been cultivated in numerous regions of the world mainly for its edible oil and colorful petals, which are valued as food coloring and flavoring agents, as well as for sources of red and yellow dyes for clothing (Rashid and Anwar 2008b). India, the United States, and Mexico are the principle producers of safflower; however, it has been cultivated in many other countries, such as Australia, Argentina, Ethiopia, Russia, Kazakhstan, and Uzbekistan. Safflower has a strong taproot that enables it to thrive in dry climates, but the plant is very susceptible to frost injury (Rashid and Anwar 2008b). Safflower has many long, sharp spines on its leaves and grows from 30 to 150 cm tall with globular flower heads that contain brilliant yellow, orange, or red flowers. Each branch usually has from one to five flower heads containing from 15 to 20 seeds per head. The seeds are also used as a component in seed mixtures for the recreational feeding of birds. Safflower seeds normally contain around 35 to 45 wt.% oil. Safflower oil, which has a very low FFA content (0.2 wt.%), primarily consists of linoleic acid (76.0 wt.%; Table 15.10), followed by oleic (14.2 wt.%), palmitic (6.9 wt.%), and stearic (2.1 wt.%) acids (Rashid and Anwar 2008b). The physical properties of safflower oil FAME include an IV of 142, CP, PP, and CFPP values of −2°C, −8°C, and -6°C, respectively, flash point of 176°C, kinematic viscosity of 4.29 mm^2/s, and a CN of 52 (Rashid and Anwar 2008b). Refer to Table 15.9 for an example of reaction parameters used in the preparation of biodiesel from safflower seed oil.

Sesamum indicum, commonly known as sesame, is an annual herbaceous flowering plant of the Pedaliaceae family that has recently been adapted to semi-arid climates and is an important oilseed crop cultivated in many regions of the world (Elleuch et al. 2007). Sesame is widely used in food, nutraceutical, and pharmaceutical applications as a result of its high oil, protein, and antioxidant content. Sesame seeds normally contain around 44 to 58 wt.% oil (Elleuch et al. 2007). Sesame seed oil primarily consists of oleic (43 wt.%; Table 15.10) and linoleic (35 wt.%) acids, with palmitic (11 wt.%) and stearic (7 wt.%) acids comprising most of the remaining constituents (Elleuch et al. 2007). The physical properties of sesame oil FAME include an IV of 98 (corrected), CN of 50, kinematic viscosity (40°C) of 4.2 mm^2/s, flash point of 170°C, and CP and PP values of –6°C and –14°C, respectively (Saydut et al. 2008). The exhaust emissions of FAME from sesame seed oil are summarized by Banapurmath et al. (2008). Refer to Table 15.9 for an example of reaction parameters used in the preparation of biodiesel from sesame seed oil.

Sclerocarya birrea, commonly known as marula, is a medium-sized (up to 18 m in height) deciduous Savannah tree belonging to the Anacardiaceae family that is found in southern and western Africa (Mariod et al. 2006). The primary fatty acids found in marula oil, which contains 2.1 wt.% FFA, are oleic (67.3 wt.%; Table 15.10), palmitic (14.2 wt.%), stearic (8.8 wt.%), and linoleic (5.9 wt.%) acids. The physical properties of the resultant FAME include a CFPP value of 5°C, an IV of 76 and a CN of 62 (Mariod et al. 2006). Refer to Table 15.9 for an example of reaction parameters used in the preparation of biodiesel from marula seed oil.

Cucurbita pepo, commonly known as pumpkin, is a member of the Cucurbitaceae family and is grown in many regions of the world for various agricultural and ornamental applications. Pumpkin seeds are a common snack food in several cultures, and pumpkin seed oil may be used as a salad oil (Schinas et al. 2009). The primary fatty acids contained in pumpkin seed oil, which is obtained in 45 wt.% yield from seeds and has a low FFA content of 0.3 wt.%, are linolenic (43.7 wt.%; Table 15.10), oleic (37.1 wt.%), and palmitic (12.5 wt.%) acids (Schinas et al. 2009). The physical properties of pumpkin seed oil FAME include a kinematic viscosity (40°C) of 4.41 mm^2/s, IV of 115, CFPP value of –9°C, and a density (15°C) of 883.7 kg/m^2 (Schinas et al. 2009). Refer to Table 15.9 for an example of reaction parameters used in the preparation of biodiesel from pumpkin seed oil.

Simmondsia chinensis, commonly known as jojoba, is a perennial shrub belonging to the Simmondsiaceae family that is native to the Mojave and Sonoran deserts of Mexico, California, and Arizona. Jojoba is unique in that the lipid content of the seeds, which is between 45 and 55 wt.%, is in the form of long-chain esters of fatty acids and alcohols (wax esters; Canoira et al. 2006; Bouaid et al. 2007) as opposed to TAG. The fatty acid component of jojoba wax esters primarily consists of eiconenoic, erucic, and oleic acids with *cis*-11-eicosen-1-ol and *cis*-13-docosen-1-ol principally composing the alcohol component (Canoira et al. 2006). As a consequence of the unique composition of jojoba wax, methanolysis affords a product that consists of a mixture of FAME and long chain alcohols, as the separation of these materials is problematic. The physical properties of this mixture do not compare favorably with biodiesel prepared from other feedstocks, as the CFPP value is 4°C and the kinematic viscosity (40°C) is 11.82 mm^2/s (Canoira et al. 2006).

Reduction of the alcohol component through laborious purification results in an improvement in CFPP to −14°C; however, the kinematic viscosity (40°C; 9.0 mm^2/s) remains considerably above accepted American (ASTM D6751) and European Union (EN 14214) limits (Canoira et al. 2006). The reaction conditions for the production of biodiesel from jojoba wax esters have been optimized by RSM (Bouaid et al. 2007).

Animal fats. Animal fats may include materials from a variety of domesticated animals, such as cows, chickens, pigs, and other animals such as fish and insects. Animal fats are normally characterized by a greater percentage of saturated fatty acids in comparison to oils obtained from the plant kingdom. Animal fats are generally considered as waste products, so they are normally less expensive than commodity vegetable oils, which makes them attractive as feedstocks for biodiesel production.

Waste salmon oil is a by-product of the fishing industry and is obtained from fish by-product material by extraction or centrifugation from either fresh salmon by-product or from hydrolysate made from salmon by-product (Wright 2004; Chiou et al. 2008; El-Mashad et al. 2008). The major fish by-products include fish heads, viscera, and frames (Chiou et al. 2008). The hydrolysate, which contains around 10 wt.% oil, is typically produced by maceration of viscera or skinny by-products followed by enzymatic digestion of protein and removal of bones (El-Mashad et al. 2008). Acid is normally added to stabilize the final product by lowering the pH to around 3.7, which in combination with other handling procedures results in an inedible oil (El-Mashad et al. 2008). Biodiesel may be prepared from oil obtained from either fresh salmon by-product (termed non-acidified oil) or from hydrolsyate (acidified oil). The FFA contents of non-acidified and acidified oils are 1.7 and 6.0 wt.%, respectively (El-Mashad et al. 2008). The fatty acid compositions of these oils is nearly identical and is characterized by a wide range of fatty acids not normally found in other materials, such as eicosapentaenoic (C20:5; 11.1 wt.%; Table 15.11), docosahexaenoic (C22:6; 13.7 wt.%), arachidonic (C20:4; 3.3 wt.%), and glupanodonic (C22:5; 3.0 wt.%) acids, along with more typical constituents such as palmitic (14.9 wt.%), linolenic (11.5 wt.%), myristic (6.8 wt.%), and palmitoleic (6.1 wt.%) acids. In addition, oleic acid, along with the 10-*cis* isomer, constitutes 15.6 wt.% of salmon oil (Chiou et al. 2008). As a result of the high FFA content of salmon oil, a two-step reaction procedure is required in which the oil initially undergoes mineral acid-catalyzed (H_2SO_4) pretreatment, followed by conventional homogenous alkaline base-catalyzed transesterification (KOH) to afford salmon oil FAME (Chiou et al. 2008; El-Mashad et al. 2008). The reaction conditions for the production of biodiesel from salmon oil have been optimized by RSM (El-Mashad et al. 2008). The physical properties of non-acidified salmon oil FAME include CP, PP, and CFPP values of −2°C, −3°C, and −8°C, respectively, and an OSI (110°C) value of 6.1 h (Chiou et al. 2008). The physical properties of acidified salmon oil FAME include PP and CFPP values of −3°C and −6°C, respectively, and an OSI (110°C) value of 12.1 h (Chiou et al. 2008).

Aspongubus viduatus, commonly known as melon bug, is an insect belonging to the Hemiptera order that is widely distributed in Sudan and other regions of Africa and is considered the primary pest of watermelon (Mariod et al. 2006). The melon bug normally contains around 45 wt.% oil. The primary fatty acids found in melon bug oil, which contains 3.0 wt.% FFA, include oleic (46.6 wt.%; Table 15.11), palmitic (30.9 wt.%), and palmitoleic (10.7 wt.%) acids (Mariod et al. 2006). As a result of the high saturated fatty acid content of melon bug oil, the corresponding FAME display a high CFPP value of 10°C and a low IV of 56. Other parameters of note include a CN of 55 and an OSI (110°C) value of 4.9 h (Mariod et al. 2006). Refer to Table 15. 9 for an example of reaction parameters used in the preparation of biodiesel from melon bug oil.

Agonoscelis pubescens, commonly known as sorghum bug, is an insect belonging to the Hemiptera order that is widely distributed in Sudan and other regions of Africa and is a pest (Mariod et al. 2006). The sorghum bug normally contains around 60 wt.% oil. The primary fatty acids found in sorghum bug oil, which contains 10.5 wt.% FFA, include oleic (40.9 wt.%; Table 15.11), linoleic (34.5 wt.%), and palmitic (12.1 wt.%) acids (Mariod et al. 2006). The physical properties of the resultant FAME include a CFPP value of 4°C, an IV of 84, a CN of 55, and an OSI (110°C) value of 1.0 h (Mariod et al. 2006). Refer to Table 15. 5 for an example of reaction parameters used in the preparation of biodiesel from sorghum bug oil.

Animal fats such as beef tallow and chicken fat are by-products of the food processing industry and represent low value feedstocks for biodiesel production. The primary fatty acids found in beef tallow (Table 15.11) include oleic (47.2 wt.%), palmitic (23.8 wt.%), and stearic (12.7 wt.%) acids. The primary fatty acids contained in chicken fat (Table 15.11) include oleic (40.9 wt.%), palmitic (20.9 wt.%), and linoleic (20.5 wt.%) acids (Wyatt et al. 2005). As a result of the very low polyunsaturated fatty acid content of beef tallow, the corresponding FAME display excellent oxidative stability, as evidenced by an OSI (110°C) value of 69 h. In addition, other physical properties of beef tallow FAME include a kinematic viscosity (40°C) of 5.0 mm^2/s, a flash point of 150°C, and CP, PP, and CFPP values of 11°C, 13°C, and 8°C, respectively (Wyatt et al. 2005). As a result of the high polyunsaturated fatty acid content of chicken fat, the corresponding FAME display poor oxidative stability, as evidenced by an OSI (110°C) value of 3.5 h. In addition, other physical properties of beef tallow FAME include a kinematic viscosity (40°C) of 4.3 mm^2/s, a flash point of 150°C, and CP, PP, and CFPP values of 4°C, 6°C, and 1°C, respectively (Wyatt et al. 2005). Combustion of B20 blends of beef tallow and chicken fat FAME results in an increase in NO$_x$ exhaust emissions of only 0.0% and 2.4% versus 6.2% for a B20 blend of SME (Wyatt et al. 2005). In addition, the exhaust emissions of beef tallow FAME are summarized by McCormick et al. (2001).

Pork lard is a by-product of the food processing industry and represents a low value feedstock for biodiesel production. The primary fatty acids found in pork lard (Table 15.11) include oleic (44.7 wt.%), palmitic (26.4 wt.%), linoleic (12.7 wt.%), and stearic (12.1 wt.%) acids (Jeong et al. 2009). As a result of the relatively high saturated fatty acid content of pork lard, the corresponding FAME exhibit a relatively high CFPP value of 8°C and a relatively low IV of 72, along with a kinematic viscosity (40°C) of 4.2 mm^2/s (Jeong et al. 2009). Another study determined that pork lard FAME have a kinematic viscosity (40°C) of 4.8 mm^2/s, a flash point of 160°C, OSI (110°C) value of 18.4 h, and CP, PP, and CFPP values of 11°C, 13°C, and 8°C, respectively (Wyatt et al. 2005). Furthermore, combustion of B20 blends of pork lard FAME results in an increase in NO_x exhaust emissions of only 3.0% versus 6.2% for a B20 blend of SME (Wyatt et al. 2005). In addition, the exhaust emissions of pork lard FAME are summarized by McCormick et al. (2001). Refer to Table 15.6 for an example of reaction parameters used in the preparation of biodiesel from refined and bleached pork lard that have been optimized by RSM. Exploration of the influence of pork lard and beef tallow FAME blended with rapeseed and linseed oil FAME in various ratios on CN and exhaust emissions is summarized by Lebedevas and Vaicekauskas (2006).

Other waste oils. Waste oils may include a variety of low-value materials such as used cooking or frying oils, vegetable oil soapstocks, acid oils, tall oil, and other waste materials. Waste oils are normally characterized by relatively high FFA and water contents and potentially the presence of various solid materials that must be removed by filtration prior to conversion to biodiesel. In the case of used cooking or frying oils, hydrogenation to increase the useful cooking lifetime of the oil may result in the introduction of relatively high melting *trans* constituents, which influence the physical properties of the resultant biodiesel fuel.

Used or waste frying or cooking oil is primarily obtained from the restaurant industry and may cost anywhere from free to 60% less expensive than commodity vegetable oils, depending on the source and availability (Predojevic 2008). The fatty acid profiles of several used or waste frying or cooking oils is presented in Table 15.11. Conditions for the preparation of biodiesel from used or waste frying or cooking oils is presented in Table 15.5 and 15.6. The physical properties of FAME prepared from used or waste cooking or frying oils include kinematic viscosities (40°C) of 4.23 (Meng et al. 2008), 4.79 (Cetinkaya and Karaosmanoglu 2004), and 4.89 mm^2/s (Phan and Phan 2008), flash points of 171 (Meng et al. 2008), 176 (Cetinkaya and Karaosmanoglu 2004), and 120°C (Phan and Phan 2008), a CN of 55 (Meng et al. 2008), an IV of 125 (Cetinkaya and Karaosmanoglu 2004), CFPP values of 1°C (Meng et al. 2008) and −6°C (Cetinkaya and Karaosmanoglu 2004), CP values of 9°C (Cetinkaya and Karaosmanoglu 2004) and 3°C, and PP values of −3°C (Cetinkaya and Karaosmanoglu 2004) and 0°C (Phan and Phan 2008). The differences in the physical property data among the various studies may be a result of differing feedstock origin or differences in product purity. The exhaust emissions of used cooking oil FAME are summarized by Meng et al. (2008), Ozsezen et al. (2008), and Kocak et al. (2007).

Soapstock, a by-product of the refining of vegetable oils, consists of a heavy alkaline aqueous emulsion of lipids that contains around 50% water, with the remaining material made up of FFA (and soaps), phosphoacylglycerols, TAG, pigments, and other minor nonpolar components (Haas 2005). The total fatty acid content, including both FFA and lipid-linked acids, is normally around 25 to 30 wt.% (Haas et al. 2001). Soybean soapstock is generated at a rate of about 6 vol.% of crude oil refined, which equates to an annual US production of approximately one billion pounds with a market value of about one fifth of the price of crude soybean oil (Haas 2005). The fatty acid composition of soybean soapstock (Table 15.11) is similar to that of crude soybean oil. The production of FAME from soybean soapstock is complicated by the presence of a substantial amount of water, which must be removed prior to conversion to biodiesel. Freeze-drying is an effective method for removal of water after hydrolysis and acidulation of TAG (Haas 2005). Soybean soapstock at this stage is entirely composed of FFA, so it is referred to as acid oil (Haas et al. 2003). Acid-catalyzed (H_2SO_4) esterification of acid oil may be used for conversion to FAME (Haas et al. 2003; Haas 2005). The physical properties of biodiesel prepared from soybean soapstock are similar to those of biodiesel prepared from soybean oil (Haas et al. 2003; Haas 2005). For instance, FAME prepared from soybean soapstock exhibit a kinematic viscosity (40°C) of 4.3 mm^2/s, flash point of 169°C, CN of 51, IV of 129, and a CP value of 6°C (Haas et al. 2001). The exhaust emissions of soybean soapstock FAME are summarized by Haas et al. (2001). Refer to Table 15.9 and 15.5 for examples of reaction parameters used in the preparation of biodiesel from soybean soapstock and acid oil, respectively.

Rendered animal fats and restaurant waste oils are known as yellow greases if the FFA level is less than 15 wt.% and brown greases if the FFA content is in excess of 15 wt.% (Canakci and Van Gerpen 2001). The fatty acid compositions (Table 15.11) of yellow and brown greases are similar, with oleic acid present in the greatest abundance, followed by palmitic, stearic, and linoleic acids. As a result of the high FFA content of greases, conventional homogenous alkaline base-catalyzed transesterification is not practical. Therefore, acid pretreatment must precede base-catalyzed transesterification to afford FAME (Canakci and Van Gerpen 2001). Alternatively, various diarylammonium catalysts may be used instead of homogenous alkaline base catalysts in the transesterification of acid-pretreated greases to provide FAME (Ngo et al. 2008). The exhaust emissions of yellow and brown grease FAME are summarized by Canakci and Van Gerpen (2003b) and McCormick et al. (2001). Refer to Table 15.5 for an example of reaction parameters used in the preparation of biodiesel from brown grease.

Tall oil is obtained from coniferous wood recovered in the Kraft pulping process as a by-product of cellulose production (Keskin et al. 2008). Chemically, tall oil is a mixture of several components, including resin acids and other terpenoids (40–60 wt.%), fatty acids and TAG (30–50 wt.%), unsaponifiables (4–10 wt.%), water (0.5–3 wt.%), and ash (0.1–1.0 wt.%; Demirbas 2008). The chemical composition of tall oil varies with the age, pine species, geographical location, and the pulping process used (Keskin et al. 2008). As a result of the high FFA content of tall oil, conventional homogenous alkaline base-catalyzed transesterification is not practical. The physical properties of a B60 blend of tall oil FAME in petrodiesel include a kinematic viscosity (40°C) of 5.3 mm^2/s, CN of 50, flash point of 88°C, and CP and CFPP values of −2°C and −6°C, respectively (Keskin et al. 2008). The exhaust emissions of tall oil FAME blended with petrodiesel are summarized by Keskin et al. (2008). Refer to Table 15.5 for an example of reaction parameters used in the preparation of biodiesel from tall oil.

Dried distiller's grains (DDG) are a by-product of dry-grind ethanol production and contain around 8–10 wt.% corn oil. Ordinarily, these grains are used as an animal feed supplement, but extraction of the corn oil may represent a sizeable feedstock for biodiesel production in addition to providing a more nutritive and valuable animal feed supplement (Balan et al. 2009; Noureddini et al. 2009). However, corn oil extracted from distiller's grains may contain 7.0 wt.% FFA or more, resulting in the need for acid-catalyzed pretreatment or other alternative methods to prepare FAME in high yield. For instance, a recent study used sulfuric acid-catalyzed pretreatment to lower the FFA content of corn oil extracted from grains to 0.25 from 7.0 wt.%, followed by classic homogenous alkaline base-catalyzed (NaOH) methanolysis to yield corn oil FAME in 98 wt.% yield (Noureddini et al. 2009).

Effects of Blending Biodiesel with Other Fuels

Biodiesel–petrodiesel blends. Biodiesel is completely miscible with ULSD and can be used as a blend component in any proportion. However, ASTM D975 and D7467 only allow up to 5 and 20 vol.% biodiesel, respectively. Biodiesel and petrodiesel are not chemically similar: biodiesel is composed of long-chain FAAE, whereas petrodiesel is a mixture of aliphatic and aromatic hydrocarbons that contain approximately 10 to 15 carbons. Because biodiesel and petrodiesel have differing chemical compositions, they have differing fuel properties. Once mixed, the blend will exhibit properties different from neat biodiesel and petrodiesel fuels. Specifically, the most important fuel properties influenced by blending are lubricity, exhaust emissions, CN, flash point, oxidative stability, low-temperature operability, kinematic viscosity, and energy content.

Lubricity of petrodiesel is positively impacted through blending with biodiesel (Geller and Goodrum 2004; Goodrum and Gellar 2005; Hu et al. 2005; Knothe and Steidley 2005b; Moser et al. 2008). Specifically, B2 and B20 blends of SME in ULSD (contains no lubricity-enhancing additives) significantly improve lubricity (60°C according to ASTM D6079) from 551 to 212 and 171 μm, respectively (Moser et al. 2008). Exhaust emissions of ULSD, with the exception of NO_x, are reduced through blending with biodiesel, as previously discussed (Chang et al. 1996; Altiparmak et al. 2007; Korres et al. 2008; Lapuerta et al. 2008). The CN of petrodiesel is increased upon blending with biodiesel (Chang et al. 1996; Altiparmak et al. 2007). For example, the CN of petrodiesel, tall oil FAME, and the corresponding B50, B60, and B70 blends are 47, 54, 52, 52, and 53, respectively (Altiparmak et al. 2007). The flash point of petrodiesel is increased upon blending with biodiesel (Alptekin and Canakci 2009). The flash points of FAME are much higher than those of petrodiesel and range from around 110 to 200°C versus 50 to 60°C for petrodiesel. When blended with petrodiesel, biodiesel does not impact flash point up to B20, but beyond B20 the flash point increases significantly (Alptekin and Canakci 2009). The oxidative stability of petrodiesel is negatively impacted upon blending with biodiesel (Mushrush et al. 2003, 2004). This is because the hydrocarbon constituents of petrodiesel are more stable to oxidation than FAME (especially in the case of unsaturated FAME). The amount of gravimetric solids formed (indicators of oxidative degradation) according to ASTM D5304 in the cases of petrodiesel and the corresponding B10 and 20 blends (SME) were 0.6, 4.2, and 9.0 mg/100 mL, respectively (Mushrush et al. 2004). The low-temperature operability of petrodiesel is negatively impacted once blended with biodiesel (Altiparmak et al. 2007; Benjumea et al. 2008; Moser et al. 2008; Alptekin and Canakci 2009). For instance, the CFPP of petrodiesel, tall oil FAME, and the corresponding B50, 60, and 70 blends were −8°C, −3°C, −7°C, −6°C, and −6°C, respectively (Altiparmak et al. 2007). In another example, the PP of ULSD and B2, 5, 20, and 100 blends (PME) were −21°C, −21°C, −18°C, −12°C, and 18°C, respectively (Moser et al. 2008b). The kinematic viscosity of petrodiesel increases upon blending with biodiesel (Altiparmak et al. 2007; Benjumea et al. 2008; Moser et al. 2008; Alptekin and Canakci 2009, 2008). For example, the kinematic viscosities (40°C) of ULSD and B1, 2, 5, and 20 (SME) blends are 2.32, 2.40, 2.48, 2.57, and 2.71 mm^2/s, respectively (Moser et al. 2008). In another example, the kinematic viscosities (40°C) of petrodiesel and B50, 60, 70, and 100 (tall oil FAME) blends are 2.60, 4.50, 4.82, 5.12, and 7.10 mm^2/s, respectively (Altiparmak et al. 2007). Lastly, the heat of combustion (energy content) of petrodiesel is reduced upon blending with biodiesel (DeOliveira et al. 2006; Altiparmak et al. 2007; Benjumea et al. 2008). Specifically, the energy contents of petrodiesel and B50, 60, 70, and 100 (tall oil FAME) blends are 43.76, 41.90, 41.51, 41.15, and 40.02 MJ/kg, respectively (Altiparmak et al. 2007). In another example, the energy contents of petrodiesel and B2, 5, 10, 20, and 100 (SME) blends are 46.65, 46.01, 45.46, 44.48, 43.75, and 39.09 MJ/kg, respectively (DeOliveira et al. 2006).

Biodiesel–alcohol blends. Low-level blends of ethanol in diesel fuel (E-diesel) significantly reduce harmful exhaust emissions such as PM, THC, and CO as a result of increased fuel oxygenation. For instance, an E20 (20% ethanol in diesel fuel) blend results in reductions of 55%, 36%, and 51% for CO, THC, and PM exhaust emissions, respectively (Ajav et al. 2000). However, drawbacks of E-diesel include reduced energy content (Can et al. 2004; Li et al. 2005), CN (Li et al. 2005), flash point (Li et al. 2005), lubricity (Fernando and Hanna 2004), and immiscibility of ethanol in diesel over a wide range of temperatures (Satge de Caro et al. 2001; Fernando and Hanna 2004; Li et al. 2005). To correct the immiscibility problem, surfactants at levels of up to 5% are required to stabilize E-diesel mixtures (Satge de Caro et al. 2001; Fernando and Hanna 2004). A recent study explored the utility of ethanol–biodiesel–diesel blends (EB–diesel) as a means to mitigate the miscibility issues of E-diesel (Fernando and Hanna 2004). The disadvantages of E-diesel are substantially reduced or eliminated in the case of EB–diesel blends prepared from 5% ethanol and 20% biodiesel (SME) in petrodiesel (Fernando and Hanna 2004). A later study (Rahimi et al. 2009) revealed that 3% ethanol, 2% biodiesel (sunflower oil FAME), and 95% low sulfur diesel (LSD, <500 ppm S) reduced the PP in comparison to neat LSD. In general, EB-diesel blends result in reduced CO and THC exhaust emissions versus neat LSD (Rahimi et al. 2009). Also discussed were the effects of blending ethanol with biodiesel (E-biodiesel) in a ratio of 6 to 4 on

PP, kinematic viscosity, and flash point. Specifically, the PP of biodiesel was reduced from −3°C to −9°C, the kinematic viscosity (40°C) was reduced from 4.22 to 1.65 mm^2/s, and the flash point was reduced from 187°C to 14°C after blending with ethanol (Rahimi et al. 2009). Analogously, a blend of ethanol and biodiesel prepared from *Madhuca indica* oil yielded lower flash point, reduced kinematic viscosity, lower PP, reduced CO emissions, slightly higher THC emissions, and lower NO$_x$ emissions versus unblended *Madhuca indica* oil FAME (Bhale et al. 2009). Other effects of blending ethanol with biodiesel that have not yet been reported but can be reasonably speculated may be reduced lubricity, energy content, and CN in comparison to neat biodiesel.

Multi-feedstock biodiesel blends. Mixed feedstock biodiesel production may be employed to provide biodiesel with improved physical properties in comparison to the individual fuels on their own. Mixed feedstock production may also arise as a result of economic considerations. For instance, it may be economically advantageous to extend the lifetime of a comparatively more expensive feedstock through blending with a less expensive feedstock. Recent examples of the influence of blending various feedstocks on biodiesel fuel properties include blends of canola, palm, soybean, and sunflower oil FAME (Moser 2008a), blends of palm, rapeseed and soybean oil FAME (Park et al. 2008b), blends of jatropha and palm oil FAME (Sarin et al. 2007), and blends of cottonseed, soybean, and castor oil FAME (Meneghetti et al. 2007). In the case of jatropha–palm blends, the objective was to obtain a biodiesel blend with superior low-temperature performance to PME and superior oxidative stability to jatropha oil FAME. Specifically, the CFPP of PME was improved from 12°C to 3°C upon blending (20:80 by volume) with jatropha oil FAME (neat: 0°C Sarin et al. 2007). The oxidative stability of jatropha oil FAME was improved from 3 to 4 h after blending (same proportion as above) with PME (Sarin et al. 2007). The objective of another study was to evaluate the kinematic viscosities, specific gravities, and IV of biodiesel fuels obtained from binary mixtures (20 and 80 vol.%) of castor, soybean, cotton, and canola oil FAME in an effort to find combinations that were satisfactory according to the EN 14214 biodiesel standard (Albuquerque et al. 2009).

Other Uses of Biodiesel

Fatty acid alkyl esters have attracted considerable interest as alternative bio-based diesel fuels for combustion in CI engines. However, a number of additional applications have been developed or discovered for these versatile oleochemical materials. For instance, biodiesel may be used as a replacement for petroleum as a heating oil (Mushrush et al. 2001). As such, the European standard EN 14213 (CEN 2003b) was established to cover the use of biodiesel for this purpose. In the United States, blends of up to 5% biodiesel in heating oils (B5 Bioheat) have recently been approved for inclusion in the ASTM heating oil standard, D396 (ASTM 2008d). Another combustion-related application of biodiesel is as an aviation fuel, although the relatively poor low-temperature properties of biodiesel restrict its use to low-altitude aircraft (Dunn 2001). Additionally, the use of biodiesel in diesel-fueled marine engines to reduce environmental impact is another important application of this biodegradable and non-toxic fuel (Nine et al. 2000). Because there are less harmful exhaust emissions from biodiesel than those from petrodiesel, the use of biodiesel to power underground mining equipment is another fuel-related application. Biodiesel may also be used as a fuel for generators and turbines for the generation of electricity (Hashimoto et al. 2008; Kalbande et al. 2008; Kram 2008a; Lin et al. 2008) or as a substitute for hydrogen in fuel cells (Kram 2008b). Finally, other niche uses such as in national parks and other environmentally sensitive locations are important fuel-related applications of biodiesel.

An important non-fuel application is as an industrial environmentally friendly solvent, since FAAE are biodegradable, have high flash points, and have very low volatilities (Wildes 2002). The high solvent strength of biodiesel makes it attractive as a substitute for a number of conventional and harmful organic solvents (Hu et al. 2004) in applications such as industrial cleaning and degreasing, resin cleaning, and removal (Wildes 2001, 2002), plasticizers in the production of plastics (Wehlmann 1999), liquid–liquid extractions (Spear et al. 2007), polymerization solvent (Salehpour and Dube 2008), and as a medium in site bioremediation of crude petroleum spills (Miller and Mudge 1997; Mudge and Pereira 1999; Glória Pereira and Mudge 2004; Fernandezalvarez et al. 2006). The strong solvent properties of biodiesel are particularly noticeable in cases where diesel engines have been operated with petrodiesel for many years or miles, which results in build-up of insoluble deposits in fuel tanks and lines. Upon switching to biodiesel, especially B100, the deposits will get released, which results in vehicle inoperability due to fuel filter clogging. Although biodiesel often gets blamed for clogging fuel filters with deposits in cases such as these, it is in fact the fault of the petrodiesel from which the deposits originated, since biodiesel is merely acting as a cleansing solvent.

Fatty acid alkyl esters can also serve as valuable starting materials or intermediates in the synthesis of fatty alcohols (Peters 1996), lubricants (Willing 1999; Moser et al. 2007; Sharma et al. 2007; Dailey et al. 2008; Padua 2008), cold flow improver additives (Moser and Erhan 2006, 2007; Moser et al. 2007; Dailey et al. 2008), cetane improving additives (Poirier et al. 1995), and multifunctional lubricity and combustion additives (Suppes et al. 2001; Suppes and Dasari 2003). Lastly, biodiesel in conjunction with certain surfactants can act as a contact herbicide to kill broadleaf weeds in turfgrass (Vaughn and Holser 2007).

Glycerol

Glycerol (or glycerin; 1,2,3-propanetriol) is produced in addition to FAAE during transesterification of vegetable oils and animal fats (Fig. 15.1). Prior to the increase in biodiesel production that occurred over the past decade as a result of the continued interest in renewable fuels, the market demand for glycerol was relatively balanced with supply. However, the emergence of the biodiesel industry has spawned numerous efforts to find new applications, products, and markets using this versatile chemical. A recently published review (Behr et al. 2008) thoroughly covers recent developments on the chemistry and utility of glycerol. In general, glycerol may be used as a chemical feedstock in the production of polyurethanes, polyesters, polyethers, and other materials. Glycerol may also be found in lubricants, wrapping and packaging materials, foods, drugs, cosmetics, and tobacco products. Applications and products that displace existing petroleum-derived materials or feedstocks are of particular interest. A recent significant advance is the development of a synthetic route to propylene glycol (1,2-propanediol) from glycerol, which represents a viable alternative to the classic petrochemical route from propylene (Dasari et al. 2005; Suppes 2006; Feng et al. 2008). Propylene glycol represents a replacement for the common toxic antifreeze component ethylene glycol.

Future Outlook for Biodiesel

Despite its many advantages as a renewable alternative fuel, biodiesel presents a number of technical problems that must be resolved before it will be more attractive as an alternative to petrodiesel. These problems include improving the relatively poor low-temperature properties of biodiesel as well as monitoring and maintaining biodiesel quality against degradation during long-term storage.

Maintaining fuel quality during long-term storage is a concern for biodiesel producers, marketers, and consumers. The most cost-effective means for improving oxidative stability of biodiesel is treatment with antioxidant additives. Care must be exercised in cleaning storage tanks before filling them with biodiesel and in monitoring storage conditions inside of the tanks such as temperature, moisture content, exposure to direct sunlight, and the atmosphere (nitrogen "blanket" is preferable) over which the fuel is stored. Biodiesel stored over long periods should be monitored regularly for signs of degradation. The National Biodiesel Board recommends that biodiesel not be stored for longer than 6 mo.

The use of additives to address a great number of fuel performance issues is ubiquitous in the biodiesel and petrodiesel industries. Unless the fuels themselves are enhanced through compositional modification, the employment of additives is likely to continue for the foreseeable future. As such, in spite of the impressive technological advances that have been made over the last 50 or more years in the field of fuel additives, a great deal of research remains to be accomplished to fully address technical deficiencies inherent in fuels, in particular the comparatively new arena of biodiesel and biodiesel blends in ULSD. With the conversion from low-sulfur ($\leq$500 ppm) no. 2 petrodiesel to ULSD in the United States in 2006, many additive treatment technologies that were previously effective with no. 2 petrodiesel may not yield similar results in ULSD. By extension, additives used for blends of SME/low-sulfur petrodiesel may not be as beneficial in blends of SME/ULSD, which once again emphasizes the need for continued research and development of fuel additives.

The primary market for biodiesel in the near to long-term future is likely to be as a blend component in petrodiesel (ULSD). As such, it is critical that a thorough understanding be developed as to how biodiesel prepared from various feedstocks influences important fuel properties of the resultant petrodiesel/biodiesel blends, such as exhaust emissions, low-temperature operability, oxidative stability, water content, kinematic viscosity, AV, lubricity, and corrosiveness.

Soybean oil, which is currently the predominant feedstock for biodiesel production in the United States, is comparatively expensive, does not have an optimal fatty acid composition, has numerous competing food-related applications, and is not obtained in high yield from the seeds from which it is obtained. Development of alternative feedstocks for biodiesel production is another important area of current and future research. Additionally, genetic modification of existing oilseed sources to yield higher oil content and optimal fatty acid compositions is another potential strategy to yield improved quantity and fuel properties of biodiesel. However, genetically modified crops require extensive testing, regulatory evaluation, and approval prior to widespread commercial production.

Biodiesel in the coming years may face competition from non-ester renewable diesel fuels such as those produced from catalytic hydroprocessing of vegetable oils or animal fats. Although many of the benefits of biodiesel may be lost in the production of hydrocarbons from vegetable oils or animal fats, some of the negative aspects of biodiesel are lost as well. From a commercial standpoint, the traditional petroleum industry is more comfortable with these non-ester renewable diesel fuels than with biodiesel, which may present a substantial challenge to the widespread deployment of biodiesel as an alternative fuel in the future. However, the many environmental benefits and applications of biodiesel will continue to ensure that a substantial market exists for this attractive alternative to conventional petroleum diesel fuel.

Further Reading

For additional recent reviews on biodiesel production and processing, *see* Sharma et al. (2008), Vasudevan and Briggs (2008), Marchetti et al. (2007), Meher et al. (2006a), and Gerpen 2005. For further reading on catalysts for the production of biodiesel, *see* Ranganathan et al. (2008; enzymatic catalysts), Narasimharao et al. (2007; general catalyst review), and Demirbas (2005; supercritical methanol). For more information on alternative feedstocks for biodiesel production, *see* Meng et al. (2009; microorganisms), Canakci, Sanli (2008; various), Li et al. (2008; microbial oils), Scholz and da Silva (2008; castor oil), Chisti (2007; microalgae), and Mohibbeazam et al. (2005; Indian oils). For additional reading on the influence of structure on the physical properties of biodiesel, *see* Knothe (2008), Knothe (2007), and Knothe (2005). Further reviews on additives (Ribeiro et al. 2007), glycerol (Behr et al. 2008), standards (Knothe 2006), methods (Monteiro et al. 2008), renewable diesel (Demirbas and Dincer 2008; Huber and Corma 2007), and progress and trends in biofuels (Demirbas 2007) are also available. Several recent books also discuss biodiesel in whole or in part, which include Hou and Shaw (2008), Nag (2008), Erhan (2005), Knothe et al. (2005), Mittelbach and Remschmidt (2004).

References

Ajav E. A.; Singh B.; Bhattacharya T. K. Thermal balance of a single cylinder diesel engine operating on alternative fuels. *Energ. Convers. Manage.* 41: 1533–1541; 2000. doi:10.1016/S0196-8904(99)00175-2.

Alamu O. J.; Waheed M. A.; Jekayinfa S. O. Effect of ethanol–palm kernel oil ratio on alkali-catalyzed biodiesel yield. *Fuel* 87: 1529–1533; 2008. doi:10.1016/j.fuel.2007.08.011.

Albuquerque M. C. G.; Machado Y. L.; Torres A. E. B.; Azevedo D. C. S.; Cavalcante C. L. Jr.; Firmiano L. R.; Parente E. J. S. Jr. Properties of biodiesel oils formulated using different biomass sources and their blends. *Renew Energ.* 34: 857–859; 2009. doi:10.1016/j.renene.2008.07.006.

Ali Y.; Hanna M. A. Alternative diesel fuels from vegetable oils. *Bioresource. Technol.* 50: 153–163; 1994. doi:10.1016/0960-8524(94)90068-X.

Alptekin E.; Canakci M. Determination of the density and the viscosities of biodiesel-diesel fuel blends. *Renew Energ* 33: 2623–2630; 2008. doi:10.1016/j.renene.2008.02.020.

Alptekin E.; Canakci M. Characterization of the key fuel properties of methyl ester-diesel fuel blends. *Fuel* 88: 75–80; 2009. doi:10.1016/j.fuel.2008.05.023.

Altiparmak D.; Keskin A.; Koca A.; Guru M. Alternative fuel properties of tall oil fatty acid methyl ester-diesel fuel blends. *Bioresource. Technol.* 98: 241–246; 2007. doi:10.1016/j.biortech.2006.01.020.

Anonymous Dictionary Section. In: Gunstone F. D.; Harwood J. L.; Dijkstra A. J. (eds) The lipid handbook. 3rd ed. CRC, Boca Raton, pp 444–445; 2007.

Antolin G.; Tinaut F. V.; Briceno Y.; Castano V.; Perez C.; Ramirez A. I. Optimisation of biodiesel production by sunflower oil transesterification. *Bioresource. Technol.* 83: 111–114; 2002. doi:10.1016/S0960-8524(01)00200-0.

ASTM Standard specification for biodiesel fuel (B100) blend stock for distillate fuels. In: Annual Book of ASTM Standards, ASTM International, West Conshohocken, Method D6751-08; 2008a.

ASTM Standard specification for diesel fuel oil, biodiesel blend (B6 to B20). In: Annual Book of ASTM Standards, ASTM International, West Conshohocken, Method D7467-08a; 2008b.

ASTM Standard specification for diesel fuel oils. In: Annual Book of ASTM Standards, ASTM International, West Conshohocken, Method D975-08a; 2008c.

ASTM Standard specification for fuel oils. In: Annual Book of ASTM Standards, ASTM International, West Conshohocken, Method D396-08b; 2008d.

Balan V.; Rogers C. A.; Chundawat S. P. S.; da Costa Sousa L.; Slininger P. J.; Gupta R.; Dale B. E. Conversion of extracted oil cake fibers into bioethanol including DDGS, canola, sunflower, sesame, soy, and peanut for integrated biodiesel processing. *J. Am. Oil. Chem. Soc.* 86: 157–165; 2009. doi:10.1007/s11746-008-1329-4.

Ban-Weiss G. A.; Chen J. Y.; Buchholz B. A.; Dibble R. W. A numerical investigation into the anomalous slight NO_x increase when burning biodiesel; A new (old) theory. *Fuel Process Technol.* 88: 659–667; 2007. doi:10.1016/j.fuproc.2007.01.007.

Banapurmath N. R.; Tewari P. G.; Hosmath R. S. Performance and emissions characteristics of a DI compression ignition engine operated on Honge, Jatropha, and sesame oil FAME. *Renew Eng* 33: 1982–1988; 2008. doi:10.1016/j.renene.2007.11.012.

Barbour, R. H.; Rickeard, D. J.; Elliott, N. G. Understanding diesel lubricity. SAE Tech Pap Ser 2000-01-1918; 2000.

Behr A.; Eilting J.; Irawadi K.; Leschinski J.; Linder F. Improved utilization of renewable resources: new important derivatives of glycerol. *Green Chem.* 10: 13–30; 2008. doi:10.1039/b710561d.

Benjumea P.; Agudelo J.; Agudelo A. Basic properties of palm oil biodiesel-diesel blends. *Fuel* 87: 2069–2075; 2008. doi:10.1016/j.fuel.2007.11.004.

Bhale P. V.; Deshpande N. V.; Thombre S. B. Improving the low temperature properties of biodiesel fuel. *Renew Energ.* 34: 794–800; 2009. doi:10.1016/j.renene.2008.04.037.

Bhatnagar A. K.; Kaul S.; Chhibber V. K.; Gupta A. K. HFRR studies on FAME of nonedible vegetable oils. *Energ. Fuel* 20: 1341–1344; 2006. doi:10.1021/ef0503818.

Bhatt Y. C.; Murthy N. S.; Datta R. K. Use of mahua oil (*Madhuca indica*) as a diesel fuel extender. *J. Institutional Eng. (India): Agric. Eng. Div.* 85: 10–14; 2004.

Bondioli P.; Cortesi N.; Mariani C. Identification and quantification of steryl glucosides in biodiesel. *Eur. J. Lipid Sci. Technol.* 110: 120–126; 2008. doi:10.1002/ejlt.200700158.

Bondioli P.; Gasparoli A.; Bella L. D.; Tagliabue S.; Toso G. Biodiesel stability under commercial storage conditions over one year. *Eur. J. Lipid Sci. Technol.* 105: 35–741; 2003. doi:10.1002/ejlt.200300783.

Boocock D. G. B.; Konar S. K.; Mao V.; Lee C.; Buligan S. Fast formation of high-purity FAME from vegetable oils. *J. Am. Oil. Chem. Soc.* 75: 1167–1172; 1998. doi:10.1007/s1746-998-0130-8.

Boocock D. G. B.; Konar S. K.; Mao V.; Sidi H. Fast one-phase oil-rich processes for the preparation of vegetable oil FAME. *Biomass. Bioenerg.* 11: 43–50; 1996a. doi:10.1016/0961-9534(95)00111-5.

Boocock D. G. B.; Konar S. K.; Sidi H. Phase diagrams for oil/methanol/ether mixtures. *J. Am. Oil. Chem. Soc.* 73: 1247–1251; 1996b. doi:10.1007/BF02525453.

Bostyn S.; Duval-Onen F.; Porte C.; Coic J. P.; Fauduet H. Kinetic modeling of the degradation of α-tocopherol in biodiesel-rape methyl ester. *Bioresource Technol.* 99: 6439–6445; 2008. doi:10.1016/j.biortech.2007.11.054.

Bouaid A.; Bajo L.; Martinez M.; Aracil J. Optimization of biodiesel production from jojoba oil. *Process Saf. Environ.* 85: 378–382; 2007. doi:10.1205/psep07004.

Breccia A.; Esposito B.; Breccia Fratadocchi G.; Fini A. Reaction between methanol and commercial seed oils under microwave irradiation. *J. Microwave Power EE* 34: 3–8; 1999.

Bringi N. V. Non-traditional oilseeds and oils of India. Oxford and IBH, New Delhi; 1987.

Can O.; Celikten I.; Usta N. Effects of ethanol addition on performance and emissions of a turbocharged indirect injection Diesel engine running at different injection pressures. *Energ. Convers. Manage.* 45: 2429–2440; 2004. doi:10.1016/j.enconman.2003.11.024.

Canakci M.; Sanli H. Biodiesel production from various feedstocks and their effects on the fuel properties. *J. Ind. Microbiol. Biot.* 35: 431–441; 2008. doi:10.1007/s10295-008-0337-6.

Canakci M.; Van Gerpen J. Biodiesel production via acid catalysis. *Trans. ASAE* 42: 1203–1210; 1999.

Canakci M.; Van Gerpen J. Biodiesel production from oils and fats with high free fatty acids. *Trans. ASAE* 44: 1429–1436; 2001.

Canakci M.; Van Gerpen J. A pilot plant to produce biodiesel from high free fatty acid feedstocks. *Trans. ASAE* 46: 945–954; 2003a.

Canakci M.; Van Gerpen J. Comparison of engine performance and emissions for petroleum diesel fuel, yellow grease biodiesel, and soybean oil biodiesel. *Trans. ASAE* 46: 937–944; 2003b.

Canoira L.; Alcantara R.; Garcia-Martinez M. J.; Carrasco J. Biodiesel from jojoba oil-wax: transesterification with methanol and properties as a fuel. *Biomass. Bioenerg.* 30: 76–81; 2006. doi:10.1016/j.biombioe.2005.07.002.

Cardone M.; Prati M. V.; Rocco V.; Seggiani M.; Senatore A.; Vitolo S. *Brassica carinata* as an alternative oil crop for the production of biodiesel in Italy: engine performance and regulated and unregulated exhaust emissions. *Environ. Sci. Technol.* 36: 4656–4662; 2002. doi:10.1021/es011078y.

Cetinkaya M.; Karaosmanoglu F. Optimization of base-catalyzed transesterification reaction of used cooking oil. *Energ. Fuel* 18: 1888–1895; 2004. doi:10.1021/ef049891c.

Chang D. Y. Z.; Van Gerpen J. H.; Lee I.; Johnson L. A.; Hammond E. G.; Marley S. J. Fuel properties and emissions of soybean oil esters as diesel fuel. *J. Am. Oil. Chem. Soc.* 73: 1549–1555; 1996. doi:10.1007/BF02523523.

Chapagain B. P.; Yehoshua Y.; Wiesman Z. Desert date (*Balanites aegyptiaca*) as an arid lands sustainable bioresource for biodiesel. *Bioresource Technol.* 100: 1221–1226; 2009. doi:10.1016/j.biortech.2008.09.005.

Chiou B. S.; El-Mashad H. M.; Avena-Bustillos R. J.; Dunn R. O.; Bechtel P. J.; McHugh T. H.; Imam S. H.; Glenn G. M.; Ortz W. J.; Zhang R. Biodiesel from waste salmon oil. *Trans. ASABE* 51: 797–802; 2008.

Chisti Y. Biodiesel from microalgae. *Biotechnol. Adv.* 25: 294–306; 2007. doi:10.1016/j.biotechadv.2007.02.001.

Chiu C. W.; Schumacher L. G.; Suppes G. J. Impact of cold flow improvers on soybean biodiesel blend. *Biomass. Bionerg.* 27: 485–491; 2004. doi:10.1016/j.biombioe.2004.04.006.

Choi C. Y.; Reitz R. D. An experimental study on the effects of oxygenated fuel blends and multiple injection strategies on DI diesel engines. *Fuel* 78: 1303–1317; 1999. doi:10.1016/S0016-2361(99)00058-7.

Committee for Standardization Automotive fuels—fatty acid FAME (FAME) for diesel engines—requirements and test methods. European Committee for Standardization, Brussels; 2003a. Method EN 14214.

Committee for Standardization Heating fuels—fatty acid FAME (FAME)—requirements and test methods. European Committee for Standardization, Brussels; 2003b. Method EN 14213.

Committee for Standardization Automotive fuels—diesel—requirements and test methods. European Committee for Standardization, Brussels; 2004. Method EN 590.

Dailey O. D.; Prevost N. T.; Strahan G. D. Synthesis and structural analysis of branched-chain derivatives of methyl oleate. *J. Am. Oil Chem. Soc.* 85: 647–653; 2008. doi:10.1007/s11746-008-1235-9.

Dantas M. B.; Almeida A. A. F.; Conceicao M. M.; Fernandes V. J. Jr.; Santos I. M. G.; Silva F. C.; Soledade L. E. B.; Souza A. G. Characterization and kinetic compensation effect of corn biodiesel. *J. Therm. Anal. Calorim.* 87: 847–851; 2007. doi:10.1007/s10973-006-7786-9.

Danzer, M. F.; Ely, T. L.; Kingery, S. A.; McCalley, W. W.; McDonald, W. M.; Mostek, J.; Schultes, M. L. Biodiesel cold filtration process. US Pat Appl 20070175091, filed 02/01/2007; 2007.

Darnoko D.; Cheryan M. Kinetics of palm oil transesterification in a batch reactor. *J. Am. Oil Chem. Soc.* 77: 1263–1267; 2000. doi:10.1007/s11746-000-0198-y.

Dasari M. A.; Kiatsimkul P. P.; Sutterlin W. R.; Suppes G. J. Low-pressure hydrogenolysis of glycerol to propylene glycol. *Appl. Catal. A-Gen.* 281: 225–231; 2005. doi:10.1016/j.apcata.2004.11.033.

Demirbas A. Biodiesel fuels from vegetable oils via catalytic and non catalytic supercritical alcohol transesterifications and other methods: a survey. *Energ. Convers. Manage.* 44: 2093–2109; 2003. doi:10.1016/S0196-8904(02)00234-0.

Demirbas A. Biodiesel production from vegetable oils via catalytic and non-catalytic supercritical methanol transesterification methods. *Progress Energ. Combust.* 31: 466–487; 2005. doi:10.1016/j.pecs.2005.09.001.

Demirbas A. Biodiesel production via non-catalytic SCF method and biodiesel fuel charactertistics. *Energ.convers.Manage.*47:2271–2282;2006.doi:10.1016j.enconomon.2005:11.019.

Demirbas A. Progress and recent trends in biofuels. *Prog. Energ. Combust.* 33: 1–18; 2007. doi:10.1016/j.pecs.2006.06.001.

Demirbas A. Production of biodiesel from tall oil. *Energ. Source Part A* 30: 1896–1902; 2008.

Demirbas A.; Dincer K. Sustainable green diesel: a futuristic view. *Energ. Source Part A* 30: 1233–1241; 2008.

DeOliveira E.; Quirino R. L.; Suarez P. A. Z.; Prado A. G. S. Heats of combustion of biofuels obtained by pyrolysis and by transesterification and of biofuel/diesel blends. *Thermochim. Acta* 450: 87–90; 2006. doi:10.1016/j.tca.2006.08.005.

Dias J. M.; Alvim-Ferraz M. C. M.; Almeida M. F. Comparison of the performance of different homogenous alkali catalysts during transesterification of waste and virgin oils and evaluation of biodiesel quality. *Fuel* 87: 3572–3578; 2008. doi:10.1016/j.fuel.2008.06.014.

Dimitrakis W. J. The importance of lubricity. *Hydrocarb. Eng.* 8: 37–39; 2003.

Doell R.; Konar S. K.; Boocock D. G. B. Kinetic parameters of a homogenous transmethylation of soybean oil. *J. Am. Oil Chem. Soc.* 85: 271–276; 2008. doi:10.1007/s11746-007-1168-8.

Domingos A. K.; Saad E. B.; Wilhelm H. M.; Ramos L. P. Optimization of the ethanolysis of *Raphanus sativas* (L. var.) crude oil applying the response surface methodology. *Bioresource. Technol.* 99: 1837–1845; 2008. doi:10.1016/j.biortech.2007.03.063.

Dorado M. P.; Ballesteros E.; Lopez F. J.; Mittelbach M. Optimization of alkali-catalyzed transesterification of *Brassica carinata* oil for biodiesel production. *Energ. Fuel* 18: 77–83; 2004. doi:10.1021/ef0340110.

dos Santos I. C. F.; de Carvalho S. H. V.; Solleti J. I.; Ferreira de Le Salles W.; Teixeira de Silva de La Salles K.; Meneghetti S. M. P. Studies of *Terminalia catappa* L. oil: characterization and biodiesel production. *Bioresource. Technol.* 99: 6545–6549; 2008. doi:10.1016/j.biortech.2007.11.048.

Drown D. C.; Harper K.; Frame E. Screening vegetable oil alcohol esters as fuel lubricity enhancers. *J. Am. Oil Chem. Soc.* 78: 679–584; 2001. doi:10.1007/s11746-001-0307-y.

Dunn R. O. Alternative jet fuels from vegetable oils. *Trans. ASAE* 44: 1751–1757; 2001.

Dunn R. O. Oxidative stability of soybean oil fatty acid FAME by oil stability index (OSI). *J. Am. Oil Chem. Soc.* 82: 381–387; 2005a. doi:10.1007/s11746-005-1081-6.

Dunn R. O. Effect of antioxidants on the oxidative stability of methyl soyate (biodiesel). *Fuel Process. Technol.* 86: 1071–1085; 2005b. doi:10.1016/j.fuproc.2004.11.003.

Dunn R. O.; Bagby M. O. Low-temperature properties of triglyceride-based diesel fuels: transesterified FAME and petroleum middle distillate/ester blends. *J. Am. Oil Chem. Soc.* 72: 895–904; 1995. doi:10.1007/BF02542067.

Dunn R. O.; Shockley M. W.; Bagby M. O. Improving the low-temperature properties of alternative diesel fuels: vegetable oil-derived FAME. *J. Am. Oil Chem. Soc.* 73: 1719–1728; 1996. doi:10.1007/BF02517978.

Dunn, R. O.; Shcokley, M. W.; Bagby, M. O. Winterized FAME from soybean oil: an alternative diesel fuel with improved low-temperature properties. SAE Tech Pap Ser 1997-01-971682; 1997.

Eathington S. R.; Crosbie T. M.; Edwards M. D.; Reiter R. S.; Bull J. K. Molecular markers in a commercial breeding program. *Crop. Sci.* 47S3: S154–S163; 2007. doi:10.2135/cropsci2007.04.0015IPBS.

El-Mashad H. M.; Zhang R.; Avena-Bustillos R. J. A two-step process for biodiesel production from salmon oil. *Biosyst. Eng.* 99: 220–227; 2008. doi:10.1016/j.biosystemseng.2007.09.029.

Elleuch M.; Besbes S.; Roiseux O.; Blecker C.; Attia H. Quality characteristics of sesame seeds and by-products. *Food Chem.* 103: 641–650; 2007. doi:10.1016/j.foodchem.2006.09.008.

Encinar J. M.; Gonzalez J. F.; Rodriguez J. J.; Tejedor A. Biodiesel fuels from vegetable oils: transesterification of *Cynara cardunculus* L. oils with ethanol. *Energ. Fuel* 16: 443–450; 2002. doi:10.1021/ef010174h.

Encinar J. M.; Gonzalez J. F.; Rodriguez-Reinares A. Ethanolysis of used frying oil. Biodiesel preparation and characterization. *Fuel Process. Technol.* 88: 513–522; 2007. doi:10.1016/j.fuproc.2007.01.002.

Environmental Protection Agency (EPA) A comprehensive analysis of biodiesel impacts on exhaust emissions. Draft Technical Report EPA420-P-02-00. National Service Center for Environmental Publications, Cincinnati, OH; 2002.

Erhan S. Z. Industrial uses of vegetable oils. AOCS, Champaign; 2005.

Feng J.; Fu H.; Wang J.; Li R.; Chen H.; Li X. Hydrogenolysis of glycerol to glycols over ruthenium catalysts: effect of support and catalyst reduction temperature. *Catal. Com.* 9: 1458–1464; 2008. doi:10.1016/j.catcom.2007.12.011.

Fernandezalvarez P. F.; Vila J.; Garrido-Fernandez J.; Grifoll M.; Lema J. M. Trials of bioremediation on a beach affected by the heavy oil spill of the Prestige. *J. Hazard. Mater. B* 137: 1523–1531; 2006. doi:10.1016/j.jhazmat.2006.04.035.

Fernando S.; Hall C.; Jha S. NO_x reduction from biodiesel fuels. *Energy Fuels* 20: 376–382; 2006. doi:10.1021/ef050202m.

Fernando S.; Hanna M. Development of a novel biofuel blend using ethanol-biodiesel-diesel microemulsions: EB-diesel. *Energ. Fuel* 18: 1695–1703; 2004. doi:10.1021/ef049865e.

Foglia T. A.; Nelson L. A.; Dunn R. O.; Marmer W. N. Low-temperature properties of alkyl esters of tallow and grease. *J. Am. Oil Chem. Soc.* 74: 951–955; 1997. doi:10.1007/s11746-997-0010-7.

Frankel E. N. Lipid oxidation. 2nd ed. The Oily Press, Bridgewater; 2005.

Freedman B.; Butterfield R. O.; Pryde E. H. Transesterification kinetics of soybean oil. *J. Am. Oil Chem. Soc.* 63: 1375–1380; 1986. doi:10.1007/BF02679606.

Freedman B.; Pryde E. H.; Mounts T. L. Variables affecting the yields of fatty esters from transesterified vegetable oils. *J. Am. Oil Chem. Soc.* 61: 1638–1643; 1984. doi:10.1007/BF02541649.

Frohlich A.; Rice B. Evaluation of *Camelina sativa* oil as a feedstock for biodiesel production. *Ind. Crop. Prod.* 21: 25–31; 2005. doi:10.1016/j.indcrop.2003.12.004.

Frohlich A.; Schober S. The influence of tocopherols on the oxidative stability of FAME. *J. Am. Oil Chem. Soc.* 84: 579–585; 2007. doi:10.1007/s11746-007-1075-z.

Geller D. P.; Goodrum J. W. Effects of specific fatty acid FAME on diesel fuel lubricity. *Fuel* 83: 2351–2356; 2004. doi:10.1016/j.fuel.2004.06.004.

Georgogianni K. G.; Kontominas M. G.; Pomonis P. J.; Avlontis D.; Gergis V. Conventional and in situ transesterification of sunflower seed oil for the production of biodiesel. *Fuel Process. Technol.* 89: 503–509; 2008. doi:10.1016/j.fuproc.2007.10.004.

Gerpen J. Biodiesel processing and production. *Fuel Process. Technol.* 86: 1097–1107; 2005. doi:10.1016/j.fuproc.2004.11.005.

Ghadge S. V.; Raheman H. Biodiesel production from mahua (*Madhuca indica*) oil having high free fatty acids. *Biomass. Bioenerg.* 28: 601–605; 2005. doi:10.1016/j.biombioe.2004.11.009.

Ghadge S. V.; Raheman H. Process optimization for biodiesel production from mahua (*Madhuca indica* L.) oil using response surface methodology. *Bioresource. Technol.* 97: 379–384; 2006. doi:10.1016/j.biortech.2005.03.014.

Glória Pereira M. G.; Mudge S. M. Cleaning oiled shores: laboratory experiments testing the potential use of vegetable oil biodiesels. *Chemosphere*. 54: 297–304; 2004. doi:10.1016/S0045-6535(03)00665-9.

Goodrum J. W.; Geller D. P. Influence of fatty acid FAME from hydroxylated vegetable oils on diesel fuel lubricity. *Bioresource. Technol.* 96: 851–855; 2005. doi:10.1016/j.biortech.2004.07.006.

Graboski M. S.; McCormick R. L. Combustion of fat and vegetable oil derived fuels in diesel engines. *Prog. Energ. Combust.* 24: 125–164; 1998. doi:10.1016/S0360-1285(97)00034-8.

Gunstone F. D. The chemistry of oils and fats. sources, composition, properties and uses. CRC, Boca Raton: 23–33 pp; 2004.

Gunstone F. D.; Harwood J. L. Occurrence and characterization of oils and fats. In: Gunstone F. D.; Harwood J. L.; Dijkstra A. J. (eds) The lipid handbook. 3rd ed. CRC, Boca Raton, pp 37–142; 2007.

Haas M. J. Improving the economics of biodiesel production through the use of low value lipids as feedstocks: vegetable oil soapstock. *Fuel Process. Technol.* 86: 1087–1096; 2005. doi:10.1016/j.fuproc.2004.11.004.

Haas M. J.; Michalski P. J.; Runyon S.; Nunez A.; Scott K. M. Production of FAME from acid oil, a byproduct of vegetable oil refining. *J. Am. Oil Chem. Soc.* 80: 97–102; 2003. doi:10.1007/s11746-003-0658-4.

Haas M. J.; Scott K. M.; Alleman T. L.; McCormick R. L. Engine performance of biodiesel fuel prepared from soybean soapstock: a high quality renewable fuel produced from a waste feedstock. *Energ. Fuel* 15: 1207–1212; 2001. doi:10.1021/ef010051x.

Hamad B.; Lopes de Souza R. O.; Sapaly G.; Carneiro Rocha M. G.; Pries de Oliveira P. G.; Gonzalez W. A.; Andrade Sales E.; Essayem N. Transesterification of rapeseed oil with ethanol over heterogeneous heteropolyacids. *Catal. Com.* 10: 92–97; 2008. doi:10.1016/j.catcom.2008.07.040.

Hancsok J.; Bubalik M.; Beck A.; Baladincz J. Development of multifunctional additives based on vegetable oils for high quality diesel and biodiesel. *Chem. Eng. Res. Des.* 86: 793–799; 2008. doi:10.1016/j.cherd.2008.03.011.

Hashimoto N.; Ozawa Y.; Mori N.; Yuri I.; Hisamatsu T. Fundamental combustion characteristics of palm methyl ester (PME) as alternative fuel for gas turbines. *Fuel* 87: 3373–3378; 2008. doi:10.1016/j.fuel.2008.06.005.

Hess M. A.; Haas M. J.; Foglia T. A. Attempts to reduce NO_x exhaust emissions by using reformulated biodiesel. *Fuel Process Technol.* 88: 693–699; 2007. doi:10.1016/j.fuproc.2007.02.001.

Hess M. A.; Haas M. J.; Foglia T. A.; Marmer W. M. The effect of antioxidant addition on NO_x emissions from biodiesel. *Energ. Fuel* 19: 1749–1754; 2005. doi:10.1021/ef049682s.

Heywood J. Internal combustion engine fundamentals. McGraw-Hill Press, New York: 572–577 pp; 1998.

Hoed V.; Zyaykina N.; De Greyt W.; Maes J.; Verhe R.; Demeestere K. Identification and occurrence of steryl glucosides in palm and soy biodiesel. *J. Am. Oil Chem. Soc.* 85: 701–709; 2008. doi:10.1007/s11746-008-1263-5.

Holman R. A.; Elmer O. C. The rates of oxidation of unsaturated fatty acids and esters. *J. Am. Oil Chem. Soc.* 24: 127–129; 1947. doi:10.1007/BF02643258.

Holser R. A.; Harry-O'Kuru R. Transesterified milkweed (*Asclepias*) seed oil as a biodiesel fuel. *Fuel* 85: 2106–2110; 2006. doi:10.1016/j.fuel.2006.04.001.

Hou C. T.; Shaw J. F. Biocatalysts and bioenergy. Wiley, Hoboken; 2008.

Hu J.; Du Z.; Li C.; Min E. Study on the lubrication properties of biodiesel as fuel lubricity enhancers. *Fuel* 84: 1601–1606; 2005.

Hu J.; Du Z.; Tang Z.; Min E. Study on the solvent power of a new green solvent: biodiesel. *Ind. Eng. Chem. Res.* 43: 7928–7931; 2004. doi:10.1021/ie0493816.

Huber G. W.; Corma A. Synergies between bio- and oil refineries for the production of fuels from biomass. *Ang. Chem. Int. Ed.* 46: 7184–7201; 2007. doi:10.1002/anie200604504.

Hughes J. M.; Mushrush G. W.; Hardy D. R. Lubricity-enhancing properties of soy oil when used as a blending stock for middle distillate fuels. *Ind. Eng. Chem. Res.* 41: 1386–1388; 2002. doi:10.1021/ie010624t.

International Grains Council Grain market trends in the stockfeed and biodiesel industries. *Australian Grain* 17: 30–31; 2008.

Issariyakul T.; Kulkarmi M. G.; Dalai A. K.; Bakhshi N. N. Production of biodiesel from waste fryer grease using mixed methanol/ethanol system. *Fuel Process. Technol.* 88: 429–436; 2007. doi:10.1016/j.fuproc.2006.04.007.

Jeong G. W.; Yang H. S.; Park D. H. Optimization of transesterification of animal fat ester using response surface methodology. *Bioresource. Technol.* 100: 25–30; 2009. doi:10.1016/j.biortech.2008.05.011.

Joshi, H. C.; Toler, J.; Moser, B. R.; Walker, T. Biodiesel from canola oil using a 1:1 mixture of methanol and ethanol. *Eur J Lipid Sci Technol.* 111: 464–473; 2009. doi:10.1002/ejlt.200800071.

Joshi H. C.; Toler J.; Walker T. Optimization of cottonseed oil ethanolysis to produce biodiesel high in gossypol content. *J. Am. Oil Chem. Soc.* 85: 357–363; 2008. doi:10.1007/s11746-008-1200-7.

Kalbande S. R.; More G. R.; Nadre R. G. Biodiesel production from non-edible oils of jatropha and karanj for utilization in electrical generator. *Bioenerg. Res.* 1: 170–178; 2008. doi:10.1007/s12155-008-9016-8.

Karmee S. K.; Chadha A. Preparation of biodiesel from crude oil of *Pongamia pinnata*. *Bioresource. Technol.* 96: 1425–1429; 2005. doi:10.1016/j.biortech.2004.12.011.

Kerschbaum S.; Rinke G.; Schubert K. Winterization of biodiesel by mirco process engineering. *Fuel* 87: 2590–2597; 2008. doi:10.1016/j.fuel.2008.01.023.

Keskin A.; Guru M.; Altiparmak D. Influence of tall oil biodiesel with Mg and Mo based fuel additives on diesel engine performance and emission. *Bioresource. Technol.* 99: 6434–6438; 2008. doi:10.1016/j.biortech.2007.11.051.

Knothe G. Dependence of biodiesel fuel properties on the structure of fatty acid alkyl esters. *Fuel Process Technol.* 86: 1059–1070; 2005. doi:10.1016/j.fuproc.2004.11.002.

Knothe G. Analyzing biodiesel: standards and other methods. *J. Am. Oil Chem. Soc.* 83: 823–833; 2006. doi:10.1007/s11746-006-5033-y.

Knothe G. Some aspects of biodiesel oxidative stability. *Fuel Process. Technol.* 88: 669–677; 2007. doi:10.1016/j.fuproc.2007.01.005.

Knothe G. "Designer" biodiesel: optimizing fatty ester composition to improve fuel properties. *Energ. Fuel* 22: 1358–1364; 2008. doi:10.1021/ef700639e.

Knothe G.; Bagby M. O.; Ryan T. A. III Cetane numbers of fatty compounds: influence of compound structure and of various potential cetane improvers. *SAE Tech. Pap. Ser.* 971681: 127–132; 1997.

Knothe G.; Dunn R. O. Dependence of oil stability index of fatty compounds on their structure and concentration in the presence of metals. *J. Am. Oil Chem. Soc.* 80: 1021–1025; 2003. doi:10.1007/s11746-003-0814-x.

Knothe G.; Matheaus A. C.; Ryan T. W. III Cetane numbers of branched and straight-chain fatty esters determined in an ignition quality tester. *Fuel* 82: 971–975; 2003. doi:10.1016/S0016-2361(02)00382-4.

Knothe G.; Sharp C. A.; Ryan T. W. III Exhaust emissions of biodiesel, petrodiesel, neat FAME, and alkanes in a new technology engine. *Energ. Fuel* 20: 403–408; 2006. doi:10.1021/ef0502711.

Knothe G.; Steidley K. R. Kinematic viscosity of biodiesel fuel components and related compounds. Influence of compound structure and comparison to petrodiesel fuel components. *Fuel* 84: 1059–1065; 2005a. doi:10.1016/j.fuel.2005.01.016.

Knothe G.; Steidley K. R. Lubricity of components of biodiesel and petrodiesel. The origin of biodiesel lubricity. *Energ. Fuel* 19: 1192–1200; 2005b. doi:10.1021/ef049684c.

Knothe G.; Van Gerpen J.; Krahl J. The Biodiesel Handbook. AOCS, Urbana; 2005.

Kocak M. S.; Ileri E.; Utlu Z. Experimental study of emission parameters of biodiesel fuels obtained from canola, hazelnut, and waste cooking oils. *Energ. Fuel* 21: 3622–3626; 2007. doi:10.1021/ef0600558.

Korres D. M.; Karonis D.; Lois E.; Linck M. B.; Gupta A. K. Aviation fuel JP-5 and biodiesel on a diesel engine. *Fuel* 87: 70–78; 2008. doi:10.1016/j.fuel.2007.04.004.

Korus, R. A.; Hoffman, D. S.; Bam, H.; Peterson, C. L.; Brown, D. C. Transesterification process to manufacture ethyl ester of rape oil, First Biomass Conference of the Americas. Burlington, VT, vol 2: 815–822; 1993.

Kotrba R. Transition period. *Biodiesel. Mag.* 512: 52–57; 2008.

Kram J. W. Gallons of megawatts. *Biodiesel. Mag.* 55: 76–80; 2008a.

Kram J. W. Power without the burn. *Biodiesel. Mag.* 53: 73–77; 2008b.

Kucek K. T.; Aparecida M.; Cesar-Oliveira F.; Wilhelm H. M.; Ramos L. P. Ethanolysis of refined soybean oil assisted by sodium and potassium hydroxides. *J. Am. Oil Chem. Soc.* 84: 385–392; 2007. doi:10.1007/s11746-007-1048-2.

Kulkarni M. G.; Dalai A. K.; Bakhshi N. N. Transesterification of canola oil in mixed methanol/ethanol system and use of esters as lubricity additive. *Bioresource. Technol.* 98: 2027–2033; 2007. doi:10.1016/j.biortech.2006.08.025.

Kumari V.; Shah S.; Gupta M. N. Preparation of biodiesel by lipase-catalyzed transesterification of high free fatty acid containing oil from *Madhuca indica*. *Energ. Fuel* 21: 368–372; 2007. doi:10.1021/ef0602168.

Kumartiwari A. K.; Kumar A.; Raheman H. Biodiesel production from jatropha oil (*Jatropha curcas*) with high free fatty acids: An optimized process. *Biomass. Bioenerg.* 31: 569–575; 2007. doi:10.1016/j.biombioe.2007.03.003.

Kusdiana D.; Saka S. Effects of water on biodiesel fuel production by supercritical methanol treatment. *Bioresource Technol.* 91: 289–295; 2004. doi:10.1016/S0960-8524(03)00201-3.

Lang X.; Dalai A. K.; Bakkshi N. N.; Reaney M. J.; Hertz P. B. Preparation and characterization of bio-diesels from various bio-oils. *Bioresource Technol.* 80: 53–62; 2001. doi:10.1016/S0960-8524(01)00051-7.

Lapuerta M.; Herreros J. M.; Lyons L. L.; Garcia-Contreras R.; Briceno Y. Effect of the alcohol type used in the production of waste cooking oil biodiesel on diesel performance and emissions. *Fuel* 87: 3161–3169; 2008. doi:10.1016/j.fuel.2008.05.013.

Lebedevas S.; Vaicekauskas A. Use of waste fats of animal and vegetable origin for the production of biodiesel fuel: quality, motor properties, and emissions of harmful components. *Energ. Fuel* 20: 2274–2280; 2006. doi:10.1021/ef060145c.

Lee I.; Johnson L. A.; Hammond E. G. Use of branched-chain esters to reduce the crystallization temperature of biodiesel. *J. Am. Oil Chem. Soc.* 72: 1155–1160; 1995. doi:10.1007/BF02540982.

Lee, I.; Mayfield, J. L.; Pfalzgraf, L. M.; Solheim, L.; Bloomer, S. Processing and producing biodiesel and biodiesel produced there from. US Pat Appl 20070151146, filed 12/21/2006; 2006.

Lee I.; Pfalzgraf L. M.; Poppe G. B.; Powers E.; Haines T. The role of sterol glucosides on filter plugging. *Biodiesel. Mag.* 4: 105–112; 2007.

Lee K. T.; Foglia T. A.; Chang K. S. Production of alkyl ester as biodiesel from fractionated lard and restaurant grease. *J. Am. Oil Chem. Soc.* 79: 191–195; 2002. doi:10.1007/s11746-002-0457-y.

Leung D. Y. C.; Guo Y. Transesterification of neat and used frying oil: optimization for biodiesel production. *Fuel Process Technol.* 87: 883–890; 2006. doi:10.1016/j.fuproc.2006.06.003.

Li D.; Zhen H.; Xingcai L.; Wu-gao Z.; Jiang-guang Y. Physico-chemical properties of ethanol-diesel blend fuel and its effect on performance and emissions of diesel engines. *Renew Energ.* 30: 967–976; 2005. doi:10.1016/j.renene.2004.07.010.

Li Q.; Du W.; Liu D. Perspectives of microbial oils for biodiesel production. *Appl. Microbiol. Biot.* 80: 749–756; 2008. doi:10.1007/s00253-008-1625-9.

Liang Y. C.; May C. Y.; Foon C. S.; Ngan M. A.; Hock C. C.; Basiron Y. The effect of natural and synthetic antioxidants on the oxidative stability of palm biodiesel. *Fuel* 85: 867–870; 2006. doi:10.1016/j.fuel.2005.09.003.

Lima J. R. O.; Silva R. B.; Moura E. M.; Moura C. V. R. Biodiesel of tucum oil, synthesized by methanolic and ethanolic routes. *Fuel* 87: 1718–1723; 2008. doi:10.1016/j.fuel.2007.09.007.

Lin Y. C.; Tsai C. H.; Yang C. R.; Jim Wu C. H.; Wu T. Y.; Chang-Chien G. P. Effects on aerosol size distribution of polycyclic aromatic hydrocarbons from the heavy-duty diesel generator fueled with feedstock palm-biodiesel blends. *Atmos. Environ.* 42: 6679–6688; 2008. doi:10.1016/j.atmosenv.2008.04.018.

Liu X.; Piao X.; Wang Y.; Zhu S. Calcium ethoxide as a solid base catalyst for the transesterification of soybean oil to biodiesel. *Energ. Fuel* 22: 1313–1317; 2008. doi:10.1021/ef700518h.

Loh S. K.; Chew S. M.; Choo Y. M. Oxidative stability and storage behavior of fatty acid FAME derived from used palm oil. *J. Am. Oil Chem. Soc.* 83: 947–952; 2006. doi:10.1007/s11746-006-5051-9.

Lotero E.; Liu Y.; Lopez D. E.; Suwannakarn K.; Bruce D. A.; Goodwin J. G. Jr Synthesis of biodiesel via acid catalysis. *Ind. Eng. Chem. Res.* 44: 5353–5363; 2005. doi:10.1021/ie049157g.

Mahajan S.; Konar S. K.; Boocock D. G. B. Standard biodiesel from soybean oil by a single chemical reaction. *J. Am. Oil Chem. Soc.* 83: 641–644; 2006. doi:10.1007/s11746-006-1251-6.

Mahajan S.; Konar S. K.; Boocock D. G. B. Variables affecting the production of standard biodiesel. *J. Am. Oil Chem. Soc.* 84: 189–195; 2007. doi:10.1007/s11746-006-1023-3.

Mao V.; Konar S. K.; Boocock D. G. B. The pseudo-single-phase, base catalyzed transmethylation of soybean oil. *J. Am. Oil Chem. Soc.* 81: 803–808; 2004. doi:10.1007/s11746-004-0982-8.

Marchetti J. M.; Miguel V. U.; Errazu A. F. Possible methods for biodiesel production. *Renew. Sustain. Energy Rev.* 11: 1300–1311; 2007. doi:10.1016/j.rser.2005.08.006.

Mariod A.; Klupsch S.; Hussein H.; Ondruschka B. Synthesis of alkyl esters from three unconventional Sudanese oils for their use as biodiesel. *Energ. Fuel* 20: 2249–2252; 2006. doi:10.1021/ef060039a.

Mbaraka I. K.; Radu D. R.; Lin V. S. Y.; Shanks B. H. Organosulfonic acid-functionalized mesoporous silicas for the esterification of fatty acid. *J. Catal.* 219: 329–336; 2003. doi:10.1016/S0021-9517(03)00193-3.

McCormick, R. L., Alvarez, J. R., Graboski, M. S. 2003 NREL Final Report. SR-510-31465.

McCormick R. L.; Graboski M. S.; Alleman T. L.; Herring A. M. Impact of biodiesel source material and chemical structure on emissions of criteria pollutants from a heavy-duty engine. *Environ. Sci. Technol.* 35: 1742–1747; 2001. doi:10.1021/es001636t.

McGeehan, J. A. Diesel engines have a future and that future is clean. *SAE Tech Pap Ser* 2004-01-1956; 2004.

Meher L. C.; Dharmagadda V. S. S.; Naik S. N. Optimization of alkali-catalyzed transesterification of *Pongamia pinnata* oil for production of biodiesel. *Bioresource. Technol.* 97: 1392–1397; 2006b. doi:10.1016/j.biortech.2005.07.003.

Meher L. C.; Sagar D. V.; Naik S. N. Technical aspects of biodiesel production by transesterification—a review. *Renew. Sust. Engerg. Rev.* 10: 248–268; 2006a. doi:10.1016/j.rser.2004.09.002.

Meneghetti S. M. P.; Meneghetti M. R.; Serra T. M.; Barbosa D. C.; Wolf C. R. Biodiesel production from vegetable oil mixtures: cottonseed, soybean, and castor oils. *Energ. Fuel* 21: 3746–3747; 2007. doi:10.1021/ef070039q.

Meneghetti S. M. P.; Meneghetti M. R.; Wolf C. R.; Silva E. C.; Lima G. E. S.; de Lira Silva L.; Serra T. M.; Cauduro F.; de Oliveira L. G. Biodiesel from castor oil: a comparison of ethanolysis versus methanolysis. *Energ. Fuel* 20: 2262–2265; 2006. doi:10.1021/ef060118m.

Meng X.; Chen G.; Wang Y. Biodiesel production from waste cooking oil via alkali catalyst and its engine test. *Fuel Process Technol.* 89: 851–857; 2008. doi:10.1016/j.fuproc.2008.02.006.

Meng X.; Yang Y.; Xu X.; Zhang L.; Nie Q.; Xian M. Biodiesel production from oleaginous microorganisms. *Renew. Energ.* 34: 1–5; 2009. doi:10.1016/j.renene.2008.04.014.

Miao X.; Wu Q. Biodiesel production from heterotrophic microalgal oil. *Bioresource. Technol.* 97: 841–846; 2006. doi:10.1016/j.biortech.2005.04.008.

Miller J. A.; Bowman C. T. Mechanisms and modeling of nitrogen chemistry in combustion. *Prog. Energ. Combust.* 15: 287–338; 1989. doi:10.1016/0360-1285(89)90017-8.

Miller N. J.; Mudge S. M. The effect of biodiesel on the rate of removal and weathering characteristics of crude oil within artificial sand columns. *Spill. Sci. Technol. B* 4: 17–33; 1997. doi:10.1016/S1353-2561(97)00030-3.

Mittelbach M.; Gangl S. Long storage stability of biodiesel made from rapeseed and used frying oil. *J. Am. Oil Chem. Soc.* 78: 573–577; 2001. doi:10.1007/s11746-001-0306-z.

Mittelbach M.; Remschmidt C. Biodiesel - a comprehensive handbook. Martin Mittelbach, Graz; 2004.

Mittelbach M.; Schober S. The influence of antioxidants on the oxidation stability of biodiesel. *J. Am. Oil Chem. Soc.* 80: 817–823; 2003. doi:10.1007/s11746-003-0778-x.

Miyashita K.; Takagi T. Study on the oxidative rate and prooxidant activity of free fatty acids. *J. Am. Oil Chem. Soc.* 63: 1380–1384; 1986. doi:10.1007/BF02679607.

Mohibbeazam M. M.; Waris A.; Nahar N. M. Prospects and potential of fatty acid FAME of some non-traditional seed oils for use as biodiesel in India. *Biomass. Bioenerg.* 29: 293–302; 2005. doi:10.1016/j.biombioe.2005.05.001.

Mondala A.; Liang K.; Toghiani H.; Hernandez R.; French T. Biodiesel production by *in situ* transesterification of municipal primary and secondary sludges. *Bioresource. Technol.* 100: 1203–1210; 2009. doi:10.1016/j.biortech.2008.08.020.

Monteiro M. R.; Ambrozin A. R. P.; Liao L. M.; Ferreira A. G. Critical review on analytical methods for biodiesel characterization. *Talanta* 77: 593–605; 2008. doi:10.1016/j.talanta.2008.07.001.

Moreau R. A.; Scott K. M.; Haas M. J. The identification and quantification of steryl glucosides in precipitates from commercial biodiesel. *J. Am. Oil Chem. Soc.* 85: 761–770; 2008. doi:10.1007/s11746-008-1264-4.

Moreira A. B. R.; Perez V. H.; Zanin G. M.; de Castro H. F. Biodiesel synthesis by enzymatic transesterification of palm oil with ethanol using lipases from several sources immobilized on silica-PVA composite. *Energ. Fuel* 21: 3689–3694; 2007. doi:10.1021/ef700399b.

Moser B. R. Influence of blending canola, palm, soybean, and sunflower oil FAME on fuel properties of biodiesel. *Energ. Fuel* 22: 4301–4306; 2008a. doi:10.1021/ef800588x.

Moser B. R. Efficacy of myricetin as an antioxidant additive in FAME of soybean oil. *Eur. J. Lipid Sci. Technol.* 110: 1167–1174; 2008b. doi:10.1002/ejlt.200800145.

Moser B. R. Comparative oxidative stability of fatty acid alkyl esters by accelerated methods. *J. Am. Oil Chem. Soc.* 86: 699–706; 2009a. doi:10.1007/s11746-009-1376-5.

Moser B. R.; Williams A.; Haas M. J.; McCormick R. L. Exhaust emissions and fuel properties of partially hydrogenated soybean oil FAME blended with ultra low sulfur diesel fuel. *Fuel Process. Technol.* 90: 1122–1128; 2009b. doi:10.1016/j.fuproc.2009.05.004.

Moser B. R.; Cermak S. C.; Isbell T. A. Evaluation of castor and lesquerella oil derivatives as additives in biodiesel and ultra low sulfur diesel fuel. *Energ. Fuel* 22: 1349–1352; 2008. doi:10.1021/ef700628r.

Moser B. R.; Erhan S. Z. Synthesis and evaluation of a series of α-hydroxy ethers derived from isopropyl oleate. *J. Am. Oil Chem. Soc.* 83: 959–963; 2006. doi:10.1007/s11746-006-5053-7.

Moser B. R.; Erhan S. Z. Preparation and evaluation of a series of α-hydroxy ethers from 9,10-epoxystearates. *Eur. J. Lipid Sci. Technol.* 109: 206–213; 2007. doi:10.1002/ejlt.200600257.

Moser B. R.; Erhan S. Z. Branched chain derivatives of alkyl oleates: tribological, rheological, oxidation, and low temperature properties. *Fuel* 87: 2253–2257; 2008. doi:10.1016/j.fuel.2008.01.005.

Moser B. R.; Shah S. N.; Winkler-Moser J. K.; Vaughn S. F.; Evangelista R. L. Composition and physical properties of cress (Lepidium sativum L.) and field penncyress (Thlaspi arvense L.) oils. *Ind. Crops Prod.* 30: 199–205; 2009. doi:10.1016/j.indcrop.2009.03.007.

Moser B. R.; Sharma B. K.; Doll K. M.; Erhan S. Z. Diesters from oleic acid: synthesis, low temperature properties, and oxidation stability. *J. Am. Oil Chem. Soc.* 84: 675–680; 2007. doi:10.1007/s11746-007-1083-z.

Mudge S. M.; Pereira G. Stimulating the biodegradation of crude oil with biodiesel. Preliminary results. *Spill. Sci. Technol. B* 5: 353–355; 1999. doi:10.1016/S1353-2561(99)00075-4.

Mushrush G.; Beal E. J.; Spencer G.; Wynne J. H.; Lloyd C. L.; Hughes J. M.; Walls C. L.; Hardy D. R. An environmentally benign soybean derived fuel as a blending stock or replacement for home heating oil. *J. Environ. Sci. Heal A* 36: 613–622; 2001. doi:10.1081/ESE-100103749.

Mushrush G. W.; Wynne J. H.; Hughes J. M.; Beal E. J.; Lloyd C. T. Soybean-derived fuel liquids from different sources as blending stocks for middle distillate ground transportation fuels. *Ind. Eng. Chem. Res.* 42: 2387–2389; 2003. doi:10.1021/ie021052v.

Mushrush G. W.; Wynne J. H.; Willauer H. D.; Lloyd C. T.; Hughes J. M.; Beal E. J. Recycled soybean cooking oils as blending stocks for diesel fuels. *Ind. Eng. Chem. Res.* 43: 4944–4946; 2004. doi:10.1021/ie030883d.

Nag A. Biofuels refining and performance. McGraw Hill, New York; 2008.

Naik M.; Meher L. C.; Naik S. N.; Das L. M. Production of biodiesel from high free fatty acid Karanja (*Pongamia pinnata*) oil. *Biomass. Bioenerg.* 32: 354–357; 2008. doi:10.1016/j.biombioe.2007.10.006.

Narasimharao K.; Lee A.; Wilson K. Catalysts in production of biodiesel: a review. *J. Biobased. Mat. Bioenerg.* 1: 19–30; 2007. doi:10.1166/jbmb.2007.002.

Nebel B. A.; Mittelbach M. Biodiesel from extracted fat out of meat and bone meal. *Eur. J. Lipid Sci. Technol.* 108: 398–403; 2006. doi:10.1002/ejlt.200500329.

Newhall, H. K.; Shahed, S. M. Kinetics of nitric oxide formation in high-pressure flames. Proceedings of the Thirteenth International Symposium on Combustion:381–390, The Combustion Institute; 1971.

Ngo H. L.; Zafiropoulos N. A.; Foglia T. A.; Samulski E. T.; Lin W. Efficient two-step synthesis of biodiesel from greases. *Energ. Fuel* 22: 626–634; 2008. doi:10.1021/ef700343b.

Nimcevic D.; Puntigam R.; Worgetter M.; Gapes R. Preparation of rapeseed oil esters of lower aliphatic alcohols. *J. Am. Oil Chem. Soc.* 77: 275–280; 2000. doi:10.1007/s11746-000-0045-1.

Nine R. D.; Clark N. N.; Mace B. E.; Morrison R. W.; Lowe P. C.; Remcho V. T.; McLaughlin L. W. Use of soy-derived fuel for environmental impact reduction in marine engine applications. *Trans. ASAE* 43: 1383–1391; 2000.

Noureddini H.; Bandlamudi S. R. P.; Guthrie E. A. A novel method for the production of biodiesel from the whole stillage-extracted corn oil. *J. Am. Oil Chem. Soc.* 86: 83–91; 2009. doi:10.1007/s11746-008-1318-7.

Ozsezen A. N.; Canakci M.; Sayin C. Effects of biodiesel from using frying palm oil on the exhaust emissions of an indirect injection (IDI) diesel engine. *Energ. Fuel* 22: 2796–2804; 2008. doi:10.1021/ef800174p.

Padua M. V. Modifying vegetable oils for industrial lubricant applications. *Fuel Lube Int.* 14: 34–35; 2008.

Palada, M. C.; Changl, L. C. Suggested cultural practices for *Moringa*. International Cooperators Guide AVRDC. AVRDC pub # 03-545:1-5; 2003.

Park J. Y.; Kim D. K.; Lee J. P.; Park S. C.; Kim Y. J.; Lee J. S. Blending effects of biodiesels on oxidation stability and low temperature flow properties. *Bioresource. Technol.* 99: 1196–1203; 2008b. doi:10.1016/j.biortech.2007.02.017.

Park J. Y.; Kim D. K.; Wang Z. M.; Lu P.; Park S. C.; Lee J. S. Production and characterization of biodiesel from tung oil. *Appl. Biochem. Biotech.* 148: 109–117; 2008a. doi:10.1007/s12010-007-8082-2.

Paulson, N. D.; Ginder, R. G. The growth and direction of the biodiesel industry. Working Paper 07-WP 448, Center for Agricultural and Rural Development, Iowa State University; 2007.

Peters R. A. Alcohol production and use. *Inform* 7: 502–504; 1996.

Peterson C. L.; Reece D. L.; Thompson J. C.; Beck S. M.; Chase C. Ethyl ester of rapeseed used as a biodiesel fuel - a case study. *Biomass. Bioenerg.* 10: 331–336; 1996. doi:10.1016/0961-9534(95)00073-9.

Pfalzgraf L.; Lee I.; Foster J.; Poppe G. Effect of minor components in soy biodiesel on cloud point and filterability. *Inform* 18Suppl 4: 17–21; 2007.

Phan A. N.; Phan T. M. Biodiesel production from waste cooking oils. *Fuel* 87: 3490–3496; 2008. doi:10.1016/j.fuel.2008.07.008.

Plessis L. M.; de Villiers J. B. M.; van der Walt W. H. Stability studies on methyl and ethyl fatty acid esters of sunflower oil. *J. Am. Oil Chem. Soc.* 62: 748–752; 1985. doi:10.1007/BF03028746.

Poirier M. A.; Steere D. E.; Krogh J. A. Cetane improver compositions comprising nitrated fatty acid derivatives. *US Patent* 5: 454, 842; 1995.

Predojevic Z. J. The production of biodiesel from waste frying oils: a comparison of different purification steps. *Fuel* 87: 3522–3528; 2008. doi:10.1016/j.fuel.2008.07.003.

Qureshi N.; Saha B. C.; Cotta M. A. Butanol production from wheat straw by simultaneous saccharification and fermentation using *Clostridium beijerinckii*: Part II—Fed-batch fermentation. *Biomass. Bioenerg.* 32: 176–183; 2008b. doi:10.1016/j.biombioe.2007.07.005.

Qureshi N.; Saha B. C.; Hector R. E.; Hughes S. R.; Cotta M. A. Butanol production from wheat straw by simultaneous saccharification and fermentation using *Clostridium beijerinckii*: Part I—Batch fermentation. *Biomass. Bioenerg.* 32: 168–175; 2008a. doi:10.1016/j.biombioe.2007.07.004.

Raccuia S. A.; Melilli M. G. Biomass and grain oil yields in *Cynara cardunculus* L. genotypes grown in a Mediterranean environment. *Field Crop. Res.* 101: 187–197; 2007. doi:10.1016/j.fcr.2006.11.006.

Raheman H.; Phadatare A. G. Diesel engine emission and performance from blends of karanja methyl ester and diesel. *Biomass. Bioenerg.* 27: 393–397; 2004. doi:10.1016/j.biombioe.2004.03.002.

Rahimi H.; Ghobadian B.; Yusaf T.; Najafi G.; Khatamifar M. Diesterol: an environmental-friendly IC engine fuel. *Renew Energ.* 34: 335–342; 2009. doi:10.1016/j.renene.2008.04.031.

Ramadhas A. S.; Jayaraj S.; Muraleedharan C. Biodiesel production from high FFA rubber seed oil. *Fuel* 84: 335–340; 2005. doi:10.1016/j.fuel.2004.09.016.

Ranganathan S. V.; Narasimhan L.; Muthukumar K. An overview of enzymatic production of biodiesel. *Bioresource. Technol.* 99: 3975–3981; 2008. doi:10.1016/j.biortech.2007.04.060.

Rashid U.; Anwar F. Production of biodiesel through optimized alkaline-catalyzed transesterification of rapeseed oil. *Fuel* 87: 265–273; 2008a. doi:10.1016/j.fuel.2007.05.003.

Rashid U.; Anwar F. Production of biodiesel through base-catalyzed transesterification of safflower oil using an optimized protocol. *Energ. Fuel* 22: 1306–1312; 2008b. doi:10.1021/ef700548s.

Rashid U.; Anwar F.; Moser B. R.; Ashraf S. Production of sunflower oil FAME by optimized alkali-catalyzed methanolysis. *Biomass. Bioenerg.* 32: 1202–1205; 2008b. doi:10.1016/j.biombioe.2008.03.001.

Rashid U.; Anwar F.; Moser B. R.; Knothe G. *Moringa oleifera* oil: A possible source of biodiesel. *Bioresource Technol.* 99: 8175–8179; 2008a. doi:10.1016/j.biortech.2008.03.066.

Reid E. E. Studies in esterification. IV. The interdependence of limits as exemplified in the transformation of esters. *Am. Chem. J.* 45: 479–516; 1911.

Retka-Schill S. Walking a tightrope. *Biodiesel. Mag.* 53: 64–70; 2008.

Ribeiro N. M.; Pinto A. C.; Quintella C. M.; de Rocha G. O.; Teixeira L. S. G.; Guarieiro L. L. N.; Rangel M. C.; Veloso M. C. C.; Rezende M. J. C.; da Cruz R. S.; de Oliveira A. M.; Torres E. A.; de Andrade J. B. The role of additives for diesel and diesel blended (ethanol or biodiesel) fuels: a review. *Energ. Fuel* 21: 2433–2445; 2007. doi:10.1021/ef070060r.

Rodrigues R. C.; Volpato G.; Wada K.; Ayub M. A. Z. Enzymatic synthesis of biodiesel from transesterification of vegetable oils and short chain alcohols. *J. Am. Oil. Chem. Soc.* 85: 925–930; 2008. doi:10.1007/s11746-008-1284-0.

Rosa C.; Morandim M. B.; Ninow J. L.; Oliveira D.; Treichel H.; Vladimir Oliveira J. Lipase-catalyzed production of fatty acid ethyl esters from soybean oil in compressed propane. *J. Supercrit. Fluid* 47: 49–53; 2008. doi:10.1016/j.supflu.2008.06.004.

Sahoo P. K.; Das L. M.; Babu M. K. G.; Naik S. N. Biodiesel development from high acid value polanga seed oil and performance evaluation in a CI engine. *Fuel* 86: 448–454; 2007. doi:10.1016/j.fuel.2006.07.025.

Saka S.; Kusdiana D. Biodiesel fuel from rapeseed oil as prepared in supercritical methanol. *Fuel* 80: 225–231; 2001. doi:10.1016/S0016-2361(00)00083-1.

Salehpour S.; Dube M. A. Biodiesel: a green polymerization solvent. *Green Chem.* 10: 321–326; 2008. doi:10.1039/b715047d.

Sarin R.; Sharma M.; Sinharay S.; Malhotra R. K. Jatropha-palm biodiesel blends: an optimum mix for Asia. *Fuel* 86: 1365–1371; 2007. doi:10.1016/j.fuel.2006.11.040.

Sarma A. K.; Konwer D.; Bordoloi P. K. A Comprehensive analysis of fuel properties of biodiesel from *Koroch* seed oil. *Energ. Fuel* 19: 656–657; 2005. doi:10.1021/ef049754f.

Satge de Caro P.; Mouloungui Z.; Vaitilingom G.; Berge J. C. Interest of combining an additive with diesel-ethanol blends for use in diesel engines. *Fuel* 80: 565–574; 2001. doi:10.1016/S0016-2361(00)00117-4.

Saydut A.; Duz M. Z.; Kaya C.; Kafadar A. B.; Hamamci C. Transesterified sesame (*Sesamum indicum* L.) seed oil as a biodiesel fuel. *Bioresource Technol.* 99: 6656–6660; 2008. doi:10.1016/j.biortech.2007.11.063.

Schinas P.; Karavalakis G.; Davaris C.; Anastopoulos G.; Karonis D.; Zannikos F.; Stournas S.; Lois F. Pumpkin (*Cucurbita pepo* L.) seed oil as an alternative feedstock for the production of biodiesel in Greece. *Biomass. Bioenerg.* 33: 44–49; 2009. doi:10.1016/j.biombioe.2008.04.008.

Scholz V.; da Silva J. N. Prospects and risks of the use of castor oil as a fuel. *Biomass. Bioenerg.* 32: 95–100; 2008. doi:10.1016/j.biombioe.2007.08.004.

Schwab A. W.; Bagby M. O.; Freedman B. Preparation and properties of diesel fuels from vegetable oils. *Fuel* 66: 1372–1378; 1987. doi:10.1016/0016-2361(87)90184-0.

Sern C. H.; May C. Y.; Zakaria Z.; Daik R.; Foon C. S. The effect of polymers and surfactants on the pour point of palm oil FAME. *Eur. J. Lipid Sci. Technol.* 109: 440–444; 2007. doi:10.1002/ejlt.200600242.

Sharma B. K.; Doll K. M.; Erhan S. Z. Oxidation, friction reducing, and low temperature properties of epoxy fatty acid FAME. *Green Chem.* 9: 469–474; 2007. doi:10.1039/b614100e.

Sharma Y. C.; Singh B.; Upadhyay S. N. Advancements in development and characterization of biodiesel: a review. *Fuel* 87: 2355–2373; 2008. doi:10.1016/j.fuel.2008.01.014.

Singh A.; Singh I. S. Chemical evaluation of mahua (*Madhuca indica* [*M longifolia*] seeds. *Food Chem.* 40: 221–228; 1991. doi:10.1016/0308-8146(91)90106-X.

Sinha S.; Agarwal A. K.; Garg S. Biodiesel production from rice bran oil: transesterification process optimization and fuel characterization. *Energ. Convers. Manage.* 49: 1248–1257; 2008. doi:10.1016/j.enconman.2007.08.010.

Song J.; Cheenkachorn K.; Want J.; Perez J.; Boehman A. L.; Young P. J.; Walker F. J. Effect of oxygenated fuel on combustion and emissions in a light-duty turbo diesel engine. *Energ. Fuel* 16: 294–301; 2002. doi:10.1021/ef010167t.

Soriano N. U.; Migo V. P.; Sato K.; Matsumura M. Crystallization behavior of neat biodiesel and biodiesel treated with ozonized vegetable oil. *Eur. J. Lipid Sci. Technol.* 107: 689–696; 2005. doi:10.1002/ejlt.200501162.

Soriano N. U. Jr.; Migo V. P.; Sato K.; Matsumura M. Ozonized vegetable oil as pour point depressant for neat biodiesel. *Fuel* 85: 25–31; 2006. doi:10.1016/j.fuel.2005.06.006.

Spear S. K.; Griffin S. T.; Granger K. S.; Huddleston J. G.; Rogers R. D. Renewable plant-based soybean oil FAME as alternatives to organic solvents. *Green Chem.* 9: 1008–1015; 2007. doi:10.1039/b702329d.

Sridharan R.; Mathai I. M. Transesterification reactions. *J. Sci. Ind. Res.* 22: 178–187; 1974.

Srivastava A.; Prasad R. Triglycerides-based diesel fuels. *Renew. Sust. Energ. Rev.* 4: 111–133; 2000. doi:10.1016/S1364-0321(99)00013-1.

Srivastava P. K.; Verma M. Methyl ester of karanja oil as an alternative renewable source energy. *Fuel* 87: 1673–1677; 2008. doi:10.1016/j.fuel.2007.08.018.

Stavarache C. E.; Morris J.; Maeda Y.; Oyane I.; Vinatoru M. Syringa (*Melia azedarach* L.) berries oil: a potential source for biodiesel fuel. *Revista de Chimie* 59: 672–677; 2008.

Stavarache C. E.; Vinatoru M.; Nishimura R.; Maeda Y. Conversion of vegetable oil to biodiesel using ultrasonic irradiation. *Chem. Lett.* 32: 716–717; 2003. doi:10.1246/cl.2003.716.

Sun H.; Hu K.; Lou H.; Zheng X. Biodiesel production from transesterification of rapeseed oil using KF/Eu$_2$O$_3$ as a catalyst. *Energ. Fuel* 22: 2756–2760; 2008. doi:10.1021/ef700778r.

Suppes G. J. Propylene glycol from glycerol. *Ind. Bioprocess.* 28: 3; 2006.

Suppes G. J.; Dasari M. A. Synthesis and evaluation of alkyl nitrates from triglycerides as cetane improvers. *Ind. Eng. Chem.* 42: 5042–5053; 2003. doi:10.1021/ie030015g.

Suppes G. J.; Goff M.; Burkhart M. L.; Bockwinkel K.; Mason M. H.; Botts J. B.; Heppert J. A. Multifunctional diesel fuel additives from triglycerides. *Energ. Fuel* 15: 151–157; 2001. doi:10.1021/ef000122c.

Szybist J. P.; Boehman A. L.; Taylor J. D.; McCormick R. L. Evaluation of formulation strategies to eliminate the biodiesel NO$_x$ effect. *Fuel Process Technol.* 86: 1109–1126; 2005. doi:10.1016/j.fuproc.2004.11.006.

Tang H.; Wang A.; Salley S. O.; Ng K. Y. S. The effect of natural and synthetic antioxidants on the oxidative stability of biodiesel. *J. Am. Oil Chem. Soc.* 85: 373–382; 2008. doi:10.1007/s11746-008-1208-z.

Thomson L. A. J.; Evans B. Species profiles for pacific island. Agroforestry. *Terminalia Catappa* 2.2: 1–20; 2006.

United States Department of Agriculture, Foreign Agricultural Service, Office of Global Analysis. Oilseeds: World Markets and Trade, Circular Series FOP 12-08, Table 03, pg. 5; 2008.

Usta N. Use of tobacco seed oil methyl ester in a turbocharged indirect injection diesel engine. *Biomass. Bioenerg.* 28: 77–86; 2005. doi:10.1016/j.biombioe.2004.06.004.

Vasudevan P. T.; Briggs M. Biodiesel production - current state of the art and challenges. *J. Ind. Microbiol. Biot.* 35: 421–430; 2008. doi:10.1007/s10295-008-0312-2.

Vaughn S. F.; Holser R. A. Evaluation of biodiesels from several oilseed sources as environmentally friendly contact herbicides. *Ind. Crop. Prod.* 26: 63–68; 2007. doi:10.1016/j.indcrop.2007.01.005.

Veljkovic V. B.; Lakicevic S. H.; Stamenkovic O. S. Todorovic, ZB.; Lazic, ML Biodiesel production from tobacco (*Nicotiana tabacum* L.) seed oil with a high content of free fatty acids. *Fuel* 85: 2671–2675; 2006. doi:10.1016/j.fuel.2006.04.015.

Vicente G.; Martinez M.; Aracil J. Optimization of *Brassica carinata* oil methanolysis for biodiesel production. *J. Am. Oil Chem. Soc.* 82: 899–904; 2005. doi:10.1007/s11746-005-1162-6.

Wang P. S.; Tat M. E.; Van Gerpen J. The production of fatty acid isopropyl esters and their use as a diesel engine fuel. *J. Am. Oil Chem. Soc.* 82: 845–849; 2005. doi:10.1007/s11746-005-1153-7.

Wang Y.; Wu H.; Zong M. H. Improvement of biodiesel production by lipase TL IM-catalyzed methanolysis using response surface methodology and acyl migration enhancer. *Bioresource Technol.* 99: 7232–7237; 2008. doi:10.1016/j.biortech.2007.12.062.

Wehlmann J. Use of esterified rapeseed oil as plasticizer in plastics processing. *Fett-Lipid* 101: 249–256; 1999. doi:10.1002/(SICI)1521-4133(199907)101:7<249::AID-LIPI249>3.0.CO;2-I.

Wildes S. Clean machines from beans. *Chem. Innov.* 5: 23; 2001.

Wildes S. Methyl soyate: a new green alternative solvent. *Chem. Heal. Saf.* 9: 24–26; 2002. doi:10.1016/S1074-9098(02)00292-7.

Williams, A.; McCormick, R. L.; Hayes, R. R.; Ireland, J.; Fang, H. L. Effect of biodiesel blends on diesel particulate filter performance. *SAE Tech Pap Ser* 2006-01-3280; 2006.

Willing A. Oleochemical esters—environmentally compatible raw materials for oils and lubricants from renewable resources. *Fett-Lipid* 101: 192–198; 1999. doi:10.1002/(SICI)1521-4133(199906)101:6<192::AID-LIPI192>3.0.CO;2-W.

Wright I. Salmon by-products. Aqua Feeds. *Formulation & Beyond* 11: 10–12; 2004.

Wu W. H.; Foglia T. A.; Marmer W. N.; Dunn R. O.; Goering C. E.; Briggs T. E. Low-temperature property and engine performance evaluation of ethyl and isopropyl esters of tallow and grease. *J. Am. Oil Chem. Soc.* 75: 1173–1177; 1998. doi:10.1007/s11746-998-0131-7.

Wyatt V. T.; Hess M. A.; Dunn R. O.; Foglia T. A.; Haas M. J.; Marmer W. M. Fuel properties and nitrogen oxide emission levels of biodiesel produced from animal fats. *J. Am. Oil Chem. Soc.* 82: 585–591; 2005. doi:10.1007/s11746-005-1113-2.

Yang F-X.; Su Y-Q.; Li X-H.; Zhang Q.; Sun S-C. Studies on the preparation of biodiesel from *Zanthoxylum bungeanum* maxim seed oil. *J. Agric. Food Chem.* 56: 7891–7896; 2008. doi:10.1021/jf801364f.

Yao L.; Hammond E. G. Isolation and melting properties of branched-chain esters from lanolin. *J. Am. Oil Chem. Soc.* 83: 547–552; 2006. doi:10.1007/s11746-006-1238-3.

Yu L.; Lee L.; Hammond E. G.; Johnson L. A.; Van Gerpen J. H. The influence of trace components on the melting point of methyl soyate. *J. Am. Oil Chem. Soc.* 75: 1821–1824; 1998. doi:10.1007/s11746-998-0337-8.

Yuan X.; Liu J.; Zeng G.; Shi J.; Tong J.; Huang G. Optimization of conversion of waste rapeseed oil with high FFA to biodiesel using response surface methodology. *Renew. Energ.* 33: 1678–1684; 2008. doi:10.1016/j.renene.2007.09.007.

Zappi M.; Hernandez R.; Sparks D.; Horne J.; Brough M.; Arora S. M.; Motsenbocker W. D. A review of the engineering aspects of the biodiesel industry. Mississippi Biomass Council, Jackson, MS: 71 pp; 2003.

Zhang J.; Jiang L. Acid-catalyzed esterification of *Zanthoxylum bungeanum* seed oil with high free fatty acids for biodiesel production. *Bioresource. Technol.* 99: 8995–8998; 2008. doi:10.1016/j.biortech.2008.05.004.

Zhang Y.; Dube M. A.; McLean D. D.; Kates M. Biodiesel production from waste cooking oil via two-step catalyzed process. *Energ. Convers. Manage.* 48: 184–188; 2003. doi:10.1016/j.enconman.2006.04.016.

Zhou W.; Boocock D. B. G. Phase behavior of the base-catalyzed transesterification of soybean oil. *J. Am. Oil Chem. Soc.* 83: 1041–1045; 2006a. doi:10.1007/s11746-006-5160-5.

Zhou W.; Boocock D. B. G. Phase distribution of alcohol, glycerol, and catalyst in the transesterification of soybean oil. *J. Am. Oil Chem. Soc.* 83: 1047–1052; 2006b. doi:10.1007/s11746-006-5161-4.

Zhou W.; Konar S. K.; Boocock D. G. B. Ethyl esters from the single-phase base-catalyzed ethanolysis of vegetable oils. *J. Am. Oil Chem. Soc.* 80: 367–371; 2003. doi:10.1007/s11746-003-0705-1.

Index

A

Aboveground net primary productivity (ANPP), 75, 87
Alcohol, biodiesel production
   butanolysis, 299
   ethanolysis
      ethyl esters, 297
      nucleophilicity, 298
      stable emulsions, 298
   methanolysis
      optimization, 296
      response surface methodology (RSM), 295
      TAG and FAME, 297
ANPP. *See* Aboveground net primary productivity

B

BESC. *See* BioEnergy Science Center
Biodiesel
   advantages and disadvantages
      exhaust emissions, 291
      petrodiesel, miscible, 290
   alcohols
      butanolysis, 299
      ethanolysis, 297–298
      methanolysis, 295–297
   alternative feedstock, production
      "advanced biofuels", 313
      animal fats, 324
      Asclepias syriaca, 320
      Balanites aegyptiaca, 318–319
      beef tallow and chicken fat, 325
      Brassica carinata, 321
      Calophyllum inophyllum, 321–322
      Camelina sativa, 321
      capacity, 312
      Carthamus tinctorius, 322
      castanhola and rubber tree, 319
      Cynara cardunculus, 322
      dried distiller's grains (DDG), 327–328
      fatty acid profile, 315–316
      Madhuca indica, 317
      Melia azedarach, 317–318
      melon bug, 324–325
      Moringa oleifera and *Nicotiana tabacum,* 318
      oilseeds, 313, 317
      pork lard, 325–326
      Raphanus sativus, 320–321
      reaction conditions, 314
      rice bran, 320
      salmon oil, 324
      Sclerocarya birrea and *Cucurbita pepo,* 323
      Sesamum indicum, 322–323
      Simmondsia chinensis, 323–324
      soapstock, 326–327
      sorghum bug, 325
      tall oil, 327
      Vernicia fordii, 319
      waste oils, 326
      yellow greases, 327
      Zanthoxylum bungeanum, 320
   aviation fuel, 330–331
   blending
      alcohol, 329–330
      multi-feedstock, 330
      petrodiesel, 328–329
   catalysts, production
      heterogeneous lipases, 294
      homogenous alkaline base catalysts, 294
      noncatalytic transesterification, 295
   composition, fuel properties
      cetane number (CN), 307–308
      contaminants, 310–311
      exhaust emissions, 306–307

Biodiesel (*cont.*)
    fatty ester, 300
    feedstock, 300–301
    heat of combustion, 308–309
    kinematic viscosity, 305–306
    low-temperature operability, 301–303
    lubricity, 309–310
    minor components, 311–312
    oxidative stability, 303–305
  cottonseed oil, 6
  definition, 46–47, 286
  diesel engine, 5
  FFA, production
    high content, 293
    homogenous alkaline base catalysts, 291
    low-cost feedstock, 292
    mineral acid pretreatment, 292
    purification, 292, 294
    soap formation, 292
  glycerol, 331–332
  India
    advantages, 195–196
    combustion properties, 181–182
    food commodities, 182
    motor fuel, 182
    oil-bearing seeds, 195
    policy options, 193–194
    technical definition, 181
  oils, feedstock, 288, 289
  petroleum, replacement, 330
  solvent strength, 331
  soybean oil, 333
  standard
    ASTM D6751, 288, 289
    ASTM D7467, 290
    EN 14214, 289, 290
  transesterification reaction, 286
  typical fatty acid composition, 287
  ultra-low sulfur diesel (ULSD) fuel, 332–333
  vegetable oils conversion, 5–6
Bioenergy Research Centers (BRCs), U.S. Department of Energy
  BESC
    commercial applications, 16–17
    education, 17–18
    partners, 11–12
    research strategy, 12–16
    team, 10
  GTL research program, 10
BioEnergy Science Center (BESC)
  characterization and modeling, 13, 16
  commercial applications, 16–17
  deconstruction and conversion, 13, 15–16
  educational and training programs, 17–18
  formation and modification, 12–15
  lignocellulosic biomass, 12
  partners, 11–12
  team, 11
Bioethanol
  Arab oil embargos, 1
  crops, production
    areas, 214
    cassava and sugarcane breeding, 215
    corn, 213–214
    feedstock classification, 213
    lignocellulosic material, 216
    sweet sorghum, 215–216
  energy crisis, 2
  fueling station, 2–3
  greenhouse gas emissions, 238
  industry, 217–218
  petroleum and ethanol industries, 2
  plants, 4
  production plant, 3–4
  raw materials and natural resources, 3
  US corn ethanol production, 5
Biofuel
  biodiesel (*see* Biodiesel)
  bioethanol (*see* Bioethanol)
  cellulosic ethanol, 104
  cobalt, 59
  feedstock
    climatic zones, 103
    features, 103–104
  rise of
    ethanol and bioethanol, 238
    first-generation technologies, 237–238
    lignocellulosic materials, 239
    sugarcane bioethanol, Brazil, 238–239
Biomass
  characterization and modeling, 13, 16
  corn ethanol, 64
  C4 plants, 115
  deconstruction and conversion, 13, 15–16
  enzymatic hydrolysis, 243–244, 250
  flowering time, 115
  formation and modification
    BESC scientific approach, 14
    biofuel yields, 13–14
    cell-wall biosynthesis, 15
    goals, 12
  introduced species, productivity
    Arabidopsis CBF2 transcription factor, 145–146
    CBF genes, 145
    DOE, 146
    eucalyptus, 144
    RFS, 146–147

lignocellulosic, 224, 271
native species, productivity
    cell wall development, 143–144
    direct genetic modification, 143
    drought-tolerant genotypes, 142–143
    loblolly pine, 144
    Populus genotypes, 142
plant, components, 242
tallgrass prairie, North America, 74–77
Blending, biodiesel
    alcohol
        ethanol, 329
        Madhuca indica, 330
    multi-feedstock, 330
    and petrodiesel
        hydrocarbon constituents, 328–329
        kinematic viscosity, 329
        lubricity, 328

C

CBP. *See* Consolidated bioprocessing
Cellulosic biofuel crops
    alternative renewable energy solutions, 114
    carbon neutral energy sources, 113
    characteristics
        biomass yield and invasiveness, 115–116
        C4 plants, 115
        ethanol production, 114
        rhizome growth, 116
        vegetative to reproductive phase, 115
    genetic manipulation
        breeding program, 116
        Miscanthus and switchgrass, 117
    germplasm characterization
        genetic diversity, 118
        higher ploidy levels, 118–119
        SSR and SNP markers, 119
    germplasm collection
        breeding program, 117–118
        genetic resources, 117
        Miscanthus accessions, 118
    MAS
        biomass yield, 122
        breeding programs, 123
        genotype selection, 122–123
        quantitatively inherited traits, 124
    trait-linked markers
        gene function, 121–122
        genetic mapping, 119
        invasiveness, 121
        Miscanthus and switchgrass, 119–120
        optimal chemical composition, 120–121

quantitative trait loci (QTLs), 120
resources, 122
transgenic approaches
    CBF transcription factors, 126–127
    cell wall degradability, 127–128
    chemical induction methods, 126
    disease resistance, breeding, 127
    herbicide and insect resistance, 124
    lignin composition, 125–126
    Miscanthus hybrid production, 125
    sorghum photoperiod-sensitive gene, 124–125
Cellulosic ethanol
    biofuels, rise of
        ethanol and bioethanol, 238
        first-generation technologies, 237–238
        lignocellulosic materials, 239
        sugarcane bioethanol, Brazil, 238–239
    biomass recalcitrance, 257
    corn stover to ethanol conversion, biochemical route
        cost comparison, 54, 58
        enzyme-based process, 52
        flow diagram, 51–52
        process economics, 53
        saccharification and fermentation steps, 51–52
    corn stover to ethanol conversion, thermochemical route
        cost comparison, 56–58
        crude synthesis gas, 56
        process flow diagram, 53, 55
    economic aspects
        China, 240
        production costs, 239
        Project Liberty, 239–240
    enzyme expression, crops
        Acidothermus cellulolyticus catalytic domain, 255–256
        cellobiohydrolase (CBH), 256
        cellulases and hemicellulases, 251–254
        Corn Event 3272, 250
        nuclear transformants, 250, 255
        thermostable cellulases and hemicellulases, 255
        total soluble protein (TSP), 256–257
    enzymes
        cellulosic hydrolysis, 246
        cellulosomes, pectinase, 247
        hemicellulases, 246, 247
    feedstock
        cellulose content and lignin, 243
        cost, 240, 242

Cellulosic ethanol (cont.)
    plant hemicelluloses and lignin, 242
    sugarcane bagasse and corn stover, 242
  green-house gas emissions, 240
  hydrolysis and fermentation
    consolidated bioprocessing, 249
    NSSF, 248–249
    Saccharomyces cerevisiae, 249
    SHF, 247–248
    SSF, 248
    strategies, 248
  pretreatment
    acids and alkalis, 244
    AFEX technologies, 244, 246
    description, 243
    enzymatic hydrolysis, 246
    features, 245
    recalcitrance, 243–244
  research areas, progress and challenges, 241
  SCSBD project, 258
CFPP. *See* Cold filter plugging point
China, biofuels
  crude oil, production and import, 212
  development
    feedstock cost, 219
    research challenges, 219–220
  energy crops, production
    biodiesel, 216–217
    bioethanol, 213–216
  energy use efficiency, 211
  industries
    biodiesel, 218–219
    bioethanol, 217–218
  matrix, energy, 212
  NDRCPRC, 213
Cold filter plugging point (CFPP)
  camelina oil, 321
  definition, 301
  non-acidified salmon oil, 324
  petrodiesel, 329
  pork lard, 325
  pumpkin seed oil, 323
  safflower oil, 322
  tall oil, 329
  tung oil, 319
"Cold soak filtration test", 312
Consolidated bioprocessing (CBP)
  biocatalysts, 13
  Clostridium thermocellum, 15
  description, 249
Consumer price index (CPI)
  cross-correlation, 26
  crude oil prices and inflation rates, 26
  production costs, 28
  soybean prices, 27
Corn butanol
  biochemical/thermochemical biorefinery, 64–65
  cellulosic potentials, 61–63
  design and process variations, 59
  energy density and transportation advantages, 59
  process
    economics comparison, 63
    flow diagram, 60
    intensification, 64
  production
    cost, 62
    economic assumptions, 61–62
    enzyme, 66
    via biomass, 58
  two-liquid modeling package, 59–60
Corn ethanol
  boiler systems, biomass, 64
  dry grind
    baseline economic assumption, 42–43
    description, 40
    operating cost, 43
    process diagrams, 40–41
    techno-economic models, 42
  production costs
    USA, Brazil and EU, 41–42
    USDA, 42
  production methods, 267
  wet milling
    description, 40
    process diagrams, 40–41
    techno-economic models, 42
CPI. *See* Consumer price index
Crude oil prices
  depressed, 205
  overall food prices, 31
  soybean
    costs, production, 27–29
    prices, 25–27
    producers, 29–31

D

DEC. *See* Dedicated energy crop
Dedicated energy crop (DEC)
  algae, 106–107
  biofuel feedstock
    climatic zones, 103
    features, 103–104
  energy cane, 106

first-generation 'conventional'
    biofuels, 98
forest resources, 104–105
industry development and projections
    biodiesel production, 100
    biofuel refineries, 101–102
jatropha, 106
Miscanthus spp., 105–106
policy and economic drivers
    ad valorem tariff, 99
    corn and soybeans, 99
    RFS renewable fuel, 98–99
switchgrass, 105

E
EBP. *See* Ethanol blending program
EISA. *See* Energy Independence and Security Act
Energy crops, Chinese biofuels production
    bioethanol
        areas, 214
        cassava and sugarcane breeding, 215
        corn, 213–214
        feedstock classification, 213
        lignocellulosic material, 216
        sweet sorghum, 215–216
    oil, biodiesel
        Pistacia chinensis, 216–217
        woody plants, 216
Energy Independence and Security Act
    (EISA), 24, 38, 98, 240, 266, 313
Energy Policy Act (EPAct), 24, 98–99
Ethanol blending program (EBP), 192

F
FFA. *See* Free fatty acids
Food price inflation
    biofuels
        agricultural commodities, 24
        corn and soybean, 24–25
        demands, 23–24
        ethanol production, 25
    climate change
        farming, growing season, 22–23
        global weather conditions, 22
    crude oil and soybean price impact
        biodiesel production, 25
        cross-correlation, 26
        international export, 27
        lead/lag regression, 26–27
    emerging economies, 23
    energy costs
        biofuels, 21
        commodity, 20
        crude oil, 21
        global liquid fuels supply, 20–21
    solutions and policy implications
        agricultural production
            and management, 33
        commodity crops supply, 31
        corn breeding methods, 31–32
        gene transfer and genomics, 32
        insect pest protection, 32–33
    soybean producers
        crude oil price, 29
        operating incomes, 29–30
    soybean production costs
        crude oil prices, 27–28
        regression model, 28–29
    US dollar devaluation, 21–22
Free fatty acids (FFA)
    biodiesel production
        high content, 293
        homogenous alkaline base catalysts, 291
        low-cost feedstock, 292
        mineral acid pretreatment, 292
        purification, 292, 294
        soap formation, 292
    bound glycerol, 311
    polanga seed, oil, 321
    pro-oxidants, 311
    safflower oil, 322
    salmon oil, 324
Fuel ethanol
    corn-based, industry, 266
    India
        bioalcohol transport fuel, 188
        fermentation, 181
        imported, Brazil, 191
        molasses and alcohol, 188–189
        production and consumption, 189
        sugar industry and oil marketing
            companies, 192
    sugar-based feedstock, 270

G
Geographic information systems (GIS), 33
Germplasm
    characterization
        genetic diversity, 118
        higher ploidy levels, 118–119
        SSR and SNP markers, 119

Germplasm (*cont.*)
  collection
    breeding program, 117–118
    genetic resources, 117
    Miscanthus germplasm, 118
GHG. *See* Greenhouse gas
GIS. *See* Geographic information systems
Global energy
  biofuel, 239
  climate change, 174
  crops production, 175
  economic growth, 173–174
"Green clearance", 180
Greenhouse gas (GHG)
  bioethanol use, 238
  cellulosic ethanol, 240
  emissions, 100
  perennial biomass crops, 104
  rise of, 174

I
India, biofuels
  biodiesel
    advantages, 195–196
    combustion properties, 181–182
    food commodities, 182
    motor fuel, 182
    oil-bearing seeds, 195
    policy options, 193–194
    technical definition, 181
  commercial initiatives
    biodiesel commercial production, 198, 201–203
    jatropha plantations, 200–202
  constraints
    ethanol manufacturing cost, 205
    fuel alcohol program, 204
    jatropha cultivation, 205–206
    sugarcane production, 204–205
  energy challenges
    annual economic growth, 175
    coal reserves, 175–176
    crude oil production, 176
    economic cost, 179
    electricity generation, 176
    oil consumption, 178
    renewable energy resources, 177
  ethanol
    production, 183
    sugar industry and sugarcane supply, 183–184
  federal initiative progress, 197–198
  fuel ethanol
    bioalcohol transport, 188
    fermentation, 181
    imported, Brazil, 191
    molasses and alcohol, 188–189
    production and consumption, 189
    sugar industry and oil marketing companies, 192
  global energy, 173–175
  land availability
    cultivation, forest, 197
    Jatropha plantation, 196
  molasses and alcohol
    consumption trends, 191
    ethanol producing units, 187–188
    potable and industrial segments, 188
    production, 187
  national coordination committee, 192–193
  policy initiatives
    air pollutants, 181
    commercial energy requirements, 179–180
    fuels utilization, 180
    integrated energy, 179
    transports, 180
  state initiatives, 198
  sugarcane and sugar production
    area, growth, 185
    consumption, 187
    utilization, 186
  sugar policy, 184–185
Integrated biorefineries, engineered microbes and high-value co-products
  EISA and RFS, 266
  ethanologenic organism, 281
  first-generation, corn
    advantages, 269
    distillers, 268
    liquefaction, 267
    saccharification and fermentation, 267–268
    yield, 267
  second-generation, technologies and limitations
    challenges, 269–270
    lignocellulosic biomass, 271
    saccharification, 271–272
    seasonal availability, sugar crops, 270–271
    sugar-based feedstock, 270
  synergies, first-and second-generation processes
    advanced biorefinery, 272, 273
    integration, 272

technology improvements
  costs and profits, comparison, 280
  desired metabolic capabilities, 278
  ferment mixed sugars, 277
  microbes, ethanologens, 275–276
  multiple products, microbe engineering, 277–278
  recombinant GMAX yeast strain, 279
  Saccharomyces cerevisiae, 274
  sequence determination, 278
  valuable co-products, 280

L

Lignin biosynthesis, genetic modification
  breeding, plant biomass feedstock, 224–225
  cellulosic ethanol production
    antisense gene constructs, 230
    cell wall recalcitrance, 231
  down-regulation, 226
  hemicellulose, 224
  lignocellulosic biofuels, 224
  monocots and
    brown-midrib *(bm)* mutants, 228
    COMT gene, 228–229
    tall fescue plants, transgenic, 229
    transgenic maize, 229
  pathway, 225, 226
  plant transformation and gene regulation methods
    Agrobacterium, 227
    RNAi and VIGS, 227
  precursors/monolignols, 225
  reduction, biofuel crops, 232
  technology, transgenic, 223
  transgenic alfalfa, 230
  wall polysaccharides, 231
Little bluestem, North America
  breeding
    biomass production, 77–78
    eastern South Dakota, 78
    tallgrass prairie, 77
  genetics
    DNA extraction, 85
    RAPD markers and variation, 84

M

Marker-assisted selection (MAS)
  biomass yield, 122
  breeding programs, 123
  genotype selection, 122–123
  quantitatively inherited traits, 124
Minimum ethanol selling price (MESP), 39, 58

N

National development and reform commission of people's republic of China (NDRCPRC), 213, 217, 218
Native herbaceous perennial feedstock
  bioenergy
    crops, 72
    landscapes, 72–73
  environmental issues, 73
  little bluestem, North America
    breeding, 77–78
    genetics, 84–85
  model biomass production system
    bluestem, 76
    LIHD mixtures, 75
    soils and cultivation, 75–76
    switchgrass, 74–75
    warm-season grass establishment, 76–77
  potential pathogens and pests
    fungi and water molds, 86
    Ischnodemus falicus, 86–87
    monoculture, 85–86
    mosaic symptoms, 87
    plant–water relations, 87–88
    prairie cordgrass and little bluestem, 89
    primary root rotting, 88
    Puccinia sparganioides, 86
    seed production and viability, 88–89
  prairie cordgrass, North America
    breeding, 81–84
    genetics, 78–81
  switchgrass, 71–72
Non-isothermal simultaneous saccharification and fermentation (NSSF), 248–249

P

Prairie cordgrass, North America
  breeding
    big bluestem and switch grass, 78
    biomass production, 79–80
    natural stands, 80
    vs. switchgrass, 81
  genetics
    autotetraploid species, 83
    biomass productivity, 81
    halophyte grass, 84
    molecular markers, 82–83
    NCBI database, 82
    preliminary characterization, 83–84
    Spartina genus, 81–82
ProAlcohol
  agriculture management procedures, 159–160
  cultivation and production, 161–162

ProAlcohol (*cont.*)
   description, 158
   drawbacks, 160
   electricity production, 161
   ethanol fueled cars, 160–161
   oil price, 159
   varieties, 165
Process economics, biofuels
   commercialized biofuels, 38
   comparative analysis
      cellulosic ethanol, 51–58
      corn butanol, 58–66
      corn ethanol, 40–43
      soybean biodiesel, 46–51
      sugarcane ethanol, 43–46
   concept design methodologies and models
      energy balance calculations, 39–40
      engineering decision making, 39
   energy policy, 38
   ethanol production
      Asia countries, 40
      Brazil, 38, 40
      European Union (EU) countries, 40
      USA, 37–38, 40
   petroleum prices, 38
   production cost, 38
Properties, biodiesel
   cetane number (CN)
      determination, 307–308
      end points, 308
      index, 308
   contaminants
      glycerol and free fatty acids, 311
      water, 310
   exhaust emissions
      oxides of nitrogen and carbon monoxide, 306–307
      Zeldovich and Fenimore mechanisms, 307
   fatty ester, 300
   feedstock, 300–301
   heat of combustion
      energy content, 308
      unsaturation, 309
   kinematic viscosity
      structural features, 306
      vegetable oils and animal fats, 305
   low-temperature operability
      chemical compounds, 301
      melting point, 301–302
      parameters, 301
      structural features, 302–303
   lubricity
      scar values, 309
      structural features, 309–310
   minor components
      quantities, 311–312
      steryl glucosides, 312
   oxidative stability
      cis *vs.* trans isomers, 304–305
      EN 14112 method, 303–304
      higher molecular weight compounds, 305
      initial lipid autoxidation, 303, 304
      oil stability index (OSI), 303
      structural features, 304

R

Renewable fuel standard (RFS)
   biodiesel fuel production, 266
   biofuels production, 98
   biomass-derived diesel, 100
   productivity and total planted acreage, 146–147
   qualification, 107
   volumes, 99

S

SCSBD. *See* Syngenta Centre for Sugarcane Biofuels Development
Separate hydrolysis and fermentation (SHF), 247–248
Short-rotation woody crops
   biofuel production
      and bioenergy applications, 141
      renewable energy, feasibility, 141
   biomass productivity, genetic improvement
      introduced species, 144–147
      native species, 142–144
   biotech trees
      adoption rate, 150
      bioenergy applications, 149–150
      biofuels production, 149
      male cone clusters, 151–152
      perennial wind pollinated species models, 151
   feedstock conversion efficiency
      ethanol production, 147
      lignin biosynthetic pathway, 147–148
      lumen-to-wall ratio, 149
      negligible negative effects, 148
      transportation and storage costs, 148–149
   genetics, silviculture and biotechnology

Simultaneous saccharification
    and fermentation (SSF)
  advantages, 267
  ethanol, 248
  variants, 248–249
Soybean biodiesel
  biofuels process economics
    alternative feedstock development, 51
    lignocellulosic materials, 50
    economical challenges, 48–49
    feedstock, production, 47
    oil price, 47–48, 50
    operating cost, 49
    petroleum refining, 46–47
    process flow diagram, 48
    transesterification, 47
SSF. *See* Simultaneous saccharification
    and fermentation
Sugarcane ethanol
  economic assumptions, 45–46
  fermentation, separation and purification, 44
  juice extraction, 45
  operating cost, 46
  process diagram, 44–45
  production
    in Brazil, 43–44
    cost, USA, 45
Sugarcane ethanol industry, Brazil
  biotechnology
    genes and traits, 167
    herbicide and insect resistance, 168–169
    microsatellite markers, 169
    SUCEST database, 167–168
    transcriptome, functional analysis, 168
  breeding programs
    average crop productivity, 167
    biotic and abiotic stresses, 165–166
    diseases control, 165
    Saccharum spontaneum, 164–165
    technological level, 164
    varieties distribution, 166
  energy and environment
    land area, agriculture, 162–163
    pollutant emissions, 162
    profile, 163

  energy sustainability, 157–158
  ethanol, car fuel, 158–159
  price control system, 158
  ProAlcohol
    agriculture management procedures,
        159–160
    cultivation and production, 161–162
    drawbacks, 160
    electricity production, 161
    ethanol fueled cars, 160–161
    oil price, 159
Syngenta Centre for Sugarcane Biofuels
    Development (SCSBD), 258

T
Transesterification
  by-product, 182
  canola oil, 298
  definition, 286
  description, 47
  methanol, phases, 297
  noncatalytic, 295
  pilot plants, 198
  triacylglycerols, 286

U
US Department of Agriculture (USDA)
  corn dry mill ethanol model, 59–60
  "cost of production" survey, 41
  dry-grind and wet mill models, 42
  ethanol production cost, 42

V
VeraSun energy, 5
Virus-induced gene silencing
    (VIGS), 227

W
Woody biomass
  description, 140
  feedstock conversion efficiency, 147